T0140302

Springer Proceedings in Mathematics & Statistics

Volume 367

This book series features volumes composed of selected contributions from workshops and conferences in all areas of current research in mathematics and statistics, including operation research and optimization. In addition to an overall evaluation of the interest, scientific quality, and timeliness of each proposal at the hands of the publisher, individual contributions are all refereed to the high quality standards of leading journals in the field. Thus, this series provides the research community with well-edited, authoritative reports on developments in the most exciting areas of mathematical and statistical research today.

More information about this series at http://www.springer.com/series/10533

Antônio Márcio Tavares Thomé ·
Rafael Garcia Barbastefano · Luiz Felipe Scavarda ·
João Carlos Gonçalves dos Reis ·
Marlene Paula Castro Amorim
Editors

Industrial Engineering and Operations Management

XXVI IJCIEOM (2nd Edition), Rio de Janeiro, Brazil, February 22–24, 2021

Challenge and Trends for Sustainability in the 21st Century

Editors
Antônio Márcio Tavares Thomé
Pontifical Catholic University of Rio de
Janeiro
Rio de Janeiro, Brazil

Rafael Garcia Barbastefano
Federal Center for Technological Education
"Celso Suckow da Fonseca"
Rio de Janeiro, Brazil

Luiz Felipe Scavarda
Pontifical Catholic University of Rio de
Janeiro
Rio de Janeiro, Brazil

João Carlos Gonçalves dos Reis ⓘ
Lusofona University and EIGeS Research
Unit
Lisbon, Portugal

Marlene Paula Castro Amorim ⓘ
GOVCOPP & DEGEIT
University of Aveiro
Aveiro, Portugal

ISSN 2194-1009   ISSN 2194-1017   (electronic)
Springer Proceedings in Mathematics & Statistics
ISBN 978-3-030-78572-7   ISBN 978-3-030-78570-3   (eBook)
https://doi.org/10.1007/978-3-030-78570-3

Mathematics Subject Classification (2010): 00A69, 90-XX, 90-06, 90Bxx, 90B06, 93Exx

Preface

We are pleased to Preface this second volume of the Proceedings to the XXVI International Conference on Industrial Engineering and Operations Management (IJCIEOM). The Conference was initially scheduled to take place on the campus of the Pontifical Catholic University of Rio de Janeiro, Brazil, in the first semester of 2021. Due to the pandemic of COVID-19, the Conference migrated to a virtual edition, held successfully between 22 and 24 February 2021. It resulted equally in the combination of the events of 2020 and 2021, each in a single Conference. The 2020–2021 Conference was jointly organized by ABEPRO (Associação Brasileira de Engenharia de Produção), ADINGOR (Asociación para el Desarrollo de la Ingeniería de Organización), IISE (Institute of Industrial and Systems Engineers), AIM (European Academy for Industrial Management), and ASEM (American Society for Engineering Management). This second volume is a testimony of the persistent and high-quality production of our Operations Management and Industrial Engineering (OM&IE) peers in difficult times.

The Conference's theme was the "Challenge and Trends for Sustainability in the twenty-first Century, looking ahead to bridge theory and practice of economic, social, and environmental sustainability." This volume assembles 48 papers selected through a double-blind review process. The content of the selected papers covers a diversified array of themes and methods in the broad fields of operations and supply chain management, industrial engineering, engineering management, and related disciplines. It is a companion to the first volume of the Proceedings, which assembled 52 papers originally submitted to the 2020 edition of the Conference. Papers submitted for the 2021 edition are now being published in this separate volume.

The topics covered in this volume can be regrouped under the general headings of innovation, sustainability, sector-specific applications and management practices, maintenance engineering, decision-making processes, and education. Innovation appears often in combination with the related areas of Industry 4.0 and the digital transformation of supply chains. This cluster of papers includes open and green innovation ecosystems, contributions of Industry 4.0 technologies to sustainable operations and functional machinery safety management in the automotive industry, and the digital transformation of supply chain management processes. It embraces also the circular economy and recycling. The concept of circular economy is analyzed concerning the digital transformation of eco-innovation, supply chain management, its measurement in companies, with exemplar applications to industrial clusters, to packaging shape models, and the segments of lubricating oils and the packaging recycling chain.

Sustainability is equally represented as a core and diversified theme. It includes papers on cleaner production in aeronautical workshops, sustainable entrepreneurship, use of photovoltaics in electrical vehicles, building construction, the use of sales and operations planning as a driver to sustainability, public transportation in city logistics, and the adaptation of engineering to a sustainable service sector. The analysis of effluents and contaminants is well represented in papers about the critical factors for effluent drainage network projects, biosafety in supermarkets, and CO_2 emissions in the soil-cement industry.

The application of management practices in specific economic sectors appears chiefly under the heading of lean and lean six-sigma applied to a diverse set of industrial engineering projects implementation, improvement of electronic components, and the application of takt-time to the production of coffee powder. Applications to the health sector are also well represented, with papers on simulation models in emergency departments and on the effects of COVID-19 on vaccine supply chains and the planning of intensive care units. Other sector-specific application covers the banking sector, with papers on benchmarking of sustainable banking under COVID-19 pandemic, and bank mergers and acquisitions. A particular application of management practices appears in papers offering system dynamic views of disasters and human migration and a review of deprivation costs in humanitarian logistics and operations. Project management is represented by two articles on dynamic project management and the integration of business process model life cycle analysis in the design thinking process. General supply chain management processes are analyzed for the segment of porcelain export.

Several papers are situated in the area of maintenance engineering. They cover topics about fuzzy criticality assessments of corrosion risks in the petroleum industry, autonomous maintenance in factories, and total preventive maintenance in lean health care. The relationships between the key indicators of the maintenance process, the dimensions of sustainability, and the use of multi-criteria decision-making in maintenance prediction are equally represented.

The decision-making process was the focus of a proposed framework for the development of competencies, an analysis of entrepreneurship resilience and gender, and a broader review of the relations between the decision-making process and sustainability in general, and more specifically in the mining industry. Education in OM&IE covered the use of games and teaching cases, the measurement of universities' commitment to sustainable development goals, and an appraisal of the relevance of the concepts of advanced manufacturing and industry 4.0 to technology development.

Altogether, the papers gathered in this second volume of proceedings of the XXVI IJCIEOM Conference are representatives of the diversity of interests and methods applied to the betterment of the theory and practice of OM&IE. The editors also acknowledge the support of Fundação de Amparo à Pesquisa do Estado do Rio de Janeiro—FAPERJ (210.368/2020—SEI-260003/002594/2020). As editors, we are proud to offer this companion edition to the first volume of the XXVI IJCIEOM proceedings and we hope you will enjoy the reading.

Rio de Janeiro, Brazil   Antônio Márcio Tavares Thomé
Rio de Janeiro, Brazil   Rafael Garcia Barbastefano
Rio de Janeiro, Brazil   Luiz Felipe Scavarda
Lisbon, Portugal   João Carlos Gonçalves dos Reis
Aveiro, Portugal   Marlene Paula Castro Amorim

Contents

Data Modelling and Validation of An Emergency Department Simulation Model—A Lean Healthcare Approach

Alexandre Castanheira-Pinto©, **Bruno Gonçalves**©, **Rui M. Lima**©, and **José Dinis-Carvalho**©

Abstract An increasing demand for the public emergency systems has been noticed in the recent past. Since public hospitals capacity, in terms of resources, doesn't followed such increase proportionally, operational problems started to emerge. To pursue better hospital operations management, lean philosophy has been adopted since the its successful cases in the industry. For this purpose, a statistical analysis was made in the hospital database in order to obtain key performance indicators (KPI), such as *Takt Time* as well as the arrival patterns per priority. Such statistical data and KPIs were posteriorly used to validate a discrete simulation model that represents the emergency system of a public hospital in Portugal. Therefore, multiple processes were modeled, like the admission, triage and the arrival of patients to the treatment rooms. Finally, it is presented the validation procedure for the simulation model. After testing it was possible to prove that the simulation model is valid, which means that is capable of representing the real system, achieving realistic results.

Keywords Simulation · Lean healthcare · Discrete event simulation · Simulation model

1 Introduction

In the mid-twentieth century, a new productive ideology was born within Toyota, capable of breaking the productive paradigms rooted at the time. Contrary to the concept of mass production developed by Henry Ford, the Toyota Production System (TPS) intends to achieve the same objective, continuous production flow, with fewer resources [1]. The benefits caused by TPS in the production department, such as the costs reduction and increasing product quality, quickly led to its expansion to different sectors [2]. This expansion process was intensified when several industrial segments, besides the automobile industry, reached prominent results with the adoption of Lean

A. Castanheira-Pinto (✉) · B. Gonçalves · R. M. Lima · J. Dinis-Carvalho
Algoritmi Centre, Production and Systems Department, School of Engineering, University of Minho, Guimarães, Portugal
e-mail: amgcpinto@gmail.com

A. M. Tavares Thomé et al. (eds.), *Industrial Engineering and Operations Management*, Springer Proceedings in Mathematics & Statistics 367, https://doi.org/10.1007/978-3-030-78570-3_1

principles in their production processes [3]. According to Hines et al. [4] delivering value-added products to customers, passed not only through the application of Lean principles at the operational level, but also at the strategic level, given rise to a new concept known as Lean Enterprise [1]. This philosophy excels in identifying the processes and tasks that truly add value to the service provided to customers. Thus, by implementing Lean Enterprise, it is expected to reduce operating costs, improve the quality of the service provided, as well as the reduction of its lead-time. Such gains provided by adopting a Lean Enterprise philosophy quickly stimulated curiosity in different sectors. There has been an increasing tendency to transport Lean principles to the public sector in order to reduce costs and simultaneously improve the efficiency of government operations [5]. Notwithstanding the above, the health sector, due to its great complexity regarding operations and resources management and entities flow control, is being seen as a very suited sector for the application of lean principles. Thus, the sector is capturing the scientific community attention, given rise to a new concept known as Lean Healthcare [6].

To advocate a management policy based on Lean Healthcare concepts means a constant search to provide more health services using a lower number of resources. It is also important to mention that Lean Healthcare presents an antagonistic perspective to traditional management policies, which is based on workers dismissal or compromising the quality of the service provided [6]. The exact beginning of the application of Lean Healthcare principles is somewhat diffuse, however a pioneering work stands out [7], in which the author transports productive technology to the hospital context in order to reduce inventories. Since then, multiple publications have emerged in the scientific community, addressing several problems from the application of SMED techniques for reducing preparation time in an operating room [8] to the application of 5S techniques [9]. Lean Healthcare has been in the spotlight of the scientific community, with an exponential growth in the number of annual publications [6]. However, the work of Proudlove et al. [10] published in 2008, reinforced in 2015 by the work of D'Andreamatteo et al. [11], show the embryonic stage that Lean Healthcare is currently at, with countless opportunities yet to be explored. Combining this with the constant technological advancement, has enhanced the development of simulation software more robust and capable of numerically translating a real scenario. This fact makes possible to test changes into existing processes without the need to implement them on a real scale. Despite representing a theme that has been cherished in academia since 1950, the use of discrete simulation as a tool to analyze issues related to production only proliferated in 1986 [12]. Since then, the adoption of simulation as an analysis technique has gained notoriety, having been published several studies in different contexts [13–15]. Just as Lean has expanded to different sectors, so has simulation. The application of this technique as a problem analysis technique that aid to the decision-making process quickly spread in the health sector, highlighting the adoption of themes such as the study of surgical procedures, intensive care unit design, among others [16].

Both the discrete event simulation technique and the Lean Healthcare principles share the goal of improving the process under analysis, but represent two isolated approaches. The first attempt at integrating these two concepts was proposed in 2012 by Stewart Robinson [17]. Thus, testing the various scenarios resulting from the adoption of measures based on Lean Healthcare concepts became possible with the use of discrete event simulation [18, 19].

Based on the succinct historical framework presented, it is clear the importance and the innovative and avant-garde nature of the theme presented for this article. Both the application of the Lean philosophy to the health sector, as well as the use of the simulation technique as an aid to the improvement of hospital services represent a recent theme in the scientific community. This work is being developed in a public hospital which is undergoing an expansion and complete remodeling of its emergency service physical plant. So, it is intended to develop studies to evaluate the capacity, use of resources and the operating mode of the hospital new emergency service, through the association of the simulation technique with the Lean Healthcare philosophy. In this way, several variant scenarios developed according to Lean Healthcare principles can be studied without the need for scale tests. As easily understood the investigation methodology is based on a case-study approach, being identified the system operation in order to develop a simulation model.

2   System Description and Parameterization

Since modern societies develop new standards of living, some public sectors started to observe an increased demand and consequently an inefficiency in the operation processes. Solving such problem, and respond more efficiently to the increasing demand, can generally be solved by adopting new management principles, as the ones preconized by Lean philosophy. This work presents an ongoing study of an emergency department in a Portuguese hospital located in the north of Portugal. In order to better understand the patient flow, as well as the tasks and phases from which the patient undergoes in the system, an illustrative scheme is presented in Fig. 1. It is important to emphasize that only the flows analyzed are outlined in the figure, with a multiplicity of flows ignored due to their weak relevance to the study.

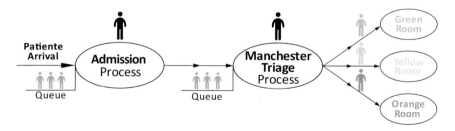

Fig. 1 Illustrative scheme of the emergency department *modus operandi*

2.1 Patient Arrival

Starting by analyzing the demand of the emergency service to identify demand patterns (very important for applying to the simulation model), the authors had access to the hospital 2018 database. Since every emergency episode (patient emergency request) is registered, the database had 92978 entries, which represents a very good sample for statistical purposes. Thus, all episodes were separated by hour for each day of week, being posteriorly determined the average of arrivals per hour. Using such information was possible to draw the emergency arrivals pattern, as it is expressed in Fig. 2. In spite of being presented only four days, the demand pattern was assessed for all weekdays. Since the patients with priority between green and orange color (Manchester Triage System) represent 98.75% of the sample, the statistical analysis only focused on these patient's groups. Therefore, some conclusions can be drawn by scrutinizing Fig. 2. Firstly, there are two periods of the day with an arrival peak, between 8–10 and 14–16 h. Another interesting aspect is the fact that in night shift, between 0 and 7 h, the arrival rate is almost the same independently of the weekday considered. Finally, Monday express itself as the more demanding day having an average arrival rate of almost 30 patients in the peak hour.

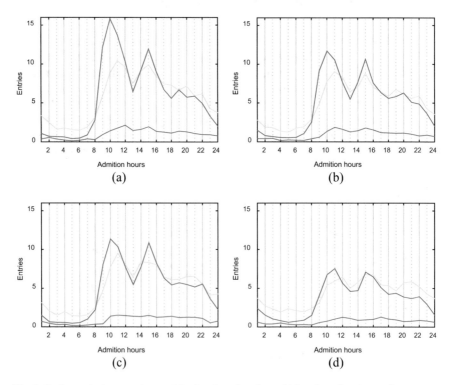

Fig. 2 Patient arrival rate per hour: **a** Monday, **b** wednesday; **c** friday; **d** sunday; (green dots—green priority; yellow dots—yellow priority; orange dots—orange priority)

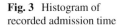

Fig. 3 Histogram of
recorded admission time

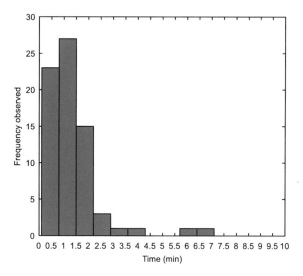

2.2 Admission Process

The admission generically represents the emergency registration, which is the first task that a patient will experience in an emergency episode. The hospital monitors the patients flow during their stay in the emergency service by registering a set of process times, event times and waiting times. However, the hospital can only start monitoring the patients after their registration in the emergency service. So, in order to obtain the process times of the registration task, an observation action was conducted in the field to collect the registration task process time (Fig. 3). With the purpose of using such information in the simulation model, the collected data was treated and fitted, conservatively, to a uniform distribution, with a minimum and maximum of 3 and 5 min respectively.

2.3 Triage System

After concluding the admission process, the patients will wait for the Manchester triage system where they are assigned with a priority level (represented by a color), and a medical specialty is specified to attend the patients based on their symptoms. Since the processing time for the triage task is monitored by the hospital, a probabilistic fit to the database provided was performed, using the maximum likelihood method [20], being expressed in Fig. 4 the cumulative frequencies for the real data, as well as for the theoretical distribution.

A Lognormal distribution was found to be the random distribution which better describes the triage process, with an average of 113.836 s and 4667.12 of variance.

Fig. 4 Cumulative frequencies of triage data (blue circles—real data, red line—theoretical distribution)

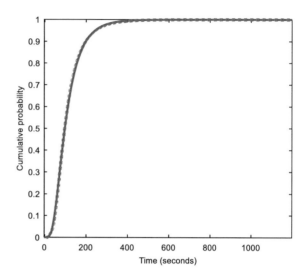

After the triage task the patient is allocated to a designated area in accordance to the priority received in the Manchester triage system.

2.4 Treatment Room Arrivals

Lack of efficiency is commonly associated to the emergency services worldwide, being characterized by long waiting queues and consequently by huge waiting times. Low medical resources as well as empirical design of the space per treatment area influence directly the unproductive behavior of an emergency system. Thus, to pursue a better performance, and properly design the requirement resources is imperative to known the exact demand of the treatments rooms, which may be assessed by measuring its *Takt Time*. *Takt Time* measure, in average, how often a patient arrives at the treatment room. Since the hospital monitors at which time the patient begins the triage system and also the triage duration, it is possible to determine at which time the patient arrives at the treatment room. *Takt Time* will be the average time between arrivals, being expressed in Table 1.

Analyzing the data of Table 1, it is possible to observe that green and yellow patients arrive almost at the same rate, which is related to the fact that they represent a similar percentage of the emergency episodes. Orange episodes, as it makes reference to more serious health diagnosis are not so common as the green and yellow, being

Table 1 *Takt Time* obtained by the hospital database

Green room (min)	Yellow room (min)	Orange room (min)
13	12	69

the *Takt Time* considerable higher. For the purpose of stablishing an efficient service, the *Takt Time* represents the needed treatment rate at which there will be no queue in the treatment rooms. Therefore, *Takt Time* can be used to design the medical staff capacity required to respond to a specific demand.

3 Simulation Model

As an attempt to predict how the new emergency department will behave, a discrete simulation software (simio) was used to model the system. In Fig. 5 it can be seen an overview of the model being superimposed all the available flows that a patient can follow. To better understand how the model was developed a stage-by-stage explanation will be followed.

Fig. 5 Overview of the simulation model with the patient flows available

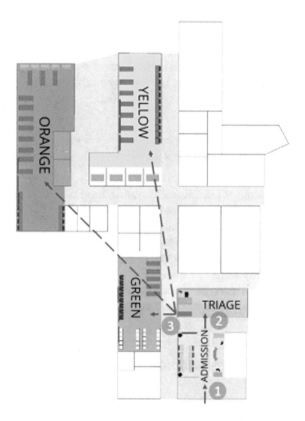

3.1 Entities Arrival

In the simulation an entity represents the object that will cross a designated process. As can easily be perceived, in this context an entity will represent a patient in the emergency system. Thus, the arrival rates, obtained in the system's parametrization, were considered in the simulation model to generate the patient's arrivals. Considering the patient arrival as an arrival rate, makes the model flexible to study the impact of an increasing demand expected to occur in the future.

3.2 Admission Process

As it was said previously, the admission process is the beginning of an emergency episode. Such task was modeled in the simulation model (processing time) using the uniform distribution expressed in the system parametrization, making the model dynamic and representative of the real scenario. When the arrival rate is less than the operations time, achieved by resources considered, queue will be experienced, being the simulation modeled able to count for such particularity. As observed *in loco* the admission process can be done by more than one person, being implemented the same feature in the simulation model.

3.3 Triage Process

Concluding the admission process the entities follows to the triage process. In a real-life event, after a patient is admitted to the emergency system, needs to wait for the triage process (which might be full or without available triage resources). Similarly, in the simulation model when the entity exits the admission process checks for available space in the triage process. If there is no available space in the triage process, then the entity waits in a queue. For triage's processing time, the lognormal distribution presented in the previous chapter was assumed. The simulation model is also flexible to considered any scenario of triage workplaces.

3.4 Treatment Rooms Arrival

Finally, the entities are forwarded for a designated room, in accordance to the priority received in the triage system. To posterior validation, the arrival time is recorded per entity at the room entrance.

Table 2 *Takt Times* (min) comparison

Triage		Green room		Yellow room		Orange room	
Real	Simulated	Real	Simulated	Real	Simulated	Real	Simulated
5.5	[5.2; 5.8]	**13**	[10.9; 12.4]	**12**	[10.1; 14.1]	**69**	[54.8; 83.1]

3.5 Model Validation

The validation procedure consisted in the direct comparison between the *Takt Time* registered in the simulation model and the real *Takt Time* obtained by the hospital database (Table 1). Since admission and triage times are dictated by probabilistic distributions several replications of the model were ran for the results to achieve statistical meaning. In the present study, and for this preliminary stage, it was assessed that 40 replications are enough for a robust statistical analysis. As the results of the simulation model are in fact a multiplicity of results, as many as the replications considered, the software expresses them as a confident interval with a confidence level of 95% (probability of being inside the interval). A comparison of both results, real and simulated are presented in Table 2 being expressed in minutes. As can be seen, all the simulated *Takt Times* are reasonable in terms of results magnitude. Another interesting aspect is the capacity of the model to simulate properly the emergency behavior since the results achieved contained, for almost every process, the real *Takt Time*, being the simulated model validated.

Despite of the green 's *Takt Time* does not contain the real value, it can be perceived that is very close, which. can be justified by the fact that the calculation of the database's *Takt Time* is merely an arithmetic average and the simulation model have the consideration of randomness in the arrivals process.

4 Conclusions

An overview of the emergency system from the studied public hospital was presented with the correspondent process parameterization, being found that the triage process time can be described by a lognormal distribution and the admission as a uniform distribution. Demand pattern assessment allowed the identification firstly of two major peaks which occurs between 8–10 and 14–16 h, and concomitantly that Monday is the most requested day. In the pursue to preview how the emergency system will behave in the future, a simulation model capable of representing the real system was developed. To properly simulate the emergency system until the treatment room's arrivals, several features needed to be considered. The dynamic behavior of the operations was one of them, with the adoption of probabilistic distribution for the processing times. The simulation model is also capable of considering the desired scenario in terms of workplaces on all the operations modeled. A direct comparison between the real *Takt Time*, obtained by treating the hospital database, with the result achieved using the simulation model was followed as a validation strategy. To achieve statistical meaning, the simulation model ran 40 replications, being the results expressed as confident intervals. It was seen that for all the priorities considered the real *Takt Time* is achieved by the simulation model. Such fact proves the validation of the model in producing accurate results, and therefore as a powerful tool to predict outcomes to a desired scenario.

Acknowledgements This work was partially supported by projects UIDCEC003192019 and POCI-01-0145-FEDER-030299, from Fundação para a Ciência e Tecnologia (FCT), Portugal.

References

1. T. Melton, "The benefits of lean manufacturing: what lean thinking has to offer the process industries," *Chemical engineering research and design,* vol. 83, no. 6, pp. 662–673, 2005.
2. M. Ballé, G. Beauvallet, A. Smalley, and D. K. Sobek, "The thinking production system," *Reflections,* vol. 7, no. 2, pp. 1–12, 2006.
3. J. P. Womack and D. T. Jones, "Lean thinking: Banish waste and create wealth in your organisation," *Simon and Shuster, New York, NY,* vol. 397, 1996.
4. P. Hines, M. Holweg, and N. Rich, "Learning to evolve: a review of contemporary lean thinking," *International journal of operations & production management,* vol. 24, no. 10, pp. 994–1011, 2004.
5. E. A. Scorsone, "New development: what are the challenges in transferring Lean thinking to government?," *Public Money and Management,* vol. 28, no. 1, pp. 61–64, 2008.
6. D. Weinstock, "Lean healthcare," *The Journal of medical practice management: MPM,* vol. 23, no. 6, p. 339, 2008.
7. S. E. Heinbuch, "A case of successful technology transfer to health care: total quality materials management and just-in-time," *Journal of Management in Medicine,* vol. 9, no. 2, pp. 48–56, 1995.
8. M. Leslie, C. Hagood, A. Royer, C. P. Reece, and S. Maloney, "Using lean methods to improve OR turnover times," *AORN journal,* vol. 84, no. 5, pp. 849–855, 2006.
9. S. Kanamori, S. Sow, M. C. Castro, R. Matsuno, A. Tsuru, and M. Jimba, "Implementation of 5S management method for lean healthcare at a health center in Senegal: a qualitative study of staff perception," *Global health action,* vol. 8, no. 1, p. 27256, 2015.
10. N. Proudlove, C. Moxham, and R. Boaden, "Lessons for lean in healthcare from using six sigma in the NHS," *Public Money and Management,* vol. 28, no. 1, pp. 27–34, 2008.
11. A. D'Andreamatteo, L. Ianni, F. Lega, and M. Sargiacomo, "Lean in healthcare: A comprehensive review," *Health policy,* vol. 119, no. 9, pp. 1197–1209, 2015.
12. L. F. McGinnis and O. Rose, "History and perspective of simulation in manufacturing," in *Simulation Conference (WSC), 2017 Winter,* 2017, pp. 385–397: IEEE.
13. B. Dengiz and O. Belgin, "Simulation optimization of a multi-stage multi-product paint shop line with Response Surface Methodology," *Simulation,* vol. 90, no. 3, pp. 265–274, 2014.
14. S. A. M. Hashim, "Simulation for reducing energy consumption of multi core low voltage power cable manufacturing system," *Journal on Technical and Vocational Education,* vol. 1, no. 2, pp. 1–10, 2017.
15. W. Terkaj, T. Tolio, and M. Urgo, "A virtual factory approach for in situ simulation to support production and maintenance planning," *CIRP Annals,* vol. 64, no. 1, pp. 451–454, 2015.
16. M. Thorwarth and A. Arisha, "Application of discrete-event simulation in health care: a review," 2009.

17. S. Robinson, Z. J. Radnor, N. Burgess, and C. Worthington, "SimLean: Utilising simulation in the implementation of lean in healthcare," *European Journal of Operational Research*, vol. 219, no. 1, pp. 188–197, 2012.
18. N. Ö. Doğan and O. Unutulmaz, "Lean production in healthcare: a simulation-based value stream mapping in the physical therapy and rehabilitation department of a public hospital," *Total Quality Management & Business Excellence*, vol. 27, no. 1–2, pp. 64-80, 2016.
19. F. Rocha, J. Queiroz, J. Montevechi, and J. Gomes, "Aplicação de value stream mapping e simulação a eventos discretos para melhoria de processo de um hospital," *Anais do XLVI Simpósio Brasileiro de Pesquisa Operacional, Salvador,* 2014.
20. G. N. Murshudov, A. A. Vagin, and E. J. Dodson, "Refinement of macromolecular structures by the maximum-likelihood method," *Acta Crystallographica Section D: Biological Crystallography,* vol. 53, no. 3, pp. 240–255, 1997.

Assessing the Influence of Circular Economy Practices in Companies that Orchestrate an Ecosystem of a Brazilian Industrial Cluster

Marcia M. C. Bacovis and **Miriam Borchardt**

Abstract The purpose of this study is to verify the presence of principles, strategies, and practices of the circular economy in a group of companies installed in an industrial cluster in the Brazilian Amazon region. The research is exploratory, using multiple case studies. Using as constructs, the eight Schools of Thought, precursors of the circular economy. Which guided the identification of 37 sustainable circular strategies and practices that guide the transition to the circular economy. The practices identified were transformed into a questionnaire, used in the interview with managers of three large companies (orchestrators of an ecosystem). As a result, it was found that the practices of schools of thought Natural Capitalism and Supply Chain in Closed Loop (CLSC) are the most influential and present in the sample companies; and the practices with less influence refer to the School of Biomimicry. The study found that companies have focused on minimizing the generation of waste, reducing the consumption of water, energy, reducing unnecessary resources, recycling packaging; but there are few efforts and strategies to keep components, materials, and products in the loop for longer. New partners are needed to collaborate in the value chain and reconsider creating value, extending the product's useful life, or looking for ways to regenerate and recover its waste.

Keywords Circular economy · School of thinking · Natural capitalism · Closed-loop supply chains · Biomimicry · Industrial cluster

M. M. C. Bacovis (✉)
Federal Institute of Education, Science and Technology of Amazonas, Manaus, AM 69075-351, Brazil
e-mail: mmbacovis@ifam.edu.br

M. Borchardt
Vale do Rio dos Sinos University (UNISINOS), São Leopoldo, RS 93022-000, Brazil

A. M. Tavares Thomé et al. (eds.), *Industrial Engineering and Operations Management*,
Springer Proceedings in Mathematics & Statistics 367,
https://doi.org/10.1007/978-3-030-78570-3_2

1 Introduction

Sustainable development has been a challenge for business organizations, which are called upon to balance their rate of exploitation of natural resources without compromising future generations [1]. In the context of sustainable development, Circular Economy (CE) emerges as a strategy that aims to improve the efficiency of materials and energy use [2]. CE is an approach to the management of restorative and regenerative industrial production systems [3–5], in which products, components, and materials that remain valuable in the greatest usefulness and for the longest possible time [2–4]. The CE theme has gained considerable attention from researchers and professionals in recent years because of its potential economic, social, and environmental benefits [3]. Research on CE acceptance by the industry remains limited [6], as there are several challenges in the transition to CE to be faced by manufacturing companies [7]. CE is not an easy concept to implement [3], and empirical evidence on CE adoption at the organizational level has been particularly critical in emerging economies, in addition to China [1, 6], where much research has been carried out on the topic. The same does not happen in Brazil [8, 9].

For the authors [10–12], there are little empirical evidences on the implementation of CE practices at the micro-level, deserving greater attention, especially in the context of companies manufacturing facilities installed in emerging countries, such as Brazil [1, 8]. For [9], the CE in Brazil still faces institutional gaps and sustainability paradoxes. The dissemination of CE in emerging economies has been hampered because the field of CE research is filled with divergent approaches, with little research on the benefits of adopting CE strategies and practices having been developed [12, 13], pointing to the need for more research on innovative practices and strategies for CE [14].

De Oliveira et al. [15] state that, in Brazil, research on the application of Circular Economy practices is still incipient and requires the identification and articulation of actions coordinated by companies. In this context, despite the literature pointing to a growing interest in CE practices, there are still few studies analyzing the adoption of circular business practices and models in emerging countries, especially in Brazil [8, 16]. Thus, this research was carried out in three companies located in the Brazilian Amazon region; a region that has an enormous amount of available natural resources, but that cannot be fully utilized without the mechanisms that promote sustainable development, such as the principles of the circular economy, being properly applied; since in underdeveloped regions, economic success is almost always achieved with the depletion of natural resources, generating a loss of biodiversity and climate change, in addition to impacts on the health of the local population. Given this context, the research questions that guided this study were: (Q1) What are the strategies and practices of CE implemented in the researched manufacturing companies? (Q2) How CE strategies and practices are adopted by companies located in emerging countries, subsidiaries of large manufacturing companies, in order to minimize the impacts of their activities in the region?

The paper is organized as follows. The Introduction in the first section. In the second section, we have a review of the literature on strategies and practices of CE Schools of Thought. The third section describes the methodology used to conduct the research and case studies. The fourth session presents the results and discussions. The fifth section summarizes the contributions, makes a conclusion, and suggests future research.

2 The CE Practices from the Circular Economy Schools of Thought

The CE is based on established concepts and theories that defend the transformation of linear economic systems into circular systems, in which there are no losses. Linear systems rely on extraction, production, consumption, and disposal processes. The general concept underlying the circular economy has been developed by many Schools of Thought (ST) [2, 17, 18], such as Industrial Ecology [19, 20], Cradle to Cradle (C2C) [21], Performance Economics [22, 23], Biomimicry [24], Blue Economics [25], Regenerative Design [26], Closed-loop supply chain [27], and Natural Capitalism [28].

Industrial Ecology (IE) is a well-established area of research, considering a systemic perspective, complex patterns of material and energy flows within and outside the industrial system and technological dynamics, which requires a systemic view [29]. An important strategy of IE is Eco-efficiency, by which the company can create value and reduce the environmental impact [12, 30]. Resource-efficient production saves them for other purposes or for use in future generations. The Industrial Symbiosis strategy involves industries traditionally separated in a collective approach of competitive advantage and the physical exchange of materials, energy, water and/or by-products, through cooperation between companies [12, 19]. In both IE and CE, there must be cooperation between actors within and between the technical and biological cycles. The CE model is systemic, focused on effectiveness in terms of systems (the entire product life cycle and its relationships with other cycles, with the environment and society) [5]. Another strategy identified in IE is investment in new technologies. For Agyemang et al. [1] technology is important for the company to achieve sustainability in advanced manufacturing; it also allows for radical transitions during the manufacturing process, allowing for reduced energy, material consumption and toxic emissions. The **Cradle to Cradle (C2C)** is based on the proposal that products can be designed so that their constituent materials circulate indefinitely in biological or technical systems when using energy from renewable sources, with a circular design and considering waste as food [2, 12, 21]. The design of the product is crucial in the design of sustainable circular systems. Efforts to close the material cycle by C2C design approaches have been researched since 1990 [29, 31]. The expansion of the return on products used, through the reverse logistics strategy, is explicitly related to companies' sustainability efforts [2, 32]. The C2C is a holistic framework with the objective of creating efficient, sustainable and waste-free systems, as the concept goes beyond the design and manufacturing processes [32].

Performance Economics emphasizes the design of durable products, guaranteeing strategies for extending the product's useful life and defending the sale of the service of a product and not the physical product itself [22]. The "design for durability" and "design for reliability" will allow you to extend the life of the products; promoting the adoption of new business models based on use-oriented services (PSS), in addition to options for reusing products and components [2, 33]. Recent thinking about CE focuses on "dematerializing" the economy, reducing material flows in production and consumption; create products and services that provide consumers with the same level of performance, but with an inherently lower environmental burden [22]. Dematerialization has promoted new forms of business models such as the economy of sharing and collaborative consumption, which are an integral part of CE [22, 31].

In **Biomimicry**, nature is the great source of inspiration for innovation [24]. It is inspired by natural ecosystems where the waste of one organism is the food or shelter of another being. In this sense, the CE can be seen as an application of biomimicry at the level of an ecosystem, as the perfect source of examples and successful models of materials and energy flow is in ecosystems [12, 24]. Eco-innovation, present in Biomimicry, is becoming a theoretical reference for strategic development, which can lead to an increase in productivity and an improvement in competitive advantage at the company level [34]. The **Blue Economy** highlights that waste does not exist: nutrients, matter and energy have a ripple effect; by-products can be used for the development of new products [25]. The use of resources from renewable sources provides compliance with the first principle of CE: preserving and improving natural capital, controlling finite stocks and balancing the flow of renewable resources. Thus, natural capital must be valued by reducing its use for the minimum necessary [35]. Circular Economy is inspired by this school, based on a model that optimizes the flow of goods, maximizing the use of natural resources and minimizing the production of waste. Competitiveness are two characteristics of the Blue Economy that serve as a motivation for CE [25, 32]. **Regenerative Design** is based on systems theory and has as premises that the recovery of materials and products should not be treated at the end of its useful life, but be contemplated from the design, through the choice of renewable materials or design for disassembly [2, 26, 29]. The **Regenerative Design**, in the context of CE, requires central skills in circular design to facilitate the disassembly of products and which allows the reconditioning, remanufacturing, reuse and recycling of each one, that is, to guarantee its use in cascade [2, 26, 31, 32]. For Moreno et al. [31], designers are responsible for defining how products and services are designed and built and that the design of most products is far from circular, as it follows the linear pattern of resource use. It is worth pointing out that the design of circular products needs to consider the chosen business model [17, 36, 37] and the selection of material [17].

The **Closed loop supply chain (CLSC)** proposes the "closing of the loops" through the retention of value promoted by the "R" frameworks and Reverse logistics. The literature points to several "R" frameworks that make it possible to retain the value of products and resources: such as the 3Rs (reduce, reuse and recycle) [12, 38], the 4R (reduce, reuse, recycle and recover), 6R (reduce, redesign, reuse, remanufacture, recycle, recover), and 9R are proposed by [39] (refuse, reduce, redesign, reuse, repair, remanufacture, recondition, recycle, recover). **Natural Capitalism** refers to the world's natural assets, such as air, water use and land use. The model argues that the interests of the environment and companies are not mutually exclusive, but overlapping [2, 28, 32]. Natural capital must be valued by reducing its use to the minimum necessary. For Bradley et al. [35] the energy and resources used must be primarily from renewable sources. Moving to CE requires organizations to innovate in their business models [2, 17, 30]. The reformulation of the business model allows traditional companies to more easily integrate sustainability into their businesses [17, 30]. The table in the Appendix, present the characteristics of the schools and the main practices and strategies recommended by the ST, as well as the main references that empirically base each body of knowledge.

3 Methodology

This research is classified as follows: as for the objectives, it is exploratory; as for nature, it is quantitative, as for the object of study it is a study of multiple cases. The analysis of the literature provided a theoretical basis for the study of multiple cases performed here. The case studies are suitable for investigating contemporary phenomena [40]. This research method was chosen due to the exploratory nature of the research. In addition to the study of multiple cases, a content analysis [41] of the selected articles was carried out to codify strategies and practices and analyze the interviews; using the NVIVO version 11 software. The strategies and practices were submitted to three Brazilian experts on the CE theme for validation. The specialists act as consultants on the subject and have publications in the area of life cycle management and circular economics.

After reviewing the literature, providing the main bibliographic reference, the next step was to choose the companies that would be part of the research. The main criteria for selecting companies was to disclose in the Sustainability Report, published on its website, that they are applying the CE principles and practices (energy efficiency, emissions reduction, reuse of materials, redistribution, remanufacturing, recycling), diversifying their models business models to close loops, delay resource loops and reduce resource flows [17] or regeneration and recovery processes [2]. The multiple case study will be conducted at three companies located in an industrial cluster in northern Brazil that are subsidiaries of global corporations. The companies included in the sample are orchestrators of the cluster's ecosystems, with environmental certification according to the ISO 14.001 family or equivalent, with environmental, social and economic sustainability objectives.

Alpha is a large manufacturer of home appliances, with more than 15 thousand employees and three manufacturing units in Brazil. The company is certified in ISO 9001, ISO 14001, OHSA 18.001 and considers aspects of environmental and social sustainability in its Mission and values, demonstrating that the environmental dimension is of great importance for its business. The company diversified its business model to offer a line of water purifiers in the subscription model (Product Service System). Beta is a supplier of metal packaging for beverages, food and household products. It is the largest manufacture of aluminum beverage cans in Brazil, with a high content of recycled material (97%). Gama, in turn, is a world leader in technology and materials for welding and offers services for the purchase and industrialization of welding residues (Blot/Oxides/Pastes) from the welding process of electronic components, with Environmental Responsibility in the welding process recovery and final destination of this Solder Waste, closing the production cycle. The three companies are corporations, headquartered outside Brazil, so many sustainability strategies are defined at the corporate level. Despite the different sectors and business models, each company chosen as a case study represents a starting point to characterize elements and functions that contribute to the CE.

Data collection was carried out mainly based on structured interviews, using the practices and strategies identified in the literature as a protocol. All visits and interviews were carried out between May and August 2019 and were carried out with key informants from the companies. In Alpha, the interviews were conducted with the Environmental Analyst and EHS Manager (Environment, Health, and Safety). At Beta and Gama companies, there is a manager and/or coordinator responsible for quality (ISO 9001) and environmental (Environmental Management System, ISO 14.001) issues. Thus, the Quality and Environment Manager and Environmental Analysts participated in the interviews. During the research, the content of the questions and the purpose of the study were clarified. Respondents confirmed that they understood the concepts. The interviews lasted more than 90 min and the researcher recorded and transcribed the content for content analysis in the NVIVO11 software. To ensure consistency between the responses and the empirical situation of the companies, the interviewees also manually filled out a spreadsheet summarizing information about each practice, to synthesize the discussion with a categorical question, evaluated on a scale that indicated the importance of that practice/strategy for the company, 1 = very high; 0.75 = high; 0.5 = intermediates; 0.25 = low; 0 = very low]. The question that the study poses is: How does the company evaluate the contribution or the importance of such CE practice/strategy in the company's environmental management strategy? In this context, each practice listed in Table 1, guided a discussion on the motivation to adopt the practice/strategy, as well as the current situation of the company regarding this practice.

Biases in the interviews and reports were considered during data analysis, as pointed out by previous qualitative research on corporate sustainability. To mitigate this research limitation, the interview data were supplemented by published documents and research sites. companies. Besides, during the interviews, concrete examples were asked to illustrate generic statements. The results, transcripts, and final information were presented at a final meeting (which lasted, on average, 90 min), in which the interviewees corrected the results found.

A case study research strategy must guarantee the internal validity and reliability of the results. The main procedures for avoiding interference are extensive review and triangulation. In this context, the study should reflect the action of the variables, not the noise of the participants' feedback (review and possible modifications), as well as the use of a logical model (to logically organize the variables) and the peer review (variables validated by specialists) [42]. Reliability in case studies refers to the consistency of the results. Multiple applications for the same object should provide similar results [29].

4   Results and Discussions

Table 2, in appendix, presents the mean and standard deviation of the answers, the ranking of the practices, and the mean and ranking of the ST. At the end respondents reviewed the answers, confirming the results.

Despite the low score (22.5 out of a maximum score of 37), Alpha has a final result (the mean) substantially higher than Beta and Gama. The research did not investigate why Alpha has an overall performance so different than the others, which would require a specific case study. The Fig. 1 shows the distribution of performance, considering the average of the general influence of the ST on the environmental strategies of the companies included in the sample. Figure 2 compares the influence of each ST on individual companies.

Regarding the practices of the **Industrial Ecology school**, the three companies manage sustainable supply chains based on Collaboration/Cooperation and a systemic approach. Alpha reported that it advises its partners on sustainable and CE practices; which performs audits and requires certifications from its suppliers. The Beta Manager stated that "everyone can win when they cooperate". Gama's interviewees stated that "ISO 14.001 assists in this cooperation, since it has mandatory issues to be carried out between partners". The results are in line with [30, 43] which states that collaboration between companies and other stakeholders has become more important with the increasing global pressures on sustainability. For Bocken et al. [30], collaboration with non-industrial actors may be the key to creating value from waste. Regarding the use of New Technologies, Alpha and Beta reported investing in new product and process technologies in order to promote greater eco-efficiency. Gamma has outdated and old process technologies, but invests in environmental technologies, such as: filters in chimneys and ovens and invests in technologies in the product (Solder Lead free, elimination of volatile organic compound). The results show that the companies in the sample have already realized the potential of using advanced technologies and information technology to boost the CE and revolutionize the way they operate; what is in line with the studies by Agyemang et al. [1], Kirchherr et al. [3], Bocken et al. [30] who pointed to technology as a driver for the CE. As for industrial symbiosis, Gama reuses the solder sludge produced in the client's process, which is the only residue it processes; Alpha and Beta said they did not practice industrial symbiosis in the manufacturing process. But it has its waste, mainly from packaging, being reused as a secondary raw material, generating new business opportunities for other companies. Although the companies are part of an industrial

Table 1 Reference framework with strategies for circularity from the schools of thought

School of thought	Practices	Description of the practice	Authors
Industrial ecology	Systemic approach	Explain how parts influence each other as a whole and the relationship of the whole to the parts	Ellen MacArthur Foundation et al. [2], Silva et al. [8], Chertow [19], Gregson and Crang [20], Geisendorf and Pietrulla [32]
	Cooperation	Collaboration and cooperation are essential in closed-loops that turn waste into useful resources	
	Industrial symbiosis	A process-oriented solution concerned with transforming the waste output from one process into raw material to another process or product line	
	Eco-efficiency	The strategy adopted by the company to promote the application of environmental, social, and economic sustainability practices throughout the product cycle	
	Technological changes	Technological changes and information technologies play an important role in the transition to the circular economy	
Cradle to cradle	C2C design	To close the cycle, products must be developed so that they can be regenerated and reused	Ellen MacArthur Foundation et al. [2], McDonough and Braungart [21], Lieder and Rashid[29], Bocken et al. [30]
	Reduce waste generation	It means reducing waste by closing the loop; reduce losses of valuable materials; and develop industrial processes with less waste	
	Reverse logistics	Return of post-sale or post-consumption waste for a new application, in the same or in another process	
	Energy recovery	Recover energy from by-products or waste by technologies such as incinerators and biogas exploitation	

(continued)

Table 1 (continued)

School of thought	Practices	Description of the practice	Authors
Performance economics	Design for durability	Enlarges the useful life of developed products	Ellen MacArthur Foundation et al. [2], Stahel [22], Tukker [23], Moreno et al. [31], Elia et al. [33]
	Design for reliability	Develops products with a design that ensures reliability, with an adequate product or component life, aims to reduce the number of failures and extend their service life	
	Dematerialization	Reduce the use of raw material in manufacturing but the performance remains unaltered	
	Product service system—PSS	A Product Service System (PSS) integrates products and services to enhance value to users	
	Sharing	Expands the efficient use of underutilized resources, which fulfills its social function	
Biomimicry	Nature inspired products	Designers should draw on organisms, biological processes, and ecosystems as a way to solve human problems	Ellen MacArthur Foundation et al. [2], Benyus [24], Moreno et al. [31], Geisendorf and Pietrulla [32], De Jesus et al. [34]
	Nature-inspired processes	In Biomimicry, nature inspires products and processes	
	Return to nature	In CE, the resources used in the production of goods must be recovered and returned to nature. The principle of regeneration aims to restore, retain, and restore the health of ecosystems	
	Eco-innovation	Green innovation assists in environmental sustainability as it promotes changes in the way we produce with less environmental impact	
Blue economy	Co-product generation	Materials produced during the primary production process of the main product from the same input (resources)	Ellen MacArthur Foundation et al. [2], Bocken et al. [17], Pauli [25], Bocken et al. [30], Geisendorf and Pietrulla [32]
	Co-products use	The use of co-products contributes to sustainability in industries	

(continued)

Table 1 (continued)

School of thought	Practices	Description of the practice	Authors
	Increased competitiveness	Investments in social and environmental initiatives are no longer an additional cost but an opportunity for innovation and competitiveness	
	Use of renewable energies	It relies on natural resources with a utilization rate lower than the renewal rate, such as solar, wind, hydro, geothermal, among others	
Regenerative design	Design for disassembly	Design that considers the need for disassembly, which facilitates the repair and remanufacturing of end-of-use returns as well as recycling	Ellen MacArthur Foundation et al. [2], Bocken et al. [17], Lyle [26], Moreno et al. [31], Geisendorf and Pietrulla [32]
	Modular design	Design that develops products composed of functional modules, so that they can be updated with new features. Modules can be individually refurbished or replaced, increasing product longevity	
	Resource reuse	It involves the use of fewer resources and less workforce to produce new products from virgin materials or even to recycle and discard products	
	Energy performance	It involves the use of less energy to produce new products from virgin materials or even to recycle and discard products	
	Material selection	This refers to the preference, in the process of acquisition and production, for pure materials, which offer easier classification at the end of life of products	
Closed-loop supply chain	Reduce	Includes input reduction and use of natural resources; reduction of emission levels; loss reduction of valuable materials	Ellen MacArthur Foundation et al. [2], Ghisellini et al. [12], Krikke et al. [27], Geisendorf and Pietrulla [32], De Jesus et al. [34], Ruggieri et al. [43]
	Reuse	It means using a product again for the purpose for which it was originally designed and produced, with little improvement or alteration	

(continued)

Table 1 (continued)

School of thought	Practices	Description of the practice	Authors
	Remanufacture	Remanufacturing ensures that products meet original performance specifications by restoring and replacing components	
	Refurbish	It consists of updating a used product, replacing parts that are failing or likely to fail soon	
	Recycle	Recycling is any recovery operation whereby waste is reprocessed into products, materials, or substances, whether for original or other purposes	
Natural capitalism	Land-use optimization	Proper management of solid waste decreases the pressure on natural resource consumption and impacts on the soil for landfill disposal.	Ellen MacArthur Foundation et al. [2], Bocken et al. [17], Hawken et al. [28], Bocken et al. [30], Moreno et al. [31], Geisendorf and Pietrulla [32]
	Water use optimization	The company must manage the use of water, making treatment and reuse	
	Use of the Atmosphere	Proper management of atmospheric emissions decreases the pollution generated by the activity and the need for technological resources to mitigation	
	Change business model	Sustainable business models allow to close and narrow loops and dematerialize products for sustainable development solutions	

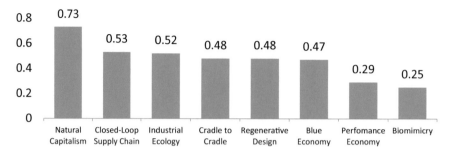

Fig. 1 Mean performance of the companies in the eight bodies of knowledge

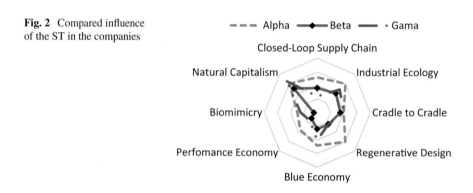

Fig. 2 Compared influence of the ST in the companies

Cluster, the exchange of waste is incipient; such results are similar to those found in the study by Ruggieri et al. [43], who identified that there is no interorganizational symbiosis in two of the three companies surveyed, even though the companies are inserted in a cluster with potential opportunities for interorganizational cooperation. Many researches point to industrial symbiosis as essential for the operationalization of the CE [19, 30, 43].

The **CLSC School** is the second school with the best influence on the companies surveyed. Recycling is the most common practice. The three companies recycle 100% of the used packaging, obtaining a financial return. In addition to recycling, Alpha reuses wooden pallets but also forwards them to reuse in the furniture industry. Gama recycles 100% of its main waste and internally recycles the tin sludge (solder sludge) returned from customers. The three companies claim to be committed to raising awareness among employees to reduce energy consumption, consumption of raw materials, and waste of materials. Alpha maps the seven losses to chemicals, waste, and water; but the results are not entirely satisfactory in terms of remanufacturing and reconditioning. Finally, only Gama has a reuse policy, involving wooden packaging. The results confirm the studies of [12], who states that the implementation of the CE, worldwide, is still in the initial stages with a focus on recycling instead of reuse, remanufacturing and repair.

Regarding the practices of the **Cradle-to-Cradle** school of thought, the three companies develop successful programs that aim to reduce and even eliminate the generation of waste. The three companies do not send solid waste to industrial landfills (zero waste). Alpha has an extensive cleaner production program with a significant reduction rate. Beta and Gamma also have successful zero defect and loss prevention programs. Reverse logistics and life cycle management are only partially developed in companies, confirming the findings of Gusmerotti et al. [44], where only 10.6% of companies surveyed in Italy focus on production and logistics practices. The line fully returns the after-sales and post-consumer leftovers, but Alpha and Beta do this only partially, returning some types of aluminum and scrap packaging, respectively. Only Alpha has some incipient initiatives to recover energy from biomass and other types of solid waste. Respondents from the three companies were unaware of the concept of Circular design and Cradle to Cradle; however, Alpha develops eco-design initiatives. These results confirm the research of [8, 45].

In relation to the **Regenerative Project**, the three companies adopt consistent practices for the selection of materials, mainly chemical raw materials. The three companies have extensively reduced the need for hazardous materials to an almost negligible level. The energy performance, the reuse of resources, and the design for disassembly show incipient, but promising results. The three companies have already implemented these practices in accordance with their environmental management systems, even though the results are still initial, which explains the intermediate assessment. Finally, none of the companies knows the concept of modular design, which explains the lack of relevance of the variable. The adoption of material selection practices and modular design are pointed out in the research by [27] as essential for the implementation of the CE.

In relation to the **Blue Economy**, companies extensively seek to increase competitiveness and minimize waste through the specific programs mentioned, such as zero defects, loss prevention, and cleaner production. These successful programs explain the high scores attributed by respondents to waste management and the strategy to increase competitiveness. Managers understand that investing in issues of environmental and social sustainability brings economic results. The three companies recognize that investment in certifications (ISO 9001, ISO 14.001, OHSA18.001), that compliance with standards have become an opportunity for innovation and increased the company's competitiveness, due to reduced fines for meeting environmental laws. As for the strategy of using renewable energies, the company Alpha has isolated and successful initiatives in the use of solar and wind energy; Beta and Gama have a project for the use of renewable energy, but they are in the approval phase. The company Gama has an isolated and successful initiative in the management of by-products in the use of solder sludge in production; this result is in agreement with the results found in the study by [43].

Regarding **Performance Economics**, the three companies widely adopt actions that increase the reliability of the product; in the design process, aiming mainly to avoid failures and losses in the production process and in the product life cycle. The main motivation for the choice is to reduce costs in the manufacturing process and technical assistance within the warranty period. Companies develop only incipient actions that influence the rest of the practices. Regarding the dematerialization/virtualization of products, in the company Alpha the product instructions manual is "delivered" in a virtual way and for the author [23] to offer performance, the sale of the product-service (PSS), which also it is characterized as dematerialization; Beta and Gama do not have dematerialization practices. The three companies are unaware of the concept of resource sharing, which explains the variable's zero score.

The practices of **Natural Capitalism** point to results that suggest that companies increase productivity in the use of natural resources (land, water and atmosphere). Alpha uses chimneys with filters in the painting and furnace processes, collecting harmful gases. The company Gama has dust collectors and controls its emissions. Beta managers reported that the company does not release particles into the atmosphere. In Beta, 95% of the inputs (aluminum) come from recycled material. Gama also uses a large percentage of recycled material in its process; demonstrating that there is a concern to optimize land use, extracting less resources. The three companies recycle 100% of industrial water and collect rainwater for industrial use. They also capture all air emissions by filters, partially recovering the heat. As for the business model adopted, they have not evolved significantly to the circular business model. Alpha has diversified its business model and offers its customers a line of products such as the sale of services (PSS), but this model is not yet well consolidated, a greater effort is needed to make people aware of the reconditioning and remanufacturing practices. This is the most effective way to close the loop; because, according to [23] maintaining product ownership covers the highest level of responsibility for it. The company Gama operates in two business segments: conventional model of production and sale of solder paste and also buys the waste, solder sludge, from its customers to recycle, closing the resource cycle, creating value with what would be waste [17].

Finally, the practices least mentioned in the interviews are those of **Biomimicry**. In the interviews and at the last meeting, all respondents said they had never heard of biomimicry, which explains the little influence of biomimicry on the environmental management of companies. Intuitively, Alpha promotes some small improvements to the packaging process using gravity to reduce labor. Beta and Gama support the cooperatives of waste pickers that collect organic waste that is returned to feed manufacturers. Alpha recognizes the importance of working with the concept of eco-innovation and develops more eco-efficient products and processes. This result is in line with the study of [34] that analyzes drivers and barriers when using eco-innovation to transition to CE.

5   Final Considerations

This study confirms previous studies that argue that CE practices have been adopted by global companies and located in developed countries that have tax regulations and incentives [12, 34], but the same is not true for companies located in emerging countries [1]. Our study confirms the result of the study by Gusmerotti et al. [44] that the CE paradigm is still far from institutionalized in all manufacturing companies. This study also confirms that the CE, in these companies installed in Brazil, is in an elementary stage, confirmed by the study by Silva et al. [8], Sehnem et al. [16]; and requires changes in the business model with the adoption of practices that promote the successful transition to CE [16–18]. These CE practices have the potential to advance the design of reverse supply chains, recycling, reuse or remanufacturing end-of-life products [16, 32, 36]. Although the companies surveyed are orchestrators of the ecosystem of an industrial cluster, managers and environmental analysts interviewed in the companies are unaware of some key concepts, such as modular design, design for durability, circular design, biomimetic and products and processes inspired by nature. This is a specific issue raised by this study. Future research must identify why this occurs in relevant companies in a relevant industrial cluster. In addition, the study shows that only 17 of the 37 practices have an influence on the sample of companies.

The study opens space for new research, which should replicate the research on the practices adopted and the effective return of the CE in companies manufacturing other clusters. In this perspective, the study found that companies have focused on minimizing the generation of waste, reducing the consumption of water, energy, reducing unnecessary resources, recycling packaging; but there are few efforts and strategies to keep components, materials and products in the loop for longer. For CE principles to be established, they need to find new partners that collaborate in the value chain and reconsider creating value by extending the product's useful life or looking for ways to regenerate and recover its waste. To solve this, new tools, technologies and ecological innovations, thought from the product design and the process design, are necessary to guide companies in their journey of transition to the CE.

The main contributions of the study corroborate the theoretical framework presented in the research, by pointing out the main sustainable practices employed in manufacturing companies in an emerging country, regardless of the sector in which the company operates. Another contribution of this study points to the need for more research on drivers and barriers that prevent companies from implementing CE. A limitation of the study was the number of companies, only three, which leads us to suggest that future research be carried out to verify the presence of these practices in a larger sample of companies, located in industrial clusters and in emerging countries; not necessarily orchestrators of the ecosystem, but in the entire set of companies, whose results can be analyzed by a model of structural equations. To meet this purpose, a new block of questions, comprising competitive dimensions such as cost, quality, flexibility, and delivery, must be added to assess the strategic relevance of CE practices.

Appendix

See (Table 2).

Table 2 Categorized answers

School of thought	Practices	Company			Practices		School of thought		
		Alpha	Beta	Gama	μ	Ranking	μ	σ	Ranking
Industrial ecology	Systemic approach	0.75	0.75	0.5	0.67	13	0,52	0.07	3°
	Cooperation	0.75	1	0.5	0.75	9			
	Industrial symbiosis	0.5	0.5	1	0.17	28			
	Eco-efficiency	0.75	0	0.25	0.33	23			
	Technological changes	1	0.25	0.75	0.67	12			
Cradle to cradle	C2C design	0.25	0	0	0.08	33	0,48	0.13	4°
	Reduce waste generation	1	1	1	1.00	1			
	Reverse logistics	0.25	0.75	1	0.67	14			
	Energy recovery	0.5	0	0	0.17	29			
Performance economy	Durability design	0.5	0.25	0	0.25	25	0,29	0.10	7°
	Reliability design	1	0.5	1	0.83	5			
	Dematerialize	0.25	0	0	0.13	31			
	Product service system (PSS)	0.5	0	0	0.25	26			
	Sharing	0	0	0	0.00	35			
Biomimicry	Nature inspired products	0	0	0	0.00	36	0,25	0.16	8°
	Nature-inspired processes	0.25	0	0	0.08	34			
	Return to nature	0.25	0.25	0.75	0.42	20			
	Eco innovation	1	0	0.5	0.50	10			
Blue economy	Waste minimization	1	0.5	0.75	0.75	6	0,47	0.11	6°
	Co-products generation	0	0	0.25	0.08	30			

(continued)

Table 2 (continued)

School of thought	Practices	Company			Practices		School of thought		
		Alpha	Beta	Gama	μ	Ranking	μ	σ	Ranking
	Co-products use	0	0	0.5	0.17	24			
	Increased competitiveness	1	0.75	1	0.92	2			
	Use of renewable energies	1	0.25	0	0.42	21			
Regenerative	Disassembly design	0.75	0	0	0.38	22	0,48	0.15	5°
Design	Modular design	0	0	0	0.00	37			
	Resource reuse	1	0.5	0.5	0.67	15			
	Energy performance	1	0.5	0	0.50	18			
	Material selection	1	0.5	1	0.83	7			
Closed-loop supply chain	Reduce	1	0.75	0.5	0.75	11	0,53	0.07	2°
	Reuse	0.25	0	0	0.13	32			
	Remanufacture	0.75	0.5	0	0.63	17			
	Refurbishing	0.5	0	0	0.25	27			
	Recycle	0.75	1	1	0.92	3			
Natural capitalism	Land use	1	0.75	0.75	0.83	8	0,73	0.13	1°
	Water use	0.75	0.75	0.5	0.67	16			
	Use of the atmosphere	0.75	1	1	0.92	4			
	Business model change	0.5	0	1	0.50	19			
	Sum	22.5	13.0	160					

References

1. Agyemang, Martin, et al. "Drivers and barriers to circular economy implementation." Management Decision (2019).
2. Ellen MacArthur Foundation, Zumwinkel, K., & Stuchtey, M. R. "Growth within: a circular economy vision for a competitive Europe". Cowes: Ellen MacArthur Foundation. (2015).
3. Kirchherr, J., Reike, D., Hekkert, M. Conceptualizing the circular economy: An analysis of 114 definitions. Resources, Conservation and Recycling, 127, 221–232. (2017).
4. Korhonen, J., Honkasalo, A., and Seppälä, J. Circular Economy: The Concept and its Limitations. Ecological Economics, 143, 37–46. (2018).
5. Webster, Ken. "The circular economy: A wealth of flows". Ellen MacArthur Foundation Publishing, (2017).
6. Stewart, Raphaëlle, and Monia Niero. "Circular economy in corporate sustainability strategies: A review of corporate sustainability reports in the fast-moving consumer goods sector." Business Strategy and the Environment 27.7: 1005–1022. (2018).
7. Parida, V., Burström, T., Visnjic, I., & Wincent, J. (2019). Orchestrating industrial ecosystem in circular economy: A two-stage transformation model for large manufacturing companies. Journal of Business Research, 101, 715–725.
8. Silva, F. C., Shibao, F. Y., Kruglianskas, I., Barbieri, J. C., and Sinisgalli, P. A. A. Circular economy: analysis of the implementation of practices in the Brazilian network. Revista de Gestão, 26(1), 39–60. (2019).
9. Jabbour, Charbel Jose Chiappetta et al. Stakeholders, innovative business models for the circular economy and sustainable performance of firms in an emerging economy facing institutional voids. Journal of Environmental Management, v. 264, p. 110416, (2020).
10. Homrich, Aline Sacchi et al. The circular economy umbrella: Trends and gaps on integrating pathways. Journal of Cleaner Production, v. 175, p. 525–543, (2018).
11. Urbinati, Andrea, Enes Ünal, and Davide Chiaroni. "Framing the managerial practices for circular economy business models: a case study analysis." 2018 IEEE International Conference on Environment and Electrical Engineering and 2018 IEEE Industrial and Commercial Power Systems Europe (EEEIC/I&CPS Europe). IEEE, 2018.
12. Ghisellini, P., Cialani, C., & Ulgiati, S.. "A review on circular economy: The expected transition to a balanced interplay of environmental and economic systems". Journal of Cleaner Production, 114, 11–32. (2016).
13. Kalmykova, Yuliya; Sadagopan, Madumita; Rosado, Leonardo. Circular economy–From review of theories and practices to development of implementation tools. Resources, conservation and recycling, v. 135, p. 190–201, 2018.
14. Merli, Roberto; Preziosi, Michele; Acampora, Alessia. How do scholars approach the circular economy? A systematic literature review. Journal of Cleaner Production, v. 178, p. 703–722, 2018.
15. De Oliveira, F. R., França, S. L. B., Rangel, L. A. D. "Challenges and opportunities in a circular economy for a local productive arrangement of furniture in Brazil". Resources, Conservation and Recycling, 135, 202–209. (2018).
16. Sehnem, Simone, et al. "Circular business models: level of maturity." Management Decision (2019).
17. Bocken, N. M. P. et al. Product design and business model strategies for a circular economy. Journal of Industrial and Production Engineering, 33(5), 308–320. (2016).
18. Lewandowski, M. Designing the business models for circular economy—Towards the conceptual framework. Sustainability 8(1), 43. (2016).
19. Chertow, Marian R. Industrial symbiosis: literature and taxonomy. Annual review of energy and the environment, v. 25, n. 1, p. 313–337 (2000).
20. Gregson, Nicky, and Mike Crang. "From waste to resource: The trade in wastes and global recycling economies." Annual Review of Environment and Resources 40, 151–176 (2015).
21. McDonough, W., & Braungart, M. "Cradle to cradle: Remaking the way we make things". New York: North Point Press. (2010).

22. Stahel, W. "The performance economy". New York: Springer. (2010).
23. Tukker, A. "Product services for a resource-efficient and circular economy - A review". Journal of Cleaner Production, 97, 76–91. (2015).
24. Benyus, Janine. "Biomimicry: What would nature do here." Nature's Operating Instructions: The true biotechnologies. Sierra Club Books, San Francisco. 3–16. (2004).
25. Pauli, G. A. The blue economy: 10 years, 100 innovations, 100 million jobs. Boulder: Paradigm Publications. (2010).
26. Lyle, J. T. (1996). Regenerative design for sustainable development. John Wiley & Sons.
27. Krikke, H., Blanc, I., & Van De Velde, S. (2004). Product modularity and the design of closed-loop supply chains. California management review, 46(2), 23–39.
28. Hawken, P., Lovins, A. B., and Hunter, L. Natural capitalism: The next industrial revolution. London: Routledge. (2013).
29. Lieder, M., & Rashid, A. (2016). Towards circular economy implementation: a comprehensive review in context of manufacturing industry. Journal of Cleaner Production, 115, 36–51.
30. Bocken, N. M. P. et al. A literature and practice review to develop sustainable business model archetypes. Journal of Cleaner Production, 65, 42–56 (2014).
31. Moreno, M. et al. "A conceptual framework for circular design". Sustainability, 8(9), 937. (2016).
32. Geisendorf, Sylvie, and Felicitas Pietrulla. "The circular economy and circular economic concepts—a literature analysis and redefinition." Thunderbird International Business Review 60.5, 771–782 (2018).
33. Elia, V., Gnoni, M. G., and Tornese, F. Measuring circular economy strategies through index methods: A critical analysis. Journal of Cleaner Production, 142, 2741–2751. (2017).
34. De Jesus, A. et al. "Eco-innovation in the transition to a circular economy: An analytical literature review". Journal of Cleaner Production, 172, 2999–3018. (2018).
35. Bradley, R. et al. A Framework for Material Selection in Multi-Generational Components: Sustainable Value Creation for a Circular Economy. Procedia CIRP, v. 48, p. 370–375, 2016.
36. Geissdoerfer, M., Morioka, S. N., de Carvalho, M. M., & Evans, S. Business models and supply chains for the circular economy. Journal of cleaner production, 190, 712–721. (2018).
37. Urbinati, A., Chiaroni, D., Chiesa, V. "Towards a new taxonomy of circular economy business models". Journal of Cleaner Production, 168, 487–498. (2017).
38. Blomsma, F., & Brennan, G. The emergence of circular economy: A new framing around prolonging resource productivity. Journal of Industrial Ecology, 21(3), 603–614. (2017).
39. Potting, J. et al. Circular economy: measuring innovation in the product chain. Sidney: PBL Publishers. (2017).
40. Yin, Robert K. Case study research and applications: Design and methods. Sage publications, (2017).
41. Bardin, L. (2006). Análise de conteúdo. Lisboa: Edições 70.
42. Kitchenham, B. (2004). Procedures for performing systematic reviews. Keele, UK, Keele University, 33(2004), 1–26.
43. Ruggieri, Alessandro et al. A meta-model of inter-organisational cooperation for the transition to a circular economy. Sustainability, v. 8, n. 11, p. 1153, (2016).
44. Gusmerotti, N. M., Testa, F., Corsini, F., Pretner, G., Iraldo, F. "Drivers and approaches to the circular economy in manufacturing firms". Journal of Cleaner Production, 230, 314–327. (2019).
45. Sellitto, Miguel Afonso, et al. "Práticas de ecodesign em um pólo moveleiro do sul do Brasil: das práticas incipientes à melhoria." Jornal de Política e Gestão de Avaliação Ambiental 19.01 (2017): 1750001.

Public Transport Systems and its Impact on Sustainable Smart Cities: A Systematic Review

Roberto Rivera⊙, Marlene Amorim⊙, and João Reis⊙

Abstract This article presents a systematic literature review that includes research papers published since 2015–2019, and addresses topics in the areas of on Public Transport Systems, Sustainability and Smart Cities. From 42 articles, 20 were documents that met the inclusion criteria and represented a diverse sample. This article also mapped 171 smart cities from 5 continents, where the transport system is most relevant. Although the results show a similar trend in terms of the close relationship between sustainability and public transport systems in terms of Information and Communication Technologies (ICT), it differs from one country to another in terms of the implementation indicators, policies and user behaviors. In light of the above, this research offers a contemporary view of the activities carried out under the theme and creates the basis for future action plans.

Keywords Public transport systems · Smart cities · Sustainability · Information and communication technologies · Urban mobility

1 Introduction

The concept of Smart City (SC) integrates the presence, application and use of ICTs as a complex system that allow citizens, business, transports, communication networks, services and utilities being interconnected to each other's. The mentioned situation allows efficiency in operations that improve the quality of life in public

R. Rivera (✉)
Research Unit on Governance, Competitiveness and Public Policies (GOVCOPP), University of Aveiro, 3810-193 Aveiro, Portugal
e-mail: r.rivera@ua.pt

M. Amorim
Department of Economics, Management, Industrial Engineering and Tourism, University of Aveiro, 3810-193 Aveiro, Portugal

J. Reis
Industrial Engineering and Management, Faculty of Engineering, Lusofona University, Campo Grande, 1749-024 Lisbon, Portugal

© The Author(s), under exclusive license to Springer Nature Switzerland AG 2021   33
A. M. Tavares Thomé et al. (eds.), *Industrial Engineering and Operations Management*, Springer Proceedings in Mathematics & Statistics 367,
https://doi.org/10.1007/978-3-030-78570-3_3

spaces, particularly with regard to urban mobility [1, 2]. This phenomenon is also facing challenges, mainly due to the population growth and climate changes. Thus, in recent years, governmental initiatives have focused on policies and programs to implement strategies and actions aimed to create sustainable urban environments, monitoring environmental impacts, economic growth and social inclusion [3].

Access to urban services are rapidly changing, as cycles of technological innovation are shorter, particularly regarding to the digitization and online communication systems that are advancing at a fast pace. Connectivity and big data have also allowed for a better evaluation of services and consequently an improvement in quality [4, 5].

In light of the above, this research lies down on a systematic literature review, which focuses on the challenges that urban mobility faces, specifically with regard to the public transport service. The previous option is justified by the relevance that this theme has for the citizens' lifestyle, especially those residing in SC [6, 7].

In the literature, the aspects related to public transport systems have been widely discussed; mainly from the point of view of the development of urban transport, with the sustainable panorama still being considered to a lesser extent [8]. As projects that involve both, urban transport and sustainability, are becoming more relevant, due to the pressure for better planning, greater sustainability and governance in transport, new approaches are also emerging. Therefore, this research, emphasizes those approaches that are integrating strategies developed by local authorities, which allow the development of more sustainable transportation alternatives. These alternatives tended to result in efficiency and system reliability, passenger comfort and the implementation of sustainability and safety policies [9].

Overall, this research tries to shed some light on the public transport systems by standing on the shoulders of those that have significantly contributed to the literature in this past 5 years. Additionally, our objective is to explain the development of the public transport system and its close connection with sustainability in SC. In particular on the development of innovative business models and the use of digital technologies that directly contributed to the economic growth, the protection of the environment and the health and safety of citizens.

The first section presents an overview of the 20 articles selected as well as data from 171 cities around the world addressed in the papers; following, we focus on the methodological process, which emphasized the inclusion and exclusion criteria; then, the results highlighted the review process and findings discussion; finally, the last section presented the conclusions and recommendations for future research.

2   An Overview

A great extent of mobility in urban areas is supported by public transport systems. However, in several cities the services offered by public transport still hold important inefficiencies as well as issues related with safety. Many systems are characterized with delays, long waiting times, embarkation and disembarkation procedures, etc. [10]. Therefore, major challenges remain to be addressed concerning the optimization of mobility through the implementation of efficient and effective urban transport services [6, 11].

According to Vanolo [12], transport systems represent a crucial factor in the structures of SC, by directly impacting in terms of sustainability, as a result of the use of ICTs and the access to citizen's data for offering emerging technologies and better services [3, 13]. This exchange of information will influence the relationship between transport companies and their stakeholders. Customer-centric services will be based on data from individual users and their needs, and this will allow, for example, real-time traffic management, increase in the frequency of public transport under specific timetable demands or during special events [6, 14].

Transport systems in urban areas involve environmental, social, economic and cultural concerns [8, 15]. Air pollution, traffic congestion, noise pollution, and loss of time in transfers, are some of the factors that affect the quality of life in cities, which, in addition to the impact on population health, also contribute to lost productivity and economic efficiency [16, 17]. For the European Commission [18], the economy of this continent loses about 100 billion euros annually due to traffic congestion as a source of pollution, accidents, productivity and efficiency in companies.

In environmental terms, the increase in population and access to improved purchasing power that allows the use of private cars as an essential element [19] has caused serious traffic problems, especially in large cities, which consequently increases pollution levels [6]. Staricco [20] estimates based on 27 European Union countries that 25% of greenhouse gases (GHG) emissions and more than 30% of the total energy consumed were due to the transport sector. When considering that more than 90% of the previous results come from non-renewable sources [15], several proposals for solutions have been considered in order to somehow reverse the negative impact that the use of private vehicles brings, such as the increase in the use of urban public services; bike path networks; shared transport services; etc. [21–23].

Although advanced economy countries are investing in the development of projects of public transport systems that focus on reducing environmental impacts, Roda et al. [15] mentions that the problem of emissions from vehicles with internal combustion engines is still relevant and caused mainly by the lack of capacity in public transport guaranteeing quality and comfort services, therefore stimulating the use of individual means of transport.

3 Research Methodology

This research draws on a systematic literature review (SLR). The SLR carried out in January 2020 focused on documents retrieved from the following electronic databases: SCOPUS and Web of Science (WoS). The documents ranged from 2015 to 2019, and the keywords used were: Sustainability, Smart Cities, Public Transport or Public Transportation, as shown in Table 1.

In the first phase 42 articles were identified, 24 from Scopus database and 18 from the Web of Science. The higher percentage of published articles ranged from 2017 to

Table 1 Literature review process

Keywords	"Smart cities" AND "sustainability" AND "public transport"
	"Smart cities" AND "sustainability" AND "public transportation"
Fields	Article title, Abstract, Keywords
Language	English
Source type	Journals and conferences
Document type	Articles and conference paper
Years	2015–2019

Table 2 Source and year of publication of articles identified

Year	"smart cities" AND "sustainability" AND "public transport"		"smart cities" AND "sustainability" AND "public transportation"		TOTAL
	SCOPUS	WoS	SCOPUS	WoS	
2015	2	0	1	1	4
2016	0	3	1	2	6
2017	3	2	4	3	12
2018	3	2	1	3	9
2019	5	2	4	0	11
TOTAL	13	9	11	9	42

2019, indicating that manuscripts based on the selected keywords are gaining greater interest in the scientific community. Table 2 shows the distribution list.

From the 42 papers identified in the first filtering criteria, 12 were duplicated, one was published in two different journals and, one document could not be accessed. From the remaining 28 articles, eight papers were partially related to the keywords, by using one or more of them only in the name of the Journal or Conference where they were published or in the name of participating institutions without any significant involvement in the investigation. Thus, 20 papers were selected, of which 6 correspond to articles published in conferences and 14 published in journals. Moreover, 43% of the journals are in Quartile 1 of the SCimargo Journal Rank (SJR). The SJR was considered ahead of Clarivate Analytics from Web of Science to calculate the reliability score since it is an open access search engine [24].

Figure 1 illustrates the process in a schematic way based on the Preferred Reporting Items for Systematic Reviews and Meta-Analyses (PRISMA) technique, a transparent process based on the evaluation of random documents focused on reducing the margins of error in the selection of publications during systematic reviews [25]. To increase the validity and reliability of the paper, the activities corresponding to each of the stages that set up the PRISMA technique were distributed among the authors. After the second author has collected and analyzed the data according to the content analysis technique, the first author reviewed the entire

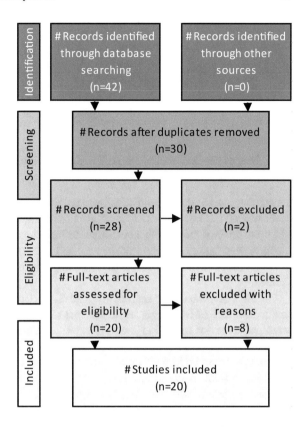

Fig. 1 Filtering process based on the PRISMA model

methodological process, the three authors were subsequently involved in the review process, in particular regarding with methodology. Once the process was finished, the triangulation of perspective was initiated in order to evaluate and interpret the data obtained in the 20 articles selected. Several aspects were discussed that would allow the categorization of the specific actions identified in the work, and, in case of disagreement, it was necessary to analyze the evidence more deeply and try to bring about the discrepancy as recommended by Stake [26].

4　Results and Discussion

4.1　Geographical Location

The information and geographical distribution on a global scale, representing the 171 cities retrieved from the reading of the selected articles are shown in Fig. 2. Based on this, it is possible to identify, in the first place, the localities that in the last 5 years have been investing in the areas of sustainability and public transport. The three most

Fig. 2 Geographic distribution of SC considered in the literature review

relevant regions in this type of initiatives are Europe, Australia and North America; nevertheless, the European Union effort is remarkable, mainly Italy and Spain that add up to 75% of the 89 cities belonging to that continent. In the case of Australia, 27 cities were mentioned in 3 of the selected articles and in North America, 67% of the 21 cities mentioned in 4 articles belong to the United States, and 7 cities to Canada.

On the other hand, the regions of Asia, Latin America and Africa included a smaller number of SC compared to the rest of the world. In Asia, the most representative countries regarding the application of sustainable policies in the public transport service in SC are China and India with four cities each. Latin America and Africa is represented in most cases only by capital cities, with the exception of two countries in Latin America and two in Africa that consider alternative cities: León and Guadalajara in Mexico, São Paulo and Rio de Janeiro in Brazil, Johannesburg in South Africa and Minna in Nigeria. Figure 3 presents the information on the 171 cities included in the 20 studies analyzed.

4.2 Specific Actions

The analysis of the selected papers aimed at identifying projects and practices concerning public transport systems in the context of SC. To this end, the papers were read in full and each project addressed in the text was attributed a classification. The coding process was iterative, and the classifications attributed were revised and refined throughout the process. The projects and cases concerning public transport systems were assigned to the proposed classification groups according to its characteristics and the degree of similarity with other projects in each group. A total of 7 distinct categories were identified and characterized with the purpose of highlighting the relationship between public transport systems and their impact on the sustainability of SC.

Transport Policies. With the application of policies in SC, it is increasingly common to notice changes in the social behavior of citizens regarding the sustainable use of resources [9]. The implementation of transport policies is associated to the strategy

AUSTRALIA	USA-Atlanta	ESP-Castellón Plana	GBR-London	NLD-Amsterdam
AUS-Adelaide	USA-Austin	ESP-Ciudad Real	GBR-Milton keynes	NLD-Rotterdam
AUS-Boroondara	USA-Boston	ESP-Córdoba	ITA-Ancona	POL-Warsaw
AUS-Cambridge	USA-Denver	ESP-Getafe	ITA-Bari	PRT-Porto
AUS-Chittering	USA-Fort Worth	ESP-Gijón	ITA-Bologna	SWE-Stockholm
AUS-Dumbleyung	USA-Los Angeles	ESP-Huelva	ITA-Bolzano	SWE-Västerås
AUS-Gingin	USA-Miami	ESP-Huesca	ITA-Brescia	TUR-Istanbul
AUS-Gr. Melbourne	USA-Minneapolis	ESP-Jaén	ITA-Cagliari	ASIA
AUS-Hornsby	USA-Nebraska	ESP-Logroño	ITA-Campobasso	ARE-Dubai
AUS-Joondalup	USA-New York	ESP-Lugo	ITA-Catania	CHN-Beijing
AUS-Kent	USA-Portland	ESP-Madrid	ITA-Catanzaro	CHN-Hong Kong
AUS-Ku-ring-gai	USA-San Diego	ESP-Majadahonda	ITA-Firenze	CHN-Shanghai
AUS-Lane Cove	USA-San Francisco	ESP-Málaga	ITA-Florence	CHN-Shenzhen
AUS-Lower Eyre P.	USA-Washington DC	ESP-Móstoles	ITA-Frosinone	IDN-Semarang
AUS-Mallala	LATIN AMERICA	ESP-Motril	ITA-Genova	IND-Almaty
AUS-Manningham	ARG-Buenos Aires	ESP-Murcia	ITA-Lecce	IND-Delhi
AUS-Melbourne	BRA-Rio de Janeiro	ESP-Oviedo	ITA-Messina	IND-Kampala
AUS-Mosman	BRA-Sao Paulo	ESP-P. de Mallorca	ITA-Milan	IND-Mumbai
AUS-Narrogin	CHL-Santiago	ESP-Paterna	ITA-Napoli	JPN-Tokyo
AUS-Nedlands	COL-Bogota	ESP-R. Vaciamadrid	ITA-Palermo	KOR-Seoul
AUS-Nillumbik	ECU-Quito	ESP-Salamanca	ITA-Piacenza	MYS-Kajang
AUS-Peppermint Gr.	MEX-Guadalajara	ESP-S. C. Laguna	ITA-Potenza	MYS-Singapore
AUS-Pittwater	MEX-Leon	ESP-St C. del Vallés	ITA-R. Calabria	PHL-Makati
AUS-Sydney	MEX-Mexico City	ESP-Santander	ITA-Rimini	RUS-Moscow
AUS-Wandering	PER-Lima	ESP-S. de Compostela	ITA-Rome	RUS-St Petersburg
AUS-Wickepin	EUROPE	ESP-Segovia	ITA-Salerno	SAL-Makkah
AUS-Willoughby	AUT-Vienna	ESP-Sevilla	ITA-Silver Coast	TWN-Taipei
AUS-Woodanilling	CHE-Zurich	ESP-Tarragona	ITA-Teramo	VNM-Haiphong
NORTH AMERICA	DEU-Berlin	ESP-Toledo	ITA-Terni	AFRICA
CAN-Montreal	DEU-Frankfurt	ESP-Valencia	ITA-Torino	KEN-Nairobi
CAN-Saint-Augustin	DEU-Munich	ESP-Valladolid	ITA-Treviso	ZAF-Johannesburg
CAN-Shawinigan	DEU-Solingen	ESP-Vitoria-Gasteiz	ITA-Turin	JOR-Amman
CAN-Surrey	DNK-Copenhagen	ESP-Zaragoza	ITA-Udine	NGA-Minna
CAN-Toronto	ESP-Alcobendas	FIN-Helsinki	IRL-Dublin	
CAN-Vancouver	ESP-Alicante	FRA-Paris	LUX-Luxembourg	
CAN-Vaughan	ESP-Barcelona	GBR-Edinburgh	LVA-Riga	

Fig. 3 Smart cities considered in the systematic review

in the management of sustainable conditions that integrate ICT in the infrastructure of the city, creating viable solutions regarding the mobility of citizens through public services [27, 28].

In recent years, government entities belonging to the European Union have prioritized the implementation of sustainable transport policies through encouraging the use of low-emission vehicles, active travel (cyclists and pedestrians), public transport and/or shared mobility; establishing a cost-benefit balance and contributing significantly to the reduction of pollution levels based on initiatives, such as using smaller vehicles or the adequacy of the offer according to the demand in rotating schedules [5, 6, 29].

The application of transport policies requires an adequate, flexible and compatible infrastructure with the challenges that the emerging SC demand. Thus, the public transport must be accessible, efficient, safe, comfortable and ecologically viable. But, at the same time, it must maintain adequate levels of service quality and be accessible to all economic sectors of the society [27, 30].

It is clear that urban mobility is largely influenced by public and private policies [29]; therefore, both sectors must work in synergy to integrate technological tools as part of smart services between conventional and intelligent transport systems. For instance, Bus Rapid Transit (BRT) and ridesharing schemes, where passengers are matched to spare seats in private car journeys [29, 31]. The latter is a controversial alternative, because it is considered an unfair competitor because it directly affects individual public transport services (taxis). For that reason, in several regions, legal efforts have been made to try to ban the existence of services such as UBER, Cabify, etc. [29].

This integration is defined by Roda et al. [15] from a technological and sustainable point of view, considering three types of infrastructure: (1) an intermodal system of sustainable mobility, (2) an energy infrastructure based on renewal sources for vehicle fleets and, (3) an ICT structure based on Internet of Things (IoT), Open Data, sensors, information security management and Geographic Information Systems (GIS).

Planned Communities. The close link between citizens and technology define the concept of SC development, since emerging technologies need to integrate and respond to the needs and habits of citizens in order to effectively improve urban sustainability [3]. The planning or regeneration of urban centers that incorporate ICT is committed to creating attractive environments based on sustainable development [32]. As a result, cities around the world rely on action plans to assess specific conditions for the proper use of their resources in sustainable terms.

In Italy, the ITACA Protocol is mainly used to support specific policies to promote sustainable construction [33], and places the accessibility of public transport as one of the main criteria in terms of the quality of the location for the design and development of urban enterprises sustainable [32]. Likewise, the ELAN project applied in cities in Slovenia, Belgium, Croatia, Czech Republic and Portugal, aims to improve the perception of cities and the quality of life by changing the behavior of mobility while the shared use of cars and improving the public transport service infrastructure through real-time data transmission [11, 28]. On the other hand, some cities in Spain rely on the Sustainable Urban Mobility Plans (SUMP), which provides multiple views of sustainable mobility when evaluated by various criteria that aim to improve the general infrastructure and service of a given location [17].

Behaviors. The applicability of ICT in public services has developed behavioral changes in citizens of SC [34]. According to Sunstein [35], some cities provide a set of personalized and persuasive interventions, in order to encourage individuals to choose public transport over their private vehicles. For example, in Durham, North Carolina, citizens were provided with personalized route maps and their options for commuting between home and work in various types of vehicles, considering that less availability of public transport is associated with greater concentration of private vehicles [3]. In Enschede, the Netherlands, an application was developed that encourages people to take different routes, in order to avoid using private vehicles and prefer the use of public transport, cycling or walking [34]. Variations between the use of public and private transport were also linked to the capacity of data transmission over broadband. Yigitcanlar and Kamruzzaman [36] research highlights the negative relationship between Internet access, sustainable travel patterns and remote work, resulting from the decrease in the use of public transport and the considerable increase in the use of private transport, possibly the cause of the fragmentation of work activities.

Another model of behavior identified in SC while using public transport is the connection of citizens with their cities, although the research by Belanche et al. [23] indicates that being connected to a city is not enough to generate a certain behavior, this emotional bond can come to positively influence the affective-evaluative perceptions of urban services, such as the use of public transport. On the other hand, Vierling and Schmuelling [29] presents a different point of view to that of users and focuses on changing the behavior of drivers when facing new technologies incorporated into more recent bus models. The BOB project applied to the city of Solingen in Germany, aims to develop an information system based on a training program that allows drivers to read the vehicle parameters effectively and adapt to the new demands required to offer quality services.

Transport Sharing. The advancements of urban mobility faithfully follow the emergence and application of ICT to urban centers, placing them in appropriate positions to promote changes that suit citizens' needs. However, current mobility systems, especially public transport services, still remain unsustainable [6]. In recent years, several research works have emerged regarding the variety of innovative and sustainable business models that allow individuals to share their mobility using technology. According with Roda et al. [15] these initiatives could contribute to the re-planning of public transport systems and, in the long term, contribute to the reduction of the motorization rate, which, in addition to the environmental advantages, these measures would allow citizens to benefit from the reduction of space occupied by private vehicles and the economic advantages when preferring the use of alternative mobility systems.

In particular, bicycle sharing systems (BSS) are attributed several advantages in the urban context. Firstly, it promotes the reduction of emissions by largely replacing any other type of motorized transport, including public transport [37], likewise, the use of these systems promotes the health benefits of citizens while the number of calories burned with each use [38]. On the other hand, car sharing services are gaining more space in urban mobility in SC. For Pinna et al. [5], the recent and rapid evolution of this service in recent years is due to the inclusion of the service through fleets of electric cars. Although it is considered as a complementary option and not as an alternative, the personalized service compared to traditional public transport plays an important factor in choosing as a means of transport.

Integrated Services. These services offer great benefits for the environment, directly impacting the CO_2 reduction of large cities. Intermodality is recognized as an integral part of improvements in public transport [17], not only focused on the means of transportation themselves, but on the infrastructure that offers a comprehensive transport service, therefore, pedestrian paths, mobility platforms, columns of mobility. recharging for electric vehicles, bike lanes and e-buses, are considered as part of this sector [15].

Indicators. In the same way that several protocols have been applied for the management of public transport services, the use of evaluation indicators is frequently being used for the management of SC. Table 3 summarizes the indicators used by several

Table 3 Indicators found in the literature review

Authors	Indicators
Abdullahi et al. [39]	Promoting public transportation facilities.
	Increase population density especially around public transportation nodes
Mozos-Blanco et al. [17]	Public transport
	Cycling
	Intermodality
	Accessibility
	Air and noise pollution
	Public participation
	Indicators
Pinna et al. [5]	Public transport
	Cycle lanes
	Bike sharing
	Car sharing
Sharma and Agrawal [7]	Km of high capacity public transport system per 100,000 Population (h)
	Km of light passenger transport system per 100,000 population (l)
	Annual number of public transport trips per capita (p)
	Percentage of commuters using a travel mode to work other than a personal vehicle
	Km of bicycle paths and lanes per 100,000 population (b)
	Greenhouse Gas (GHG) Emissions in tonnes per capita
	Noise Pollution
Shmelev and Schmeleva [40]	Number of underground stations per million inhabitants
	CO_2 emissions per person per year (tonnes)
	Citizens walking, cycling or taking public transport to work (%)
Vassileva et al. [41]	Transport performance in public transport
	Energy demand in public transport
	CO_2 emissions in public transport
	Cost of a monthly ticket for transport
Wu et al. [42]	Provide a variety of transportation choices

authors and which only correspond directly to those concerning urban mobility, specifically to public transport systems.

The indicators above are the result of several analysis and case studies applied to different cities around the world. Abdullahi et al. [39] presents a series of indicators resulting from the application of geographic information systems and radar remote sensing technology and synthetic aperture radar (SAR) data that analyzes the urban sustainability of the city of Kajang in Malaysia, these indicators are defined in three categories: density, mixed and intensity; the latter considered as the main parameter to determine the degree of compaction of an urban area making it more sustainable, and this category includes the two specific indicators referring to public transport services, mentioned by the author. On the other hand, the 7 indicators mentioned by Mozos-Blanco [17] are part of the Sustainable Urban Mobolity Plans (SUMP) mentioned in the "planned communities" section, indicators selected from a total of 15 and used in the evaluation of the general mobility plan applied to 38 cities in Spain.

In the case of Italy, 22 cities were selected to be evaluated using the indicators presented by Pinna [5] who, in addition to the general criteria shown in Table 3, collected data on the density of the bus network per square kilometer; the demand for public transport for passengers per year; the density of cycle paths for every 100 km^2 and for every ten thousand inhabitants; and, the density of bicycle stations by the number of stations per 100 km^2 and per ten thousand inhabitants. Sharma and Agrawal [7] use a convenience sample of 30 SC in order to assess transport and environmental pollution parameters from public data. Shmelev and Schmeleva [40] are based on the results obtained through a multiple criteria approach using 20 indicators applied to 57 cities, of which three referred directly to public transport services. Vassielela [41] focuses on Sweden with the application of 4 indicators in public transport that considers the impact of enabling technology on energy efficiency indicators; finally, Wu et al. [42] considers the city of Zurich in Switzerland to apply one of the ten indicators that assesses the smart growth of the city and that is related to the public transport services of this location.

Data. The data transfer is one of the essential elements in SC. The public transport systems of these urban centers are opportunely considering the opinion of users when planning and creating technological tools that contribute to improving services through the analysis and exchange of information regarding traffic reports, transportation schedules, travel preferences and online ticket purchase platforms [43]. These initiatives allow users to understand the selection of a particular type of transport, as long as access to data related to habits and preferences, as well as specific objectives and needs is allowed. On the other hand, the citizens´ feedback on improving services are essential for the development of an efficient and sustainable city [6].

Currently, the use of social networks has facilitated the sharing of information between users and public transport providers, which allows a more effective interaction and allows taking advantage of collaborative data technologies such as crowdsourcing [44]. In the University of Nebraska Omaha (UNO) for example, pressure from the student community contributed to the creation of a public and sustainable mobility plan within the campus through algorithms that models and identifies potential areas for optimizing traffic routes, relieves tension of the user and reduces the carbon footprint by reducing CO_2 emissions by avoiding the use of private cars within campuses [45].

5 Conclusions

This paper gathers relevant insights from the literature on public transport systems and its sustainable impact in 171 cities around the world. This paper evidences that the research developed in the last five years has focused on ICT applications in urban mobility services and its influence on the quality of life of people living in SC.

The improvement of transport systems in those locations has basically depended on seven applications: transport policies; planned communities; evaluation of citizens behavior regarding to the use of public transport systems; transport sharing and integrated mobility; transport indicators and the analysis of data collected through ICT. The results also focused on the application of indicators, which showed a comprehensive approach by involving sectors not only related to public transport but also various factors that interfere with the environmental impact of SC. Those indicators are such as the development of public policies, territorial planning and the increase in citizens' awareness of environmental issues [7, 17].

This research is based on a systematic review of the literature considering documents published in two of the main databases; therefore, only secondary data were used. Since this paper addresses a relatively recent theme, the object of study was limited to an exploratory methodology through the identification of areas of investigation and application of several projects, for this reason, there was no calculation of publications bias.

Nonetheless, the results suggest that the relationship between the increase in terms of public transport usage and the decrease in the use of private cars was considered as one of the most significant results with significant environmental impact in SC. However, little is known about the advantages of specific type of public transport, such as railway infrastructure or fleets of electric buses. Therefore, it is intended that future research will emphasize even more specific links with regard to environmental variations and sustainable impacts when using electric public transports in cities that significantly involve the use of ICT in their operations.

Acknowledgements This work was financially supported by the research unit on Governance, Competitiveness and Public Policy (project POCI-01-0145-FEDER-008540), funded by FEDER funds through COMPETE 2020—Programa Operacional Competitividade e Internacionalização (POCI)—and by national funds through FCT—Fundação para a Ciência e a Tecnologia.

References

1. Caragliu, A., Bo, C. de, and Nijkamp, P. Smart cities in Europe. 3rd Central European Conference in Regional Science – CERS (2019).
2. Neirotti, P., Marco, A. de, Cagliano, A. C., Mangano, G., and Scorrano, F. Current trends in Smart City initiatives: Some stylised facts. Cities, 38, 25–36 (2014). https://doi.org/10.1016/j.cities.2013.12.010.
3. Papa, R., Gargiulo, C., and Russo, L. The evolution of smart mobility strategies and behaviors to build the smart city. 5th IEEE International Conference on Models and Technologies for

Intelligent Transportation Systems, MT-ITS 2017 - Proceedings, 409–414 (2017). https://doi.org/10.1109/MTITS.2017.8005707.

4. Hofhuis, P., Luining, M., and Rood, J. EU Transition towards green and smart mobility. Action Toolbox to Unleash Innovation Potentials. The Hague (2016).

5. Pinna, F., Masala, F., and Garau, C. Urban policies and mobility trends in Italian smart cities. Sustainability, 9(4), (2017). https://doi.org/10.3390/su9040494.

6. Cruz, R. Smart Rail for Smart Mobility. 16th International Conference on Intelligent Transportation Systems Telecommunications, 1(7) (2018).

7. Sharma, N., and Agrawal, R. Prioritizing environmental and transportation indicators in global smart cities: Key takeaway from select cities across the globe. Nature Environment and Pollution Technology, 16(3), 727–736 (2017).

8. Patlins, A. Improvement of Sustainability Definition Facilitating Sustainable Development of Public Transport System. Procedia Engineering, 192, 659–664 (2017). https://doi.org/10.1016/j.proeng.2017.06.114.

9. Haque, M.M., Chin, H.C., and Debnath, A.K. Sustainable, safe, smart—three key elements of Singapore's evolving transport policies. Transport Policy, 27, 20–31 (2013). https://doi.org/10.1016/j.tranpol.2012.11.017.

10. Kamau, J., Ahmed, A., Rebeiro-H, A., Kitaoka, H., Okajima, H., and Ripon, Z. Demand responsive mobility as a service. IEEE International Conference on Systems, Man, and Cybernetics (SMC), 001 741–001 746 (2016).

11. European Commission. European Urban Mobility: Policy Context. Technical Report, (2017). [Online]. Available: http://civitas.eu/document/.

12. Vanolo, A. Smartmentality: The smart city as disciplinary strategy. Urban Studies, 51(5), 883–898 (2014).

13. Rivera, R., Amorim, M., and Reis, J. Robotic Services in Smart Cities: An Exploratory Literature Review. IEEE 15th Iberian Conference on Information Systems and Technologies (CISTI) (2020). Forthcoming.

14. Kitchin, R. The Real-Time City? Big Data and Smart Urbanism. GeoJournal, 79, 1–14 (2014).

15. Roda M., Giorgi, D., Joime, G. P., Anniballi, L., London, M., Paschero, M., and Mascioli, F.M.F. An integrated methodology model for smart mobility system applied to sustainable tourism. IEEE 3rd International Forum on Research and Technology for Society and Industry. Conference Proceeding, 0–5 (2017). https://doi.org/10.1109/RTSI.2017.8065912.

16. Benevolo, C., Dameri, R.P., and D'Auria, B. Smart mobility in smart city. In T. Torre, A. M. Braccini, and R. Spinelli, Eds. Cham: Springer International Publishing. Empowering Organizations: Enabling Platforms and Artefacts 11, 13–28 (2016).

17. Mozos-Blanco, M. A., Pozo-Menéndez, E., Arce-Ruiz, R., and Baucells-Aletà, N. The way to sustainable mobility. A comparative analysis of sustainable mobility plans in Spain. Transport Policy, 72, (October) 45–54 (2018). https://doi.org/10.1016/j.tranpol.2018.07.001.

18. European Commission. Towards a New Culture for Urban Mobility. Directorate-General for Energy and Transport, 1–6 (2007).

19. Gakenheimer, R. Urban mobility in the developing world. Transportation Research Part A: Policy and Practice, 33(7), 671 – 689 (1999).

20. Staricco, L. Smart Mobility Opportunities and Conditions. Jornal of Land Use, Mobility Environment, 6 (3), 341–354 (2013). https://doi.org/10.6092/1970-9870/1933.

21. Tiwari, R., Cervero, R., and Schipper, L. Driving CO2 reduction by Integrating Transport and Urban Design strategies. Cities, 28(5), 394–405 (2011). https://doi.org/10.1016/j.cities.2011.05.005.

22. Mulley, C., and Moutou, C.J. Not too late to learn from the Sydney Olympics experience: Opportunities offered by multimodality in current transport policy. Cities, 45, 117–122 (2015). https://doi.org/10.1016/j.cities.2014.10.004.

23. Belanche, D., Casaló, L.V., and Orús, C. City attachment and use of urban services: Benefits for smart cities. Cities, 50, 75–81 (2016). https://doi.org/10.1016/j.cities.2015.08.016.

24. Scopus. How is SJR (SCImago Journal Rank) used in Scopus? (2019). [Online]. https://service.elsevier.com/app/answers/detail/a_id/14883/supporthub/scopus/kw/scimago/.

25. Liberati, A., Altman, D. G., Tetzlaff, J., Mulrow, C., Gøtzsche, P. C., Ioannidis, J. P. A., Clarke, M., Devereaux, P. J., Kleijnen, J., and Moher, D. The PRISMA statement for reporting systematic reviews and meta-analyses of studies that evaluate health care interventions: explanation and elaboration. Journal of Clinical Epidemiology, 62, e1–e34 (2009).
26. Stake, R.R. Qualitative Research: Studying how Things Work. New York, (2010).
27. Budi, P.A., Buchori, I., Riyanto, B., and Basuki, Y. Smart mobility for rural areas: Effect of transport policy and practice. International Journal of Scientific and Technology Research, 8 (11), 239–243 (2019).
28. Evangelinos, C., Tscharaktschiew, S., Marcucci, E., and Gatta, V. Pricing workplace parking via cash-out: Effects on modal choice and implications for transport policy. Transportation Research Part A: Policy and Practice, 113, 369–380 (2018). https://doi.org/10.1016/j.tra.2018.04.025.
29. Vierling, D., and Schmuelling, B. Driver Information System for sustainable public transportation. 3rd International Conference on Sustainable Information Engineering and Technology, 278–282 (2018). https://doi.org/10.1109/SIET.2018.8693186.
30. Shi, Y., Arthanari, T., Liu, X., and Yang, B. Sustainable transportation management: Integrated modeling and support. Journal of Cleaning Production, 212, 1381–1395 (2019). https://doi.org/10.1016/j.jclepro.2018.11.209.
31. Hyland, M.F., and Mahmassani, H.S. Taxonomy of Shared Autonomous Vehicle Fleet Management Problems to Inform Future Transportation Mobility. Journal of the Transportation Research Board, 2653, 26–34 (2017).
32. Marino, F.P.R., Lembo, F., and Fanuele, V. Towards more sustainable patterns of urban development. IOP Conference Series: Earth and Environmental Science, 297(1), (2019). https://doi.org/10.1088/1755-1315/297/1/012028.
33. Moro, A. Transnational comparison of instruments according to ecological evaluation of public buildings. Italy (2011).
34. Ranchordás, S. Nudging citizens through technology in smart cities. International Review of Law, Computers and Technology, 0(0), 1–23 (2019). https://doi.org/10.1080/13600869.2019.1590928.
35. Sunstein, C.R. The Ethics of Nudging. Yale Journal of Regulation, 32, 413–451 (2015).
36. Yigitcanlar, T., and Kamruzzaman, M. Smart Cities and Mobility: Does the Smartness of Australian Cities Lead to Sustainable Commuting Patterns?. Journal of Urban Technology, 26(2), 21–46 (2019). https://doi.org/10.1080/10630732.2018.1476794.
37. Fishman, E., Washington, S., and Haworth, N. Bike Share: A Synthesis of the Literature. Transport Reviews, 33(2), 148–165 (2013). https://doi.org/10.1080/01441647.2013.775612.
38. Médard de Chardon, C. The contradictions of bike-share benefits, purposes and outcomes. Transportation Research Part A: Policy and Practice, 121(January), 401–419 (2019). https://doi.org/10.1016/j.tra.2019.01.031.
39. Abdullahi, S., Pradhan, B., and Jebur, M.N. GIS-based sustainable city compactness assessment using integration of MCDM, Bayes theorem and RADAR technology. Geocarto International, 30(4), 365–387 (2015). https://doi.org/10.1080/10106049.2014.911967.
40. Shmelev, S.E., and Shmeleva, I.A. Global urban sustainability assessment: A multidimensional approach. Sustainable Development, 26(6), 904–920 (2018). https://doi.org/10.1002/sd.1887.
41. Vassileva, I., Campillo, J., and Schwede, S. Technology assessment of the two most relevant aspects for improving urban energy efficiency identified in six mid-sized European cities from case studies in Sweden. Applied Energy, 194, 808–818 (2017). https://doi.org/10.1016/j.apenergy.2016.07.097.
42. Wu, H., Yin, L., Zhou, T., and Ye, S. City Smart-Growth Evaluation System. IEEE International Conference on Smart Grid and Smart Cities, 1, 293–297 (2017).

43. Bell, S., Benatti, F., Edwards, N. R., Laney, R., Morse, D. R., Piccolo, L., and Zanetti, O. Smart Cities and M3: Rapid Research, Meaningful Metrics and Co-Design. Systemic Practice and Action Research, 31(1), 27–53 (2018). https://doi.org/10.1007/s11213-017-9415-x.
44. Olaverri-Monreal, C. Intelligent technologies for mobility in smart cities. Hiradtechnika Journal, 71, 29–34 (2016).
45. Nelson, Q., Steffensmeier, D., and Pawaskar, S. A Simple Approach for Sustainable Transportation Systems in Smart Cities: A Graph Theory Model. IEEE Conference on Technologies for Sustainability, SusTech 2018, 1–5 (2019). https://doi.org/10.1109/SusTech.2018.8671384.

Eco-Innovation and Digital Transformation Relationship: Circular Economy as a Focal Point

Adriane Cavalieri◉, Marlene Amorim◉, and João Reis◉

Abstract The objective of the study is to understand the relationship between Eco-innovation (EI) and Digital Transformation (DT) to environmental sustainability in manufacturing sector. In this sense, the central question is: what are the knowledge domains that integrate EI and DT in manufacturing sector? What are their inter-relationships? The study is a qualitative research methodology using PRISMA process "Preferred Reports Items for Systematic Reviews and Meta-Analyzes". Two computer programs support the development of this study: VOSviewer and ZOTERO. The research is a Tertiary study carried out through Web of Science (WoS) database, which papers should be in English language to avoid definitions ambiguity, and the period is limited from 2015 until 2020 (August). DT is believed a facilitator to Circular Economy (CE), although the discourse is emphasizing its main principles. This indicates that, DT to boost the environment sustainability should be aligned and fit the EI definitions, even diverse, to guarantee successful CE implementation, not a run to digitalized way of life. Research was based on one scientific database Web of Science (WoS). The contribution of this paper is a presentation of diverse applications of new digital technologies with a single objective of CE implementation. The originality is the application of VOSviewer for approaches which concepts are constantly developing as EI, and the study of the relationship between EI and DT to environmental sustainability in manufacturing industry, embodied by the concept of Circular Economy.

A. Cavalieri (✉) · M. Amorim · J. Reis
Department of Economics Management, Industrial Engineering and Tourism, GOVCOPP, Aveiro University, Campus Universitário de Santiago, 3810-193 Aveiro, Portugal
e-mail: adriane.cavalieri@int.gov.br

M. Amorim
e-mail: mamorim@ua.pt

J. Reis
e-mail: reis.joao@ua.pt

A. Cavalieri
National Institute of Technology (Avaliações e Processos Industriais, Instituto Nacional de Tecnologia/Ministério da Ciência, Tecnologia e Inovações), Av. Venezuela 82, 20081-312 Rio de Janeiro, Brazil

© The Author(s), under exclusive license to Springer Nature Switzerland AG 2021   49
A. M. Tavares Thomé et al. (eds.), *Industrial Engineering and Operations Management*,
Springer Proceedings in Mathematics & Statistics 367,
https://doi.org/10.1007/978-3-030-78570-3_4

Keywords Eco-Innovation · Digital transformation · Circular economy ·
VOSviewer · PRISMA

1 Introduction

Reis et al. [1] define Digital Transformation as "(…) the use of new digital technologies that enables major business improvements and influences all aspects of
customers' life" (p. 418). The overarching umbrella concept of new digital technologies include, for example, the "Cyberphysical Systems", "Cloud Computing", "Big
Data and Analytics", "Artificial Intelligence", "Internet of Services", "Internet of
Things", "Internet of Contents and Knowledge" and "Internet by and for People".
They allow for the interconnectedness of human beings, computers, knowledge,
services and objects in a highly integrated system [2]. Organizations may have to
endure profound changes during the implementation of Digital Transformation, especially when it involves technologies that connect the physical and the real world in
real time [3].

Digital Transformation may contribute to increase sustainability of operations
by improving the utilization of resources such as water, energy and raw materials.
Such innovations that significantly reduce the impact of production activities in the
environment have been labeled as Eco-Innovation [3].

The term Eco-Innovation was coined in 1996 as a strategy to include environmental issues in the process of developing new products and services leading to both
environmentally efficient and economically profitable results [4]. Nowadays, Eco-
Innovation has been considered a critical element in the transition to a low-carbon
economy [5] and has produced social, economic and environmental impacts [6].

Ketata et al. [7] mention that earlier studies focused on Environmental, Green or
Eco-Innovation and Environmental Management Systems didn´t include the social
dimension. Gotsch et al. [8] affirm that the term "Eco-Innovation" refers to the same
type of innovation as "Green innovation" and "Environmental innovation", that is,
innovations that reduce or prevent the environmentally harmful impacts related to
the use or disposal of the innovative object. de Jesus et al. [9] explain that Eco-
Innovation is defined, today, as a way to enable economic performance without
impeding sustainable development, a definition positively seen by the European
Commission [9].

The scope of Eco-Innovation is extensive, and may include public environmental policies; supply, technology and demand management; and performance
measurement methodologies [4]. Several models have been proposed to support
the integration and implementation of Eco-Innovation initiatives in the Organization
[10].

In this sense, Eco-Innovation and Digital Transformation seem to have many
intersections. Kuo and Smith [4] offer examples of the relevance of new digital
technologies for sustainability, namely trough the adoption of Internet of things

(IoT) to reduce energy consumption. Likewise, big data is being implemented to help enterprises understand their sustainability performance.

It would be useful to synthesize and consolidate the research work that address their interrelationship, identifying gaps that could be addressed in future research efforts. Notwithstanding, as Abedinnia et al. [11] affirm "reviewing the entire domain in a single literature review is often prohibitive. Tertiary studies (i.e., reviews of literature reviews) may support structuring and synthesizing a research area in this case, as their object of analysis are (fewer) literature reviews instead of a prohibitively large number of primary research papers" (p. 404).

This work presents the results of a systematic review, building on secondary articles [12], commonly labelled as a tertiary study, to understand the relationship between Eco-Innovation and Digital Transformation. The study identifies common knowledge domains across Eco-Innovation and Digital Transformation, in manufacturing. The central question of this research can be put forward as follows: what are the knowledge domains that integrate Eco-Innovation and Digital Transformation in manufacturing sector? What are their inter-relationships?

Despite the variety in Eco-Innovation definitions [13], this study subscribes to the environmental sustainability orientation. The study also contributes to the academic context by the application of VOSviewer for addressing approaches which concepts are constantly evolving that allow for the condensation in an article on different views, while offering a contribution for managerial practice with the presentation of diverse applications of new digital technologies with a single objective of Circular Economy.

This paper is organized as follows. Section 2 introduces the main concepts concerning this paper. Section 3 addresses the research methodology. Section 4 presents the qualitative synthesis of the relationship between Eco-innovation and Digital Transformation in the context of manufacturing sector. Section 5 summarizes the results and discussions, and Sect. 6 presents the conclusions, limitations and directions for future research.

2 Background

The Sustainable Development paradigm was proposed in 1987 by UN's World Commission for Environment and Development, as "development that meets the needs of the present without compromising the ability of future generations to meet their own needs". Around that occasion, the debate on the adoption of Eco-Innovation (EI) had emerged as an enabler of corporate sustainability [14]. Presently, EI is one of the targets of the United Nations' 2030 Agenda for Sustainable Development and Sustainable Development Goals [15].

EI gave rise to an extensive literature, and its definition was reformulated by several scholars [17], making it difficult to reach a consensual understanding of the concept [4].

de Jesus et al. [9] define Eco-Innovation as any innovation incremental or radical that has positive environmental impacts while avoiding damage to natural capital. According to this perspective, EI has implications for cost efficiency, market improvement or regulatory considerations, in order to deliver new or improved goods and services, technological and non-technological processes, marketing or organizational schemes.

Hazarika and Zhang [13] argue that empirical work about EI remains disintegrated and limited, despite being discussed by academics, and addressed by the lenses of several theories. They believe that EI is "best understood from the co-evolutionary approach, putting into perspective the process of variation, selection, and retention. An integration of Eco-innovation and sustainability transition literature would be most beneficial." (p. 76). They [13] explain that co-evolutionary approach is "inclusive of all sub-systems interacting back and forth without any hierarchy or ranking of importance of any one of the sub-systems, thus preventing the occurrence of technological bias" (p. 74).

EI has been perceived as a core driver for a transition to sustainability, indeed considered central for enabling Circular Economy (CE) [9]. The CE strategy first appeared in literature in the early nineties [16].

Reuter et al. [17] argue that numerous definitions and interpretations concerning CE can be found in the literature. Reviewing the extensive literature about the subject, de Jesus et al. [9] proposed CE as an integrative concept for achieving "clean congruence" ("a link between CE and EI bodies of work") (p. 3000) to guide new institutional arrangements that consider environmental and socioeconomic aspects, promoting technical and economic development without depending on the consumption of finite resources. CE is a multilevel structure that reforms and redirects production and business models towards resilience and sustainability, in order to minimize resource extraction, maximize reuse, increase efficiency and improve waste recycling.

CE focus, mainly, in minimizing negative environmental and societal impacts by the design of processes and products [9], based on closed-loops like a biological life cycle [18], following the "Rs" loops, which replace the "end-of-life" concept [16].

Martins and Pato [12] affirm that CE is an emerging concept in the sustainable supply chain literature. Sustainable supply chain management (SSCM) consists on product, services, finances and information shared across the value chain following sustainability principles, as economically managed and integrated, connecting supplier, manufacturers, customers, several layers of supplier in upstream side to obtain optimum productivity, and end users who are benefited from the value of the product or services [19]. The literature reveals that various organizations are taking measures towards sustainability, demonstrating that they realize SSCM as a pressure. Eco-friendly factors expand the SSCM into a 'Green SCM' (GSCM), considering the environmental influences on SSCM. Digital supply chain should be considered as key driver of GSCM [22].

Extant research also raises questions about digitalization and innovation to explore the relationship between Eco-Innovation and Digital Transformation phenomena. Thus, the concepts of innovation and digitalization should be clarified.

Innovations can be incremental, radical or disruptive. Incremental innovations maintain the status quo, where changes should focus on resource efficiencies, while radical innovations introducing distinct set of characteristics to existing companies, products, technologies and customers, but disruptive innovations incorporate exceedingly new changes in technology to offer substantially benefits to customer [20].

Digital Transformation in the context of manufacturing represents the transition from previous industrial stages towards an Industry 4.0 concept [21]. Industry 4.0 involves business dimensions such as manufacturing, product development, supply chain and working processes, which are integrated and interconnected, and whose adoption involves the "front-end technologies" (smart manufacturing, smart products, smart supply chain and smart working) and the "base technologies" (internet of things, cloud services, big data and analytics) [22].

Chalmeta and Santos-de Leon [3] states that experts believe that sustainable industrial value creation will be boosted by Industry 4.0, promoting an efficient allocation of resources, as water, energy, raw materials and other products, based on data collected in real time, resulting in new sustainable green practices.

3 Research Methodology

3.1 Method

The study presented is a qualitative research contribution building on a Systematic Literature Review [23] through PRISMA process "Preferred Reports Items for Systematic Reviews and Meta-Analyzes" to guide the fulfillment of the proposed research objective. The phases of PRISMA are: (1) Identification Phase: papers identified through the databases, (2) Screening Phase: abstracts checked, (3) Eligibility Phase: suitable full texts, and (4) Inclusion Phase: publications selected for enclosure in qualitative synthesis [24]. The study resorted to the use of two software tools: (1) VOSviewer [25] a tool for constructing and visualizing bibliometric networks, used in the "Screening phase"; and (2) ZOTERO for organizing and grouping the references.

The research issues under study are fairly new, and their limits are not yet well defined, so authors opted for a qualitative research methodology as a means to identify patterns and inter-relationships among the concepts [26]. The Systematic Literature Review assisted the development of the theoretical framework, contributing to representativeness, allowing replication of the study and avoiding conceptual ambiguities [27].

There is an important number of literature reviews dedicated to topics like "Eco Innovation", "Green Innovation", "Environmental Innovation", "Digital Transformation", "Industry 4.0" and "Digitalization", therefore authors opted to perform a Tertiary study (systematic review of secondary papers) [12] aiming to structure and synthesize the research about those themes, to identify the knowledge domains which are linked to them and to understand their interrelationship, integrating different academic points of view.

The PRISMA "Identification phase" was carried out using Web of Science (WoS) database. The research was based on one scientific database because it focuses on understanding the relation between two themes, not a comparison from databases [16]. Web of Science (WoS) database is considered one of the main academic databases due to its rigor and by ensuring the inclusion of widely peer-reviewed literature, however, it can be considered a limitation of the study [28].

Digital Transformation (DT) phenomenon research was based using the boolean search expression "'digital transformation' or 'industry 4.0' or 'digitaliz*'" [29]. The same criteria was used to the Eco-Innovation (EI) phenomenon research, but using boolean search expression "'eco innovation' or 'green innovation' or 'environmental innovation'" [8].

The PRISMA "Screening phase" was applied on peer-reviewed papers published in indexed journals to ensure paper's quality, and originally published in English, to avoid translation issues and reduce problems of ambiguity in fundamental concepts. The search period ranged from 2015 to 2020 (August), covering the body of review literature papers published about the subjects of interest.

The bibliographic data obtained were exported from WoS to VOSviewer.

The reason to apply VOSviewer is that, EI definition was reformulated by several scholars [17] making it difficult to reach a consensual understanding of the concept. In addition, accordingly to Hazarika and Zhang [13], an integration of Eco-innovation and sustainability transition literature would be most beneficial, interacting back and forth without any hierarchy, thus preventing the occurrence of technological bias, as explained on "Background Section". In this sense, in order to understand the Eco-Innovation (EI) and Digital Transformation (DT) relationship, the VOSviewer was applied to identify the knowledge domains that integrate Eco-Innovation and Digital Transformation, related to environmental sustainability in the manufacturing sector. The applications of VOSviewer at EI and DT phenomenon were developed independently from each other preventing any hierarchy or technological bias.

That step was regarded at the present study as a premise to the convergence of the interrelationship between Eco-Innovation and Digital Transformation phenomena. From an operational perspective, meaning that the VOSviewer was applied with the purpose of analyzing the terms which clusters should be similar between the Digital Transformation and Eco-Innovation phenomena, both identified as "environmentally friendly" [4, 9].

VOSviewer—a freely available software [30] allows for creating, visualizing and exploring bibliometric maps of science, and provides support for creating term maps based on a corpus of documents. They explain that a "term map is a two-dimensional map in which the terms are located in such a way that the distance between two terms can be interpreted as an indication of the relationship of the terms (…) the smaller the distance between two terms, the stronger the terms are related to each other." [25] (p. 51). VOSviewer offers three types of co-occurrence analysis by "Author

Keywords", "Keywords Plus" and both. The result is a co-occurrence network of terms with high relevance, that is grouped into clusters as a topic.

The bibliographic WoS data of the papers was analyzed at VOSviewer using exclusively the "Keywords Plus" option (defined by WoS) following Zhang et al. [31] They affirm that WoS includes two types of keywords: "Author Keywords", provided by the original authors, and "Keywords Plus", words or phrases generated by an automatic computer algorithm. They argue that to analyze the structure of a field of knowledge it is preferable to use only "Keywords Plus" option, because it covers most of the keywords contained in the "Author Keywords" option.

The option "minimum number of occurrences of a term" was also selected. The total strength of the occurrence links with other terms was calculated. The terms with the greatest total link strength were applied to generate the map. Authors identified the terms and their clusters, which were analyzed and selected. The results of these analysis supported the "PRISMA screening phase" leading to the identification of the "environmentally friend" terms that make explicit the knowledge domains related to the Digital Transformation and Eco-Innovation phenomena.

The terms analyzed and selected on VOSviewer were the knowledge domains that were applied as a restriction in WoS over the researches about the Eco-Innovation and Digital Transformation phenomena. In sequence, the papers were selected based on their titles and abstracts to ensure that their content was related to the environmental sustainability subject, the manufacturing sector and the knowledge domains focal point.

The PRISMA "Eligibility phase" consisted in reading the full texts, discarding some full-text papers out of the research scope.

The following was the PRISMA "Inclusion phase", which involved the relevant collection of papers implicated in qualitative synthesis.

3.2 PRISMA & VOSviewer Applications

PRISMA "Screening phase" was carried out refining the research by language, documents review type and published years, as explained, resulting in four hundred and eight-five papers for DT and sixty-six papers for EI phenomena. VOSviewer identified one hundred and ninety-seven terms for DT and forty-eight terms for EI phenomena.

In relation to DT, authors selected clusters numbers one and number three, associated with the terms "environmentally friend" (Fig. 1). The knowledge domains identified were circular economy, renewable energy, climate change and energy efficiency to be evaluated by similarity with the Eco-Innovation phenomenon and restrict both WoS research.

Concering EI, authors selected clusters numbers two and three, associated with the terms "environmentally friend" (Fig. 2). The knowledge domains identified were

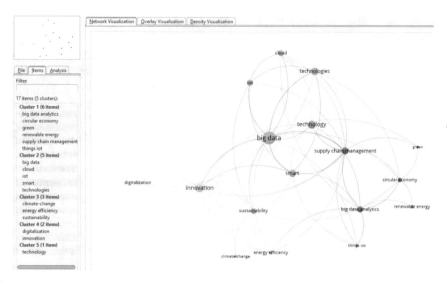

Fig. 1 VOSviewer terms map for DT phenomenon

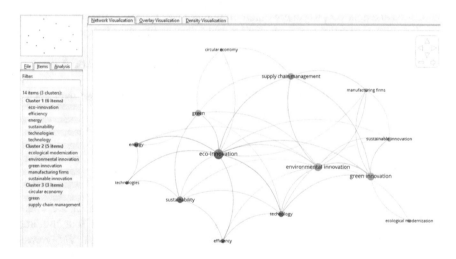

Fig. 2 VOSviewer terms map for EI phenomenon

ecological modernization, sustainable innovation, and circular economy to be evaluated by similarity with the DT Innovation phenomenon and restrict both WoS research.

The result of the similarity evaluation of the "environmentally friend" knowledge domains between DT and EI was "circular economy", for the set value of three as the minimum number of term occurrences on VOSviewer. The WoS research restriction followed the criterion that "Circular Economy" (CE) on topic for the DT research and

Table 1 Publications included in qualitative synthesis

DT phenomenon	
Bag and Pretorius [22], Sarc et al. [32]	Nobre and Tavares [33], Okorie et al. [16], Rosa et al. [34]
EI phenomenon	
de Jesus et al. [9], Kuo and Smith [4]	Saidani et al. [35]

for EI research. In total twenty-four papers corresponded to DT research, but sixteen publications out of the scope were excluded; in total nine papers corresponded to EI research, while six publications were excluded.

PRISMA phase "Eligibility phase" followed involving the reading of full texts, and the discarding of papers. Papers without a direct relation with Circular Economy were excluded.

The next PRISMA phase "Inclusion phase" led to the relevant collection of papers to be included in qualitative synthesis (Table 1).

The phases of PRISMA are illustrated in Fig. 3. The research resulted in eleven full-text papers assessed for eligibility, while three were excluded. Eight reviewed papers followed to the final PRISMA "phase for qualitative synthesis.

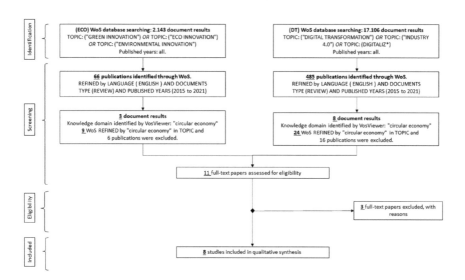

Fig. 3 PRISMA phases

4 Qualitative Synthesis: Circular Economy and the EI-DT Relationship

The interest in Eco-Innovation (EI) is emergent both in the academy and among practitioners [36]. It has been considered a key factor for competitive advantage [37] as it improves the organization image, while reducing its negative impact on the environment [4], and promoting a greener economy [13]. Effective EI adoption requires that organizations address changes to face a dynamic environment [37]. On the other hand, experts of Industry 4.0 believe that its contribution for sustainable industrial value creation will enable more efficient allocation of resources, building on insights from data collected and analyzed in real time, resulting in new sustainable green practices [3].

The main purpose of this research is the identification of the interrelationships between Eco-Innovation and Digital Transformation with focus on environmental sustainability in manufacturing sector.

4.1 Concepts and Definitions

The Eco-innovation (EI) concept is thought a key element in the development of competitive technologies and institutional forms (including new business models) that allow for 'environmental benefits', including greater efficiency in the consumption and use of resources. In the political arena, EI is a key component in the transition from a linear to a circular system of production and consumption. Not all dimensions of CE require innovation, although, when necessary, it should be consistent with CE models in a context of technical-economic change [9].

The Kuo and Smith [4] study focused on the technology aspect of EI, according to which the dimensions are design, user and product service (governance was excluded). Based on literature, they affirm that technological challenges involving EI are critical, when enterprises move towards sustainability. EI is considered as an important pathway towards sustainable development in the business sector. Previous studies have shown that EI can be measured quantitatively and used to explain the sustainability of a group. Therefore, sustainability can be seen as a result of the implementation of multidisciplinary technologies involving EI. The development and implementation of new technologies is the central driving force of EI.

Kuo and Smith [4] observe that CE has been implemented by means of several governmental policies, with Japan and Europe at the forefront. While the first, introduced material-cycle society vision, based on the 3R principle (reduce, reuse, and recycle); the second, emphasizes the strategies for "closing the loop" on linear product life cycles, and transforming them into differing loops of re-use, repair, refurbish and recycle.

Bag and Pretorius [22] identified that there are distinct forms of barriers, drivers, challenges, and opportunities to implement CE and Industry 4.0 technologies.

Regarding CE, authors affirmed that, for example, linear economy with "take-make-dispose" principles forced the society to search for sustainable ways as CE. However, manufacturers confronted several issues while adopting CE, including, setup costs; supply chain complexity; business-to-business non-cooperation, and skill gaps. Collaboration is an important dimension to CE among supply chain members for influencing environmental performance, and among other companies to support developing innovations for circularity. Environmental regulations and policies are key drivers of innovation. The literature indicates that one of the CE's main challenges is the financial unfeasibility. Sustainability challenges on the CE micro-level, or firm, for example, can be suppressed by developing a green supply focused on ecological analysis to the socio-economic system.

Saidani et al. [35], based on literature, explain that indicators have the characteristics to summarize, focus and condense the complexity of information into a simplified and useful manageable amount of meaningful knowledge. In CE, C-indicators is a facilitator for a transition toward CE practices, providing a standardized language to simplify the exchange of information, understanding, and thus facilitating this transition.

Bag and Pretorius [22] developed a research framework focused on workforce skills based on the integration among Industry 4.0 technologies, sustainable manufacturing and CE capabilities. A key CE motivation is the institutional pressure from government agencies, regulations, trade departments and industry policies, industry associations that compel manufacturing companies to acquire resource sets and configure them to support the adoption of I4.0 technologies. Institutional pressures lead to manufacturing companies to update workforce skills for the successful adoption of I4.0 technologies. Adoption of I4.0 technologies has a positive relationship with CE capabilities.

Nobre and Tavares [33] study focuses on the technology attributes present on the Ellen MacArthur Foundation [38, 39]—Regenerate, Share, Optimise, Loop, Virtualise and Exchange (ReSOLVE) Framework. They believe that CE proposals can be economically viable with applications of technologies such as big data and the Internet of things (IoT), but the necessary IT capabilities still lacks.

4.2 Technological Driven CE-EI-DT

The Technological EI emerges at CE macro level, or city, for example, as a critical as well as an incremental mechanisms based on the redesign of existing products and production methods, focusing particularly on increasing resource productivity. The technological EI is considered a facilitator of change and is perceived as essential in the creation of CE, even if the transition is recognized as requiring non-technological dimensions [9].

de Jesus et al. [9] affirm that at CE macro level is more than just about incremental change of existing components; it is also about changing the whole system of supply and demand, by technological and non-technological innovations to enhancement the resource efficiency, production and waste minimization.

Okorie et al. [16] affirm that technological CE by application of digital technologies have driven CE to field of research in engineering and computer science, and less in environmental science, business and management, materials science and chemical engineering. The focus in the literature is on recycling, recovery and reuse in the context of I4.0, while approaches like remanufacturing seem to be more scarce. They affirm that the possible reasons for this is that devices used in the energy sector are increasingly equipped with sensors and data collection devices; recycling is still seen as important by policymakers and, therefore, research continues in academia, while remanufacturing is still traditionally carried out, since disassembly, cleaning, component renovation and reassembly are labor-intensive processes.

Rosa et al. [34] affirm that studies around the digitalization of the CE do not select a specific I4.0 technology, preferring to maintain the greatest number of potential developments in the future. Only in some cases, for example, "additive manufacturing" is widely discussed among researchers, followed by "big data and analytics" and IoT.

Two concepts for a hybrid category between CE and I4.0 emerge from the literature named "Circular I4.0" and "Digital CE". Depending on the focus of the analysis, the literature indicates that, whether motivated by the principles of I4.0 or based on the CE's fields, integration CE with I4.0 can be described in a different way. While the focus is on I4.0 technologies, the intersection with CE is addressed as "Digital CE". However, if the focus is on the areas which the CE operates, the integration is called "Circular I4.0". This means that the difference lies in the type of relationship that drives CE and I4.0 integration. I4.0 is known as an CE facilitator, and not vice versa, however, there is a gap in the literature on the type of contribution that I4.0 can offer to the CE [34]. Nobre and Tavares [33] also mention a specific Internet of Things for environmental reasons (Environmental Internet of Things (EIoT)).

Sarc et al. [32] explain that digitalization in waste management is based on macroeconomic growth. They explain that systems can be used in waste treatment plants or machines in the future to make treatment of waste more efficient. Technologies are based on large amounts of data that can contribute to increasing efficiency within factories. Automated waste analysis, and material detection have advanced using sensors, artificial intelligence and access to databases that can be captured promptly without any problems.

Innovation. Different proposals of definitions that link innovation to environmental issues, still present in studies on EI and CE connections, allow for the integration of the EI-CE from various perspectives. Researchers consider the connection between EI and CE as a support to realize the potential of a new clean and coherent techno-economic paradigm. However, the relationship between CE and the notion of innovation is still not obvious. CE is an integrative multi-actor approach whereas EI (technology-based and non-technological) is a transition tool. Likewise CE strategies are drivers for EI, while EI is a capability towards CE [9].

EI is considered essential in the development of eco-industrial parks and development of institutional, technological and business model innovations; it improves a product's life cycle through the supply chain that increases resource efficiency, reuse and recycling; it reinforces the combinations of socio-technological changes that permit transition to a CE at meso level, or supply chain, for example. Innovation at CE micro level is very specific, including, for example patent data, cleaner production, eco-efficiency, eco-design and new business models. At this level, the replacement of the "take-make-dispose" business model implies new products, services, resource pool and marketing concepts, with EI as a tool to support the durability and quality of the product, in the design of efficient products [9].

Bag and Pretorius [22] affirm that Industry 4.0 (I4.0) technologies can autonomously detect the health of the system, enhance the decision-making power of humans and machines, facilitating the planning of warranty and maintenance steps of remanufactured items. However, there are various challenges in I4.0 implementation. Notably, Digital transformation will require profound changes in the human technical skills [22].

5   Results and Discussion

Eco-Innovation, in the case of sustainable manufacturing, intends to encourage the application of technological solutions that enable the substitution of toxic materials by non-toxic alternatives and the reduction of material consumption and waste [40]. However, "While incremental optimization through energy saving and eco-efficiency measures are playing an important role in widely diffusing greener practices, 'it's the disruptive end of the Eco-Innovation spectrum that is the most promising in the long term'" [41] (p. 3).

Radical alterations are understood as necessary to conduct the transition for new sustainable ways: "Certainly not all EI is linked to a CE, and not all dimensions of CE require innovation. However, a zone of overlap is bound to exist" [9] (p. 3000). At the CE micro level, incremental EI is prevalent, in terms of increasing resource efficiency, energy savings and pollution reduction and design.

EI and CE are a support to realize the potential of a new clean and coherent techno-economic paradigm. However, the relationship between CE and the notion of innovation is still not obvious, CE strategies are addressed as drivers for EI, while EI is addressed as a capability towards CE [9]. Very limited contributions are available about what kind of technologies can support the implementation of CE [34]. Nevertheless, a hybrid category seems to emerge between CE and I4.0 as "Circular I4.0" and "Digital CE", depending on the focus of the analysis. Although much has been discussed about the potential contribution offered by I4.0 technologies at CE, only in some cases has the environmental benefit (the circularity) been assessed through the adoption of I4.0 technologies [34].

6   Conclusions, Limitations and Future Research

The relationship between EI-DT and CE is perceived as a potential driver for a coherent techno-environment-economic-social paradigm. Some studies aim to identify the reduction of environmental problems as a result of the application of Industry 4.0 technologies. However, there are various challenges in the implementation of I4.0. In what concerns the social dimension, institutional pressures lead to manufacturing companies to update workforce skills for the effective adoption of I4.0 technologies.

Eco-Innovation is perceived as a phenomenon to support the sustainable environment, social and economic development, and CE focus mainly at the micro-level, minimizing negative environmental and societal impacts by the design of processes and products [9]. On the other hand, the Digital Transformation is considered a facilitator to the implementation of a Circular Economy. This indicates that, in order for Digital Transformation to boost particularly the environment sustainability, it should be aligned and fit the Eco-Innovation definitions, even if they are diverse, so to guarantee successful CE implementation, rather than a digitalized way of life.

This study contributes to the literature offering a Tertiary review, addressing Eco-innovation and Digital Transformation in the industrial context. For future researchers, the recommendations are exploring the topic by means of case studies in order to further evaluate the "Circular I4.0" and "Digital CE" movements, as well as advancing the knowledge to support the design of a framework to support the implementation of "Circular I4.0" at micro and meso levels.

References

1. Reis, J., Amorim, M., Melão, N., Matos, P.: Digital Transformation: A Literature Review and Guidelines for Future Research. In: Rocha, Á., Adeli, H., Reis, L.P., and Costanzo, S. (eds.) Trends and Advances in Information Systems and Technologies. pp. 411–421. Springer International Publishing, Cham (2018). https://doi.org/10.1007/978-3-319-77703-0_41.
2. Yao, X., Jin, H., Zhang, J.: Towards a wisdom manufacturing vision. Int. J. Comput. Integr. Manuf. 28, 1291–1312 (2015). https://doi.org/10.1080/0951192X.2014.972462.
3. Chalmeta, R., Santos-de León, N.J.: Sustainable Supply Chain in the Era of Industry 4.0 and Big Data: A Systematic Analysis of Literature and Research. Sustainability. 12, 4108 (2020). https://doi.org/10.3390/su12104108.
4. Kuo, T.-C., Smith, S.: A systematic review of technologies involving eco-innovation for enterprises moving towards sustainability. J. Clean. Prod. 192, 207–220 (2018). https://doi.org/10.1016/j.jclepro.2018.04.212.
5. Polzin, F.: Mobilizing private finance for low-carbon innovation – A systematic review of barriers and solutions. Renew. Sustain. Energy Rev. 77, 525–535 (2017). https://doi.org/10.1016/j.rser.2017.04.007.
6. Xavier, A., Reyes, T., Aoussat, A., Luiz, L., Souza, L.: Eco-Innovation Maturity Model: A Framework to Support the Evolution of Eco-Innovation Integration in Companies. Sustainability. 12, 3773 (2020). https://doi.org/10.3390/su12093773.
7. Ketata, I., Sofka, W., Grimpe, C.: The role of internal capabilities and firms' environment for sustainable innovation: evidence for Germany: Internal capabilities and firms' environment. RD Manag. 45, 60–75 (2015). https://doi.org/10.1111/radm.12052.

8. Gotsch, M., Kelnhofer, A., Jäger, A.: Environmental product innovations and the digital transformation of production: Analysing the influence that digitalising production has on generating
   environmental product innovations. Work. Pap. Sustain. Innov. No S072019 Fraunhofer ISI
   Karlsr. 61 (2019).
9. de Jesus, A., Antunes, P., Santos, R., Mendonça, S.: Eco-innovation in the transition to a circular
   economy: An analytical literature review. J. Clean. Prod. 172, 2999–3018 (2018). https://doi.
   org/10.1016/j.jclepro.2017.11.111.
10. Xavier, A.F., Naveiro, R.M., Aoussat, A., Reyes, T.: Systematic literature review of eco-
   innovation models: Opportunities and recommendations for future research. J. Clean. Prod.
   149, 1278–1302 (2017). https://doi.org/10.1016/j.jclepro.2017.02.145.
11. Abedinnia, H., Glock, C.H., Grosse, E.H., Schneider, M.: Machine scheduling problems in
   production: A tertiary study. Comput. Ind. Eng. 111, 403–416 (2017). https://doi.org/10.1016/
   j.cie.2017.06.026.
12. Martins, C.L., Pato, M.V.: Supply chain sustainability: A tertiary literature review. J. Clean.
   Prod. 225, 995–1016 (2019). https://doi.org/10.1016/j.jclepro.2019.03.250.
13. Hazarika, N., Zhang, X.: Evolving theories of eco-innovation: A systematic review. Sustain.
   Prod. Consum. 19, 64–78 (2019). https://doi.org/10.1016/j.spc.2019.03.002.
14. He, F., Miao, X., Wong, C.W.Y., Lee, S.: Contemporary corporate eco-innovation research: A
   systematic review. J. Clean. Prod. 174, 502–526 (2018). https://doi.org/10.1016/j.jclepro.2017.
   10.314.
15. Szopik-Depczyńska, K.: Innovation in sustainable development_ an investigation of the EU
   context using 2030 agenda indicators. Land Use Policy. 12 (2018).
16. Okorie, O., Salonitis, K., Charnley, F., Moreno, M., Turner, C., Tiwari, A.: Digitisation and
   the Circular Economy: A Review of Current Research and Future Trends. Energies. 11, 3009
   (2018). https://doi.org/10.3390/en11113009.
17. Reuter, M.A., van Schaik, A., Gutzmer, J., Bartie, N., Abadías-Llamas, A.: Challenges of the
   Circular Economy: A Material, Metallurgical, and Product Design Perspective. Annu. Rev.
   Mater. Res. 49, 253–274 (2019). https://doi.org/10.1146/annurev-matsci-070218-010057.
18. Smol, M., Kulczycka, J., Avdiushchenko, A.: Circular economy indicators in relation to eco-
   innovation in European regions. Clean Technol. Environ. Policy. 19, 669–678 (2017). https://
   doi.org/10.1007/s10098-016-1323-8.
19. Manavalan, E., Jayakrishna, K.: A review of Internet of Things (IoT) embedded sustainable
   supply chain for industry 4.0 requirements. Comput. Ind. Eng. 127, 925–953 (2019). https://
   doi.org/10.1016/j.cie.2018.11.030.
20. Prabhu, J.: Frugal innovation: doing more with less for more. Philos. Trans. R. Soc. Math.
   Phys. Eng. Sci. 375, 20160372 (2017). https://doi.org/10.1098/rsta.2016.0372.
21. Frank, A.G., Mendes, G.H.S., Ayala, N.F., Ghezzi, A.: Servitization and Industry 4.0 convergence in the digital transformation of product firms: A business model innovation perspective.
   Technol. Forecast. Soc. Change. 141, 341–351 (2019). https://doi.org/10.1016/j.techfore.2019.
   01.014.
22. Bag, S., Pretorius, J.H.C.: Relationships between industry 4.0, sustainable manufacturing and
   circular economy: proposal of a research framework. Int. J. Organ. Anal. ahead-of-print, (2020).
   https://doi.org/10.1108/IJOA-04-2020-2120.
23. Webster, J., Watson, R.T.: Analyzing the Past to Prepare for the Future: Writing a Literature
   Review. MIS Q. 26, xiii-xxiii/June (2002).
24. Moher, D., Liberati, A., Tetzlaff, J., Altman, D.G., for the PRISMA Group: Preferred reporting
   items for systematic reviews and meta-analyses: the PRISMA statement. BMJ. 339, b2535–
   b2535 (2009). https://doi.org/10.1136/bmj.b2535.
25. van Eck, N.J., Waltman, L.: Text mining and visualization using VOSviewer. 5 (2011).
26. Reis, J., Amorim, M., Melão, N.: Multichannel service failure and recovery in a O2O era:
   A qualitative multi-method research in the banking services industry. Int. J. Prod. Econ. 215,
   24–33 (2019). https://doi.org/10.1016/j.ijpe.2018.07.001.
27. Fink, A.: Conducting research literature reviews: from the internet to paper. SAGE, Thousand
   Oaks, California (2014).

28. Iñigo, E.A., Albareda, L.: Understanding sustainable innovation as a complex adaptive system: a systemic approach to the firm. J. Clean. Prod. 126, 1–20 (2016). https://doi.org/10.1016/j.jcl epro.2016.03.036.
29. Zangiacomi, A., Pessot, E., Fornasiero, R., Bertetti, M., Sacco, M.: Moving towards digitalization: a multiple case study in manufacturing. Prod. Plan. Control. 31, 143–157 (2020). https://doi.org/10.1080/09537287.2019.1631468.
30. van Eck, N.J., Waltman, L.: Software survey: VOSviewer, a computer program for bibliometric mapping. Scientometrics. 84, 523–538 (2010). https://doi.org/10.1007/s11192-009-0146-3.
31. Zhang, J., Yu, Q., Zheng, F., Long, C., Lu, Z., Duan, Z.: Comparing keywords plus of WOS and author keywords: A case study of patient adherence research: Comparing Keywords Plus of WOS and Author Keywords. J. Assoc. Inf. Sci. Technol. 67, 967–972 (2016). https://doi.org/10.1002/asi.23437.
32. Sarc, R., Curtis, A., Kandlbauer, L., Khodier, K., Lorber, K.E., Pomberger, R.: Digitalisation and intelligent robotics in value chain of circular economy oriented waste management – A review. Waste Manag. 95, 476–492 (2019). https://doi.org/10.1016/j.wasman.2019.06.035.
33. Nobre, G.C., Tavares, E.: Assessing the Role of Big Data and the Internet of Things on the Transition to Circular Economy: Part I : An extension of the ReSOLVE framework proposal through a literature review. Johns. Matthey Technol. Rev. 64, 19–31 (2020). https://doi.org/10.1595/205651319X15643932870488.
34. Rosa, P., Sassanelli, C., Urbinati, A., Chiaroni, D., Terzi, S.: Assessing relations between Circular Economy and Industry 4.0: a systematic literature review. Int. J. Prod. Res. 58, 1662–1687 (2020). https://doi.org/10.1080/00207543.2019.1680896.
35. Saidani, M., Yannou, B., Leroy, Y., Cluzel, F., Kendall, A.: A taxonomy of circular economy indicators. J. Clean. Prod. 207, 542–559 (2019). https://doi.org/10.1016/j.jclepro.2018.10.014.
36. Pacheco, D.A. de J., Caten, C.S. ten, Jung, C.F., Navas, H.V.G., Cruz-Machado, V.A.: Eco-innovation determinants in manufacturing SMEs from emerging markets: Systematic literature review and challenges. J. Eng. Technol. Manag. 48, 44–63 (2018). https://doi.org/10.1016/j.jengtecman.2018.04.002.
37. Salim, N., Ab Rahman, M.N., Abd Wahab, D.: A systematic literature review of internal capabilities for enhancing eco-innovation performance of manufacturing firms. J. Clean. Prod. 209, 1445–1460 (2019). https://doi.org/10.1016/j.jclepro.2018.11.105.
38. MacArthur, E., others: Towards the circular economy. J. Ind. Ecol. 2, 23–44 (2013).
39. Foundation, E.M.: Towards a circular economy: Business rationale for an accelerated transition. Ellen MacArthur Foundation Cowes (2015).
40. OECD: Eco-innovation in industry: Enabling green growth. OECD (2010).
41. OECD: The future of eco-innovation: The role of business models in green transformation. Presented at the (2012).

Open Innovation and User-Community as Enhancers of Sustainable Innovation Ecosystems

Joana Costa®, **Pedro Freire, and João Reis**®

Abstract Rapidly changing environments place different players at the vortex of the innovation process. Therefore, in the digital age, strong businesses are sometimes built on perceptions and on the approval of the community. The shift from linear value chains to ecosystems is likely to occur in 4.0 organizations adopting service or customer orientations, according to their participation in networked ecosystems. Moving from organization-centered innovation to ecosystem co-creation will approach individuals and institutions thus enhancing sustainable and smart product development along with trust. Embedded innovation is a self-sustained process in which firms and stakeholders interact in a common environment creating a common identity. Empirical results reinforce the role of open innovation strategies and the user community as pillars of sustainable innovation ecosystems. Policy actions need to reinforce these ecosystems as they will generate employment encompassing innovative and inclusive growth, fostering the resilience of societies and environmental sustainability.

Keyword Open innovation · User-community · Innovation ecosystem · Community innovation survey

---

J. Costa · P. Freire · J. Reis (✉)
Department of Economics, Management Industrial Engineering and Tourism, GOVCOPP, Aveiro University, Campus Universitário de Santiago, 3810-193 Aveiro, Portugal
e-mail: reis.joao@ua.pt

J. Costa
e-mail: joanacosta@ua.pt

P. Freire
e-mail: pedro.freire@ua.pt

J. Costa
Institute for Systems and Computer Engineering, Technology and Science (INESC TEC). Campus da Faculdade de Engenharia da Universidade do Porto, 4200-465 Porto, Portugal

J. Reis
Industrial Engineering and Management, Lusófona University, Campo Grande, 1749-024 Lisbon, Portugal

© The Author(s), under exclusive license to Springer Nature Switzerland AG 2021   65
A. M. Tavares Thomé et al. (eds.), *Industrial Engineering and Operations Management*,
Springer Proceedings in Mathematics & Statistics 367,
https://doi.org/10.1007/978-3-030-78570-3_5

1 Introduction

We are living in a digital age, where technologies are changing the way of doing business. This transformation creates value by monetizing data collection and management; this trend requires multi-firm collaboration, outside from the borders of existing organizations and industries. Vertically integrated value chains are giving place to new frameworks, leveraging distributed ecosystems to co-create value and promote sustainability. At present, rather than focusing on existing technologies, firms co-create core business value for their customers, committing to jointly create human-centric value, through innovation ecosystems under the sustainability paradigm [1, 2].

The user community can be involved in different parts of the innovation process. In the beginning, the process is typically creative through surveys. In the development phase, these agents help in the generation ideas for new products and services. Upon reaching the market, multidisciplinary teams continue to follow the cycle, by collecting the feedback from consumers to introduce new resources to their production [3, 4].

In the past, firms have pursued innovation through internal laboratories or research and development (R&D) centers; however, an important part of those projects or ideas have not made it out to the market, being abandoned or postponed, as they did not fit he companies' core businesses [5].

At present, customers can bring valuable insights to innovation promoting new business ideas and models, developed together with thinkers and practitioners. This process is the journey towards of Open Innovation 2.0, which relies in an ecosystem-based multiparty collaborative system. In light of the above, firms-universities-governments use to combine their efforts to deliver valuable outcomes to the society, which are reinforced by digital platforms [6]. The innovation ecosystem is a community of trust, reuniting people and ideas, funds and technologies. On the other hand, sustainability is a broader concept, used to identify initiatives and actions aimed at the preservation of a particular resource. Therefore, it normally relates to areas such as human, social, economic and environmental [7]. The digitization of innovation also has broad implications in both geographical, political and entrepreneurial domains. In this regard, digitization can promote productivity gains during the innovation process, reinforce regional entrepreneurial initiative, along with other economic and social gains such as the preservation of resources [8].

Extant literature in these fields does not provide a structured view of how and why open innovation, along with the user community, are used to manage innovation processes in a digital atmosphere to build sustainable innovative ecosystems. This paper, therefore contributes to the research of open innovation and co-creation stream by discussing the potential and the implications of these strategies in the innovative propensity. It also contributes to the sustainable innovation ecosystem stream by identifying its drivers and process levers that policy makers can adopt to foster and nurture the framework through user-communities in the era of digital transformation.

The paper also explored innovative strategies and behavior exploring a set of 13.701 firms, which differs in size, technological regime, innovation strategy and behavior. Data from these firms were collected from the Portuguese CIS (Community Innovation Survey) and where the outcome is twofold: first, it casts light on the empirical relation between open innovation strategies and the user-community in the innovative propensity; second, it draws some insights from the policy perspective to reinforce the importance of this binomial in the emergence of sustainable innovation ecosystems.

The paper is structured into five sections. In Sect. 2, we highlight the relevance of the topic from a theoretical perspective, evidencing the shifts in open innovation frameworks (Sect. 2.1). Thereafter, the challenges of the digital transitions are highlighted (Sect. 2.2) and the role of the user community is debated (Sect. 2.3). In Sect. 2.4, the previous theoretical debate is intertwined with the sustainable innovation ecosystem. Section 3 describes the database, the rationale and methodology in use for the empirical analysis. Section 4 reports the results and the discussion of the econometric estimation. Finally, Sect. 5 draws conclusions, summarizes the main theoretical and empirical implications weaving considerations and suggesting future research avenues.

2 Literature Review

2.1 Shifts in the Open Innovation Framework

In the last century, firms focused on their internal capabilities to generate and develop innovative ideas, so their concerns were on hiring the smartest people. The more innovative companies are, the greater their capacity for investment and, consequently, become leaders in their sector. While reaping most of their profits, they strongly protected their intellectual property, preventing their competitors from using it [9].

Open Innovation blurred the firm boundaries towards a more open framework to share and commercialize ideas and opening from outside-in, inside-out or both the innovation process; the organizations will benefit from the combination of internal and external knowledge, such that the innovation process can be more efficient and less costly to both parts [10]. The more usual, inbound OI (outside-in), describes the use of external knowledge to obtain new advantages in the internal processes from other companies, partners, customers, universities, and others. The outbound OI (inside-out) allows these new pathways to the market, allow firms to commercialize their ideas, selling "false positives" that were already do in the closed, and other companies to profit from "false negatives" ideas, no longer a firm lock their intellectual property, so they can find way to profit from them [11]. External R&D can now add a new value to companies, which can focus on their business models. Also, some of the means to reach this, include start-ups and licensing agreements, where all the players can benefit from the knowledge sharing and the creation of networks and partnerships [9, 12].

In 2013, Open Innovation was updated to its 2.0 version with the need for better solutions in global domains, bringing the opportunity to create new shared value through innovation and a society challenge [6]. The new framework will relate open innovation with the Quadruple Helix, where civil society will join the usual players (Firms, Government and University) [6, 13]. Under this new paradigm, civil society was working together with the other players to co-create promoting structural changes beyond what no one could do alone, promoting a shared vision which will add value to innovation at a broader scale and promote competitive advantage. The importance of the collaboration and networks need to establish a trusted relation aligned between the communities, so the creation of the ecosystem can benefit all the players and boost the innovation process [6, 14].

A new conceptual approach of innovation was proposed in 2010, Innovation 3.0, called "Embedded Innovation" [15]. A dynamic and more digital environment should be the main resource for firms to survive benefiting from the circulating knowledge flows. So, ecosystem networks and flows are pivotal, and the more the firms are capable to substantially integrate with their surrounding communities the more they will be able to absorb valuable knowledge [15]. The need of a sustainable and ecological approach to the products and processes is claimed by the user. The "Embedded innovation" also forges new mind-sets and consumer enrolment, creating a long term strategy and value added as self-sustained process, this framework proposes innovation from inside out.

The fourth industrial revolution, had a core milestone proposed in 2011-Industry 4.0. It required advances in technology, a more digital environment, along with automation, robotizations and artificial intelligence. Physical, digital and biological sphere, are approached in a combined way, and machines and humans' connections are reassessed [16]. Open innovation organizations are in the frontline of this shift benefiting from and the established network to maximize the innovation outcome.

This technological changeover pushes firms towards new innovation processes and open innovation will play a crucial role in clustering this evolution. With global business models, enlarged communication, knowledge exchange the allocative efficiency will be maximized, resources will be preserved, the environment respected and transforming these players a sustainable ecosystem [17].

2.2 Digital Transformation

The digital transformation has come into wide use in firms demanding for transformational or disruptive changes encompassing digital technologies [18]. Open science, as a public good, enhances possibilities for economic development, accelerating the innovation cycle and collaborative construction, which help knowledge massification. Digital transformation is a key enabler of innovation and diffusion as evidenced by the numerous firms using digital technologies to manage their innovation processes [19]. This exchange will enrich both the public and the private spheres. Entrepreneurs need to rethink products, processes, and business models making them sustainable, with a positive social impact allowing long term survival and economic prosperity [19–21].

Digitalization started with products and services, with the Internet of Things and Smart Devices, but this being insufficient in the long run, companies face the challenge of the digital transformation as defined "the process of using digital technologies to create new business processes and customer experiences that can meet the changing business and market requirements" [22, p. 1264]. The digitalization of processes and decisions, with Industry 4.0 and Big Data, the automation of processes, artificial intelligence and all that Industry 4.0 means and if every process can be digitalized it will be [21, 22].

Adapting to digital transformation is presently a new imperative strategy so firms can gain competitive advantages. A successful digitalization starts with understating the customer problem and to think how you can model your business model and make it more efficient and less costly. Identify not only where you can improve your already business, but point out new growth areas that can create digital business models Identify where you can make the bigger impact with the lowest resources [16, 22].

The process of creating value add to consumer in a dynamic market with continuously improves and novel products/processes that drives to hard competition, companies have limited resources and skills [23] that address companies to reach for knowledge sharing [22] and not only outsourcing, as crowdsourcing, the companies need to exploit external knowledge as the same time they liberate their internal [23]. Through digitalization, open innovation has changed, the new possibilities of communication, interaction and mechanism has been possible because big data analysis and data interaction, the open collaboration, as in software development, can bring value add, because allows support voluntaries and third parties in developing in open platforms, the more common are platforms with open application programme (APIs) [23].

As Travaglioni et al. [24] argue "companies digitized with fourth industrial revolution should create an ecosystem of innovation that provides opportunities for dialogue with all parties involved". Creating ecosystems that involve all the value chain of firms, from clients to stakeholders, making count every interaction and been able to absorb the knowledge to increase efficiency and effectiveness of new projects [23]. Because drivers that facilitate open innovation are dynamic capabilities (intuitive and absorptive that have the more weigh) and the know-how, your human resource need to be in full possession of the knowledge to execute their functions as consideration of all the organization, as product and process to facilitate the open innovation approach [24]. Sharing in open innovation can not only be in direct connection in the value chain, as can be created "knowledge brokers" who can facilitate the transfer of knowledge between their consumers/clients and external source that aren't directly related [22]. The (in)direct effect of digital trust combined with conventional capabilities (human-centric trust), will enhance organization's innovation abilities [25]. The digital transformation inside the open innovation framework promotes an accelerated innovation cycle responding to the volatile consumer expectations, switching to automatization, digitization and digital security. Fostering cooperation and networking among businesses inside the ecosystem will generate virtuous innovation cycles [26].

2.3   Open Innovation Ecosystems and the User Community

Open Innovation was proposed and re-adjusted, connecting to the Triple, the Quadruple Helix and recently to the Quintuple Helix, adding the civil society to the usual players (Government, University and Firms). The ignition of this process was the user-oriented innovation models; the pace of innovation is accelerated as in this mindset the different phases co-exist and developed in real world context [6].

As we erode the boundaries between the firm and the ecosystem will innovations will appear in different marketplaces, knowledge flows circulating all around feeding the networks. External sources of knowledge inside the value chain, user-community, and the environment are now equally important [12, 27].

In this new era, the Helix plays an enlarged effect and the user-community and the civil society will promote the adoption of responsible practices in innovative strategies [28]. This becomes feasible under permeable organizations, and, managers should focus on the interactions, performing constant adjustments, exploiting internal endowments of resources with the external co-operators, thus promoting sustainability.

2.4   Sustainable Innovation Ecosystems

Worldwide, nations are challenged with structural changes trying to appraise global innovation trends and technological shifts. Innovation ecosystems gained increased popularity over the last decade, due to its particular link to open innovation. Tansley [29] proposed the concept based on the ecological element embedding the creatures and their environment, Moore [30] proposes a framework of coopetitive players, focusing on the geographic dimension of knowledge spillover sharing. Presently, innovation ecosystems are a multilayer framework connecting institutions to develop and share relevant knowledge feeding the innovation processes [1].

It combines complementary and substitute connections of actors and objects and the environment promoting sustainability [2], being dynamic, interactive and embedded in an innovation mindset. They can be virtual due to the digital transformation still being a grounded hub as members need to physically meet to interact and co-create [1]. Present models combine human resources leveraging initiatives bidirectionally, as intelligence facilitators. External knowledge and contributors will feed the system. Orchestrated interactions feed the ecosystem, which should rely on trust, creating a sense of identity which will consolidate the network based on shared values which will enhance sustainable practices [28].

Human capital is the basis of collective knowledge which will generate income reinforcing competitive advantages. Endowments of qualified workers will raise firms' absorptive capacity, embedded in processes and benefit the ecosystem [21].

Traditionally, R&D labs, licensing and patenting are more likely to appear globalized innovation hubs, however, smaller ecosystems should commit to identify innovative potential to fully exploit their resources.

3 Database and Methodology

The appraisal of the role of open innovation and the user community in innovation and the promotion of sustainable innovative ecosystems were performed relying upon binary logit estimation. As a consequence, the empirical evidence will support eventual impacts of the open innovation strategy along with the user community, along in the propensity to innovate, controlled for firms' structural characteristics and the use of innovation sources.

3.1 Database Description

The empirical analysis will rely on data from the Portuguese CIS 2018, covering the biennia of 2016–2018. The database includes 13.701 firms operating in Portugal with diversified structural characteristics as well as innovative profiles. This survey is the most comprehensive concerning innovation related issues, providing robust evidence for the relevant variables in use.

Changing the innovative mindset and including new players in the system will produce delayed effects; moreover, innovative cycles are addressed as long waves, being therefore difficult to appraise the immediate results of the actions taken. However, with the consistent efforts made over the last decades many organizations are now more close to these practices. Addressing the case of Portugal in this time frame is of particular interest as it is amongst the regions which have made the upturn in the recent years, and labeled by the European Innovation Scoreboard as "strong innovator" in 2019 and 2020, something that had never happened before. The crossover to the leading stage was reached by seven countries. This period allowed the greatest "leaps" as the country as classified as "moderate innovator" during more than a decade; the leading capacity on spot was "innovation in SMEs (small and medium-sized companies)".

Table 1 Variable description

Variables	Description	Measurement
INNOV (1)	Having performed at least one type of innovation	Binary
COM (2)	Performing any co-creation with the user community	Binary
OPEN_INNOV (3)	Performing inbound and outbound innovation	Binary
CHANNELS (4)	Channels used to obtain knowledge	1–8°
FUNDS (5)	Beneficiary of funds	Binary
TECH_REG (6)	Technological regime of the firm (according to Boliacino and Pianta [32])	1 = supplier dominated 2 = scale intensive 3 = specialized supplier 4 = science based
SIZE (7)	Firm dimension	1–3°
EMPUD (8)	Human capital intensity	1–6°

3.2 Exploratory Analysis

A detailed description of the variables and their measurement is made in Table 1. Whenever possible, the CIS' original scale was maintained, in others, simple mathematical transformations were implemented. To address the impact of the technological regimes, firms were divided into four categories, with increasing technological intensities. Concerning firm dimension, firms were split into small, medium and large, according to the European Commission methodology, and the European Innovation Scoreboard [31]. Human capital intensity is a multinomial variable split accordingly to the CIS methodology and the EIS [31]. The role of the user-community was appraised by means of co-creation behaviors.

Table 2 provides the descriptive statistics and the correlations for each of the variables in use. In respect to correlation, it appears as significant for all the pairings, with moderate intensity. Statistical testing based on VIF (Variance Inflation Factor) did guarantee the inexistence of multicollinearity, further reinforcing the results presented in the correlation table.

4 Econometric Analysis

4.1 Econometric Estimations

In order to appraise the determinants of the innovative performance for each innovation type a logit model was run being presented in the following Table 3. It encompasses having performed any of the innovation types, which means product, process,

Table 2 Descriptive and correlations

Variables	Mean	S.D	Min	Max	(1)	(2)	(3)	(4)	(5)	(6)	(7)	(8)
INNOV (1)	0.36	0.48	0	1	1.00							
COM (2)	0.22	0.41	0	1	0.37	1.00						
OPEN_INN (3)	0.45	0.20	0	1	0.26	0.20	1.00					
CHANNELS (4)	2.31	2.10	0	8	0.39	0.28	0.25	1.00				
FUNDS (5)	0.23	0.57	0	4	0.23	0.19	0.25	0.25	1.00			
TECH_REG (6)	1.83	1.07	1	4	0.04	0.10	0.06	0.06	(0.02)	1.00		
SIZE (7)	1.36	0.58	1	3	0.16	0.06	0.19	0.23	0.13	0.02	1.00	
EMPUD (8)	3.39	1.86	1	7	0.21	0.15	0.14	0.36	0.14	0.17	0.16	1.00

All correlations are significant at the 1% level ($p < 0.01$)

Table 3 Econometric estimation—Marginal effects after logit estimation

Variables	INNOV
COM	0.345*** (0.012)
OPEN_INNOV	0.509*** (0.026)
CHANNELS	0.063*** (0.002)
FUNDS	0.922*** (0.009)
TECH_REG	0.008* (0.004)
SIZE	0.056*** (0.008)
EMPUD	0.017*** (0.003)
Constant	−2.330*** (0.072)
Observations	13.701

Standard errors in parentheses *** $p < 0.01$, ** $p < 0.05$, * $p < 0.1$

service, organizational or marketing. The mainstream empirical research in open innovation relies on product and process innovations; however, given the emergent importance of the other types the analysis was further extended proposing a broader proxy for the dependent variable. Given the binary nature of the dependent variable (whether or not having performed any type of innovation), binary regressions such as the logit model were implemented. Consequently, the coefficients of the Logit were omitted as they are of poor importance given that they are uninterpretable. Table 3, presents the marginal effects evidencing the impact on the innovative propensity caused by changes in the predictors.

The present estimation includes explanatory variables co-creation with user community and open innovation; both play a positive impact in probability of observing at least one type of innovation. Involving the user community raises the probability to innovate by 34,5 pp (percentage points); and, 50,94 pp when adopting open innovation practices compared to their counterparts.

All other predictors also appear with a positive impact, leveraging innovative performance; being "funds" the one that represent a higher impact of in the probability to innovate. This reinforces the importance of traditional instruments in the innovation policy, as, when firms obtain funds the probability to innovate raises by 92,2 pp. All coefficients present high statistical significance with the exception of technological regimes that appears a positive impact but only in a significance level of $p > 0,10$.

4.2   Results and Discussion

How and with whom should firms integrate their innovation processes is a crucial point of view, nowadays, economies need to understand market and social forces' needs and requirements. Since the integration of the Quadruple and Quintuple Helix innovation concept new players were added to the framework challenging firms to rebuild their strategies. The role of social and environmental involvement in the innovation practices takes into account the civil society, its opinions and needs as well as product redesign considering the inclusion of responsible practices in the innovation process [33]. In this line, the empirical results evidence the influence of the user-community in the innovation strategies, which has showed a positive and significant impact in the model tested, as expected, when the user and others are involved in the co-creation process the effect is twofold: first, the demand is naturally satisfied as they were involved in the development of new innovations. Secondly, insights and criticisms from the user-community will feed the innovation cycle, which will consolidate the existence of sustainable ecosystems.

The impact of open innovation on the development of innovations as shown positive results since the first time the concept was introduced [9]. Using inflows and outflows of knowledge, the capacity to commercialize ideas and open the firms' barriers to the technological progress and good practices can flow between all users will enhance firm performance [12]. Here again, the variable has showed positive influence in innovation and in this model being no exception, with positive and significant impact in innovation strategy. In the same line of open innovation, the increase in the number of channels that a firm utilizes in an efficient way, raises the probability to innovate, here, this variable shows high significance.

Regardless of the founding source as well as its application in innovation processes; the variable shows a positive and significant impact and demonstrates the need of subsidization to exploit innovation activities [11, 34]. Firms that have a higher internal R&D intensity are expected to have the internal resources to exploit radical innovations in high-tech, since low-tech industries possess limited resources and less capacity to internally innovate [35]. In this model, the technological regimes appear with weaker significance (only at the 10% level) as an enhancer to innovate. The size of the firm is positively related to innovation propensity as shown in several studies [36]. In the present, this variable also appears having a positive and very significant impact in innovation.

Finally, the human capital intensity which has showed a crucial contribution on innovation processes, since it contributes to its underlying foundations; and firms that foster the education and the absorptive capacity will have an increased potential to innovate [37, 38]. In line with extant literature, human capital plays a positive role in innovation propensity.

5 Implications and Policy Recommendations

Digital technologies, platforms, networks and infrastructures have transformed the innovation process. Recently, internet and the entire virtual world have made open science a reality. Academia has contributed with both theoretical and empirical measures to appraise the impacts of digital innovation.

Concerning the public policy perspective, and traditional instruments such as funding, individualized grants promoted by closed innovation models do not ensure welfare maximization due to intellectual property. As a consequence, the promotion of collaboration and knowledge share must me embedded in policy actions, promoting open knowledge ecosystems.

Further efforts need to be made in terms of the public policy to include the user community in the ecosystem. Digital transformation relies on connections and networks. Given the need of faster and better innovation cycles, and the avoidance of its abandon, the inclusion of users will enhance the probability of marketplace acceptance.

In general, firms are increasingly using external sources of knowledge rather than confining themselves to internal resources. Still, uncertainty and complexity of the legal framework are putting away firms' open innovation networks. Appropriability problems affect specially SMEs.

Policy makers should address the traits of their ecosystems along with their weaknesses to develop a more effective legal system protecting property rights. Actions should modernize SME business models, promoting the digital transition.

It is of worth mentioning that the present results emerge from a sectional analysis and despite the vastness, diversity and robustness of the respondent sample they may represent an exceptional coordination for the variables in use describing a singular result rather than a long term trend. Running the same empirical analysis in a diachronic perspective would reinforce the findings, which is an open avenue of research for future works.

Considering the present challenges of the uprising economic crisis with unprecedented consequences, given the simultaneous downturn in the demand and supply side urges emerging knowledge based model of development.

Exalting regional endowments and capabilities will increase opportunity identification and promotion of the emergence of self-sustained ecosystems. Governments must promote innovations exploiting latent capabilities and focusing on relevant problems to the ecosystem, promoting societal mind-changing actions. Open innovation ecosystems will promote sustainable and responsible practices monetizing the importance of the user-community. Digital transformation is commonly believed as de-humanizing, however, this framework is prover to be human-centered including new life styles and individual organization of with different connections in the job market, in the academia, in firms and in Governmental institutions.

References

1. Granstrand, O., Holgersson, M.: Innovation ecosystems: A conceptual review and a new definition. Technovation, 90, 102098 (2020).
2. Adner, R.: Match your innovation strategy to your innovation ecosystem. Harvard Business Review, 84(4), 98–107 (2006).
3. Bauer, M., Schlund, S., Vocke, C.: Transforming to a Hyper-connected Society and Economy – Towards an "Industry 4.0". Procedia Manufacturing, 3, 417–424 (2015).
4. Kaggermann, H.: Change Through Digitazation-Value creation in the Age of the Industry 4.0. In Management of Permanent Change (pp. 23–45). Springer Gabler, Wiesbaden (2015).
5. Stanislawski, R.: Open innovation as a value chain for small and medium-sized enterprises: Determinants of the use of open innovation. Sustainability, 12(8), 3290 (2020).
6. Curley, M., Slamelin, B.: Open innovation 2.0: A new Paradigm-White Paper. EU Open Innovation Strategy and Polocy Group (2013).
7. Brundtland, G.: Report of the World Commission on Environment and Development: Our Common Future. United Nations General Assembly Document A/42/427 (1987).
8. Burtch, G., Carnahan, S., Greenwood, B.: Can you gig it? An empirical examination of the gig economy and entrepreneurship. Management Science, 64(12), 5461–5959 (2018).
9. Chesbrough, H.: The era of open innovation. MIT Sloan Management Review, 44(3), 35–41 (2003).
10. Bogers, M., Chesbrough, H., Moedas, C.: Open innovation: Research, practices, and policies. California Management Review, 60(2), 5–16 (2018).
11. Greco, M., Grimaldi, M., Cricelli, L.: Hitting the nail on the head: Exploring the relationship between public subsidies and open innovation efficiency. Technological Forecasting and Social Change, 118, 213–225 (2017).
12. Chesbrough, H.: Managing open innovation. Research-Technology Management, 47(1), 23–26 (2004).
13. Leydesdorff, L., Park, H., Lengyel, B.: A Routine for measuring synergy in university-industry-government relations: Mutual information as a triple-helix and quadruple-helix indicator. Scientometrics, 99(1), 27–35 (2014).
14. Curley, M. Twelve principles for open innovation 2.0. Nature, 533(7603), 314–316 (2016).
15. Hafkesbrink, J., Schroll, M.: Innovation 3.0: embedding into community knowledge-collaborative organizational learning beyond open innovation. Journal of Innovation Economics Management 1(7), 55–92 (2011).
16. Schwab, K.: The Fourth Industrial Revolution. New York: Crown Business, 2017. ISBN 9781524758868.

17. Gerlitz, L.: Design Management as a Domain of Smart and Sustainable Enterprise: Business Modelling for Innovation and Smart Growth in Industry 4.0. Entrepreneurship and Sustainability Issues, 3(3), 244–268 (2016).
18. Boutetiere, H., Reich, A.: Unlocking Success in Digital Transformations. McKinsey Digital October (2018).
19. Nambisan, S., Wright, M., Feldman, M.: The digital transformation of innovation and entrepreneurship: Progress, challenges and key themes. Research Policy, 48(8), 103773 (2019).
20. Reis, J., Amorim, M., Melão, N., Matos, P.: Digital transformation: A literature review and guidelines for future research. In World Conference on Information Systems and Technologies (pp. 411–421). Springer, Cham.
21. Bogers, M., Foss, N., Lyngsie, J.: The "human side" of open innovation: The role of employee diversity in firm-level openness. Research Policy, 47(1), 218–231 (2018).
22. Crupi, A., Del Sarto, N., Di Minin, A., Gregori, G., Lepore, D., Marinelli, L., Spigarelli, F.: The digital transformation of SMEs – a new knowledge broker called the digital innovation hub. Journal of Knowledge Management 24(6), 1263–1288 (2020).
23. Burchardt, C., Maisch, B.: Digitalization needs a cultural change – examples of applying Agility and Open Innovation to drive the digital transformation. Procedia CIRP, 84, 112–117 (2019).
24. Travaglioni, M., Ferazzoli, A., Petrillo, A., Cioffi, R., De Felice, F., Piscitelli, G.: Digital manufacturing challenges through open innovation perspective. Procedia Manufacturing, 42, 165–172 (2020).
25. Mubarak, M., Petraite, M.: Industry 4.0 technologies, digital trust and technological orientation: What matters in open innovation? Technological Forecasting and Social Change 161, 120332 (2020).
26. Kaggermann, H.: Change through Digitization-Value creation in the Age of the Industry 4.0. In Management of Permanent Change (pp. 23–45). Springer Gabler, Wiesbaden (2015).
27. Chesbrough, H., Schwartz, K.: Innovating business models with co-development partnerships. Research-Technology Management, 50(1), 55–59 (2007).
28. Liu, Z., Stephens, V.: Exploring Innovation Ecosystem from the Perspective of Sustainability: Towards a Conceptual Framework. Journal of Open Innovation: Technology, Market, and Complexity, 5(3), 48 (2019).
29. Tansley, A.: The use and abuse of vegetational concepts and terms. Ecology, 16(3), 284–307 (1935).
30. Moore, J.: Predators and prey: A new ecology of competition. Harvard Business Review 71(3), 75–86 (1993).
31. European Commission. European Innovation Scoreboard, 2020. https://ec.europa.eu/docsroom/documents/42981 Accessed 16 Oct 2020.
32. Boliacino, F., Pianta, M.: The Pavitt Taxonomy, revisited: patterns of innovation in manufacturing and services. Economia Politica, 33(2), 153–180 (2016).
33. Carayannis, E., Grigoroudis, E., Stamati, D., Valvi, T.: Social Business Model Innovation: A Quadruple/ Quintuple Helix-Based Social Innovation Ecosystem. IEEE Transactions on Engineering Management, 1–14 (2019) https://doi.org/10.1109/TEM.2019.2914408.
34. Cano-Kollmann, M., Hamilton III, R., Mudambi, R.: Public support for innovation and the openness of firms' innovation activities. Industrial and Corporate Change, 26(3), 421–442 (2017).
35. Zouaghi, F., Sánchez, M., Martínez, M.: Did the global financial crisis impact firms' innovation performance? The role of internal and external knowledge capabilities in high and low tech industries. Technological Forecasting and Social Change, 132, 92–104 (2018).
36. Fang, X., Paez, N., Zeng, B.: The nonlinear effects of firm size on innovation: an empirical investigation. Economics of Innovation and New Technology, 1–18 (2019).
37. Alexy, O., Bascavusoglu-Moreau, E., Salter, A.: Toward an aspiration-level theory of open innovation. Industrial and Corporate Change, 25(2), 289–306 (2016).
38. Salampasis, D., Mention, A.: Open Innovation: Unveiling The Power Of The Human Element. World Scientific (2017).

Impact of Lean Tools on Companies During Industrial Engineering Projects Implementation: A Correlation Study

Ângela Silva⊙ **and Ana Cristina Ferreira**⊙

Abstract Lean production is an integrated methodology aiming the elimination of all activities that do not add value to products and services. To get this objective, a set of tools are typically applied regarding waste elimination. Industrial companies from different activity sectors propose the implementation of Lean tools in the development of Industrial Engineering Projects, since the focus is to solve their real problems. In this work, 22 final-year projects developed during the graduation and master degrees in Industrial Engineering were analysed. The impact of Lean tools implementation was studied and a correlation analysis with projects output indicators was performed. It was observed that the most used tools were Time and Methods, 5S, Standard Work and Visual Management. Time and costs reduction, productivity increase, human factors and layout modifications are the five indicators most used to measure the impact and the success of Industrial Engineering projects. It was obtained a good correlation between the implementation of Times and Methods and 5S tools with the productivity increase and reduction of time (*Pearson coefficient* equal to 0.98).

Keywords Lean production · Industrial engineering projects · Lean tools

Â. Silva (✉)
Centro de Investigação em Organizações, Mercados e Gestão Industrial, Universidade Lusíada, Lisbon, Portugal
e-mail: angela.a@esce.ipvc.pt

A. C. Ferreira
ALGORITMI Centre, University of Minho, Guimarães, Portugal

Mechanical Engineering and Resource Sustainability Center, University of Minho, Braga, Portugal

Â. Silva
Escola Superior de Ciências Empresariais, Instituto Politécnico de Viana Do Castelo, Valença, Portugal

© The Author(s), under exclusive license to Springer Nature Switzerland AG 2021
A. M. Tavares Thomé et al. (eds.), *Industrial Engineering and Operations Management*,
Springer Proceedings in Mathematics & Statistics 367,
https://doi.org/10.1007/978-3-030-78570-3_6

1 Introduction

Over the past few years, production has become increasingly focused on the customer and on product customization, challenging organizations to adopt new methods and tools and allowing them to remain competitive in the market. The concept of Lean Production has emerged in this context, having as main objective the production of goods or services with less time and cost, through the waste reduction [1].

1.1 Lean Production

The Lean Production concept, first used by John Krafcik to designate the Toyota Production System philosophy [2], became popular after the publication of the book "The Machine that Changed the World" by Womack et al. [3]. According to Womack and Jones [4], this concept comes from the recognition that only a small fraction of the productive time and effort add value to the customer. Companies widely recognize that to achieve and sustain a competitive advantage, they must become more customer-focused.

Womack et al. [3] defines Lean Production as an approach that pursues a better way to organize and manage a company's relationships with its stakeholders, product development and production operations, according to which it is possible to do more and more with less (less equipment, less human effort, less time, less cost, etc.). Hines et al. [5] consider Lean as a philosophical or strategic perspective considering value creation and focus on understanding customer value. In terms of Lean, the value should always be considered from a customer perspective.

Lean Thinking is underlying the Lean philosophy and, according to Hines et al. [6], it can be seen as the antidote for waste. Lean thinking is based on 5 principles (Fig. 1) that are fundamental for waste elimination and should serve

Fig. 1 Five principles of Lean thinking

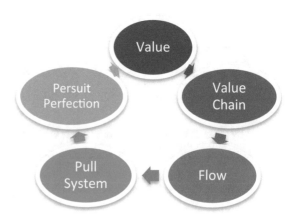

as a basis for companies that intend to adopt this Lean transformation [5]. Thus, each principle can be defined as: **1. Value**: value from the consumer's perspective. **2. Value chain**: all actions and activities necessary to bring the product to the customer, as well as those that do not add value to it and should be eliminated. **3. Flow**: all tasks that add value to the product without interruption or creating inventories. **4. Pull system**: execute only what is requested by the customer when requested, avoiding excess inventory. **5. The pursuit of perfection**: maintain the results obtained and seek a continuous improvement approach to reduce waste.

Amaro et al. [7] discuss the relevance of Lean Thinking principles implementation around the world, both in industry and services, based on the growing number of published case studies and surveys. They concluded that Lean Thinking is a global and transversal approach to improve organizations' performance (all types of industries and services). Nevertheless, several organizations are not yet fully aware of the Lean principles as they do not apply them to an entire value stream, but only to parts of value streams (i.e. to sectors or areas of the company).

The first step when applying Lean Thinking is to understand what value has the product or service and what activities or resources are needed to create that value. To do this, it is necessary to eliminate everything that is not of value, with 3 major areas to be taken into account: Muda et al. [5]. The word Muda, of Japanese origin, means waste, and it refers to any activity, from the beginning to the end of the production process, which does not add value to the product. This concept can be divided into seven main types of waste [5, 8]: **Overproduction**: production too much or too early, resulting in excess of inventory and consumption of materials before they are really needed. **Inventory**: excess inventory resulting from overproduction, which implies costs with storage and stock management. **Defect**: imperfections in products, resulting from quality problems or errors in their processing, which interrupt production and require rework or cause waste of material. **Movement**: weak organization of the workspace, which results in unnecessary movement by employees. **Incorrect processing**: production using inappropriate tools and/or procedures, often resulting from the lack of synchronization and combination of operations. **Waiting**: long periods of pause on the part of the operator, whether due to lack of work, information, material or machine breakdowns, which result in a weak production flow and high lead time. **Transport**: handling of materials or products, promoting damage and waste of time and resources.

However, reducing these seven wastes is not enough to achieve the success of Lean philosophy [5]. Alongside Muda, it is also important to take into account Mura and Muri, where Mura means variation or inequality and Muri means overload. In the case of Muri, this refers to the overload of the worker or the machine, which can be solved by standardizing the work, with the aim of making the processes more predictable and controllable. Mura refers to instabilities in production, for example when a worker expects work because the previous one is slower. This type of waste can be eliminated with the adoption of a production system pulled by the customer [8].

1.2   Lean Tools

To improve productivity and competitiveness in companies, there are many different lean tools that could be implemented. These tools are most effective if they are implemented together. However many can be used on their own to solve specific issues.

Some of the most popular tools applied in the companies are 5S, Standard Work, Visual Management, Value Stream Mapping (VSM), Just in Time (JIT) production, Kaban, Kaizen, Heijunka, Root-cause analysis, 5W2H, Ishikawa Diagram, Single minute exchange of dies (SMED), Poke-yoke, Spaghetti Diagram, Critical Path Method (CPM), Business Process Model and Notation (BPMN), One Point Lesson, Competency/Skill Matrix and ABC analysis. Almost all of them are based on Times and Methods study. These tools are the ones which are implemented in the Industrial Engineering projects. A brief description of the tools will be presented in the next paragraphs, as well as, some works which, have used these tools in their research.

Several works describing the application of these tools can be found in the literature. Antoniolli et al. [9] implemented the **standard work** tool to decrease or eliminate various activities that do not create value to the product, generating great achievements in productivity and efficiency, increasing the Overall Equipment Effectiveness (OEE) from 70 to 86%. Later, applying the same method, Rosa et al. [10] optimized the production process of an assembly line dedicated to the manufacture control cables for the automotive industry and the results were very positive: increase 43% in the line output and reduction on 30% on the line use allowing the introduction of new products on the assembly line.

The **5S methodology** is a Japanese workspace organization tool, in an efficient and safe manner, in order to achieve a productive work environment. Is one of the basic tools of Lean Manufacturing and one of the first tools applied by the companies to implement Lean methodology. This simple tool is not just for production processes, it is also applied to service operations and office implementations (lean office), promoting good results on productivity, efficiency and the overall performance of production on different companies [11–15].

Visual management is directly related to communication. It's an easy way to communicate the performance indicators, the standards, the objectives and the warnings in different scenarios. The aim of this tool is obtaining as much information as possible, with a small-time of observation. Kurpjuweit et al. [16] present a research work where the barriers, the success factors and the best practices on the implementation of visual management for continuous improvement, are investigated. Several works can be found in the literature, in services and industry [13, 17].

The continuous improvement is a pillar of Lean Production which implies the implementation of a set of tools in a cyclic way, namely **Kaizen, PDCA, VSM, One Point Lesson, Kaban and Heijunka**. All these tools suggest the great importance of the constant control of the activities that add value to the products or services and the others that do not add value, eliminating the last ones [11, 12].

Another important factor of Lean Production implementation is the process quality. To control these aspects, the quality tools such as **Ishikawa Diagram, Root-cause analysis, Poke-yoke, 5W2H, Spaghetti Diagram and ABC analysis** are implemented to identify the origin of the problems, defects, errors, wastes, etc. [13, 17].

In some specific projects, tools such as **SMED, CPM, BPMN and skill matrix** are also implemented to support the improvement of productivity and the efficiency of the projects, in order to eliminate a specific type of waste [10, 11].

1.3   Impact of Lean Tools Implementation

The impact of Lean tools applications could be observed in many studies made in different companies and areas. The impact of five essential lean methods on operational performance, i.e., cost, speed, dependability, quality and flexibility, using a linear regression analysis was studied by Belekoukias et al. [18]. The study established a correlation and impact of these lean practices on the operational performance of 140 manufacturing organisations around the world. Recently, Abu et al. [19] made a review and analysis of the motives, barriers, challenges, and the application of Lean based on a survey of 148 companies in Malaysia. Upon validation of the analyses, the results revealed that most of the lean companies agreed that the reasons for lean implementation are to increase efficiency, to clean up and organize the workplace, and to increase utilization of space. Non-lean companies believe that issues related to knowledge are the reasons for not undertaking lean implementation applications.

The layout of the shop floor is also an important factor that has a great influence on the companies' performance, as well as the workspace design, since a bad layout could promote unnecessary movements and unnecessary material flows and, consequently, the waste of time and cost which will reduce the productivity of companies. Due to this, some researchers apply the Lean tools to improve productivity making the layout optimization [20, 21].

The contribution of this study is to understand the importance of Lean tools when employed in Industrial Engineering Projects, by identifying the correlation between tools implementation and their impact on the company's productivity performance.

2   Research Methodology

enlargethispage-12ptThe relevant information was systematized considering a quantitative and qualitative content analysis. Content analysis can be defined as a research technique for making replicable and valid inferences of a meaningful matter for a certain context and it can be used on all types of written texts no matter where the material comes from [22]. This approach requires several phases: (1) planning; (2)

Fig. 2 Process of a qualitative content analysis from planning to results presentation

data collection; (3) data analysis; and (4) presentation of results. Figure 2 presents the process phases of qualitative content analysis.

The planning phase includes the definition of the scope, objective and the sample size under study. Data collection and analysis allows to transform information into organized and written as meaningful results. Nonetheless, each stage must be performed several times to maintain the quality and trustworthiness of the analysis.

2.1 Projects Characterization

The objective of the present work is to investigate the impact of Lean tools' on companies during the implementation of final-year Industrial Engineering projects. The sample considered in the study was based on a set of 22 finalized projects in the scientific fields of Industrial Engineering, Production Management and Lean Production.

These projects were developed in the context of final graduation degree projects (six months duration) and master dissertations (one-year duration) carried out by students during of 1st and 2nd cycles of higher education courses in Industrial Engineering and Management. This sample was chosen because it was considered representative of industrial training programs in Portuguese companies between the year of 2015 and 2019. These real-world experiences create opportunities to prepare both undergraduate and graduate engineering students for complex challenges in different industry sectors. All the projects collected for the analysis were supervised by the co-authors of this paper. Thus, the supervisor's perception within the training visits to the company's facilities contributed to the global analysis and discussion in order to identify the improvement opportunities.

Table 1 summarizes the specificities of final-year projects under analysis, identified as "P#" for graduation projects and "M#" master dissertations. A brief description of the project's objectives as well as the respective companies' industrial sector is disclosed. These projects were developed in 18 different companies from 10 sectors of industry. Thirteen of them correspond to final graduation degree projects (6 months) and nine were developed during master dissertations (1 year).

Table 1 Characterization of the final-year projects on industrial engineering and management

ID	Objective description	Industrial sector	Year
P1	– Plan, organize and manage the stocks in a health unit – Organize the physical space of the warehouse	Healthcare	2015
P2	– Definition of normalized times for production processes – Identify cycle time and consequent forecast of supply times – Study possible layout improvements	Textile	2015
P3	– Understand the organizational structure – Study the production process with a special focus on planning and weaving – Learn to use SAP System and propose improvements	Textile	2015
M4	– Study the materials flows during the production process – Improve the performance of the transport system for process materials	Mechanical components	2015
M5	– Implement lean tools in the ergonomic design of workstations – Increase the efficiency of a production line and reduce the assembly time – Reduce the production process costs	Mechanical components	2015
P6	– Identify the key work processes and tasks that add value to the company – Define the normalized times of the production process – Determine/suggest a new layout – Identify the cycle time and the forecast of supply timing	Textile	2016
P7	– Identify the areas and points to intervene in production – Present improvement solutions for production activities	Injection of plastics	2016
P8	– Measure production cycle times – Analyse the production scheduling and lines balancing – Evaluate risks within repetitive movements based on objective and deterministic principles – Propose challenging organizational strategies to abandon obsolete and empirical methods	Food supplements	2016
M9	– Reduce search time in work activities – Reduce or eliminate unnecessary movements – Reduce employee travels and eliminate waiting time	Electrical components	2016
P10	– Study the loading operations in of the company – Study the internal transportation and storage on the shop floor (buffers) and warehouse – Find economically viable solutions for truck loading in the expedition, while safeguarding employees integrity, both during the loading operation	Textile	2016
M11	– Characterize the way that Lean tools and technics allow the reduction of costs and wastes associated, leading to continuous improvement – Monitor the implementation of continuous improvement in the industrial environment	Automotive	2016
P12	– Create a database to assists in decision-making processes of integrated planning – Identify the actions to quantify the time and resources needed to introduce a new manufacturing order – Implement continuous improvement solutions – Improve the organizational production strategy	Mechanical components (Moulds)	2017

(continued)

Table 1 (continued)

ID	Objective description	Industrial sector	Year
P13	– Estimate the normalized and cycle times at production and consequent forecast of supply needs – Identify the waste reduction opportunities and implement the 5S methodology and visual management – Define layout changes and improvements – Suggest measures to reduce the impacts of absenteeism and increase the motivation of workers	Textile	2017
P14	– Study the flow of materials during the production process – Propose an alternative solution to improve the process itself and increase the efficiency of the company	Mechanical components	2017
P15	– Reorganize the procedures and materials location of a water supply warehouse – Suggest metrics to measure productivity increase – Reduce operating costs within the warehouse facilities	Services	2017
M16	– Improve the company's naval fleet maintenance management	Naval fleet	2017
M17	– Define a Computer-Aided Process Planning (CAPP) tools and its implementation procedure, using Methods-Time Measurement; a system of predetermined times (MTM) – Implement a usability procedure that contributes to the more frequent use of the developed tools	Automotive	2018
P18	– Study and analyse the productive operations of the company – Identify all production processes and analyse the results – Implement a time and motion study – Define and implement proposals to improve production – Create one point lessons for standard work – Define a maintenance manual for all the machinery	Mechanical components (Moulds)	2019
P19	– Understand the supply chain of the company's operation – Document the procurement processes – Propose improvements for implementation in the company's purchasing and procurement department – Apply a questionnaire to identify the improvement opportunities in the purchasing company's department	Services (Communications Electricity, IT, and Energy)	2019
M20	– Reduce waste of movement and unproductive times that do not add value to the final products – Implement a time and motion study – Create intermediate storage areas (supermarkets) near to the production sections – Implement alert methods for stock management – Optimize the industrial layout	Electrical components	2019
M21	– Analyse the synthetic material flow – Improve the related informatics process tools	Fashion accessories	2019
M22	– Improve the process of supplying production lines – Monitoring production activities – Eliminate wastes in order to reduce distances and the time spent on activities	Fashion accessories	2019

Fig. 3 List of Lean tools applied in the final-year projects under study

2.2 Lean Tools Applied and Performance Indicators

From the methodological perspective, the data content of the projects and master dissertations set was analysed, by counting and comparing several terms and predefined keywords. These keywords (Fig. 3) identify the most important tools from Lean applied during the projects accomplishment and the key indicators that were estimated to quantify the tools success implementation.

The outcomes of the analysis were reported by descriptive summaries and presented in tables and graphs with the frequencies of each of the identified indicator, used lean tools and its combination. For each project, the identified lean tools and the key indicators of performance were correlated taking into account the association between some of the Lean tools and the type of metrics that can be used to estimate the benefits of their implementation.

The application of Lean tools during all the projects requires the identification of several output indicators to assess the success of the improvement measures. During the projects, Action Research methodology was applied to all of the projects under analysis. Action Research is a close cycle methodology where the research is concerned with the resolution of organisational issues together with those who experience the issues directly. It's a methodology characterized by an iterative nature for diagnosing, planning, taking action and evaluating the work processes. Data were interpreted and validated according to the identification of clear results, which can be used to enlighten other opportunities to increase organization productivity.

During the final graduation degree projects and master dissertations, quantitative output indicators are estimated to assess the success of their implementation. Thus, projects were analysed considering the quantification of: **I1**—Reduction of Costs; **I2**—Reduction of Time; **I3**—Productivity Increase; **I4**—Human Factors; **I5**—KPIs Definition; **I6**—Cycle Time; **I7**—OEE; **I8**—Tack-time; **I9**—Layout Modification and/ or Improvement.

Costs reduction and time-saving (lead time) are the indicators that are deeply related to the company's productivity. Thus, regarding the productivity increase as an indicator, it will be accounted for when directly stated in the analysis of the results of each project. Human factors concern the identification of all aspects that may result in an improvement in ergonomic criteria for company workers.

3   Results and Discussion

After conducting the Engineering projects content analysis, all the Lean tools implemented were listed as well as the improvement indicators (I1–I9). With this data, two results analysis matrices were built. Through a Pareto analysis (Fig. 4), it was possible to identify the most frequently applied Lean tools.

Time and Methods, 5S, standard work, visual management, Ishikawa and Spaghetti diagrams, Kaizen and PDCA, VSM, Kanban are the 9 most applied Lean tools in a total of 19 tools used in the Engineering projects under study. Most of these tools represent the first step when implementing continuous improvement solutions in small and medium companies that are reluctant to invest in any manufacturing strategies, without foreseeing benefits due to monetary and resource constraints. Most of the tools are chosen because they represent an analytical instrument in providing figures to make decisions on improvements to increase the company's productivity.

Reduction of time (22.1%), productivity increase (20.8%), human factors (14.3%), reduction of costs (13.0%) and layout modifications (11.7%) are the five indicators most used to measure the impact and the success of industrial engineering projects.

These indicators are mostly based on tools that provide tangible results and benefits in a shorter period of time-related to process planning, layout changes, quality control, production planning, human resources management and supply chain optimization. According to the results, there is a correlation between the duration of the final-year projects and the number of identified indicators, showing that projects with a shorter duration limit the quantification of output indicators. On average,

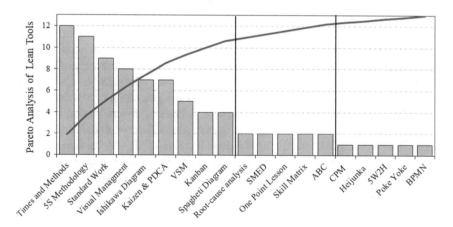

Fig. 4 Pareto analysis considering the Lean tools applied to the engineering problems

about 3 indicators were determined per final graduation degree project, which have a duration of 6 months, whereas, for master dissertations about 5 indicators were determined. In turn, it is not possible to make the same measurement between the type or quantity of Lean tools and the duration of the projects. A correlation matrix between the implemented Lean tools and the output indicators from the 22 final-year projects was built as presented in Table 2.

Results show that the successful implementation of lean tools usually requires standardising the work processes and reducing, or even eliminating wastes and activities that do not add value to the company productivity. This outcome was also observed in Belekoukias et al. [18] where a similar study was performed. Both studies provide a better insight into the relationship and impact that most important lean methods have on productivity, allowing the managers to make better and more effective decisions about the implementation of lean methods. Alves et al. [23] studied the relationship between Lean Production and Ergonomics applied to 41 lean-related projects developed in companies from different industrial sectors. The analysis disclosed a close relationship between the implementation of 5S, Visual Management and standard work and the reduction of distances travelled by the workers, as well as, the reduction of time spent in transport tasks and workers productivity.

Table 2 Correlation matrix of Lean tools and output indicators from projects implementation

Lean Tools & Indicators	Reduction of Costs	Reduction of Time	Productivity Increase	Human Factors	KPIs Definition	Cycle Time	OEE	Tack-time	Layout Modification
Times and Methods	6	10	11	7	2	4	2	3	6
5S Methodology	7	11	9	7	2	1	2	2	3
Standard Work	5	9	9	5	3	2	1	2	4
Visual Management	6	8	7	6	2	2	2	2	3
Ishikawa Diagram	5	6	6	5	2	0	2	1	3
Kaizen & PDCA	5	6	5	5	1	0	3	1	4
VSM	3	5	5	3	2	0	1	1	1
Spagheti Diagram	1	2	3	2	1	1	0	1	4
Kanban	1	4	4	2	1	0	0	0	0
Root-cause analysis	2	2	2	1	1	0	1	1	1
ABC	1	1	1	1	0	0	1	1	2
Heijunka	1	1	1	1	0	1	0	1	1
One Point Lesson	1	2	2	0	2	0	0	0	0
Skill Matrix	1	2	2	0	2	0	0	0	0
CPM	1	1	1	1	0	1	0	1	1
SMED	1	1	1	0	1	1	0	0	0
5W2H	1	1	1	1	0	0	0	0	0
Poke Yoke	1	1	1	1	0	0	0	0	0
BPMN	0	1	1	0	1	0	0	0	0

4 Conclusions

The concept of Lean Production is related to developing tools to acquire competitiveness by waste reduction and value addition approaches. Time and Methods, 5S, standard work, visual management, Ishikawa and Spaghetti diagrams, Kaizen and PDCA, VSM, Kanban are the 9 most applied Lean tools during the 22 final-year Industrial Engineering projects under analysis. Projects with shorter duration difficult the quantification of output indicators. Tools such as times and methods and 5S have a good correlation with the productivity increase and reduction of time.

Although the study was carried out in companies from different activity sectors, the research should be extended to a larger number of projects. Also, the perception of all stakeholders (company managers, employees and engineering students) should be added to develop a comprehensive point of view about the trade-off between lean tools implementation and performance indicators.

Acknowledgements This work has been supported by FCT within the R&D Units Project Scope UIDB/04005/2020 (COMEGI).

References

1. Bhamu, J., & Sangwan, K. S. Lean manufacturing: literature review and research issues. International Journal of Operations & Production Management 34(7), 876–940. (2014).
2. Krafcik, J. Triumph of lean production system. Sloan Manag. Review 30(1), 41–52. (1988).
3. Womack, J., Jones, D. & Roos, D. The machine that changed the world: the story of lean production. in world. (1990).
4. Womack, J., & Jones, D. Lean thinking: banish waste & create wealth in corporation (1996).
5. Melton, T. The benefits of lean manufacturing: what lean thinking has to offer the process industries. Chemical Engineering Research and Design 83(A6), 662–673. (2005).
6. Hines, P., Found, P., Griffiths, G., & Harrison, R. Staying lean: thriving, not just surviving. Lean Enterprise Research Centre (2008).
7. Imai, M. Gemba Kaizen: A common sense approach to a continuous improvement strategy (2nd). McGraw-Hill Education (2012).
8. Amaro, P., Alves, A.C. & Sousa, R.M. Lean thinking: a transversal and global management philosophy to achieve sustainability benefits. In book: Lean Engineering for Global Development, 1–32. (2019).
9. Antoniolli, I., Guariente, P., Pereira, T., Pinto Ferreira, L. & Silva, F.G.J. Standardization and optimization of an automotive components production line. Procedia Manufacturing 13, 1120–1127. (2017).
10. Rosa, C., Silva, F.J.G., Ferreira, L.P., Campilho, R. SMED methodology: The reduction of setup times for Steel Wire-Rope assembly lines in the automotive industry. Procedia Manufacturing 13, 1034–1042. (2017).
11. Resende, V., Alves, A.C., Batista, A. & Silva, A. Financial and Human Benefits of Lean Production in the Plastic Injection Industry: an Action Research Study. International Journal of Industrial Engineering and Management 5 (2), 61–75. (2014).
12. Jiménez, M., Romero, L., Dominguez, M., Espinosa, M. 5S Methodology implementation in the laboratories of an industrial engineering university school. Safety Science 98, 163–172. (2015).

13. Monteiro, J., Carvalho, A.C., Carvalho, M.S. Processes improvement applying Lean Office tools in a logistics department of a car multimedia components company. Procedia Manufacturing 13, 995–1002. (2017).
14. Veres, C. Marian, L., Moica, S. & Al-Akel, K. Case study 5S impact in an automotive company. Procedia Manufacturing 22, 900–905. (2018).
15. Neves, P., Silva, G., Ferreira, L. Pereira, T., Gouveia, A., Pimentel, C. Implementing lean tools in manufacturing process of trimmings products. Procedia Manufacturing 17, 696–704. (2018).
16. Kurpjuweit, S., Reinerth, D., Schmidt, C.G., Wagner, S.M. Implementing visual management for continuous improvement: barriers, success factors and best practices International Journal of Production Research, 57:17, 5574–5588. (2019).
17. Cardoso, N., Alves, A.C., Figueiredo, M., Silva, A. Improving workflows in a hospital through the application of lean thinking principles and simulation. Proceedings of International Conference on Computers and Industrial Engineering (2017).
18. Belekoukias, I., Garza-Reyes, J.A. & Kumar, C. The impact of lean methods and tools on the operational performance of manufacturing organisations. International Journal of Production Research 52(18), 5346–5366. (2014).
19. Abu, F., Gholami, H., Saman, M.Z.M., Zakuan & N. Streimikiene, D. The implementation of lean manufacturing in the furniture industry: A review and analysis on the motives, barriers, challenges, and the applications. Journal of Cleaner Production 234, 660–680. (2019).
20. Maganha, I, Silva, C., Ferreira, L.M.D.F. The layout design in reconfigurable manufacturing systems: A literature review. The International Journal of Advanced Manufacturing Technology 105, 683–700. (2019).
21. Siregar, I., Tarigan, U, Nasution T H. Layout design in order to improve efficiency in Manufacturing. IOP Conference Series: Materials Science and Engineering 309, 012001. (2018).
22. Bengtsson, M. How to plan and perform a qualitative study using content analysis. NursingPlus Open, 2, 8–14. (2016).
23. Alves, A.C., Ferreira, A.C., Maia, L., Leão, C.P., Carneiro, P. A symbiotic relationship between Lean Production and Ergonomics: insights from Industrial Engineering final year projects. *International Journal of Industrial Engineering and Management*, Vol. 10 No. 4, pp. 1–14. ISSN 2683-345X. (2019).

Mapping of the Porcelain Export Process in Brazil Based on the SCOR Methodology

Leonardo Melo Delfim⬛, Caio de Araújo Pereira Gadelha,
Jonas Figuerêdo Silva, Gabriella Gambarra Moreira,
and Maria Silene Alexandre Leite⬛

Abstract The export process is a great way to seek a larger consumer market in order to increase the company's revenues, in addition to providing a surplus in the country's trade balance. In this sense, this article aims to map the main export processes from the point of view of the exporting company, which is one of the leaders in the production of porcelain tiles in Brazil, based on the application of the Supply chain operations reference model (SCOR) logistics approach, which aims to improve the supply chain, including interactions with the customer, physical material transactions and interactions with the market. This research adopts a qualitative approach to identify and know the export process of the focus company, being conducted through semi-structured interviews with the agents responsible for the export process of the porcelain tile, afterwards, the diagram and application of the chosen logistics approach. Thus, it was possible to identify the processes that involve the fulfillment of requirements for the export to be carried out, allowing the manager a better understanding of the process and the sequence of activities.

Keywords Export · Process · SCOR

1 Introduction

The main activities of foreign trade stand out as the most important means of exchange of goods between countries and become largely responsible for the economic development of a nation [1]. In those cases, when considering the high interrelation of agents, a high cost is involved [2], requiring in-depth studies aimed at their optimization.

In these processes, exports highlight due to the high volume of sales compared with imports, mainly in Brazil [3]. In the literature, exports are considered the most common strategies of internationalization of companies due its flexibility and its cost benefits, as well as improving the implementation of internalization strategies

L. Melo Delfim (✉) · C. de Araújo Pereira Gadelha · J. Figuerêdo Silva · G. Gambarra Moreira ·
M. Silene Alexandre Leite
Universidade Federal da Paraíba – João Pessoa (PB), João Pessoa, Brazil

of companies, have led to a rapid growth of international trade [4]. Moreover, there are still many studies on the behavior of exports, which utilize a variety of theories, providing a fragmentary view of the process [5].

Several papers analyze this process and try to find variables that have great impact on your performance and that can bring improvements, highlighting the knowledge of managers in exporting companies [4], the use and improvement of the Internet [6, 7] and implementation of data analysis systems [8].

Still, Junior et al. [9] emphasize that Brazilian economy is based on producing and exporting different kinds of goods such as minerals and agricultural products, and import manufactured and semi-manufactured products, especially in bovine protein scenarios, as claimed by Neto [10]. However, there are still many challenges and difficulties in this process, as well as major risks in some sectors due to lack of studies [7].

Some researchers have been conducted in the foreign trade sector taking into account the import process [11–14] and the recognition of the logistical costs involved in all studied chain [2], however there is still a gap in the literature considering the export supply chain as a starting point for optimization of the costs of the process.

Due to the great interrelatedness of companies involved in these processes, it can be considered the export process as a supply chain and used techniques of this field as a mean of analysis and future optimization. The Council of Supply Chain Management Professionals, the main organization responsible for providing supply chain management methods proposed the SCOR methodology (Supply Chain Operations Reference) as a key point for performance analysis of supply chains, giving a portfolio of steps and processes business, as well as metrics to be applied for the lifting of possible performance indicators. The SCOR model describes the business activities associated with all phases of meeting the demand of a customer. The model itself is organized around six main processes of Plan, Source, Make, Deliver, Return and Enable. Using these components, the SCOR model can be used to describe very simple supply chains or very complex using a common set of definitions in different sectors.

Also, to be widely used such methodology, has a number of applications in various fields, highlighting the use for management of sustainable practices [15], Performance analysis and performance [16–20], quality management [21], universities management [22], and analysis risks [23].

Thus, aiming to help managers in decision making in their export process and to fill the literature gap regarding the lack of work and studies on the performance analysis in area export chains, this paper aims to get an overview of the main export processes from the point of view of the exporter company, based on the application of the SCOR logistics approach.

2 Objectives

Aiming to assist managers in decision making in their export process, this article aims to get an overview of the main export processes from the point of view of the exporter company, based on the application of the SCOR logistics approach. The exporting company is one of the leading producers of porcelain in Brazil, with a manufacturing capacity of 2 million square meters per month, and the export process of porcelain is the object of study.

In this way, the specific objectives are: (1) mapping the process from the point of view of the exporter company, and (2) applying the SCOR methodology for modeling of the company's business processes.

3 Methods

This work constitutes as a case study and is intended to meet the objective. In this sense, this research adopts a qualitative approach to identify and meet the export process focus company, belonging to the field of ceramic tiles, porcelain tiles and cement. In order to do so, we identified four (4) major players in the export process of the investigated company, which are represented by: the commercial representative who makes contact with the customer for selling the products, the export sector, which manages all the export process, PCP sector (production planning and control), responsible for the production and the shipping sector, responsible for packing and shipping the product. Porcelain it's the focus product in this export, with more than $60,000 \text{ m}^2$ to be exported to the USA.

Thus, to map the processes aimed at export, it was used the SIPOC tool (Supplier, Input, Process, Outputs and Customer), which analyzes both suppliers, inputs, process itself, outputs and customers in order to clarify the process steps and identify the responsibilities of each agent, necessary documents, and how the process is developed. The interviews were semi-structured, in which the respondent has more flexibility to expose information about the process, and such were recorded and transcribed later. The activities of the sales representative were raised by the responsible for the export sector, the interviews lasted around 1 h and 30 min. Finally, the interviews with the PCP and shipping sector, were held as technical visits in the production of the product and the storage and loading of the same (Table 1).

Table 1 Respondents
characterization

Respondents characterization		
Agent	Function	Time in the company (years)
Exporting company	Sales executive	2
	PPC analyst	3
	Dispatcher	2

With processes properly validated and outlined, next step was the application of the SCOR model, following the rules of APICS [24]. The SCOR model describes a complex set of definitions from different sectors in order to use it as support for projects to improve supply chain. The model contains four main sections: performance, practices, people and processes, the last (processes) the focus of this article, applying the first hierarchical level of the process section, because it is where you have initial descriptions of management processes and relations.

Then, the applied mapping method BPMN (Business Process Model and Notation), as it is highlighted by Silveira et al. [25], the mapping process allows the visualization and understanding of the activities that are performed by the organization, with the BPMN notation, a graphical representation easy to interpret (diagrams), containing basic information showing the hierarchy of activities. Such data are outlined through the Bizagi Modeler software for better understanding.

4 Results

The exporter company has more than three decades of tradition in ceramic tiles, being a major producer in Brazil and one of the leaders in the production of porcelain. In order to increase its visibility in the market and consequently its revenues, the company seeks increasingly gaining ground in the international market, mainly in the USA, place of origin of the importing company.

The export process starts in the negotiation through the sales representative, that contact the customers through events or in person. Thus, there is the initial negotiation with the request samples of the most similar products to those already marketed in the country of the importer. This process is followed by the exporter company in the country of origin and the trade representative, until the approval of the product by the customer.

Parallel to the process, there is the need of the exporter to comply with some requirements of the importer through audits and a number of forms in the security context, such as anti-terrorism action, internal security and absence of slave labor, and demands for market service guarantee.

At the end of the process of auditing and approval of products, begins the negotiation of specifics values, as the volume to be asked, then is mapped the requirements to carry out the business, in this case the export process is by DDP (Delivered Duty

Paid) on the purchase and sale agreement, that is the exporter bears all costs and risks in the logistics activity from the factory, shipping and international road transport to arrive at the agreed place with the customer, in this case the distribution centers.

Set these parameters, mapping costs related to this process, as the costs of production, which includes packaging and pallets with due fumigation certificate, and the logistics costs, with the dispatcher, with the national road transport, with stuffing of pallets in the container, there are also negotiations with shipowners, these refer to the marine, and negotiations with loads of agents, they manage the entire export process from the national port to the international port and also responsible for the international road transport to the arrival at the agreed place, analyzing the profitability of the operation.

Thus, we reach the final negotiation process with the customer, the exporter company and the commercial representative deal with the client the amounts to be charged. A soon as the customer approves such values, the sales representative creates the Proforma Invoice which is a document that formalizes the purchase and sale, which contained the products, volume, shipment and port of destination, among other information, it must be signed by the importer. After signing, the document is sent to the exporter company, thus confirming the customer's request, completing the process of negotiation/sale. Every process can be visualized in the following diagram (Fig. 1).

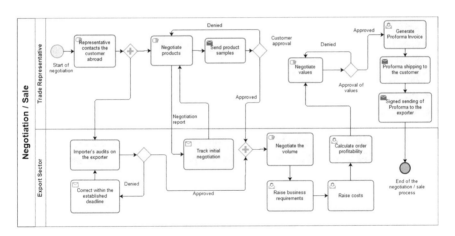

Fig. 1 Negotiation/sale

The process of negotiation/Sale is presented without applying the SCOR model, because it does not address the sales and marketing processes (demand generation), as well as the development and research, but it is presented as it is important for the understanding of the process as a whole.

Upon receipt of the customer's request, in which are described their data, the required products and volumes begins the process of planning and production control

(PCP) and registrations up applications in the system to generate the information necessary for purchases of inputs.

Later begins training the teams to meet the special requirements of customer service, afterwards is production, and the PCP reports dates and forecasts of production. Along with production, the quality control sector makes reviews of lots and informs the availability of service requests, monitoring the drafting process of the package layout. The expedition begins with the loading process as scheduled by the carrier which in this case is the same as the responsible for the stuffing.

Simultaneously with the process described above, the export sector is responsible for the preparation of the necessary documentation, and makes up the reserves of batches produced respecting the volumes proformas, reserves spaces on the ship respecting the dates stipulated by customers, revenues (first invoice), shipping documents as: packing list and Invoice, the certificate of purge (fumigation, which is needed to be stamped on all the pallets), the DUE (electronic unified statement) which is a tax document in the export process and has data export, form of export, country, net weight, value in real, value of foreign currency and ports. Once you receive the booking of the cargo agent, the material is sent to the stuffing terminal that is responsible for receiving the goods and removing the empty containers and therefore sending the filled container to the port. All the documentation is sent to the dispatcher so that it starts the export's formalities, which it is responsible for registering the proforma with the IRS and consenting bodies.

So after stuffing the container and loading the ship, the cargo agent issues the Bill of Lading, which is the load presence on the ship, knowing whether the departure and arrival dates as part of the feedback process and sends-if the Bill of lading to the customer.

The process of planning and production control starts with the receipt of the request from the export sector, from this point, the product is registered on the system and the requirements for the export process are identified, as well as the specificities of the customer order. It is then necessary to draw up the layout of the package, which is a responsibility of the marketing sector, but also managed by the export sector.

The planning to buy the raw materials it's parallel with the process described above. The fumigation is carried out on the pallet that is specific to export it, a service provided by a company hired by the export sector. Then, finally, check the requirements for export.

With such tasks performed, the planning of the lots according to the used machines is carried out. Finalizing the PCP process, the entire production is accompanied by, so to check that there are no failures, as there is feedback from the quality sector until the sending of the lots for the shipping industry.

The dispatch process starts with the receipt of the volume of information required on the customer's request for export. Thus, it is planned the location where the volume will be stored. To begin the production and finalize the lots, the shipping industry receives the pallets and stores.

In parallel, the carrier hired by the exporter sector, informs the sector the number of the note, the Purchase Order number, the proforma number, the board of the vehicle and the driver's name, which will transport the lots the next day. Thus, preparing the entire proforma, print labels and also plans the packing list to load the vehicles in the amount and correct proforma.

When the truck arrives in the shipping sector, enveloping up the pallets according to the proforma, which will be loaded into the truck. At the end of the lot, cut up truck photos, gives again the proforma with lots loaded and asks the signature of the driver on your packing list. Before the driver leaves the factory, again there is a conference throughout loading with a copy of the packing list. The feedback if the truck arrived is given the next day.

The loading is responsibility of the exporter company, however, after the cargo leaves the company, it becomes responsibility of the carrier. The stuffing in containers is the responsibility of the carrier and the maritime and international road transport, cargo agent.

After applying the SCOR model, is showed the business activities proposed by itself. This division assists to observe the nature of activities, being planning, supply, make and deliver, identifying the activities associated with the phases of the customer demand satisfaction. The beginning steps in each sector refer to the activities of planning and preparation of documents, then the supply of raw materials or services to production, ending in the attendance of the request.

5 Conclusion

Thus, this article aims to give an overview of the main export processes from the point of view of the exporter company, based on the application of the SCOR logistics approach. Thus, the mapping of the main export processes was carried out through interviews. The main processes consists of the processes of negotiation/sales, management, planning and production control (PCP) and shipping.

The use of the SCOR logistics approach assists in the performance analysis of supply chains, providing business processes, metrics and performance indicators. As the business process description is the focus of this article. We identified in Figs. 2, 3 and 4 activities that make up business processes, planning, activities related to the determination of requirements and corrective actions to achieve supply chain objectives, supply, processes associated with application delivery, receipt and transfer of raw material items, goods and/or services, which are activities associated with the conversion of materials or the creation of content to services and deliver, which are processes associated with the implementation of order management and order fulfillment activities.

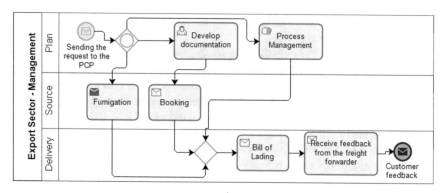

Fig. 2 Management

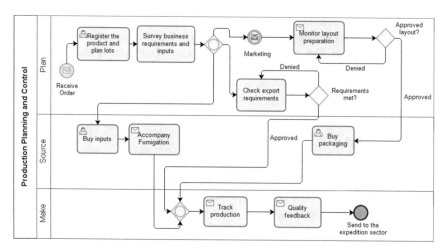

Fig. 3 PCP process

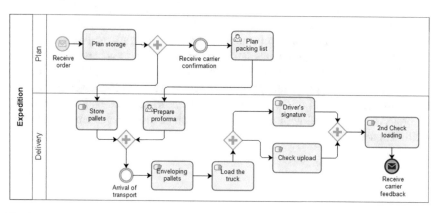

Fig. 4 Expedition

According to the Brazilian government [26], the export of goods containerized takes around 13 days, with an average of 2.200 US dollars per container costs. Therefore, the identification of the main processes involved in the export of a product, seeks a better understanding of these in order to improve customer satisfaction and increase business performance helping the company to clearly see the strengths and weaknesses.

Thus, this mapping was observed first, the number of cases involving the fulfillment of requirements for the export to be carried out, allowing the manager a better understanding of the process and sequence of activities, and as a basis for new exports. Subsequently the application of the SCOR model that aids in clear view of the responsibilities of each activity, allowing again the manager to analyze what processes can be improved, such as the creation of the layout, which comprises three sectors, PCP, marketing and finally the export sector that relates to the customer, this may end up delaying the process. The export process can be complex and exhaustive, so the mapping of all the activities that compose it helps the company to standardize it, enabling a competitive advantage in the market.

References

1. Nagyová, L.; Horácová, M.; Moroz, S.; Horská, E.; Polakova, Z.; The Analysis of Export Trade between Ukraine and visegrad Countries. Marketing and Trade. 2, XXI (2018).
2. Medeiros, M.; Measurement of logistics costs in an operation with a solid charge in a public port in northeastern Brazil. Dissertation (Post Graduate Program in Production Engineering - UFPB). João Pessoa (2018).
3. Ministry of Economy, Industry, Trade and Services (MDIC); Trade balance (2019). Available at: <http://www.mdic.gov.br/>. Last accessed: 22 apr. 2019.
4. Kotorri, M.; Krasniqi, B.A.; Managerial Characteristics and Export Performance - Empirical Evidence from Kosovo. South East European Journal of Economics and Business Volume 13 (2), 32–48 (2018).
5. Chen, J. Sousa, C.M.; He, X.; The determinants of export performance: A review of the literature from 2006 to 2014. International Marketing Review 33 (5): 626–670 (2016).
6. Blazquez, D.; Domenech, J.; Web data mining for business monitoring export orientation. Technological and economic development of economy. 2018 Volume 24 (2): 406–428 (2018).
7. Serrano, R.; Acero, I.; Rethinking Entry Mode Choice of Agro-Exporters: The Effect of the Internet. International Food and Agribusiness Management Review. Volume 18 Issue 3 (2015).
8. Wang, C.N.; Le, A.P.; Application in International Market Selection for the Export of Goods: A Case Study in Vietnam. Sustainability 2018, 10, 4621 (2018).
9. Junior I.C.L; Guimaraes, V.A; Guabiroba, R.S; Fonseca, R.R.; Comparative study of brazilian public container terminals: technical efficiency ranking and forecasts for 2052. Brazilian Journal of Operations & Production Management 13, pp. 386–398 (2016).
10. Neto, O.A.; Brazil is the world beef market: analysis of competitiveness of production and export of Brazilian logistics. Geographic studio - Goiânia-GO, vol. 12, no. 2, pp. 183–204 (2018).
11. Moreira, G.G.; Silva, J.F.; Leite, M.S.A.; Discussion about the agents Involved in the process of acquisition, and clearance operation with loads of wheat in a public port in the Northeast. IV International Congress of port performance. Florianópolis, Santa Catarina (2017).

12. Silva, J.F; Moreira, G.G; Leite, M.S.A.; Qualitative analysis of the use of the resources used in the solid bulk cargo unloading process in a public port Northeast (br). In: International Congress of port performance, Florianópolis (2018).

13. Silva, J.F.; Moreira, G.G.; Medeiros, M.; Ribeiro, W.J.R.; Leite, M.S.A.; Mapping of import's logistics chain of the solid charge drained by the brazilian public port. The 9th International Conference on Production Research - Americas 2018. Bogota, Colombia (2018).

14. Silva, J.F; Moreira, G.G; Silva, P.H.M; Medeiros, M.; Leite, M.S.A.; Acquisition of materials in the process of importing solid cargoes in a public port in the Northeast: Description of the main forms of acquisition. IV International Congress of port performance. Florianópolis, Santa Catarina (2017).

15. Azmi, N.J.M; Rasi, R.Z.R.; Ahmad, M.F.; Review of Enviropreneurial Value Chain (EVC) based on SCOR Model Theory and NRBV. Global Conference on Business & Social Science-2014 GCBSS-2014, 15th & 16th December, Kuala Lumpur (2014).

16. Bukhori, I.B; Widodo, K.H; Ismoyowati, D.; Evaluation of Poultry Supply Chain Performance in XYZ Slaughtering House Yogyakarta SCOR and using AHP Method. Agriculture and Agricultural Science Procedia, vol. 3, pp. 221–225 (2015).

17. Guritno, A.D; Fujianti, R.; Kusumasari, D.; Assessment of the supply chain and classification factors of inventory management in level suppliers of fresh vegetables. Agriculture and Agricultural Science Procedia, vol.3, pp. 51–55 (2015).

18. Spina, M.E; Rohvein, C.; Urrutia, S.; Roark, G.; Paravié, D.; Corres, G.; Application of the model in metalworking SMEs SCOR from Olavarría. CUC INGE, vol. 12 in. 2, pp. 50–57 (2016).

19. Wibowo, M.A.; Sholeh, M.N.; The analysis of supply chain performance measurement at construction project. Procedia Engineering, vol.125, pp. 25–31 (2015).

20. Wibowo, M.A.; Sholeh, M.N.; Adji, H.S.; Supply chain management strategy for recycled materials to support sustainable construction. Procedia Engineering, vol. 171, pp. 185–190 (2017).

21. Anggrahini, D.; Karningsih, P.D.; Sulistiyono, M.; Managing Quality in a frozen shrimp risk supply chain: a case study. Procedia Manufacturing, vol. 4, pp. 252–260 (2015).

22. Gopalakrishnan, G.; How to apply Academic Supply Chain Management: The Case of an International University. Management, Vol. 20, pp. 207–221 (2015).

23. Astuti, R.; Silalahi, R.L.R.; Rosyadi, R.A.; Risk Mitigation Strategy for Business Using Mangosteen House of Risk (HOR) Methods: (A Case Study in "Wijaya Buah," Blitar District, Indonesia). The 3rd International Conference on Agro-Industry, pp. 17–27 (2016).

24. APICS. Supply chain operations reference (SCOR) model. Versão 12.0. 2017. Chicago: APICS (2017). Available in: https://www.apics.org/apics-for-business/frameworks/scor. Last accessed: 10 abr. 2020.

25. Silveira, L.S.; Longaray A.A.; Tondolo, V.G.; Sarquis, A.; Munhoz, P.; Tondolo, R.P.; Proposal Process Mapping using BPMN: A Case Study on a shipbuilding industry. In: International Performance Port Congress, 3, 2016, Florianópolis. Proceedings III International Performance Port Congress, Florianópolis, Santa Catarina (2016).

26. SERPRO; Mais agilidade para o comércio exterior (2017). Available in: http://intra.serpro.gov.br/tema/noticias-tema/tecnologia-proporciona-mais-agilidade-ao-comercio-exterior-brasileiro, last accessed 25 jan. 2020.

Circular Economy for Lubricating Oils in Brazil

Marcelo Guimaraes Araújo, Giancarlo Lovón-Canchumani, and Lilian Bechara Elabras Veiga

Abstract Circular Economy—CE concept is starting to gain attention in Brazil. In developing countries, the background for CE development is quite different from developed countries, where there is an established waste management structure and environmental policy is robust. This paper aims to propose a CE evaluation model for Brazil—ECBR, based on CE concepts, origins, supporting legislation and applications in countries that are ahead in its implementation. ECBR seeks to make circularity possible in different economic sectors, encompassing the various dimensions needed for its effective implementation. The ECBR model is applied to lubricating oil re-refining in Brazil with the support of Life Cycle Assessment tool and Simapro software. The circular chain of lubricating oil is analyzed. Primary data was collected from a major Brazilian re-refining company and secondary data was obtained from Ecoinvent database. The results revealed environmental impacts reduction and economic gains from loop closing. It is also shown that pervasiveness of products and its waste can be a shortcoming for CE initiatives due to the impacts from waste collection and transport to a recycling unit.

Keywords Circular economy · Brazil · Resource efficiency · Lubricating oil

1 Introduction

In Nature, cycles are frequent and, differently from anthropogenic actions, there are no residues. That means, in Nature everything is transformed and used. What is sought through Circular Economy (CE) is this circularity, in a similar way to natural

M. Guimaraes Araújo (✉)
Oswaldo Cruz Foundation, Rio de Janeiro, RJ, Brazil
e-mail: marcelo.araujo@ensp.fiocruz.br

G. Lovón-Canchumani
Federal University of Paraná, Jandaia do Sul, PR, Brazil

L. Bechara Elabras Veiga
Federal Institute of Rio de Janeiro, Rio de Janeiro, RJ, Brazil

© The Author(s), under exclusive license to Springer Nature Switzerland AG 2021 103
A. M. Tavares Thomé et al. (eds.), *Industrial Engineering and Operations Management*,
Springer Proceedings in Mathematics & Statistics 367,
https://doi.org/10.1007/978-3-030-78570-3_8

ecosystems, where the by-products of one process become raw material in another process and the concept of waste ceases to exist [1]. Thus, CE seeks to migrate from the linear industrial system (extraction, manufacture, use and disposal) to a circular process (extraction, manufacture, use, extension of useful life, reuse, recycling and many others process). In addition to these physical processes, new business models such as product sharing, servitization and virtualization of products and services, and others are associated with CE which allow these cycles to become more efficient.

Circular Economy is complex and multidisciplinary and requires a structured approach considering the specificities of each country or economic sector. In this sense, CE assessment models at the meso and micro levels have been developed [2, 3, among others]. The EEA—European Union Environmental Agency [3], proposed a conceptual model to be followed by member countries in a systemic and comprehensive perspective, from the macro to the micro level.

In developing countries, the background for CE development is quite different from developed countries, where there is an established waste management structure and environmental policy is robust. Based on the above, this article aims to propose a model for evaluating CE in the Brazilian context, from the discussion of the fundamentals on which it is based, analyzing the existing actions and the correlated variables in the country, in addition to the challenges and opportunities for its dissemination.

Initially, a literature review is carried out on some CE models developed by academia and economic sectors. A CE analysis model for Brazil was built, considering the country's specificities, within four dimensions: resource characteristics; organizational and social structure of the chains; economic and governance aspects and impacts on the environment and human health. These four dimensions were defined based on a literature review and existing legislation, where several variables were identified and grouped considering the national context. The re-refining of lubricating oil is used as a case study due to its relevance and since it is a well established chain in the country.

2   Circular Economy Models

The models presented in this section have different approaches to CE and consider, to a greater or lesser extent, different variables. These models have been the object of analysis and revisions by researchers and professionals in the CE field, some of which have been applied by countries and/or productive sector chains.

Some authors have developed conceptual models to analyze the circularity of specific products or services (micro level), such as, desalination membrane [4], or even, economic sectors (meso level), such as, Built Environment [5, 6] or iron and steel [7]. Lewandowski [2], presented a model focused on the business vision, discussing how a company should structure itself to foster CE, highlighting that the existing CE models were limited and specific. In Brazil, the CNI—National Confederation of Industry [8] suggests a CE business model based on concentric circles, in which the inner circle presents circular business models, and is circumscribed by value chains, and this, in turn, underlies the facilitating aspects of the circular economy, which include education, public policy, infrastructure, technology and innovation. Table 1 summarizes CE models found in literature, highlighting their main components.

It is worth to mention that many of the CE models found in the literature focus mainly on the chain management bias, not including environmental and social aspects, as highlighted by Geissdoerfer et al. [16]. Moreau et al. [12] consider that the social and institutional dimensions are relevant and should be included in the CE analysis. Organizational, cultural and political aspects are also highlighted by Korhonen et al. [17]. Iacovidou et al. [15] emphasize that CE approaches should be evaluated simultaneously in the technical, social, economic, and environmental domains.

The network of European Environmental Agencies [13] presents a model for evaluating the transition to CE based on three levels of detail, from the most general along the entire chain, including all actors, to specific product indicators/material. The European Union Environmental Agency [14], in turn, presents a conceptual framework for the analysis of CE in a systemic perspective, much more comprehensive, from technical to organizational aspects, from micro to macro level, with several indicators suggested to monitor circularity.

Some authors point out barriers to an effective implementation of circularity, instead of presenting detailed models that define the variables. At the macro level, Korhonen et al. [17] highlighted limitations of thermodynamics, of economic scale,

Table 1 Conceptual models for circular economy

	Level	Model components
Moreno et al. [9]	Micro	Project for material circularity in production chain
Scheepens et al. [10]	Meso	Sustainable technologies, innovative products, tax incentives, subsidized infrastructure
Ness and Xing [11]	Meso	Closed cycle of processes, encompassing networks of actors, resources and instruments; strategies and synergies for the management of these elements
Ometto et al. [8]	Meso	Value chains and education, public policy, infrastructure, technology and innovation
Moreau et al. [12]	Macro	Social and institutional dimensions must be included
EPA Network [13]	Macro	Three levels of analysis from the most general to the specific level of product, along the chain
EEA [14]	Macro	Systemic perspective, from technical to organizational aspects, from micro aspects to the macro level
Iacovidou et al. [15]	Macro	Three phases: synthesis, analysis and refinement of the components involved in the chain, respectively: academy; professionals of the productive sectors; political actors and citizens

of temporal and physical limits, of marketing, management, governance and socio-cultural aspects. Kirchherr et al. [18] summarized the barriers to EC in four types: cultural, market, regulatory and technological factors. Ritzen & Sandstrom [19] investigated barriers in companies: economic measurement of EC benefits, exchange of information, obscure distribution of responsibilities, inadequate supply chain management, lack of perception of sustainability, risk aversion, product design and integration between productive processes.

3   A Model of Circular Economy for Brazil

This section presents and analyzes the proposed CE model for Brazil. This model was developed from four central dimensions, based on the CE model depicted by the EEA European Environmental Agency [14]. The model's criteria is inserted in the context of the Brazilian reality, therefore the four dimensions are part of an envelope formed by national legislation, as shown in Fig. 1.

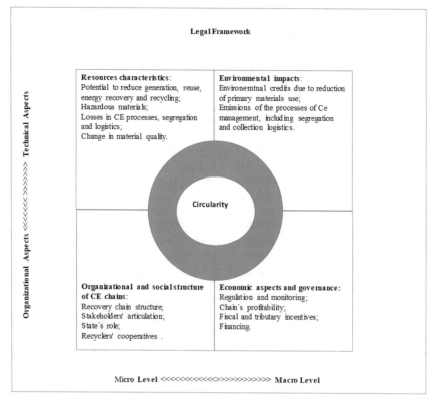

Fig.1 Circular economy model for Brazil—ECBR

The model's first layer, the envelope, refers to the legal framework, and encompasses all other dimensions. The four internal quadrants were built from two main axes: micro level to macro level and from organizational aspects to technical aspects, as described below:

Dimension 1—technical aspects at the micro level—refers to the characteristics of the materials or waste, such as: volumes generated by waste types, degree of hazardousness of the materials (toxicity, teratogenicity, pathogenesis and others), assessments of the feasibility of reduction generation, reuse, recycling and energy recovery technologies, changes in material quality, losses in processing and logistics.

Dimension 2—organizational aspects at the micro level—refers to the organizational and social structure of the chain, encompassing the role of institutions, particularly the public or private waste management company, possible economic uses for secondary materials, cooperative certifications recycling and articulation between the actors.

Dimension 3—organizational aspects at the macro level—refers to economic, regulatory and governance aspects, such as regulation (agencies), transparency and inspection of entities in the chain (state control bodies), financing, taxation, tax and tax incentives for use secondary materials and new business models.

Dimension 4—technical aspects at the macro level—refers to the infrastructure required for logistics and treatment of materials, training and capacity building, impacts on the environment such as emissions and effluents from treatment and logistics management processes, environmental credits for reducing impacts in the production of raw materials and others.

It should be noted that the boundary between these quadrants is not necessarily boldly defined. Quite the contrary, it is a nuance. However, the non-limitation of this typology does not necessarily affect the analysis in question. These dimensions are presented in more detail in the following item for the oil redefining case.

4   ECBR Application to Lubricating Oils Re-refining

Few initiatives classified as CE were successfully established in Brazil. Examples are lubricating oil and its packaging, pesticide packaging, aluminum beverage cans, waste tires, among other recovery chains. The used oil chain may be considered as a meso level cycle. Lubricating oils production and its waste management processes result in major environmental impacts that can be reduced with a CE approach.

4.1   Legal Framework for Used Oil in Brazil

Brazil has a legal framework that requires producers and importers of lubricating oils to properly dispose used oils. Used lubricating oils are considered by the Brazilian regulations as a hazardous substance. Regulations from the Brazilian National Environmental Council (CONAMA), determine that used or contaminated lubricating oils should be disposed properly forbidding any discharge into soil, surface water, groundwater, sea, sewer systems or storm drains [20].

CONAMA Resolution n° 362/2005 [21] regulates the recycling of used lubricants oils, requiring its regeneration and recovery to avoid contamination. This resolution establishes the polluter pays principle, holding the manufacturer or importer liable for collection and final disposal of used or contaminated lubricating oils and determining that all lubricants must be collected and sent for recycling by re-refining. It also sets joint and several liability of the agents (producer and collector) for environmental damage, establishes a minimum collection of 30% of the oil sold and holds the producer or importer responsible for the cost of the collection program. Finally, it prohibits the combustion and incineration of used or contaminated oils and creates mechanisms for control by the federal environmental agency—IBAMA.

In 2010, the National Waste Management Policy was enacted, Federal Law n° 12,305 [22], establishing products shared responsibility over their life cycle and a reverse logistics system for solid waste collection in many sectors, among which lubricating oils, reinforcing the previous legislation.

4.2 Dimension 1—Technical Aspects at the Micro Level

Unlike other petroleum derivatives, lubricating oils are not consumed during their useful lifetime. This creates particular responsibilities to minimize the environmental impacts they cause, through reuse or adequate disposal when no further use is possible [23]. The Brazilian market for lubricating oils is the sixth largest in the world, with an annual production of approximately 1.05 billion litters. Table 2 presents the detailed data of used oil chain in Brazil for 2017 [24].

Table 2 Used oil chain data, Brazil, 2017 [24]

Installed Capacity for used oil re refining	500,568 m³/year
Storage capacity of used oil in re refining unities	32,158 m³
Volume of re-refined oil in 2017	260,379 m³
No. of vehicles in collection	1,043
Points of collection in 2017	111,188
Used oil collection in 2017	422,977.72 m³
No. of municipalities with used oil collection in 2017	4,186

4.3 Dimension 2—Organizational Aspects at the Micro Level

The productive chain is composed of five main groups of actors: producers, resellers, consumers, collectors and re-refiners. Figure 2 shows the flows of the productive chain.

According to the trade association representing makers and blenders of refined petroleum products [26], the lubricant oil cycle involves five categories of actors, each one with a different responsibility over the product life cycle:

- Producers and importers—the companies that introduce lubricants in the market. They have the legal obligation to pay for collection and to inform consumers of their obligations and the environmental risks of inappropriate disposal.
- Resellers—the companies that sell lubricants in the wholesale and retail markets. Among their obligations are to receive used or contaminated oils from consumers and to store them in adequate installations.
- Consumers—the individuals and companies that use lubricants. They have the obligation to deliver the used products to the collection point (reseller or authorized collector).
- Collectors—companies licensed by the competent environmental authority and the National Petroleum, Natural Gas and Biofuels Agency (ANP) to engage in collection of used or contaminated lubricating oils and to deliver them to the re-refiner.
- Re-refiners—companies authorized by the ANP and licensed by the competent environmental authority to engage in re-refining. They are required to remove the contaminants and produce basic oil according to the ANP's specifications.

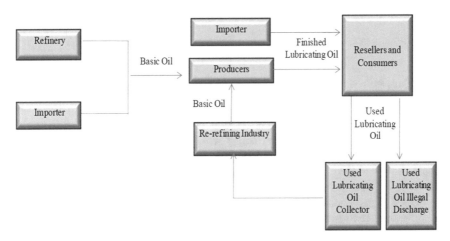

Fig. 2 Productive chain of the lubricating oils sector in Brazil—adapted from [21, 25]

4.4 Dimension 3—Economic and Governance Aspects

In 2007, the Ministry of the Environment (MMA) and the Ministry of Mines and Energy (MME) jointly issued MME/MMA Ordinance 464, containing regional and national targets.

Besides issuing the Resolution, in 2005 CONAMA created the Permanent Monitoring Group to verify the application of the Resolution, under the coordination of the MME and with the participation of representatives of the MME, state and municipal environmental agencies, businesses and nongovernmental organizations.

It can be seen from the above discussion that Brazilian environmental legislation's focus is recycling by re-refining. Since it is an economic sector with few players, its governance was easily established. The oil sector regulation in the whole country is done by a single agency—National Petroleum Agency—ANP. Only 23 companies are allowed by ANP for the collection of wasted oil and 14 for oil re-refining [24].

4.5 Dimension 4—Technical Aspects at the Macro Level

Hazards of used lubricating oil are heavy metals, metalloid particles, coloured compounds (such as polychlorinated biphenyls, or PCBs, polychlorinated dibenzo-dioxins and coloured solvents), PAHs (polycyclic aromatic hydrocarbons) and others [27]. Various authors have mentioned that efficient management of used lubricating oils can significantly reduce both the environmental impacts and the consumption of natural resources [28–30]. Technological options for the treatment of used oils include recycling to make new lubricants and its use as a fuel to generate energy or heat.

5 Discussion

Considering legislation, Brazil has enacted specific laws for the lubricating oil sector, establishing the Polluter´s Pays Principle where waste responsibility rest with the producers and importers of the sector. This rule is also adopted in countries like Spain, Italy and Portugal, among others, where producers and importers also bear the economic responsibility on managing lubricating oils over their life cycle [31].

Over the past decade, Brazil has made major advances in the sector regulation, establishing a reverse logistics system, goals for used lubricating oils collection and for the re-refining, with regional and national collection targets, which have been increasing gradually.

From 2006 to 2015, the volume of waste oils collected increased from 271 million litters to 445 million litters, an increase of 64.6% in nine years [25]. In 2017, the volume of waste oils collected was 431 million litters corresponding to 40.91% of production [24] surpassing the minimum level of 39.2% for collection imposed by the law for this year. This is a very high achievement considering the losses that occur during the use phase and in re-refining processes.

Regarding the reverse logistics structure for lubricating oils, the increase in the number of collection centers and re-refining units has boosted the collection of waste oils. Stakeholders of the lubricating oil chain implemented its circularity by mixing the primary material to the secondary material, manufacturing products with high quality and that are subjected to severe control by Brazilian authorities.

6 Final Remarks

The model presented, ECBR, encompasses different dimensions allowing the assessment of circularity in Brazil, based on a structure of four quadrants and an envelope. This envelope is composed by the Brazilian legal framework and the quadrants, organized in four dimensions, namely: characteristics of resources, supply chain social and organizational structure, recovery, economic and governance aspects and environmental impacts along the chain.

For the case study, lubricating oil re-refining, the model shows the detailed steps of the oil cycle and the interaction between them. This success case is a demonstration of the adequate application of regulation and enforcement of the laws that resulted in the adhesion of the economic actors.

This model is intended to help decision makers develop Circular Economy policies, programs and actions, promoting the circularity of the different product and material chains, reducing resources consumption and associated environmental and social impacts. It can be a basis for circularity programs and strategies development in Brazil, both by public and private sectors.

What is sought with the concept of Circular Economy is much more than just the recovery of materials as recommended by current Brazilian legislation, but rather the detachment of economic growth from the use of resources. As recommendations, for future research, it is suggested the development of analysis models for the most relevant economic sectors in the country, such as the civil construction sector and/or the agricultural sector. New business models are recent and require efforts for their analysis.

References

1. Ghisellini, P., Cialani, C., Ulgiati, S.: A review on circular economy: the expected transition to a balanced interplay of environmental and economic systems. Journal of Cleaner Production 114, 11–32 (2016).
2. Lewandowski, M.: Designing the Business Models for Circular Economy—Towards the Conceptual Framework. Sustainability, 8, 43 (2016).

3. EEA - European Environmental Agency. Circular by Design: products in the circular economy. Copenhagen: EEA report no. 06 (2017).

4. Landaburu-Aguirre, García-Pacheco, R.; Molina, S.; Rodríguez-Sáez, L.; Rabadán, J.; García-Calvo, E.: Fouling prevention, preparing for re-use and membrane recycling. Towards circular economy in RO desalination. Desalination, 393, 16–30 (2016).

5. Kilkis, S. & Kilkis, B.: Integrated circular economy and education model to address aspects of an energy-water-food nexus in a diary facility and local contexts. *Journal of Cleaner Production* xxxx, 1–15 (2017).

6. Ness, D.A. & Xing, K. (2017) Toward a Resource-Efficiency Built Environment: a literature review and conceptual model. Journal of Industrial Ecology, vol. 21, no.3, 572–592 (2017).

7. Ma, S., Wen, Z.Z., Chen, J., Wen, Z.Z.: Mode of circular economy in China's iron and steel industry: a case study in Wu'an city. Journal of Cleaner Production 64, 505–512. https://doi.org/10.1016/j.jclepro.2013.10.008 (2015).

8. Ometto, A.; Amaral, W.A.N.; Iritani, D.; Castro, L.R.K.: Economia Circular: oportunidades e desafios para a indústria brasileira. Brasilia: CNI – Confederação Nacional da Indústria (2018).

9. Moreno, M.; de los Rios, C.; Rowe, Z; Charnley, F.: A Conceptual Framework for Circular Design. Sustainability, 8, 937 (2016).

10. Scheepens, J.G.; Vogtlander, J.G.; Brezet, J.C.: Two life cycle assessment (LCA) based methods to analyse and design (regional) circular economy systems. Case: making water tourism more sustainable. Journal of Cleaner Production, 1–2, (2015).

11. Ness, D.A. & Xing, K.: Toward a Resource-Efficiency Built Environment: a literature review and conceptual model. Journal of Industrial Ecology, vol. 21, no.3, 572–592 (2017).

12. Moreau, V.; Sahakian, M.; Van Griethurssyssen, P.; Vuille, F. Coming full circle: Why social and institucional dimensions matter for the Circular Economy. Journal of Industrial Ecology, 21–3, 497–506 (2017).

13. EPA Network. European network of the heads of Environmental Protection Agencies: Input to the European Commission of the transition towards a circular economy in the European Union. Discussion paper, May (2017).

14. EEA - European Environmental Agency: More from less — material resource efficiency in Europe 2015 overview of policies, instruments and targets in 32 countries. Copenhagen: EEA report no. 10, (2016).

15. Iacovidou, E.; Millward-Hopkins, J.; Busch, J.; Purnell, P.; Velis, C.A.; Hahladakis, J.N.; Zwirner, O.; Brown, A.: A pathway to circular economy: Developing a conceptual framework for complex value assessment of resources recovered from waste. Journal of Cleaner Production 168, 1279–1288 (2017).

16. Geissdorfer, M.; Savaget, P.; Bocken, N.M.P.; Hultink, E.J.: The Circular Economy: A new sustainability paradigm? Journal of Cleaner Production, 143, 757–768 (2017).

17. Korhonen J.; Honkasalo, A.; Seppala, J.: Circular Economy: The concept and its limitation. Ecological Economics, 143, 37–46, (2018).

18. Kirchherr, J., Reike, D., Hekkert, M.: Conceptualizing the circular economy: an analysis of 114 definitions. Resources. Conservation. Recycling. 127, 221–232 (2017).

19. Ritzen, S. & Sandstrom, G.O.: Barriers to the Circular Economy – integration of perspective and domains. Procedia CIRP, 64, 7–12 (2017).

20. ABNT. Associação Brasileira de Normas Técnicas NBR 15.792:2010. Embalagem — Índice de reciclagem — Definições e método de cálculo, (2010).

21. CONAMA Conselho Nacional do Meio Ambiente (National Council of Environment) Resolução 362 de 23 de junho de 2005. Available at: www.planalto.gov.br (2005).

22. República Federativa do Brasil: Política Nacional de Resíduos Sólidos. Lei no.12.305/2010. Brasília, (2010).

23. Hsu, YL., Liu C.: Evaluation and selection of regeneration of waste lubricating oil technology, Environ Monit Assess v. 176, pp. 197–212, (2001).

24. MMA – Ministério do Meio Ambiente. Instituto Brasileiro do Meio Ambiente e dos Recursos Naturais Renováveis – IBAMA. Coleta de óleo lubrificante usado ou contaminado –2018 (ano base 2017), Brasília, (2018).

25. ANP – Agencia Nacional de Petróleo Gás Natural e Biocombustíveis: Balanço de Produção e Coleta de Óleos Lubrificantes por Região 2015. São Paulo, Brazil (2016).
26. SIMEPETRO. Sindicato Interestadual de Indústrias Misturadoras e Envasilhadoras de Produtos Derivados de Petróleo. Relatório Técnico dos Óleos Lubrificante usados. Available at: www.simepetro.com.br, (2010).
27. Baderna D., Boriani E., Giovanna F. D., Benfenati, E. Lubricants and Additives: A point View. In: Bilitewski B, Darbra RM, Barcelo M, ed. The Handbook of Environmental Chemistry. Global Risk-Based Mangement of Chemical Additives I. Production Usage and Environmental Ocurrence, Berlin, Springera-Verlag, (2011).
28. Audibert, F. Waste Engine Oils: Rerefining and energy recovery. Oxford, Elservier, (2006).
29. El-Fadel M., Khoury R. (2001) Strategies for vehicle waste-oil management: a case study, Resources, Conservation and Recycling v. 33, n. 2, pp. 75–91, (2001).
30. Hamada, A., Al-Zubaidya, E., Muhammad, E. F. Used lubricating oil recycling using hydrocarbon solvents, Journal of Environmental Management v. 74, pp. 153–159, (2005).
31. Lovón-Canchumani, G.: Óleos lubrificantes usados: um estudo de caso de avaliação de ciclo de vida do sistema de re-refino no Brasil. 2013. Tese (Doutorado em Planejamento Energético) - COPPE/UFRJ (2013).

Sustainability in Building Construction: A Comparison Between Certified and Non-certified Companies

Gabriel Aires Moreira, Rômulo Soares de Lima Filho, Maria de Lourdes Barreto Gomes, Josilene Aires Moreira, and Ricardo Moreira da Silva

Abstract Building construction is one of the most important sectors in Brazil. According to the data from the Brazilian Institute of Geography and Statistics (IBGE in Portuguese), its participation in the Gross National Product (GNP), between 2010 and 2013, were about 5% and employed about 9% of the Brazilian population. This high importance, the sector allowed new players to enter the market, increasing competitiveness. In this context, a strategy of the construction companies is to incorporate, in their management, certification systems and also, the search for sustainability, which should include the economy, equity and ecology. This article analyzes business sustainability in Building Construction (residential) in the city of João Pessoa in six major construction companies, having as variables the seven revolutions to sustainable: markets, values, transparency, life cycle technology, partnerships, timing and corporate governance. The field research (methodology) was carried out through interviews guided by a questionnaire, elaborated from the bibliographic. It can be concluded that companies that have certifications in Building Construction (residential) have a more sustainable management than those that do not have certification, because (a) they work with processes to mitigate environmental impacts, certified companies have management plans waste and non-certified companies do not have; (b) certified companies are able to maximize their profits, more effectively, through the application of transparent management; (c) there are better health and safety conditions for employees. In terms of governance and timing, all companies need to be better organized, independent of certification. It follows that certified companies seek and are closer to sustainability than non-certified.

Keywords Sustainability · Building construction · Certified companies

G. Aires Moreira · R. Soares de Lima Filho · M. de Lourdes Barreto Gomes · J. Aires Moreira · R. Moreira da Silva (✉)
Federal University of Paraiba- Cidade Universitária, João Pessoa 58051-900, Brazil

M. de Lourdes Barreto Gomes
e-mail: marilu@ct.ufpb.br

A. M. Tavares Thomé et al. (eds.), *Industrial Engineering and Operations Management*,
Springer Proceedings in Mathematics & Statistics 367,
https://doi.org/10.1007/978-3-030-78570-3_9

1 Introduction

Currently, there is a need for Building Construction (residential) companies to develop sustainability practices. This is an undeniable fact, in which, according to Philippi defines it as "the capacity to support sustainably, to maintain oneself" [1]. A sustainable activity is one that can be maintained for a long period of time (indeterminate), despite the unforeseen events that may occur.

In a context of sustainability, the triple bottom line [2] directs all business decisions to be faced with the capitalist market. The Building Construction companies, seeking survival, within sustainable development standards, have "certifications", such as High Environmental Quality (AQUA), Leadership in Energy and Environmental Design (LEED), International Organization for Standardization [3] and Occupational Health and Safety Assessment Services OHSAS [4], which are defined standards in order to standardize construction management control, ranging from environmental to social aspects.

Such certifications should guarantee: (I) consumer satisfaction, (ii) fulfillment of the contract, (iii) not pollution, (iv) eco efficiency and (v) survival, for companies. Thus, the hypothesis is that non-certified companies are less likely to be sustainable than certified companies. In this sense, the objective of this paper is to analyze corporate sustainability in Building Construction (residential) companies, between certified and non-certified companies.

2 Methodology

This was a research with a quantitative and qualitative approach. The Building Construction (residential) Companies (BCC), with and without certification interviewed were based in the city of João Pessoa—Paraiba—Brazil. The analysis variables were the seven revolutions [2]: markets, values, transparency, life cycle technology, partnerships, timing and corporate governance.

A bibliographic search was made through the Science direct using the words "Building", "Construction" and "ISO". In addition, a multi-case study was chosen. The sample consisted of six largest certified companies and non-certified companies, using the SEBRAE (Serviço Brasileiro de Apoio às Micro e Pequenas Empresas—in Portuguese—"Brazilian Business Support Service of Small and Medium Companies") size criterion.

In the data collection, a questionnaire elaborated with the reference the seven revolutions [2] was used, with the addition of a script for direct observation. Each respondent assigned a rating from 0 to 10 to each revolutions (0 represented that the company did not carry out the revolutions and the score of 10 meant that the company was fully engaged with technological tools, integrative attitudes, sustainability, environment, eco-efficiency, clean energy, etc.

In addition, for the revolution "timing", in loco observation was also used to identify whether there was over exploitation of resources. The layout of the construction sites, their remains of buildings, time management in production were analyzed.

To collect data on the revolution "corporate governance", documentary research was used in primary papers in the companies and it was also verified if they had an entrepreneurial profile. All the data collected were analyzed and displayed on the Radar graph that compares the aggregated values of the seven series of revolution.

3 Analysis of Sustainability Between Certified and Non-certified Building Construction (Residential) Companies (BCC)

The principle of sustainability in fact was introduced in 1972 with the publication of the Club of Rome report "limits of growth" [5]. In 1987 the World Commission on Environment and Development (Brundland Report) [6] published "our common future". It introduced the principle of self-responsibility, not passing environmental problems to elsewhere or the future. In 1997, John Elkington introduced the triple bottom line "people-planet-profit" that made business and sustainability no longer a contradiction [7].

But, how did these developments influence the corporate sustainability? In fact, discussing "sustainability" based on the whole planet is necessary, however, starting macro towards the specific, will often become something utopian. On the contrary, we know that the world is made up of organizations, which are supported by people, who live in society, therefore, all organizations, including market ones, need to be evaluated. And when understood individually, we can create simulations to understand the totality.

BCC can be considered an incremental sector for the development of any country. There is constant growth, the BBC market in João Pessoa-Paraiba-Brazil is predominantly made up of small and medium-sized construction companies, generally family members, however, (there are only, at most, 10 large companies) and therefore the sample of this paper was representative [8].

Large companies explore the construction of homes for the highest social class (located on the city's waterfront) and small and medium-sized companies explore the residential construction of the less economically favored population.

Within this context, the results obtained in the research are shown in Graph 1, where the score was established by the interviewees for each revolutions.

(a) **markets** - There was a change in the market because of sustainability, where it starts to become less predictable, with companies aiming at long-term returns. The research showed that there is a significant difference between certified and non-certified companies. The average grades obtained in the survey were: 6.82 for certified companies and 5.72 for non-certified, so, Certified companies are more sustainable than non-certified companies.

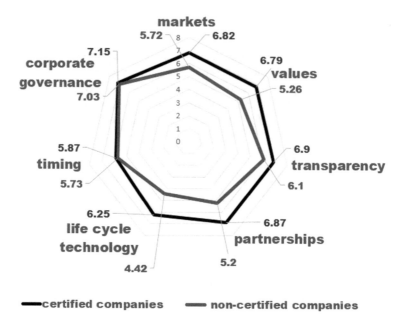

Graph 1 Comparison of sustainability in building construction (residential) companies. *Source* Survey data

It was asked if: (I) the company had prices that were lower or similar to those in the market; (ii) the products had an innovative design; (iii) the company responded quickly, the changes required by customers, (iv) quality standards were compared with other players, (v) delivery times were met, (vi) post-delivery technical assistance was performed with success, (vii) the company analyzed its competitors (viii) there were contracts with suppliers and (ix) there were plans, agreements and financing to attract customers.

In this revolution markets, it was observed that certified companies started to stabilize in the market, when they practiced lower prices or similar to the market, while non-certified companies were unable to lower prices and it leave the market more quickly.

It was noted that part of the non-certified companies worked with much longer deadlines than would be necessary (for the size of their ventures), but this was a strategy used to deliver the product to the customer within the agreed deadlines. The way companies see their competitors is similar for the two types of companies analyzed.

Regarding suppliers, both certified (and non-certified) companies, all have informal contracts with their suppliers. There is no exclusive contract.

As for attracting customers, certified companies used the strategy of attracting customers to the construction site (they usually have a sales stand on site), while in non-certified companies, they used advertising in communication vehicles.

In relation to the financing of the product themselves, all companies (certified and non-certified) received other goods and commercial government program agreements, but only certified companies, financed part of the debt, with resources own.

(b)   **values** - With the emergence of new technologies, corporate values are not rigid and become malleable. For Elkington [2], much of the technology introduced during the twentieth century, gave the worker more physical freedom, greatly affecting companies. There was so much an impact on personal mobility, as there was also an impact on lifestyles, (affecting values within organizations).

The average grades obtained in the survey in the "values" revolution were: 6.79 in certified companies and 5.26 in non-certified companies, showing that certified companies better and more efficiently incorporate the needs of their employees. An indisputable proof of this fact is that certified companies are able to pay wages above the union ceiling.

In the interviews, was it assessed whether: (I) there was a bonus or productivity bonus; (ii) there was an incentive for the worker to study (in the construction workplace); (iii) there was participatory management; (iv) there was development of regional cultural values and rites; (v) there were workers with special needs (vi) there were marketing programs against child labor; against physical violence among employees; against bullying; against moral and sexual harassment; (vii) there was a program to achieve gender equality and (viii) there was a socio-environmental responsibility program.

As a result, we find that certified companies have implemented bonus systems, such as production awards. On the contrary, non-certified companies did little to do so.

As for education and the issue of developing regional values—all surveyed companies (certified or non-certified) offer educational support for their employees in the construction workplace, however, non-certified companies offer this support to a lesser extent and lesser frequently. Also, both companies enable development regional cultural values freely, with the proviso that such rites do not interfere in the work process. Participatory management only exists in certified companies.

When it comes to programs to curb (a) child labor, (b) violence between employees, (c) bullying, (d) moral and sexual harassment, all companies develop, however, at certified companies do so with higher quality.

There is no gender equality in companies, which present in the construction workplace a male predominance (with practically 100% male workers), but, in office work, there is gender equality.

Finally, in this revolution "value", although there are small differences between the companies surveyed, as well as in the score obtained in the survey, it cannot be said that there is work differentiation (due to the fact that they are certified or non-certified companies), as it was noted the strength of regional culture, the customs of the northeast region of Brazil, in all companies.

(iii) **transparency** - Only one certified company presented a program that aimed at transparency [9]. This company was motivated by the need to generate responsibility reports, for the participation of the stock exchange—negotiations for the purchase and sale of securities.

The average grades obtained in the survey were: 6.90 for certified companies and 6.10 for non-certified companies, so there is not a big difference between companies in this revolution. The survey assessed: (I) the visibility of the financial process, (both in the corporate structure, as well as internally, among employees), (ii) visibility of goals, (iii) visibility of deadlines, (iv) disclosure of reports, (v) existence of programs to monitor customers, (vi) possibility of customization, and (vii) compliance with governmental legal requirements.

In relation to the company's goals, the employees of the certified companies, as well as the employees of the non-certified companies, have knowledge about these goals, because the workers were charged with daily goals, for the companies to reach the goals and guidelines company.

For this reason, employees of all companies knew the deadlines in depth, because of their day-to-day activities, in the workplace. In addition, they received bonus bonuses, because of productivity. However, these deadlines were not properly met.

Regarding the visibility given to customers, certified companies had a greater connection with their customers (through programs to monitor the deadlines provided), in addition, they made it possible to customize their products, meeting customer requirements, impacting more financial expense (on behalf of customers).

As for transparency for the external society, neither certified companies nor non-certified companies release reports to the population. So, they did not allow any person, other than the buyer, to access the enterprise's data, nor did they allow access to the company's data.

Studies by Isatto showed that when companies show transparency to workers, customers and society, through visibility, there is an increase in the commitment of workers in the development of improvements and also, facilitating the visualization of errors in the production; on the contrary, companies that hide and hinder information will be discarded by customers, because these seeking for a reliable company, for investments [10]. The certified companies have a broad knowledge of the financial process and the goals to be achieved, but workers have only superficial knowledge.

(iv) **partnerships** - Most companies know that business, community and social partnerships are needed. In fact, partnerships must define expectations, concepts in common and conflict, according to the principle of socio-environmental responsibility, covering the roles of the organization and the government [11].

Despite this knowledge almost worldwide, these environmental and sustainability partnerships were not found, in the companies surveyed. Only commercial supply partnerships were found.

Although in the companies surveyed do not have environmental and sustainability partnerships, certified companies have partnerships with suppliers. For this reason, the average grades obtained in the survey were: 6.87 for certified companies and 5.20 for non-certified companies.

The questionnaire evaluated the importance that the company attached to the partnership with (I) suppliers, (ii) real estate, (iii) competitors and (iv) with labor and services outsourcing companies.

The partnership between company-supplier existed in certified companies, which observe, in this relation, benefits to the enterprise. There was practically no formal partnership between real estate companies, which was surprising, because there is usually a link between the client (who buys the residence) and the construction company.

There was a partnership in two certified companies of the type company-competitors, in a kind of cooperation-competition, where companies directed their customers to evaluate the product of their competitor-partner and vice versa, with an exchange of experiences, (favoring innovation and the search for solutions to the challenges perceived by competing managers).

Partnerships were also found with outsourcing companies, both for labor and services, only in the companies surveyed certified. Evidencing that, in this case, there was a greater quality control.

Finally, the results showed that most companies are reasonably comfortable, with the idea of establishing only company-supplier business partnerships, although they all state that "there is a need for community and social partners", however, environmental and sustainability partnerships were not found, which is a mistake.

(e) **life cycle technology** - The life cycle of a sector is the result of the sum of the life cycles of the companies that are inserted in it [12]. In turn, the life cycle of a company, is composed of the cycle of their products. And how does this thinking apply to Building Construction (Residential) Companies?

To think about buildings with a sustainable life cycle is to think about sales turnover and decreasing inventory, so as not to generate obsolescence [13]. For this reason, the construction life cycle is not restricted to environmental impacts, but, the ability of a company to recycle its products, reflects directly on its profits and this recycling has to do with both Whole Process of the Construction—WPC, regarding Building Material and Component Combinations—BMCC [14].

In this paper, the data show a great difference between Certified and Non-Certified Building Construction (Residential) Companies (BCC), in fact, among the revolution surveyed, this was the one with the greatest difference in average (6.25 and 4.42 respectively) showing that certified companies rotate products more than non-certified companies. At the time of the interview, it was found that certified companies maintain a minimum stock of traditional materials.

The questionnaire assessed (I) the issue of obsolescence of the material used, (ii) the turnover of the ceramic, crockery and similar stock and (iii) whether the company seeks innovation, including non-traditional materials in the construction process (plaster wall, styrofoam, retractors etc.).

The biggest problem detected was the use of ceramics, which by itself does not represent a difficulty, however, such ceramics were purchased at the low price offered in the market and this causes many small, unused leftovers.

All companies controlled inventories, with greater or lesser rigor, but only certified companies had process control including standardized life cycle technology systems. In fact, what generates such a great differentiation in average scores is the process of selling your products by company. The marketing implanted for using the certifications, generates in the customer greater confidence, which makes the turnover of the offered products be greater.

(f) **timing** - There are at least three reasons for companies to worry about timing. (I) short-term overexploitation of natural resources, for Viola [15] this encourages the competitive exploitation of scarce resources; (ii) lack of commitment to future generations causing persistent environmental degradation and; (iii) the way in which timing itself is perceived, valued and managed where it is the main revolution in the organization of projects, which has a major impact on Building Construction (Residential) Companies [16].

The survey assessed (I) the short, medium and long term planning time, (ii) the question of the construction company's insertion in low-income government programs such as "my house, my life" and (iii) whether the deadlines were met.

The final averages for these revolutions obtained in the survey were: 5.87 for certified companies and 5.73 for non-certified companies, so we can conclude that "timing" are very close to each other. A small difference was found between certified companies and non-certified companies when the issue dealt with short, medium and long term, as non-certified companies had a short and medium term vision, and certified companies besides taking care of the "here and now", took care of the long term.

The certified companies have planning and schedules with visible and apparently effective controls, in another direction, non-certified companies have inadequate, informal planning, where time controls are in the hands of the overseers.

All companies sought to be in the right place, at the right time, taking advantage of opportunities in a JIT way. All surveyed companies had coupling with sales programs, such as real estate fairs, but because they are large, they did not fit perfectly with government housing programs such as the "my house, my life" project, which was for a public with less purchasing power.

(g) **corporate governance** - The pillars of corporate governance are (I) the owners (ii) the board of directors (iii) the management (iv) the independent audit (v) the fiscal council and (vi) conduct and conflicts of interest. In all the companies surveyed, the owners, the board of directors and the fiscal council were confused with the owners of the companies. The only difference between certified and non-certified companies was in management.

The certified companies had formal and standardized management practices, which did not happen with the non-certified ones, which presented only departmental structures, (but did not present formal production processes). Thus, in these companies studied, there is no significant difference between certified and non-certified Civil Construction companies.

The research involved (I) the worker's experience and autonomy (ii) the worker's qualification (types of training) and (iii) the construction process (organization chart, planning, maintenance, deadlines, planning and operation of the construction site and office).

In the surveyed construction companies, no company has shares traded on stock exchanges and none has an organizational structure within international standards, with collegiate systems and contracted CEOs. There was a difference in the management of certified companies with a greater concern with environmental issues, so much so that all had a waste management plan, which was non-existent in non-certified companies. The notes obtained were 7.15 and 7.03 certified and non-certified respectively, which shows that the difference in corporate governance is small.

4 Conclusion

The companies that have certifications in Building Construction (residential) sector have a more sustainable management than those that do not have certification, because:

(a)   In Ecology, they minimize the environmental impacts generated by the entire construction process, because they have waste management plans, there is reuse of materials used, cleaner and more efficient energy use and proper disposal of waste.
(b)   In Economy, companies are able to maximize their profits through the application of transparent and corporatist management;
(c)   In Equity, there are health and safety conditions at work for employees, as it was possible to observe through the use of safety equipment and application of technical standards such as the Construction Conditions and Environment Program in Civil Construction.

In fact, this difference in corporate sustainability between certified and non-certified companies was observed in Fig. 1, which is confirmed by the analysis of all research, where only in the revolution "timing" and "corporate governance", there was an approximation between companies.

Timing, regardless of certification, companies that are not proactive and do not react by taking advantage of the "here and now", disappear from the market, as quickly as they enter. In corporate governance, all companies need to be organized, since the restructuring of the globalized market, because it does not forgive disorganized companies, who do not seek eco-efficiency, (although none had a formal corporate governance structure).

With the research carried out in the large construction companies in the Building Construction (residential) sector in João Pessoa, we concluded that there is a difference between companies having or not being certified, and this was possible with an analysis of the seven revolutions of Elkington (2012). We can conclude that certified companies seek and come closer to sustainability than non-certified companies.

References

1. Philippi, Luiz Sérgio. Building Sustainable Development. In.: LEITE, Ana Lúcia Tostes de Aquino; MININNI-MEDINA, Naná. Educação Ambiental (Curso básico à distância) Questões Ambientais – Conceitos, História, Problemas e Alternativa. 2. ed, v. 5. Brasília: Ministério do Meio Ambiente, 2011
2. Elkington, John, Cannibals with forks, capstone publishing ltd, Oxford, UK 1997, in Brazil by Ed. Mbooks, 2012.
3. ISO International Organization for Standardization (2020). In http://www.iso.org/iso/home. html. Accessed jun 2020.
4. OHSAS 18001. In http://www.iso9000.com.br/consultoria-sgq-oshas-18000.html. Accessed abr 2020
5. Meadows et al, limits of growth, Cambridge (US), Universe press, New York, 1972.
6. Brundtland, G. H. "Report of the World Commission on Environment and Development: Our Common Future. Transmitted to the General Assembly as an Annex to document A/42/427-Development and International Co-operation: Environment." United Nation. Oslo, Noruega (1987).
7. Jacobs E. Seven revolutions to sustainable urban drainage. 11th International Conference on Urban Drainage, Edinburgh, Scotland, UK, in https://www.susdrain.org/delivering-suds/using-suds/background/sustainable-drainage.html. Accessed in may 2020.
8. Sindusconjp (2020)– João Pessoa Civil Construction Industry Syndicate. In https://sindusconjp. com.br/. Accessed in may 2020
9. Santos, A. dos (2020). Application of flow principles in the production management of Construction sites, in http://www.administradores.com.br/artigos/negocios/saiba-a-import ancia-da-transparencia-na-relacao-com-os-investidores/83038/. Accessed jan 2020
10. Isatto, E. L. et al. Lean Construction: Diretrizes e ferramentas para o controle de perdas na construção civil. Porto Alegre: SEBRAE, 2000.
11. Furtado, João Salvador. Sustentabilidade Empresarial. Guia de práticas econômicas, ambientais e sociais, 2012. Disponível em: http://www.intertox.com.br/index.php/biblioteca-dig ital/category/32-publicacoes?download=57:sustentabilidade-empresarial-guia-de-praticas-economicas-ambientais-e-sociais. Accessed may 2020.
12. Kayo, Eduardo Kazuo. Ativos intangíveis, ciclo de vida e criação de valor, 2006. in: http://www. scielo.br/scielo.php?pid=S141565552006000300005&script=sci_arttext. Accessed jun 2020.
13. Chaves, Helena de O. (2014) Sustainable Guidelines in Civil Construction: Life Cycle Assessment, Monografia Engenharia Civil, UFRJ.
14. Ortiz, Oscar, et al. "The environmental impact of the construction phase: An application to composite walls from a life cycle perspective." Resources, Conservation and Recycling 54.11 (2010): 832–840.
15. Viola, Eduardo J. (2020) O Movimento Ecológico no Brasil (1974–1986): Do Ambientalismo à ecopolítica, 1986. in: http://www.anpocs.org.br/portal/publicacoes/rbcs_00_03/rbcs03_01. htm. Accessed may 2020.
16. Thamhain, Hans J. (2014) Managing Technology Based Projects: Tools, Techniques, People, and Business Processes. John Wiley & Sons, Inc., ISBN 978 0470 40254–2 Hoboken, New Jersey. EUA.

S&OP as Driver for Sustainability

**Bruno Duarte Azevedo, Christian Kalla, Tobias Kreuter,
Luiz Felipe Scavarda, and Bernd Hellingrath**

Abstract Both Sales and Operations Planning (S&OP) and sustainability are enjoying growing interest in the area of Operations Management. However, the association between them still lacks in academic research and industrial applications. This paper aims to develop a first analysis of the potential relationship between S&OP and sustainability and develops a comprehension of the possibility of managing sustainability through the lens of S&OP. The researchers conducted a systematic literature review to integrate the different and fragmented contents available in the literature associated with S&OP, especially to its tactical planning perspective, and sustainability. The absence of in-depth discussions on the feasibility of using tactical planning as a driver to improve sustainability can be understood as evidence that, in general, the academic literature already accepts this idea. The current discussion seems to be in the "How to do" level, which confirms the great opportunity to adapt S&OP to better achieve goals related to sustainability. Integrating S&OP and sustainability is crucial due to the essential role of S&OP as a tactical planning approach, in linking strategic and operational activities as well as its integrative nature among different functional areas within the firm and its supply chain. To reinforce the feasibility of managing these two subjects, three propositions are also presented within this research. Finally, the development of a framework for sustainable S&OP is suggested for future works.

Keywords Tactical planning · Supply chain · Triple bottom line

B. Duarte Azevedo · L. F. Scavarda
Pontifical Catholic University of Rio de Janeiro, Rua Marquês de São Vicente, 225 sala: 952L, Rio de Janeiro 22453-900, Brazil

C. Kalla (✉) · T. Kreuter · B. Hellingrath
Westfälische Wilhelms-Universität Münster, Leonardo-Campus 3, 48149 Münster, Germany
e-mail: christian.kalla@wi.uni-muenster.de

© The Author(s), under exclusive license to Springer Nature Switzerland AG 2021
A. M. Tavares Thomé et al. (eds.), *Industrial Engineering and Operations Management*,
Springer Proceedings in Mathematics & Statistics 367,
https://doi.org/10.1007/978-3-030-78570-3_10

1 Introduction

Sales and Operations Planning (S&OP) is a topic of growing importance in the area of Operations Management (OM) that has been receiving considerable attention from academics and industry professionals [1, 2]. This is reflected in the increase of publications on the subject in the academic literature and the increase of implementations in industry [3–5]. S&OP is situated at the tactical level and seeks to balance demand and supply by integrating various business plans into one final set of plans [6]. In this sense, this practice seeks a vertical integration on the one hand, serving as a bridge between its strategic and operational levels. On the other hand, it strives for a horizontal integration, seeking a single and integrated plan aligned between the various functional areas of the company (e.g., sales, marketing, production, product development) within the company and throughout its supply chain [6–9]. The theme of S&OP is interdisciplinary but is predominantly associated with the traditional OM literature [6, 10].

Sustainability has emerged as another interdisciplinary concept of increasing interest in OM [11, 12]. It is associated with the vision of the "triple bottom line" (TBL), i.e., integrating economic, environmental, and social issues [13]. In the context of OM, sustainability can be related to generating competitive advantage through the operations of the various productive organisations [14, 15]. The sustainability approaches applied in OM require the expansion of its limits, the creation, and integration of new performance objectives, in addition to the adoption of new criteria to assist in decision-making [16].

According to Wallace and Stahl [17], integrating sustainability in S&OP is crucial due to the essential role of S&OP in linking strategic and operational activities as well as its integrative nature. Nevertheless, efforts to include sustainability in S&OP are still rare [18]. Broadly, S&OP has only been studied from the economic perspective, emphasising its impact on profit optimisation [6, 7, 19]. Despite the growing importance of both S&OP and sustainability within the OM literature, the association between them still lacks academic research and applications in the industry [18]. This scenario opens a promising future research opportunity for the area of OM, which is reinforced by the known existing association of the term "sustainability" to several OM concepts [20]. This goes from a broad concept perspective of sustainable OM [21] to more narrow ones as green supply chain management (GrSCM) [22]; sustainable supply chain management (SSCM) [23]; sustainable manufacturing practices [24], and sustainable lean production [25].

To address this research-practice gap, this paper aims to develop a first analysis of the potential relationship between S&OP and sustainability. It develops a comprehension of the possibility of managing sustainability through the lens of S&OP. Based on the premise that S&OP—similar to other OM practices—can benefit from the association to sustainability, the authors outline the following research question to guide this study: How can S&OP embrace the TBL view and be a driver for sustainability? To achieve this goal, the researchers have conducted a systematic literature review (SLR) to integrate the different and fragmented contents available in the literature associated with S&OP, especially to its tactical planning perspective, and sustainability. Then, supported by the SLR, three research propositions are presented attesting the feasibility of using S&OP as a driver to sustainability.

2 Theoretical Background

This section first provides a background on sustainability from an OM perspective and second on S&OP from a tactical planning perspective. Finally, it offers a view from the missing link on S&OP and sustainability.

2.1 Sustainability and Operations Management

Sustainability is a multi-dimensional concept composed of three distinct dimensions: economic, environmental, and social, often referred to as the "triple bottom-line" [13]. Pojasek [26, p. 94] considers sustainability as "the capability of an organization to transparently manage its responsibilities for environmental stewardship, social wellbeing, and economic prosperity over the long-term while being held accountable to its stakeholders."

In OM, the term "sustainability" has been used in several different practices to express this complex combination of concepts [20]. In the beginning, sustainability initiatives have mainly focused on environmental issues [27]. GrSCM, for instance, has not directly incorporated the social pillar of the "triple bottom line", consisting in the action of integrating environmental concepts into Supply Chain Management, starting from the design phase to managing the product at the end of its lifecycle [22, 28]. However, there is an increase in the number of publications with a larger scope of sustainability research [29, 30], "moving from environmental concerns within the confines of the firm in the late 1990s to a more inclusive economic, social and environmental definition of sustainability, both internally to the firm and externally in the supply chain" [12, pp. 10–11]. Sustainable OM allows a company to obtain competitive returns on its capital assets, "without sacrificing the legitimate needs of internal and external stakeholders and with due regard for the impact of its operations on people and the environment" [21, p. 489]. SSCM has emerged with a broader definition when compared to GrSCM, including social aspects to embrace all three dimensions of the TBL [13], combining sustainability with efficient supply chain management, and being able to integrate the concept of GrSCM as a part of its field [23, 31, 32]. Sustainable manufacturing practices embrace the complete lifecycle of a product and encompass both internal and external organisational management, integrating economic, environmental, and social aspects into operational activities [12, 24]. Herein, one can also highlight sustainable lean production. It is another relevant topic, in which there is a consensus among different authors that its success involves more than the use of tools and methods [33]; as the efforts support the development of a lean culture, where leaders and employees can achieve a truly continuous improvement [25] from the triple bottom-line lenses.

2.2 S&OP and Tactical Planning

Originally, the term 'sales and operations planning' has been used in the context of manufacturing resource planning (MRP II). Since then, it has been used synonymously with aggregated production planning (APP) [19]. According to the APICS definition, S&OP "brings together all the plans for the business (sales, marketing, development, manufacturing, sourcing, and financial) into one integrated set of plans" [34, p. 154]. S&OP seeks vertical and horizontal integration to balance supply and demand as well as strategic plans to operational plans [9, 10, 35]. Usually, it is conducted in a monthly cycle and has five steps as follows: (1) Data Gathering, in which data is updated; (2) Demand Planning, in which marketing and salespeople jointly generate the new consensus-based demand forecast; (3) Supply Planning, in which the operations department creates a supply plan, based on the forecast from step two; (4) Pre-Meeting, in which a cross-functional team discusses adjusts, and validates supply and demand plans; (5) Executive Meeting, in which executives meet to review all decisions from the pre-meeting, including those on which the pre-meeting team could not reach consensus, or which entail significant costs [17, 19, 36].

S&OP is a tactical planning process which planning horizon can vary between less than three to more than 18 months [2, 6]. Tactical or mid-term planning builds the bridge between strategic decisions and operational activities. As Fleischmann et al. [37, p. 72] state, it thereby focuses on "rough quantities and times for the flows and resources". Thus, it coordinates and organises the activities of operational units for a pre-defined time horizon [38]. Common activities of tactical planning include forecasting future demand as well as capacity, personnel, and material requirements planning [37].

2.3 Sustainability and S&OP: A Missing Link

The concept of sustainability has been deeply discussed in the academic community. However, it is still "far too abstract" among other stakeholders and decision-makers [39, p. 655]. Still, the OM perspective is being transformed to adjust organisations to the achievement of economic objectives not only from an environmentally but also socially responsible perspective [40]. Despite the tensions and conflicts that the inclusion of sustainability goals can bring to a given company, Magon et al. [12] attest the overall positive effects of sustainability on performance, such as lower costs, better delivery and product quality, enhanced volume and mix flexibility. To achieve a truly sustainable supply chain, it is of the utmost importance that all the stakeholders involved, find a joint vision of sustainability [41].

In the literature, S&OP is usually considered from the economic perspective, emphasising its impact on profit optimisation [6, 7, 19]. While many other areas in OM (e.g., SSCM; Sustainable and Lean Production) already address all three dimensions of the TBL in their practices, the S&OP literature does not yet recognise and embrace environmental and social aspects [18]. This research-practice gap is aimed to be addressed in this paper, aided by an SLR as described in the next section.

3 Research Methodology

The research methodology was based on the SLR approach to integrate the fragmented findings available in the literature towards developing a first and novel analysis of the potential relationship between sustainability and S&OP trough the perspective of tactical planning.

The SLR followed the methodology offered by Tranfield et al. [42], guided by the step-by-step approach of Thomé et al. [43] for conducting an SLR in OM, as done in Cotta Fonatinha et al. [44]. For the first step, the authors defined the research question, the scope of the review, the outcomes, and the research strategy, as described in the introduction section of this paper. In the second step, Scopus and Web of Science were selected as search engines and indexing systems, as both are well known as being good sources for extensive peer-reviewed literature and as being complementary [12]. Extensive discussions among the authors were carried out to select search keywords with different attempts using keywords and combinations as follows: ("Sales and Operations Planning" AND "Sustainab*") OR ("S&OP" AND Sustainab*") OR ("Sales and Operations Planning" AND "Green") OR ("S&OP" AND "Green"). The search process yielded just three documents. As a new attempt, the authors considered the keywords ("tactical plan*" AND sustainab*"), due to the tactical nature of S&OP. This search resulted in 50 papers after eliminating the duplicates in both databases. The inclusion criteria for determining which studies need to be considered from the scientific literature were papers that considered tactical planning through the lens of sustainability over the TBL perspective as well as papers written in English. To improve the debate in the categorisation of studies, and facilitated the data analysis and interpretation process, the authors also classified all the abstracts according to the vision of sustainability brought in the abstracts. The content analysis approach was adopted herein [31, 45]. Three authors read all the abstracts individually to avoid a biased evaluation. In the first round, they agreed that nine papers attend to the inclusion criteria, 30 did not, and three needed a full-text analysis for a final decision. On nine papers, they did not achieve unanimity, so they held a round of discussion until a consensus was reached. After this step, four papers were added, which generated 13 papers that went to full-text analysis. After this step, one more paper was excluded, as it was out of the scope. As recommended in Thomé et al. [43] and Hollmann et al. [46], to go beyond the search keywords and database limitations, one more paper was added through a snowball search (backwards and forward search). The total number of studies retrieved was 13. The presentation of the results and discussions, corresponding to step 7 of the SLR, are presented in Sect. 4.

4 Results and Discussion

This section presents the main results of the SLR. First, a discussion regarding the different views related to sustainability in the OM discipline is presented. Then, the studies related to tactical planning and TBL vision are briefly analysed. Finally, a discussion about the feasibility of using S&OP as a potential driver to reach goals related to sustainability is offered.

4.1 TBL Perspective as a Challenge

Of the 51 abstracts selected, the authors agree on the TBL vision in fourteen of them. Two abstracts present sustainability over the financial and social perspective, twelve over the financial perspective, and fifteen over the financial and environmental perspective. Additionally, in six articles the authors do not see any relation to sustainability. Many papers focus on logistical strategies for reducing greenhouse gas emissions (e.g., [47, 48]), which highlight a missing consensus on the terms green and sustainability [49]. For example, Sawadogo and Anciaux [50] present the TBL perspective in their abstract; however, the decision support system presented by them focuses on reducing the environmental impact for trip selection. This persistent confusion is probably one of the reasons for the view brought by Boukherroub et al. [51], which attest the existence of a lack in the literature regarding approaches considering the three dimensions of sustainability performance when planning the supply chain.

On the contrary, despite being one of the oldest studies analysed, Sheppard [52] already considers all three dimensions of sustainability. He introduces his work attesting that Sustainable forest management (SFM) involves balancing economic, environmental, and social values to meet society's objectives over the long term.

4.2 Tactical Planning as a Driver for Sustainability

Given the absence of papers specifically discussing S&OP and sustainability, the authors analysed papers that could answer the feasibility of using the implementation of tactical planning to improve sustainability. However, the majority of the studies focus on developing quantitative models to solve multi-objective and/or multi-criteria problems and do not deeply discuss tactical planning as a feasible strategy to improve sustainability goals (e.g., [49, 51, 53–55]).

Jensen et al. [53] describe how an Ecosystem Management Decision Support System could be successfully used in the development of logic models for integrated evaluations of economic, ecological, and social information to support strategic forest planning. Focusing on the tactical level, Boukherroub et al. [51] propose an integrated approach for the optimisation of the sustainable performance of a wood supply chain. The model presented has had to meet customer demand at the lowest cost while reducing greenhouse gas emissions and promoting local employment. They select twelve objectives to cover the TBL view: five objectives are related to the economic dimension, four to the environmental dimension, and three to the social dimension. However, after omitting the objective function related to social performance in the model test, they conclude that its analysis is one limitation of their work and should have more experiments. Fattahi et al. [54] take the uncertainty related to the renewable energy resources into account and present a novel cost-efficient multi-stage stochastic program in which strategic and tactical planning decisions are integrated. Besides the economic goal, the greenhouse gas emissions from transportation are mitigated, and the social impact of the supply chain is considered using a social life cycle assessment methodology. They conclude that the supply chain cost is sensitive to policies related to the supply chain sustainability and *"to separately reduce the network's environmental cost and improve its social impact by 5%, the expected of total [supply chain] cost increases about 2% and 1.8%, respectively"* [54, p. 12]. They also assume that the framework presented can help decision-makers to balance the economic goal, the environmental cost, and the social responsibility impact to the supply chain. Laguna-Salvadó et al. [49] present a multi-objective Master Planning Decision Support System for managing sustainable humanitarian supply chains. They attest that the tactical planning decision level is a lever to improve the performance of the sustainable supply chain, as it defines the gross operations that will take place according to the assessed needs. According to them, it enables the optimisation of the supply chain flows, and consequently, improves the operational performance. Their results show that *"managers can use the proposed model to prioritize the three sustainability dimensions and to fix a tolerance that would enable them to obtain an acceptable balance (trade-off) between the three sustainability performance objectives"* [49, p. 33]. This conclusion is aligned with Fattahi et al. [54] and deserves specific attention. Saravi et al. [55] propose a comprehensive Decision Support Tool (DST) integrating Artificial Neural Network (ANN), mathematical modelling, and solution approach to design and optimise the sustainable second-generation Bioethanol Supply Chain (BSC) while considering economic, environmental, and social aspects. Their results demonstrate that adapting an appropriate ethanol distribution policy could result in a 33% reduction in total costs. It is also suggested that a 10% increase in the total costs could be beneficial from environmental and social perspectives.

Bringing the risk management perspective, Mashaqbeh et al. [56] suggest a modified Failure Modes and Effect Analysis (FMEA) for understanding the non-technical risk comprehensively. Its main objectives are to improve awareness of risk management in power plants and to help top management having a better understanding of the organisation than lower-level managers who are close to day-to-day life.

Two documents play a more active role in demonstrating the feasibility of using tactical planning implementation to improve the sustainability of different supply chains. Reinforcing the need to better connect strategic goals with practical actions, Sizo et al. [57] present an analytical approach to support decision-making in the strategic environmental assessment, based on an application to the implementation of urban wetland conservation policies. Their focus is to strengthen the link between the strategic and the operational context. Using three sets of criteria to assess the different objectives of wetland conservation policy (economic well-being, environmental sustainability, and quality of life), they conclude that their approach is valuable for examining 'what if' strategic options and for guiding how to comply with high-level strategic policy targets. Loureiro et al. [58] bring the implementation of tactical plans as a strategy to improve the water-energy loss management. According to them, to ensure the sustainability of water supply systems, which has a direct impact on economic, environmental, and social issues, the problems should be addressed from a strategic to an operational point of view, including the tactical decision level. Their results show that the participating utilities which can define tactical measures are taken to a more efficient and sustainable service.

4.3 Discussion

The absence of in-depth discussions on the feasibility of using tactical planning as a driver to improve sustainability can be understood as evidence that, in general, the academic literature already accepts this idea. The current discussion seems to be in the "How to do" level with the majority of the studies focusing on developing quantitative models to solve multi-objective and/or multi-criteria problems (e.g., [49, 51, 53–55, 59]). Given that S&OP is a tactical planning process [6, 19, 34], this fact confirms Wallace and Stahl [17] who emphasize the need of integrating sustainability to S&OP. In this sense, there is a great opportunity to add the sustainability plan of the company to the S&OP methodology to better achieve goals related to sustainability. This scenario leads to the first proposition:

- As it is located at the tactical planning level, S&OP can be a key factor in improving sustainability goals because it can translate strategic plans into specific sustainability metrics, which are operationalised in practical activities.
  Even though many models reviewed in the literature have difficulties in embracing social performance (e.g., [51]), results show that they can already help decision-makers to balance the economic objective, the environmental cost, and the impact of social responsibility. It is important, however, to highlight that the order in which the three sustainability dimensions are prioritised has some impact on the performance measures [49, 54]. Consequently, each company needs to build a methodology that would enable to obtain an acceptable balance (trade-off) between the three sustainability objectives. Also, there is a need to align both short-term needs

and long-term responsibilities to achieve the principles of sustainable development [59]. This scenario opens a great opportunity to use the S&OP methodology as a driver to help decision-makers, and leads to the second proposition:

- Given its vertical integration dimension, the S&OP methodology can be used as an approach to help decision-makers better balancing the conflicting dimensions surrounding sustainability and translate the long-term vision to the short-term actions.

The horizontal integration view offered by S&OP [6–9] is particularly important to ease the different functional areas of a given company to find its better balance regarding sustainability goals. For example, the marketing department must have a channel to explain the existence of an increasing demand related to more sustainable products and the need for the company to respond to that, even if it increases the production costs. These multiple connections and channels provided by the S&OP methodology are a great opportunity to increase the commitment of the employers and the agility of the company to respond to external pressures related to more sustainable practices and products. This leads to the third proposition:

- Given its horizontal integration dimension, the S&OP methodology can be used to standardise the sustainable goals of the company, enhancing the commitment of all functional areas and improving its agility to respond to different scenarios.

Lastly, the persistent misunderstanding relating to the vision of sustainability is indicative that authors must pay more attention to its correct use [49]. There is a clear need to standardise the speech and avoid the use of the terms "sustainability" or "sustainable" merely because it is a hotspot in the current days. It is important to highlight the TBL vision [28, 41], which attests that to be truly sustainable, the answer to a given problem needs to consider the economic, environmental, and social aspects [52, 54].

5   Conclusion

This paper offers a first analysis of the potential relationship between sustainability and S&OP. Considering tactical planning as an important tool to bring to the operational level the strategic plans of a given company, an SLR was conducted to integrate S&OP and tactical planning literature on the one hand, and sustainability literature on the other hand. Based on the findings, it can be concluded that the academic literature already accepts the idea of using tactical planning as a driver to improve sustainability. The current discussion seems to be in the "How to do" level, with the majority of the stud-ies focusing on developing quantitative models to solve multi-objective and/or multi-criteria problems. These findings confirm the great opportunity to adapt the S&OP methodology to better achieve sustainability-related goals. Integrating sustainability and S&OP is crucial due to the essential role of S&OP as a tactical plan, in linking strategic and operational activities as well as its integrative nature among different functional areas within the firm and its supply chain. To reinforce the feasibility of managing these two subjects, three propositions describing the links between them are presented. Accordingly, including environmental and social plans into the S&OP methodology clearly offers potential benefits. This movement will help decision-makers to understand the holistic vision of sustainability without disconnecting from the need to translate a subjective concept into practical actions in the short and medium-term.

This paper addresses one major challenge of companies nowadays, i.e. facing the pressure of becoming more sustainable. Therefore, it offers interesting managerial implications. Findings indicate that S&OP could help organizations to provide a platform for discussing and integrating sustainability concerns into the business. Sustainability managers could attend S&OP meetings for aligning their specific plans and goals with the operational activities of the sales and production departments. As an example, the attendance at S&OP meetings could ensure that the production areas does not only consider the economic pilar, but also environmental and social issues. At the same time, production representatives can share their opinion and influence the process of defining sustainability goals in a way that they do not become too unrealistic. By considering all three dimensions of the triple bottom line in the S&OP meetings, executives are able to make better decisions and adjust the strategic direction so that the company becomes sustainable from an economic, environmental, and social point of view.

For future research, the development of a framework for sustainable S&OP is suggested towards guiding practitioners in the comprehension and implementation of sustainability in their operations from a tactical level perspective. Additionally, although the SLR was designed to minimise the risk of missing relevant papers, there is no guarantee that no such papers have been missed, as relevant keywords or databases may have been overlooked, that could have provided additional insights may have been excluded. A different choice of keywords and databases (e.g., specific ones for grey literature) might result in different papers retrieval.

Acknowledgements This work was supported by the following research agencies: German Academic Exchange Service – DAAD (PROBRAL Grant Number 57447177), Coordination for the Improvement of Higher Education Personnel (Coordenação de Aperfeiçoamento de Pessoal de Nível Superior – Brazil – CAPES) (Finance Code 001) & (Grant Number 88881.198822/2018-01), Brazilian National Council for Scientific and Technological Development (Conselho Nacional de Desenvolvimento Científico e Tecnológico – CNPq) (Grant Numbers 311757/2018-9).

References

1. Kristensen, J., Jonsson, P.: Context-based sales and operations planning (S&OP) research: A literature review and future agenda. International Journal of Physical Distribution & Logistics Management 48(1), 19–46 (2018).
2. Kreuter, T., Scavarda, L.F., Thomé, A.M.T., Hellingrath, B., Seeling, M. X.: Empirical and theoretical perspectives in Sales and Operations Planning. Review of Managerial Science. (2021a). Available in https://doi.org/10.1007/s11846-021-00455-y.
3. Kaipia, R., Holmström, J., Småros, J., Rajala, R.: Information sharing for sales and operations planning: Contextualized solutions and mechanisms. Journal of Operations Management 52, 15–29 (2017).
4. Michel, R.: Smoother execution through S&OP. Modern Materials Handling, 46–49 (2018).
5. Goh, S.H., Eldridge, S.: Sales and Operations Planning: The effect of coordination mechanisms on supply chain performance. International Journal of Production Economics 214(C), 80–94 (2019).
6. Thomé, A.M.T., Scavarda, L.F., Fernandez, N.S., Scavarda, A.J., 2012. Sales and operations planning: a research synthesis. International Journal of Production Economics 138 (1), 1–13 (2012).
7. Grimson, J.A., Pyke, D.F.: Sales and operations planning: an exploratory study and framework. The International Journal of Logistics Management 18(3), 322–346 (2007).
8. Tuomikangas, N., Kaipia, R.: A coordination framework for sales and operations planning (S&OP): Synthesis from the literature. International Journal of Production Economics 154, 243–262 (2014).
9. Hulthén, H., Näslund, D., Norrman, A.: Framework for measuring performance of the sales and operations planning process. International Journal of Physical Distribution & Logistics Management 46(9), 809–835 (2016).
10. Thomé, A.M.T., Sousa, R.S., Scavarda, L.F.: Complexity as contingency in sales and operations planning. Industrial Management & Data Systems 114(5), 678–695 (2014).
11. Gunasekaran, A., Subramanian, N.: Sustainable operations modeling and data analytics. Computers & Operations Research 89, 163–167 (2018).
12. Magon, R.B., Thomé, A.M.T., Ferrer, A.L.C., Scavarda, L.F.: Sustainability and performance in operations management research. Journal of Cleaner Production 190, 104–117 (2018).
13. Elkington, J.: Partnerships from cannibals with forks: the triple bottom line of 21st-century business. Environmental Quality Management 8(1), 37–51 (1998).
14. Drake, D.F., and Spinler, S.: Sustainable Operations Management: An Enduring Stream or a Passing Fancy?. Manufacturing & Service Operations Management 15(4), 689–700 (2013).
15. May, G., Stahl, B.: The significance of organizational change management for sustainable competitiveness in manufacturing: exploring the firm archetypes. International Journal of Production Research, 55(15), 4450–4465 (2016).
16. Machado, C.G., de Lima, E.P., da Costa, S.E.G., Angelis, J.J., Mattioda, R.A.: Framing maturity based on sustainable operations management principles. International Journal of Production Economics 190, 3–21 (2017).
17. Wallace, T., Stahl, B.: Sales and Operations Planning: The How-To Handbook, T.F. Wallace & Company (2008).
18. Noroozi, S.: S&OP related key performance measures with integration of sustainability and decoupling points: A case study approach. In: Proceedings of the 5th World Conference on Production and Operations Management, Havana, Cuba (2016).
19. Wagner, S.M., Ullrich, K.K., Transchel, S.: The game plan for aligning the organization, Business Horizons, 57(2), 189–201 (2014).
20. Ahi, P., Searcy, C.: An analysis of metrics used to measure performance in green and sustainable supply chains. Journal of Cleaner Production 86, 360–377 (2015).
21. Kleindorfer, P.R., Singhal, K., Van Wassenhove, L.N.: Sustainable operations management. Production and Operations Management 14(4), 482–492 (2005).

22. Srivastava, S.K.: Green supply-chain management: a state-of-the-art literature review. International journal of management reviews 9(1), 53–80 (2007).
23. Carter, C.R., Rogers, D.S.: A framework of sustainable supply chain management: moving toward new theory. International Journal of Physical Distribution and Logistics Management 38(5), 360–387 (2008).
24. Golini, R., Longoni, A., Cagliani, R.: Developing sustainability in global manufacturing networks: the role of site competence on sustainability performance. International Journal of Production Economics 147, 448–459 (2014).
25. Dombrowski, U., Mielke, T.: Lean Leadership – 15 Rules for a sustainable Lean Implementation Variety Management in Manufacturing. In: Proceedings of the 47th CIRP Conference on Manufacturing Systems, Procedia CIRP 17, 565–570 (2014).
26. Pojasek, R.B.: Understanding sustainability: An organizational perspective. Environmental Quality Management 21(3), 93–100 (2012).
27. Singh, A., Trivedi, A.: Sustainable green supply chain management: trends and current practices. Competitiveness Review: An International Business Journal 26(3), 265–288 (2016).
28. Azevedo B.D., Scavarda L.F., Caiado R.G.G, Fuss, M. Improving urban household solid waste management in developing countries based on the German experience. Waste Management, 120, 772–783 (2021).
29. Scavarda, A., Daú, G., Scavarda, L.F.; Caiado, Rodrigo, G.G.: An Analysis of the Corporate Social Responsibility and the Industry 4.0 with Focus on the Youth Generation: A Sustainable Human Resource Management Framework. Sustainability, 11, 5130 (2019).
30. Scavarda, A., Daú, G., Scavarda, L.F.; Azevedo, B.D., Korzenowski, A.L. Social and ecological approaches in urban interfaces: A sharing economy management framework. Science Of The Total Environment, 713, 134407 (2020).
31. Seuring, S., Müller, M.: From a Literature Review to a Conceptual Framework for Sustainable Supply Chain Management. Journal of Cleaner Production 16(15), 1699–1710 (2008).
32. Ansari, Z.N., and Kant, R.: A State-of-Art Literature Review Reflecting 15 Years of Focus on Sustainable Supply Chain Management. Journal of Cleaner Production 142, 2524–2543 (2017).
33. Jørgensen, F., Mathiessen, R., Nielsen, J., Johansen, J.: Lean maturity, lean sustainability. Advances in production management systems, Springer, Boston, MA, 371–378 (2007).
34. Cox, J.F. and Blackstone, J.H.: APICS Dictionary, 10th edition, Alexandria: APICS (2002).
35. Scavarda, L.F., Hellingrath, B., Kreuter, T., Thomé, A.M.T., Seeling, M.X., Fischer, J.-H., Mello, R.: A case method for Sales and Operations Planning: a learning experience from Germany. Production 27(spe) (2017).
36. Kreuter, T., Kalla, C. Scavarda, L.F., Thomé, A.M.T., Hellingrath, B.: Developing and implementing contextualised S&OP designs - an Enterprise Architecture Management approach. Journal of Physical Distribution & Logistics Management 51(6), 634–655 (2021b).
37. Fleischmann, B., Meyr, H., Wagner, M.: Advanced Planning. In: Stadtler, H., Kilger, C., Meyr, H. (eds.) Supply Chain Management and Advanced Planning – Concepts, Models, Software, and Case Studies, 5th edition, Springer, Berlin Heidelberg, 71–95 (2015).
38. Hübner, A., Kuhn, H., Sternbeck, M.: Demand and supply chain planning in grocery retail: an operations planning framework. International Journal of Retail & Distribution Management 41(7), 512–530 (2013).
39. Fuss, M., Barros, R.T.V., Poganietz, W.R.: Designing a framework for municipal solid waste management towards sustainability in emerging economy countries-An application to a case study in Belo Horizonte (Brazil). Journal of Cleaner Production 178, 655–664 (2018).
40. Buil, M., Aznar, J.P., Galiana, J., Rocafort-Marco, A.: An explanatory study of MBA students with regards to sustainability and ethics commitment. Sustainability 8(3), 280 (2016).
41. Azevedo B.D., Caiado R.G.G., Scavarda L.F.: Urban solid waste management in developing countries from the sustainable supply chain management perspective: A case study of Brazil's largest slum. Journal of Cleaner Production 233, 1377–1386 (2019).
42. Tranfield, D., Denyer, D., Smart, P.: Towards a methodology for developing evidence-informed management knowledge by means of systematic review. British journal of management 14, 207–222 (2003).

43. Thomé, A.M.T., Scavarda, L.F., Scavarda, A.J.: Conducting systematic literature review in operations management. Production Planning & Control 27, 408–420 (2016).
44. Cotta Fonatinha, T., Leiras, A., de Mello Bandeira, R., Scavarda, L.F.: Public-Private-People Relationship Stakeholder Model for disaster and humanitarian operations. International Journal of Disaster Risk Reduction 22, 371–386 (2017).
45. Ceryno, P., Scavarda, L.F., Klingebiel, K., Yuzgulec, G.: Supply Chain Risk Management: A Content Analysis Approach. International Journal of Industrial Engineering and Management 4, 141–150 (2013).
46. Hollmann, R.L., Scavarda, L.F., Thomé, A.M.T.: Collaborative planning, forecasting and replenishment: a literature review. The International Journal of Productivity and Performance Management 64, 971–993 (2015).
47. Saxena, L.K., Jain, P.K., Sharma, A.K. Tactical supply chain planning for tyre remanufacturing considering carbon tax policy. International Journal of Advanced Manufacturing Technology 97(1–4), 1505–1528 (2018).
48. Attia, A., Ghaithan, A., Duffuaa, S.: A Multi-Objective Optimization Model for Tactical Planning of Upstream Oil & Gas Supply Chains. Computers & Chemical Engineering 128, 216–227 (2019).
49. Laguna-Salvadó, L., Lauras, M., Okongwu, U., Comes, T.: A multicriteria Master Planning DSS for a sustainable humanitarian supply chain. Annals of Operations Research 283(1), 1303–1343 (2018).
50. Sawadogo, M., Anciaux, D.: Sustainable supply chain by intermodal itinerary planning: A multiobjective ant colony approach. International Journal of Agile Systems and Management 5(3), 235–266 (2012).
51. Boukherroub, T., Ruiz, A., Guinet, A., Fondrevelle, J.: An integrated approach for the optimization of the sustainable performance: A wood supply chain. IFAC Proceedings Volumes 46(9), 186–191 (2013).
52. Sheppard, S.R.J: Participatory decision support for sustainable forest management: a framework for planning with local communities at the landscape level in Canada. Canadian Journal of Forest Research 35(7), 1515–1526 (2005).
53. Jensen, M., Reynolds, K., Langner, U., Hart, M.: Application of logic and decision models in sustainable ecosystem management. In: 42nd Hawaii International Conference on System Sciences, pp. 1–10, IEEE (2009).
54. Fattahi, M., Mosadegh, H., Hasani, A.: Sustainable planning in mining supply chains with renewable energy integration: A real-life case study. Resources Policy (2018).
55. Saravi, N., Mobini, M., Rabbani, M.: Development of a comprehensive decision support tool for strategic and tactical planning of a sustainable bioethanol supply chain: Real case study, discussions and policy implications. Journal of Cleaner Production 244 (2020).
56. Mashaqbeh, S.M., Hernandez, J.E., Khan, K.M.: Developing a FMEA Methodology to Assess Non-Technical Risks in Power Plants. In: Proceedings of the World Congress on Engineering 2018, London, UK (2018).
57. Sizo, A., Noble, B., Bell, S.: Connecting the strategic to the tactical in SEA design: an approach to wetland conservation policy development and implementation in an urban context. Impact Assessment and Project Appraisal 34(1), 44–54 (2016).
58. Loureiro, D., Alegre, H., Silva, M.S., Ribeiro, R., Mamade, A., Poças, A.: Implementing tactical plans to improve water-energy loss management, Water Supply 17(2), 381–388 (2017).
59. Herazo, B., Lizarralde, G., Paquin, R.: Sustainable Development in the Building Sector: A Canadian Case Study on the Alignment of Strategic and Tactical Management. Project Management Journal 43(2), 84–100 (2012).

Industry 4.0 Contributions in Sustainable Operations

Yasmin Pires Gonçalves and Etienne Cardoso Abdala

Abstract Industrial activity has already undergone three revolutions and each one brought with it its particularities and benefits. Currently, we are going through the Fourth Industrial Revolution, also called Industry 4.0, which will bring about changes in various aspects of society, with the development of studies in this area being of great importance. Thus, the objective of this work is to analyze how Industry 4.0 can contribute to develop more sustainable operations practices. The methodologies used were the Systematic Literature Review, which consists of strategies for searching for published works, reading them and removing information, and the Content Analysis Method, which allows research, exploration and interpretation of materials, in this case, the use of reports for such analysis. After applying the methods, it was found that the technologies could be used in favor of companies not only in the economic aspect, but also in the social and environmental aspects, promoting advances and modifying the structure of organizations. The results point that there are several economics benefits, however the Fourth Industrial Revolution will contribute positively in environment, while others are still difficult to measure, as in the social pillar, in which some functions will no longer exist.

Keywords Industry 4.0 · Sustainability · Operations practices

1 Introduction

The Sustainable Development Goals 2030 Agenda by United Nations pressure organizations for cleaner production and contributes to a positive relationship between industry and sustainability, seeking to improve environmental conditions since product design and even after consumption [1]. In order to achieve better levels

Y. Pires Gonçalves
Enviromental Engineering Student - Federal University of Uberlandia, Uberlandia, Brazil

E. Cardoso Abdala (✉)
Federal University of Uberlandia, Uberlandia, Brazil
e-mail: etienne@ufu.br

of sustainability, organizations are looking for technologies capable of guaranteeing significant results at the three levels of the Triple Bottom Line (economic, social and environment).

We have experienced a technological revolution since the end of the twentieth century that promotes changes in several areas, in life, in human relationships and in work. This revolution is the Fourth Industrial Revolution or Industry 4.0 [2]. Industry 4.0 is the term used to refer to the high-tech strategy developed by Germany and which is being applied to industry. Advanced technologies are able to integrate the physical, the digital and the biological, forming a system. Machines are able to complete complex tasks due to self-optimization, self-configuration and artificial intelligence, in order to be efficient and provide better quality goods or services [3].

Industry 4.0 is a step in the direction of creating more sustainable industrial value, as it offers alternatives that allow the minimization of waste, generation of waste water and atmospheric emissions, contributing to the efficient use of raw materials and energy, providing environmental and economic benefits for companies [4].

However, each industrial revolution brings new challenges and determine new approaches within organizations, bringing with it an uncertainty in the consequences of the transformations generated [5]. Therefore, the purpose of this paper is to identify and analyze the interaction between sustainability and Industry 4.0 and thus indicate what are the main contributions that Industry 4.0 can bring to organizations so that they develop more sustainable operations. For that, a systematic review was developed with the most relevant publications that relate the two themes, in order to bring the interaction between them, and also a documental analysis through company reports from websites, in order to understand how companies are adopting industry 4.0 procedures and technologies to turn your operations more sustainable.

2 Theoretical Background

The industry 4.0 approach permeates technological advances in search of the digitalization of industrial processes [6] and develops in the implementation of strategic areas of the industry such as logistics and supply chain management, bringing benefits of these advances such as decentralization, the more accurate assessment of demand, reduction of the bullwhip effect [7] and the more integrated and transparent management of supply chains [8]. According to Stock and Seliger [9] adaptations to computerized mechanical machines can reduce losses of raw materials and energy, as well as contribute to the provision of data on quality and productivity.

Another very popular issue is sustainability in the operations of an Organization. The fulfillment of the UN Sustainable Development Goals agenda until 2030 [10, 11] and the pressure suffered by different stakeholders for cleaner production contribute to changes in the management of operations [1] since product design to post-consumption.

Organizations in general need to develop actions that result in sustainable operations, that is, activities related to the production process of goods or services that minimize the exploitation of natural resources, propitiating the reduction of waste generated at the end of the process and the possible reuse of these. Elkington [12] believes that it is possible to consider the capitalist system as something that can generate sustainability and highlights the existence of a transformation in the environmental aspect from the 1980s.

In order to achieve better levels of sustainability, organizations seek the development of technologies capable of bring results at the three levels of the Triple Bottom line: economic, social and environmental. The advent of a new 'Industrial Revolution' is a relevant element for achieving sustainability goals, as it will enable new technologies to be created in order to intensify the breadth of actions considered sustainable.

Basically, the main objective of Industry 4.0 is to reinforce and expand organizational competitiveness in the long term, by increasing the flexibility and efficiency of production through communication, information and intelligence. This 'fourth' industrial revolution leads to impacts within the organization, on the technology of production factors, organization and employees; and also outside it, in the environment and society [13]. Stock and Seliger [9] argue that in addition to the environmental contribution, industry 4.0 presents a great opportunity to add value creation in industrial sustainability in the three dimensions: Economics, environmental and social.

The benefits of industry 4.0 can be more comprehensive than just making the organization more competitive. Erol [14] argues that the fourth industrial revolution is not only aimed at increasing competitiveness, but is also built on the concept of sustainability as a key factor for economic prosperity and quality of life, and information systems are like the greatest facilitators of this vision, since they provide the assessment of ecological impact of a production process.

The discussion about industry 4.0 directly involves what is called 'internet of things', which allows full integration and connection between all systems into a value chain. As highlighted by Kagerman et al. [15] in a world 'smart and connected' to the Internet of things and services will be an element present in all key areas of an organization. This change is leading to the emergence of smart grids in the field of sustainable energy supply, mobility strategies (smart mobility, smart logistics) and smart health in the health domain. The manufacturing environment, vertical network, end-to-end engineering and horizontal integration across the entire network of product and system values, in a way increasingly intelligent is prepared to launch the fourth stage of industrialization.

Especially for logistics, industry 4.0 has contributed to the optimization of functions based on trends such as individualization, servitization, accessibility, autonomy and global networking [16] based on software and databases from which shared information is provided, generating a high degree of automation [17]. At the same time there is a vast field of new technologies to integrate physical tools with virtual tools in order to improve the integration and performance of industrial ecosystems [18] the management of these technologies applied to the processes of organizations [19] of medium logistics technology in Brazil still remains unclear.

3 Method

The systematic review is a form of research that uses the literature on a given topic as a data source. This type of research is important to integrate information from studies published separately, and from there, perform the analysis of the results obtained from all of them and posterior obtain guidance for future investigations.

Initially, a clear question is established, which serves to guide the research. The established question in this paper was: what are the contributions that technological change in industry 4.0 can bring to organizations for the development of a more sustainable operation?

In order to develop the literature research, the Periodicos Capes Brazil's website was consulted. When entering the website of the Periodicos Capes, "Search subject" was selected, the search was made in English and the chosen keywords were "sustainability and industry 4.0". The results were filtered with Business and Economics, and were refined in this current way: Publication period between 2014 to 2019 (until August); Journal Papers; Higher Level: Peer-reviewed journals; English language.

The exclusion criteria for Journal Papers define which will be part of the research and which will not. Therefore, it was defined that the two keywords determined was essential in this research and not dealing with operations, makes the study not suitable for this methodology, as the focus of the research is on sustainable operations practices.

To analyze the data collected in the systematic review, the PRISMA methodology was used [20]. This method consists of establishing certain specific items in a checklist when reporting the systematic review, recognizing that authors can review the studies or diagnosis, modifying or incorporating additional items. Thus, the PRISMA method [20] considers in a pre-defined checklist that following to evaluation of the systematic review: title of the article, structured summary, objectives, method, synthesis of results, discussion of results and research funding,

For this specific paper, all these elements of the PRISMA method were considered, but the results were presented in a synthetic way, since the studies present extensive discussions that would go beyond the limits established for publication of the paper. The method does not directly or in detail deal with conducting systematic reviews, for which other guides are available.

After applying the filters and the exclusion criteria, 22 papers were selected, from a total of 11.802 publications, they were read to verify their relevance and, subsequently, the following information was colletcted from each of the publications: title, authors, year, newspaper, goal, methodology and results. Based on these data, it was analyzed which Journal and which year had the largest number of publications, which type of research was most used and which methodology was the most usual, as shown in the results.

The second part of the data collected involves selecting and collecting reports on websites of companies that use new technologies associated with Industry 4.0, portraying situations in which organizations have adopted or developed technologies for sustainable actions and understand what were the contributions of these technologies to turn operations more sustainable. For this stage of research thirteen articles were selected from sources of specialized Brazilian business magazines and economics newspapers. All selected reports are related to sustainability in industry 4.0.

The data were analyzed using the Content Analysis method defined by Bardin [21]. Content analysis is a communication analysis technique in which what is said in interviews or observed by the researcher is analyzed. Thus, this method has been disseminated and used in order to analyze qualitative data.

The analyzed data are organized into themes or categories, in order to better understand the information passed by them. Bardin's technique has steps that are divided into: (1) pre-analysis; (2) exploration of the material, and (3) treatment of results, inference and interpretation.

In the first step, called pre-analysis, the selected material is read so that it can then move on to the next phases. The second step is called exploration of the material. All selected texts are cut out as record units (words, phrases, paragraphs) and, the words repeated or related to the bibliographic reference will raise the initial categories. When, the initial categories are grouped by categories it raises the intermediate categories and at last when combined, it results on the final categories. Finally, the third step is the treatment of results, inference and interpretation. This last step consists of capturing the contents of all the collected material. The comparative analysis highlights the aspects considered similar and those that were different [22].

This study starts from secondary data taken from *Google* and the words defined for the search were "Industry 4.0 and Sustainability". When typing these words, several reports appeared, so Business-related Brazilian Magazines were selected for analysis. As the number of results on Google after doing the research was very high, reports were selected up to page 8. When reading the texts, the most important parts that have a relationship with industry 4.0, technologies and sustainability were separated, then they were described in detail and summarized.

4   Results

4.1   Systematic Review of Literature

For the systematic review of literature five filters were used. The first filter was "Business and Economics" in "Personalize your results". The second was the date of publication between the year 2014 and the year 2019. The third filter applied was to select the type of resource "Journal Papers". The fourth filter was "peer-reviewed journals". Finally, the "English" filter. The filtering resulted in 7.322 articles. Subsequently, the selection of papers to be reviewed depended on having the two defined keywords (Sustainability and Industry 4.0) and dealing with operations, which ended up reducing the number of 7.322 to 22 articles. This is summarized in Table 1:

Table 1 Papers journals

Journals title	Number of papers found
Annals of the faculty of engineering Hunedoara	1
Benchmarking	1
Computers in industry	1
IEEE access	1
International journal of environmental research and public health	1
International journal of innovation management	1
Journal of cleaner production	1
Process safety and environmental protection	4
Social sciences	1
Sustainability	8
Technological forecasting and social change	1
The international journal entrepreneurship and sustainability	1

The most paper´s publications was in the 2018, by 12 papers. Until August 2019 were published 6 papers. It´s possible to realize that there are still few publications that deal with sustainable operations and industry 4.0 and the research that exists is recent. Therefore, in the year 2018 it can be said that there was an advance in published studies dealing with the two subjects mentioned.

Most of the research on the reviewed papers is qualitative, around 18 papers. The methodologies adopted in the 22 papers revised were various, with interviews, case studies and systematic literature reviews being the most frequent in the reviewed studies, probably due to the greater ease of seeking information on how companies are placing in practice technologies. Some researches presents more than one type of methodology adopted, and some of them uses data triangulation (quantitative and qualitative analysis). Table 2 presents the types of methodologies adopted:

By the results of the systematic review it was possible to conclude that most of results of papers investigated suggests that technologies coming from industry 4.0 associated with sustainable practices will provide competitive advantages, production will be more environmentally friendly, products can be customized, there will be improvement in product quality, continuous monitoring of production, safer work environment, less intense workload, among other factors as development of ecological products, more effectiveness in processes and inventory reduction.

The Table 3 presents the main results of papers researched in the Systematic Review.

Table 2 Types of methodologies

Methodology	Number
Surveys	3
Content analysis	2
Interviews	4
Case study	4
Systematic review	5
Literature review	4
Theory analysis	1
Operational application	1

It is possible to comprehend that among the aspects presented in Table 3 there are issues associated with the three dimensions of sustainability, as economic, social and environmental, special in rational use of resources and waste management, understanding sustainability as a strategic element of competitive advantage. In this sense, it is noted that, based on the results published in the most relevant research carried out in the last five years, that the new technologies of industry 4.0 can contribute to addition of sustainability in productive operations, despite a greater need for the inclusion of social aspects such as the valorization of human capital.

4.2 Organizational Reports Results

There are many technologies that the Fourth Industrial Revolution promotes, among them there are the internet of things (Internet of Things—IoT), big data, data analytics, autonomous robots, integrated systems, cyber-physical systems (Cyber-physical Systems—CPS), cloud computing, 3D printing, among others. Drones are examples of technological advances that bring many advantages, among them that is easy to observe, to map and take serial photos of places such as legal reserve, permanent preservation area and harvest. Real-time images make it possible to anticipate adverse environmental conditions and pests in the field, for example.

According to the Brazilian Industrial Development Agency (ABDI), by the beginning of 2018, about 2% of companies are inserted within the concept of Industry 4.0 [23]. By reading the reports presented in Table 2, it was possible to identify about 20 companies that have some technology related to the Fourth Industrial Revolution, which are: Mining Company Vale; automotives companies like Mercedes-Benz, Volkswagen, Jeep, Ford, Fiat, General Motors, Nissan; to MWM auto parts (Engines); the heavy vehicle manufacturer CNH (Case New Holland); the aircraft manufacturer Embraer; the Ambev brewery; the software and electronics company Microsoft; the elevator company Thyssen krupp; the aluminum companies Alcoa and Novelis; the aluminum and renewable energy company Hydro; the manufacturer of automotive

Table 3 Main results

Main results	Number of papers
Competitive advantage	3
Faster customer response with quality and flexibility	1
Providing detailed information on each point in the process	2
Use of resources and energy optimization	2
Increases flexibility	1
Greater efficiency	1
Waste reduction	2
Reduction in pollutant emissions	1
More autonomous decision making	1
Improved productivity	2
Improved product quality	1
Safer working environment	2
Less intense workload	1
Professional enrichment	1
Precision and accuracy in the manufacture of products	2
Development of ecological products	2
Ecological manufacturing processes	1
Improved supply chain management	1
Continuous monitoring of energy consumption	3
Cost reduction	2
Time optimization	2
Better work-life balance	1
Simplification of jobs by technological systems	1
Inventory reduction	1
Customization	2
Management waste	1
Circular economy model	1

structural components Iochpe-Maxion; and companies in the glass segment, such as Cebrace, the Saint-Gobain group and NSG, among others.

According to Silva and Fossá [22] the initial categories are the first impression of the material defined for reading. After analyzing the reports, a total of 19 categories resulted from the codification process. From the main characteristics observed during the material, it was possible to define the initial categories, described as:

Table 4 Summarized categories

First	Intermediary	Finals
1 – Cost reductions	I—Reduced negative factors	I—Management of resources and production systems
2 – Waste reductions		
3 – Error reductions		
4 – Reduction in the emission of pollutants		
5 – Inventories reduction		
6 – More accuracy	II—Productive advantages	
7 – More efficiency		
8 – Time optimization		
9 – Energy saving		
10 – Improved management of natural resources		
11 – Anticipate adverse conditions	III—Company management	
12 – Constant monitoring		
13 – Operations broad view		
14 – Real time diagnosis		
15 – Customization		
16 – Product quality	IV—Product characteristics	II—Product characteristics and human conditions
17 – Customer-oriented production		
18 – Changing work methods	V—Working and human conditions	
19 – Quality of life at work		

cost reductions, waste reductions, errors reductions, inventories reduction, reduction in the emission of pollutants, more efficiency, more accuracy, time optimization, energy saving, improved management of natural resources, anticipate adverse conditions, constant monitoring, operations broad view, real time diagnosis, customization, product quality, customer-oriented production, changing working methods, quality of life at work.

In order to gather the categories and facilitate the understanding of how the progression occurred, Table 2 presents all the categories defined. It started from 19 initial categories, which gave rise to 5 intermediate categories, which resulted in 4 final categories (Table 4).

After the application of Content Analysis by the methodology of Bardin [21] publications were analyzed based on the three pillars of sustainability: economic, social and environmental. For Industry 4.0 to promote intelligent environments is necessary to use resources consciously and attend the effects of industrial activity, aiming not only at financial gains, but also reaching the other pillars.

The automation and digitalization of industrial processes saves energy, optimizes time, reduces errors and, consequently, reduces costs. Therefore, about the economic pillar, the Fourth Industrial Revolution allows for greater accuracy of processes and a broad view of operations, avoiding material losses. The negative point is in the implementation of the technology, as high investment is required. However, after its installation and use, the production process will become economical.

There are several barriers for companies to becoming sustainable, such as the high cost of investment and the time it takes to obtain a return, lack of a tool to assist in the assessment of environmental impact, limitations in technology, among others [24]. With Industry 4.0, enterprises may be more sustainable, as this industrial revolution will bring new production technologies, organizational culture and impacts outside organizations [13].

One of the pillars of sustainability is the social one, which is still much discussed, as many functions will be replaced, mainly by robots, while others will be created. However, it is not yet possible to say how negative or positive it will be, nor whether there will be more unemployment or employment. It is understood that it will be necessary to improve the workforce qualification, since the human being remains essential in the process, but in a more intelligent way, as they will be more responsible for the management of companies. Another important factor is the improvement in the quality of life within organizations, as there will be less physical effort and less risk, since the machines will do the most stressful jobs.

According to Tartarotti et al. [25] the Internet of Things (IoT) together with Artificial Intelligence will promote a rapid advance of Industry 4.0, since together it is possible to make decisions without human interference. According to Shules [26] horizontal integration is related to the entire supply chain, while vertical integration is the integration of actuators at different levels with the sensors, until reaching the level of resource planning (ERP), which makes production more flexible and reconfigurable.

While industrial activity will be more efficient, with fewer failures during the process and less waste generated, the machinery that is responsible for all these advantages, will still have raw materials as a source and, as Industry 4.0 develops, there will be production of these technologies and spending on natural resources. To compensate all this technology, companies should be even more concerned with the environmental issue, with the use of natural resources and be much more sustainable. The concept of industry 4.0 allows advantages such as less waste of materials, less waste, reduced emission of pollutants, less stock, use of alternative energy sources with reduced consumption, among others.

5   Conclusion

Industry 4.0 will be very important for organizations in general. After reading articles and reports, it was possible to see that this industrial revolution will show that the production associated with sustainability can bring benefits not only for companies, but also for the environment. Natural resources can be used intelligently, reducing waste and reusing materials if companies invest in it and recognize that there will also be economic benefits.

With regard to sustainability, the Fourth Industrial Revolution will contribute positively in several aspects, while others are still difficult to measure, as in the social pillar, in which some functions will no longer exist and others may come to exist. Therefore, it is necessary to qualify the workforce so that people can work in the midst of technology. In the environmental pillar, smart technologies will make it easier for companies to make their operations more sustainable, as the production process will be optimized, there will be constant monitoring and less waste will be generated in production, through 3D printing, for example. And finally, the economic pillar, in which technologies will allow greater efficiency and optimization of time, reducing losses within companies.

The economic benefits of Industry 4.0 include several factors. When companies try to trying to optimize and work with these value factors, there is a high potential for improvement in the economy and in the commercial and manufacturing processes. All of these impacts make possible an analysis of how Industry 4.0 technologies can contribute to find sustainable manufacturing solutions or how they can establish a circular economy [27].

References

1. Lozano, R. Towards better embedding sustainability into companies' systems: An analysis of voluntary corporate initiatives. Journal of Cleaner Production. 25, 14–26 (2012).
2. Schwag, K. A Quarta Revolução Industrial. 1ª edição. Editora: Edipro (2016).
3. Bahrin, M.; Othman, F.; Azli, N.; Talib, M. Industry 4.0: A review on industrial automation and robotic. Journal Teknologi. 78 (6),137–143. (2016).

4. Carvalho, N.; Chaim, O.; Cazarini, E.; Gerolamo, M. Manufacturing in Fourth Industrial Revolution: a positive prospect in sustainable manufacturing. Procedia Manufacturing. 21. 671–678. (2018)
5. Perez, C. Technological revolutions and techno-economic paradigms. Cambridge Journal of Economics. 34(1).18–202. (2010).
6. Rojko, A. Industry 4.0 Concept: Background and Overview. International Journal of Interactive Mobile Technologies (iJIM). 11 (5), 77–90 (2017).
7. Hofmann, E.; Rüsch, M. Industry 4.0 and the current status as well as future prospects on logistics. Computers in Industry, 89, p. 23–34 (2017).
8. Witkowski, K. Internet of Things, Big Data, Industry 4.0–Innovative Solutions in Logistics and Supply Chains Management. Procedia Engineering, 182, 763–769, (2017).
9. Stock, T.; Seliger, G. Opportunities of Sustainable Manufacturing in Industry 4.0. ScienceDirect, 40, 536–541 (2016).
10. OECD. Organisation for Economic Co-operation and Development. Sustainable Development Goals. Available at: https://sustainabledevelopment.un.org/?menu=1300 (2015).
11. Sachs, J. D. From millennium development goals to sustainable development goals. The Lancet, 379 (9832), 2206–2211 (2012).
12. Elkington, J. Canibais com garfo e faca. São Paulo: Makron Books (2001).
13. Gabriel, M.; Pessl, E. Industry 4.0 and sustainability impacts: critical discussion of sustainability aspects with a special focus on future of work and ecological consequences. Annals of Faculty Engineering Hunedoara. International Journal of Engineering. 14, (2), 131–136. (2016).
14. Erol, S. Where is the green in Industry 4.0? or How information systems can play a role in creating intelligent and sustainable production systems of the future. First Workshop on Green (Responsible, Ethical, Social/Sustainable) IT and IS – the Corporate Perspective (GRES-IT/IS), Volume: Working Papers on Information Systems, Information Business and Operations, Department of Information Systems and Operations, WU Vienna, 2/(2016).
15. Kagermann, H.; Wahlster, W.; Helbig, J. Recommendations for implementing the strategic initiative Industrie 4.0. Acatech, 13–78 (2013).
16. Strandhagen, J. O., Vallandingham, L. R., Fragapane, G., Strandhagen, J. W., Stangeland, A. B. H., & Sharma, N. Logistics 4.0 and emerging sustainable business Advances in Manufacturing. 5 (4), 359–369. (2017).
17. Barreto, L.; Amaral, A.; Pereira, T. Industry 4.0 implications in logistics: an overview. Procedia Manufacturing, 13, 1245–1252, (2017).
18. Xu, Li Da; Xu, Eric L.; Li, Ling. Industry 4.0: state of the art and future trends. International Journal of Production Research. 1–22 (2018).
19. Jabbour, A. B. L. S.; Jabbour, C. J. C.; Godinho Filho, M. Roubaud, D. Industry 4.0 and the circular economy: a proposed research agenda and original roadmap for sustainable operations. Annals of Operations Research, 1–14 (2018).
20. Liberati, A., Altman, D. G.; Tetzlaff, J.; Mulrow, C.; Gøtzsche, P. C.; Ioannidis, J. P. A.; Clarke,M.; Devereaux, P. J.; Kleijnen, J.; Moher, D. The PRISMA statement for reporting systematic reviews and meta-analyses of studies that evaluate healthcare interventions: explanation and elaboration. BMJ. 339: b2700. (2009).
21. Bardin, L. Análise de Conteúdo. Editora: Presses Universitaires de France. (1977).
22. Silva, A. H.; Fossá, M. I. T. Análise de Conteúdo: Exemplos de Aplicação da Técnica para Análise de Dados Qualitativos. Qualit@s Revista Eletrônica, Vol.17. No 1 (2015).
23. Dino. Menos de 2% das empresas atuam no conceito da Indústria 4.0. Revista Exame (2018). Avaiable in: https://exame.abril.com.br/negocios/dino-menos-de-2-das-empresas-atuam-no-conceito-da-industria-40/ >. Acess: nov. (2019).
24. Esmaeilian, B.; Behdad, S.; Wang, B. The evolution and future of manufacturing: A review. Journal of Manufacturing Systems. 39, 79–100 (2016).
25. Tartarotti. L.; Sirtori, G.; Larentis, F. Indústria 4.0: Mudanças e Perspectivas. Universidade de Caxias do Sul. (2018).

26. Shules, M. V. Proposta de Diagnóstico para Adoção das Tecnologias da Indústria 4.0 em um Processo Produtivo com Base em Indicadores de Sustentabilidade: Um Estudo de Caso. Universidade Federal do Paraná. Curitiba. (2018).

27. Blunk, E. Werthmann, H. Industry 4.0 – an opportunity to realize sustainable manufacturing and its potential for a circular economy. DIEM: Dubrovnik International Economic Meeng. 3 (1) 644–666. Listopad (2017).

Fuzzy Criticality Assessment of Systems External Corrosion Risks in the Petroleum Industry—A Case Study

Rodrigo Goyannes Gusmão Caiado⬭, Marina Polonia Rios⬭,
Gabrielle Cordeiro Martins⬭, Paulo Ivson Netto⬭,
and Fernanda Ramos Elmas⬭

Abstract From the perspective of offshore installations, corrosion is the most common form of damage and the main factor affecting the longevity and reliability of assets. Maintenance plans based on risk and reliability are modern approaches. For a complex technical system such as a platform, maintenance management can use different approaches, such as reliability centered maintenance (RCM) that uses failure mode and effect analysis (FMEA) in determining the criticality of failure modes by the ranking of risks (e.g. related to corrosion rates and remaining life of the asset). However, in some practical cases, the implementation of the traditional FMEA may not be realistic due to the subjectivity of the dimensioning the risk factors. In this sense, this research aims to propose a more robust and practical methodology to assess the criticality of risks of external corrosion failures by the system. This methodology combines fuzzy scale method, risk space diagram and weighted Euclidean distance and is applied based on the perception of specialists in external corrosion. From an academic point of view, this research highlights the importance of using RCM for corrosion management in an empirical case. From the practitioner's perspective, we seek to provide a tool, based on the RCM and the combination of fuzzy logic and multicriteria approach to rank systems to be inspected. Therefore, this new FMEA model can help to save time, cost and reduce fieldwork, considering gains related to quality, safety, technical efficiency in inspection and maintenance processes.

Keywords Corrosion risks · Soft computing · Oil and gas industry

R. G. G. Caiado (✉) · M. Polonia Rios · G. Cordeiro Martins · P. Ivson Netto · F. Ramos Elmas
Pontifical Catholic University of Rio de Janeiro (PUC-Rio), Marquês de São Vicente Street, 225 - Gávea, Rio de Janeiro, RJ 22451-900, Brazil
e-mail: rodrigocaiado@tecgraf.puc-rio.br

R. G. G. Caiado
Federal Fluminense University (UFF), Passo da Pátria Street, 156, Niterói 22451-900, Brazil

F. Ramos Elmas
Rio de Janeiro State University (UERJ), São Francisco Xavier Street, 524 - Maracanã, Rio de Janeiro 20550-013, Brazil

1 Introduction

In the Oil and Gas (O&G) industry, maintenance costs represent 40% of total costs and most are the result of inadequate or unscientific planned maintenance activities [1]. Maintenance strategies have gained importance as their benefits such as reducing security risk and preventing costly interventions have become clear to industries [2]. In the case of offshore platforms, maintenance gains even more importance, as its activities present a high environmental, social and financial risk. Maintenance shutdowns, for example, represent an important strategic process for planning maintenance within an industrial plant, thus being fundamental in the management of equipment that requires more prolonged and in-depth inspections [3].

The detection and characterization of damages are of great importance for the development of a good maintenance plan. From the perspective of offshore installations, corrosion is the most common form of damage due to the metallic character of the materials and the aggressive environment of the sea atmosphere. Corrosion is the main factor that affects the longevity and reliability of offshore assets, consuming up to 80% of the total maintenance cost of the oil and gas exploration industry. Studies estimate that 20–30% of this cost could be saved if good corrosion management practices were adopted, including inspection and prevention strategies [4].

In this context, for a complex technical system such as a platform, maintenance management can use different approaches, such as reliability centered maintenance (RCM), to improve the planning of inspection and maintenance activities [5]. RCM can assist in the identification of functional failures, causes of failures and risk ranking that are related to corrosion rates and remaining life span of the asset [6]. The RCM is supported by three types of maintenance: risk-based maintenance (RBM); time-based maintenance (TBM); and condition-based maintenance (CBM) [7]. According to [7], at RBM, which is supported by risk-based inspection (RBI), failure mode and effect analysis (FMEA) is often applied to predict inspection targets, determine maintenance targets and select appropriate maintenance tasks for the targets. FMEA is a significant procedure and widely applied throughout the reliability engineering process [8], determining the criticality ranking of failure modes is a vital issue [9].

However, in some practical cases the implementation of the RCM may not be realistic and FMEA applications are criticized for reasons such as [9–12]: the results of the risk priority number (RPN) calculation depend heavily on the small variation in the risk factors for S, O and D, with this, the same RPN can result in different values of S, O and D; and the same importance of these parameters must be considered when calculating the RPN. In view of these reasons, different techniques for calculating criticality emerged, among the main ones, identified by [13], the use of: multi-criteria decision making models (MCDM) as AHP (analytic hierarchy process) [14]; artificial intelligence as fuzzy logic [12]; simulation like Monte Carlo [15]; and integrated approaches such as AHP-TOPSIS-VIKOR [16] and Fuzzy-AHP-TOPSIS [17]. Although academics and engineers have done research to overcome the subjectivity of the traditional FMEA method when scaling the risk factors integrated with a fuzzy inference or approximate reasoning, a substantial amount of inference rules has complicated the computational process [8].

Therefore, this research aims to propose a methodology based on the RCM strategy, combining fuzzy logic and MCDM approach to assess the criticality of external corrosion failures by system. This methodology is an adaptation of the FMEA model using Fuzzy Logic proposed by [8, 9], which is composed of a fuzzy scale method, risk space diagram and weighted Euclidean distance, and further can be used as an input to determine, the best strategy or mix of maintenance strategies.

2 Background

2.1 Reliability-Centered Maintenance (RCM)

In the past few decades, many philosophies or maintenance strategies have been developed. These include: corrective maintenance, preventive maintenance, predictive maintenance and reliability-centered maintenance (RCM) [9]. RCM is considered a structured maintenance technology to optimize the acceptable maintenance strategy (for example reactive, preventive, condition-based or proactive) for each component of a system [9]. An important step in the RCM is to determine whether a failure can be prevented or predicted and whether an investment should be made, considering its consequences [18]. Among the benefits obtained with the use of RCM and, thus, the choice of the effective maintenance strategy, there are: reduction of the severity of the impact of failures, financial gains by reducing the time and cost of repair/maintenance and possibility of increased satisfaction due to the increased reliability and availability of facilities and assets [18].

The RCM methodology can be divided into six basic steps [19]: (I) system selection and data collection; (ii) division of the system and identify the functionally significant item (FSI); (iii) determination of FMEA; (iv) criticality analysis; (v) RCM logical decision; and (vi) selection of the maintenance strategy. The initial stage, and critical point of the RCM, is to preserve the function of the system, which initially does not deal with preserving the operation of the equipment, in order to know what the expected output is and preserve that output (function). This allows maintenance engineers to systematically decide, in the later stages of the process, only which equipment relates to which functions. Thus, unlike classic preventive maintenance, in the RCM it is not assumed that "*every item of equipment is equally important*" [20]. The reliability of plants can be established by maintaining the reliability of each piece of elementary equipment [7]. The final stage of the RCM is the selection of the ideal maintenance strategy. The use of FMEA with the criticality analysis is considered the main tool of the RCM, as it helps to direct maintenance in the desired failure modes and to avoid the critical causes of failure [9]. Determining the critical classification of failure modes is a vital issue of the method, which classifies failure modes using RPNs [19]. The RPN is calculated by the product of evaluation criteria such as severity (S), occurrence (O) and detection (D) of each failure mode, according to the following Eq. (1):

$$RPN = S \, x \, O \, x \, D \tag{1}$$

2.2 Corrosion Management

The corrosion process is a complex physical–chemical phenomenon, which is influenced by several factors (environmental, climatic conditions, material characteristics and compositions, among others) [21] and from the point of view of asset management, many of the damages that affect the integrity of the equipment throughout a project is related to corrosion [22]. Corrosion management represents an operational process and organizational configuration to ensure that plans for the control, inspection, monitoring and evaluation of corrosion are implemented to optimize the integrity of assets [22]. Corrosion monitoring and inspection are essential processes in a general corrosion management system that manages corrosion risks through a combination of *proactive measures*—where the requirements and implementation of the monitoring system or inspection programs are identified and are in place before any corrosion or deterioration has been observed based, for example, on the results of a corrosion risk assessment—and *reactive measures*—implemented after the identification of a problem, as a result of proactive monitoring or because of an incident or the identification of a problem as a result of the inspection [23].

Regarding the tools used to assess corrosion risk, [24] states that this choice depends on the complexity of the situation, which can be basic, using for example a Risk Matrix, with linear or logarithmic scales, to prioritize risks and then concentrate efforts on the most influential [22] or alternatively with the attribution of a numerical value based on a criticality index (RPN), which is suggested for the case of data insertion in computer systems. In general, the information required for corrosion assessment includes a list of items of interest in the asset registry (structures, vessels, pipes, storage tanks, etc.), historical data (inspection, monitoring, maintenance), theoretical analysis (new systems based on published data/models), and informed opinions [24]. The risk classification tools seek to focus attention on critical areas [25, 26], allowing an assessment team to focus on items in a plant or processes that have varying levels of corrosion risks. Corrosion rate data, models or field information, can even be used to assess the risk of corrosion on an asset or components, based on, for example, RBI [22].

In a review covering research programs related to fatigue, as well as investigations of fatigue in specific civilian and military aircraft (DSTO-TN-0747), [27] presented a pilot study of corrosion-based reliability centered maintenance in which thirty structural and maintenance items of the C-130 J aircraft were inspected for corrosion, that is, inspections related to failure condition. According to the authors, the RCM process is designed to develop inspection intervals that detect corrosion precursors and, thus, allow the application of corrosion prevention measures, thus avoiding costly and time-consuming replacement or repair.

3   Methods

3.1   Case Study and Research Steps

This section presents the case study of the criticality assessment of systems in industrial plants according to risk factors [28] associated with external corrosion failure. In the O&G industry, systems criticality assessment aims to optimize the annual inspection process, as components with low criticality do not need to be inspected with the same frequency as components that have high criticality, being possible to adjust inspection intervals, generating a time-based inspection plan. For the evaluation of the criticality of the external corrosion of the systems, the application of the FMEA methodology is proposed, using fuzzy logic and a MCDM approach, adapted from [29], based on the evaluation of external corrosion risk experts. A FMEA team of three engineers from different areas (chemistry, civil and production) was formed to decide the severity (S), occurrence (O) and detection (D) classification of each failure mode, represented by the system to be inspected. The criticality of each system can be analyzed together with maintenance costs to define the painting plan for the following year, aiming to maximize the reliability of the asset as a whole, following the principles of the RCM, presented in Sect. 2. The proposed decision-making structure comprises eight steps (see Fig. 1).

3.2   Alternatives (Systems) and Criteria (Risk Factors) Used for MCDM

Firstly, it is necessary to define the systems (alternatives) that will be evaluated according to the key criteria (risk factors) of the FMEA. As illustrated in the hierarchical tree (see Fig. 2), a MCDM methodology based on the logic of RCM is presented.

The system is understood to mean the physical division of components into areas to be painted, carried out on platforms/industrial plants. Based on the definitions of the business unit of the studied company (case study), nine systems were selected: ceiling; bulkhead; floor; Pipes, valves and flanges; equipment; metallic structures; stairs; supports; and guardrails. In this study, the definition of [12] was adopted, in which severity: indicates the gravity of the effects of a corrosion failure affecting the system; occurrence: indicates the probability that a corrosion failure will occur; detection: measures the visibility of a corrosion failure which is the attitude of a failure mode to be identified by controls or inspections.

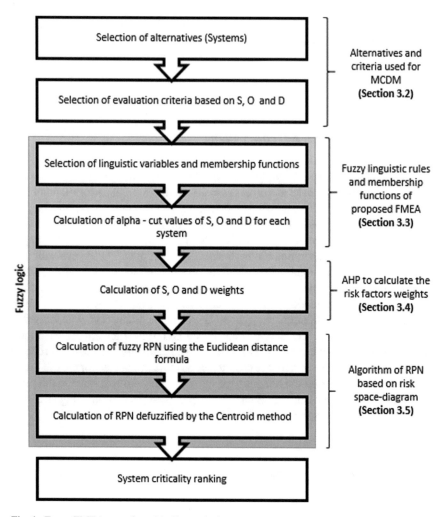

Fig. 1 Fuzzy FMEA steps for criticality analysis

3.3 Fuzzy Linguistic Rules and Membership Functions of Proposed FMEA

Each system (failure mode) is numbered sequentially to assess the risk priority of each one. The influence of the three failure parameters (S, O, D) is considered to assess the criticality RPN of a system. The fuzzy theory is adept at treating uncertain events and is widely applied to deal with the imprecision of human perception [30]. These parameters are measured on the five-point fuzzy linguistic scale V = {R = remote, L = low, M = moderate, H = high, VH = very high} in the universe [0.10] and the evaluation criteria for each these parameters are shown in Table 1. To measure

Fig. 2 Hierarchical tree adapted from [20]

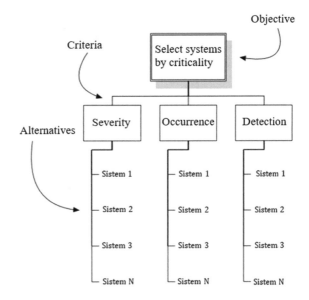

the average of the linguistic scale values in the aggregate (V), the association of these risk factors is defined by the triangular fuzzy number.

Thus, the evaluation of the classification of S, O and D for each system was based on the perception of an FMEA team of three engineers from different backgrounds, in order to avoid the influence of the experience and knowledge of the experts [8]. Through the evaluation score for each system, the triangular fuzzy number of S_I, O_I e D_I is given by the following equations:

$$S_I = (S_{iL}, S_{iM}, S_{iR}) = \left\{ \sum_{j=1}^{m} (S_{ijL}, S_{ijM}, S_{ijR}) \right\} / m \tag{2}$$

$$O_I = (O_{iL}, O_{iM}, O_{iR}) = \left\{ \sum_{j=1}^{m} (O_{ijL}, O_{ijM}, O_{ijR}) \right\} / m \tag{3}$$

$$D_I = (D_{iL}, D_{iM}, D_{iR}) = \left\{ \sum_{j=1}^{m} (D_{ijL}, D_{ijM}, D_{ijR}) \right\} / m \tag{4}$$

$$i = 1, \ldots, n; \ j = 1, \ldots, m$$

where S_{ij} represents the fuzzy scores of the ith failure mode; j represents the jth expert and the total number of experts is represented by m.

Table 1 Selection criteria for evaluating systems and membership functions

Linguistic variable	Value	Description [9]	Triangular fuzzy numbers
S	R	Insignificant effect, immediate correction by maintenance;	(1, 1, 2)
	L	Minor effect, the system suffers from a gradual degradation if not repaired;	(2, 3, 4)
	M	Moderate effect, the system does not perform its functions, but maintaining the failure requires stopping the operations;	(4, 5, 6)
	H	Critical effect, maintenance demands that the system operations be stopped;	(6, 7, 8)
	VH	Very critical effect, the failure abruptly interrupts system functions	(8, 9, 10)
O	R	Failure probability is zero;	(1, 1, 2)
	L	Failure is likely to occur once in the past 2 years;	(2, 3, 4)
	M	Failure probability is moderate (3–5 failures) in the last 2 years;	(4, 5, 6)
	H	Failure probability is high (6–8 failures) in the last 2 years;	(6, 7, 8)
	VH	Failure probability is extremely high (9–10 failures) in the past 2 years;	(8, 9, 10)
D	R	Failure indicated directly by the operator;	(1, 1, 2)
	L	Failure identified by the maintenance team during daily inspections;	(2, 3, 4)
	M	Failure identified by abnormal noises;	(4, 5, 6)
	H	Failure identified by the inspection team and cannot be performed by the operator;	(6, 7, 8)
	VH	Hidden fault, impossible to be identified by the operator or maintenance staff	(8, 9, 10)

3.4 AHP to Calculate the Risk Factors Weights

The classic AHP method is used to calculate the priority vector w_x, that is, the importance of each risk (S, O and D) in comparison with all other risks, defining the weight of the risk factors ($x = S, O, D$) of external corrosion failure. The objective is to transform the judgment consensus of experts (preference expressed by the group of engineers) in weight and then classify the risks in decreasing order of significance, according to the values of the decision-making group.

According to [31], the problem is divided into hierarchical levels, which facilitates its understanding and evaluation. The steps of the AHP method are: (i) construction of a hierarchical structuring model of criteria, (ii) construction of a pairwise comparison matrix in which the criteria (risk factors) are compared in square decision matrices using the fundamental scale developed by Saaty, (iii) standardization of the judgment matrix, (iv) summation of the judgment matrix line for obtain the relative priority of the criteria, (v) calculation of the normalized priority vector, (vi) derivation of the maximum eigenvalue and its eigenvector, and (vii) consistency check, by calculating the consistency index and, after the Consistency Ratio (CR). It is crucial for the resulting classification that the judgments be consistent, since the CR found must be less than 10% for the paired comparison matrix [32].

3.5 Algorithm of RPN Based on Risk Space-Diagram

Then, the risk priority number (RPN) that represents the measurement of systems with several S, O and D in the traditional FMEA is calculated. To avoid the dilution phenomenon, a risk space diagram is applied [8]. The risk space diagram (RSD) is used to calculate the α-level of S_i, O_i e D_i and the values on the right and left side of S_i, O_i e D_i calculated using the Zadehl extension principle [33] are expressed by the following equations [9]:

$$S_{iL}^{\alpha} = S_{iL} + \alpha(S_{iM} - S_{iL}) \tag{5}$$

$$S_{iR}^{\alpha} = S_{iR} - \alpha(S_{iR} - S_{iM}) \tag{6}$$

$$O_{iL}^{\alpha} = O_{iL} + \alpha(O_{iM} - O_{iL}) \tag{7}$$

$$O_{iR}^{\alpha} = O_{iR} - \alpha(O_{iR} - O_{iM}) \tag{8}$$

$$D_{iL}^{\alpha} = D_{iL} + \alpha(D_{iM} - D_{iL}) \tag{9}$$

$$D_{iR}^{\alpha} = D_{iR} - \alpha(D_{iR} - D_{iM}) \tag{10}$$

where S_{iL}^{α} e S_{iR}^{α} represent the value to the left and right of the S interval of the *i*th failure mode (system) by the α-level. $[O_{iL}^{\alpha}, O_{iR}^{\alpha}]$ e $[D_{iL}^{\alpha}, D_{iR}^{\alpha}]$ represent the range O e D, respectively. RSD was applied based on S, O and D at the α-level, where $0 < \alpha \leq 1$ shows the configuration and the interrelationships of the risk factors and RPN in a coordinate system [9]. As seen in [8] each failure mode has its RSD, the starting point O denotes the minimum index and the point of the vertex G denotes the maximum index, respectively, and the higher the RPN, the closer it is to the G point and the sequencing of RPN is more advanced. Subsequently, the RPN values on the left and right sides of each system are calculated using the weighted Euclidean distance formula, which is given by Eqs. (11) and (12) [9]. Considering the discrepancy in risk factors, weights (w_x) are indicated for S, O and D by the AHP method (indicated in Sect. 3.4) for calculating the RPN [8]. According to the Euclidean distance weighting, RPNi is expressed in a multiplicative weighted format, given by:

$$RPN_{iL}^{\alpha} = \sqrt{\sum_x w_x^2 \left(x_{iL}^{\alpha} - x_{imin}^{\alpha}\right)^2} / \sqrt{\sum_x w_x^2} \qquad (11)$$

$$RPN_{iR}^{\alpha} = \sqrt{\sum_x w_x^2 \left(x_{iR}^{\alpha} - x_{imin}^{\alpha}\right)^2} / \sqrt{\sum_x w_x^2} \qquad (12)$$

In the previous equations, $x_{i\,min}$ represents the minimum value of x_i, which is equal to 0. Finally, the centroid method is used to decide the critical system classification. Thus, the criticality analysis (RPN) of systems aimed to define a ranking to prioritize the systems that would be inspected, but in a complementary way it can serve as an input to determine the reliability of the systems.

4 Results

This section describes the results of applying the multicriteria decision making model to assess the criticality of systems considering external corrosion failures. It is challenging to make an accurate analysis of the risk factors for external corrosion of systems due to the lack of failure data and the various causes associated with external corrosion. Thus, the FMEA fuzzy model is employed to solve these issues. The scores of the team of multifunctional FMEA experts involved in the analysis are shown in Table 2.

Then, the values of the membership function of the risk factors deduced by Eqs. (2)–(4) are shown in Table 3.

In addition, the priority vector $w_x = [S, O, D]$ calculated for risk factors by the AHP method, considering the consensus of experts, was [0.5484, 0.2409, 0.2107], with an RC of 0.0261. Finally, from the weighted Euclidean Distance formula and

Table 2 Evaluation of each system by FMEA specialist

		Expert 1			Expert 2			Expert 3		
Id	System	S	O	D	S	O	D	S	O	D
1	Ceiling	L	H	R	L	H	L	H	L	H
2	Bulkhead	L	L	L	L	H	M	L	L	VH
3	Floor	H	M	H	M	R	M	H	L	VH
4	Pipes, valves and flanges	VH	VH	VH	VH	VH	H	VH	VH	H
5	Equipment	VH	VH	VH	VH	VH	VH	VH	VH	VH
6	Metallic structures	H	H	VH	VH	L	VH	H	H	VH
7	Stairs	H	VH	L	L	VH	H	H	H	VH
8	Supports	H	VH	H	H	VH	H	VH	VH	H
9	Body Guard	R	M	VH	VH	VH	H	H	H	L

Table 3 S, O and D membership functions for each system

Id	S	O	D
1	(3.333, 4.333, 5.333)	(4.667, 5.667, 6.667)	(3.000, 3.667, 4.667)
2	(2.000, 3.000, 4.000)	(3.333, 4.333, 5.333)	(4.667, 5.667, 6.667)
3	(5.333, 6.333, 7.333)	(2.333, 3.000, 4.000)	(6.000, 7.000, 8.000)
4	(8.000, 9.000, 10.000)	(8.000, 9.000, 10.000)	(6.667, 7.667, 8.667)
5	(8.000, 9.000, 10.000)	(8.000, 9.000, 10.000)	(8.000, 9.000, 10.000)
6	(6.667, 7.667, 8.667)	(4.667, 5.667, 6.667)	(8.000, 9.000, 10.000)
7	(4.667, 5.667, 6.667)	(7.333, 8.333, 9.333)	(5.333, 6.333, 7.333)
8	(6.667, 7.667, 8.667)	(8.000, 9.000, 10.000)	(6.000, 7.000, 8.000)
9	(5.000, 5.667, 6.667)	(6.000, 7.000, 8.000)	(5.333, 6.3337.333)

centroid defuzzified method, the RPN centroid values and criticality ranking of each system calculated using Eqs. (5–10), (11) and (12) are expressed in Table 4.

Table 4 shows that the systems have the following criticality order: Equipment = Piping, valves and flanges > Support > Structures > Stairs > Guardrails > Floor = Bulkhead = Ceiling. In this sense, this step seeks to contribute to the choice of the system to be inspected, considering the impacts of the failure generated by corrosion (S), the probability of such failure (O) and the difficulty of detecting this failure (D). This step seeks to address some of the problems of FMEA. Among them, the

Table 4 Fuzzy RPN and criticality ranking

Id	Alfa = 0		Alfa = 0,5		Alfa = 1		Centroid	Critical ranking
	RPNiL	RPNiR	RPNiL	RPNiR	RPNiL	RPNiR		
1	3.7427	5.7044	4.2269	5.2080	4.7124	4.7124	4.7178	8
2	2.7021	4.6462	3.1817	4.1552	3.6667	3.6667	3.6698	9
3	4.7342	6.6117	5.1889	6.1266	5.6442	5.6442	5.6583	7
4	7.9122	9.9107	8.4118	9.4110	8.9114	8.9114	8.9114	2
5	8.0000	10.0000	8.5000	9.5000	9.0000	9.0000	9.0000	1
6	6.2742	8.2539	6.7680	7.7580	7.2626	7.2626	7.2632	4
7	5.6033	7.5698	6.0929	7.0764	6.5840	6.5840	6.5851	5
8	7.0305	9.0238	7.5285	8.5252	8.0267	8.0267	8.0269	3
9	5.3281	7.1203	5.7261	6.6222	6.1244	6.1244	6.1742	6

difficulty of a specialized maintenance team (e.g., inspectors specialized in corrosion) to carry out a "*direct*" and exact assessment of the (intangible) quantities of the S, O and D attributes [11]. From the engineering perspective, we seek to provide a tool based on the RCM methodology and perfected by the combination of fuzzy multicriteria approach to generate a ranking of systems to be inspected. Therefore, this FMEA model combined with expert knowledge and experience can save time, cost and fieldwork.

5 Conclusions

From an academic point of view, a methodology was developed to assess the criticality of systems considering levels of severity, probability of occurrence and difficulty in detecting corrosion. As in [34], this approach was defined through a mixed research method with several sources of information: empirical knowledge of specialists, theoretical knowledge of scientific articles and international standards, using RCM concepts, MCDM and fuzzy logic, and proved to be still be a fertile area for research. Finally, this research highlights the importance of integrating RCM for corrosion management, considering gains related to quality, safety, technical efficiency, inspection processes [35].

From a practical point of view, this article highlights the following contributions: (i) definition of a methodology to assess the criticality of systems due to external corrosion failure in industrial plants; and (ii) combination of fuzzy logic and MCDM approach (e.g. AHP) at FMEA. As a sequence of this work, adjustments and increments in the proposed methodology are foreseen with the development of new evaluation criteria, arising from factors influencing corrosion based on the field experience of inspectors of the contractor, in standards and in the literature. Testing rounds are also planned, through interviews and a focus group with inspectors to validate the knowledge base involved and calibrate the models.

References

1. Mobley, T.: An introduction to predictive maintenance, (2002).
2. Nascimento, D.L.D.M., Sotelino, E.D., Lara, T.P.S., Caiado, R.G.G., Ivson, P.: Constructability in industrial plants construction: a BIM-Lean approach using the Digital Obeya Room framework. J. Civ. Eng. Manag. 23, (2017). https://doi.org/10.3846/13923730.2017.1385521.
3. Muniz, M.V.P., Lima, G.B.A., Caiado, R.G.G., Quelhas, O.L.G.: Bow tie to improve risk management of natural gas pipelines. Process Saf. Prog. 37, (2018). https://doi.org/10.1002/prs.11901.
4. Gerhardus H. Koch, Brongers, M.P.H., Virmani, Y.P., Payer, J.H.: Corrosive costs and preventive strategies in the US. Int. J. Commun. Syst. (2002). https://doi.org/10.1002/dac.3772.
5. CEN: Risk-based inspection and maintenance procedures for European industry, (2008).
6. Majid, M.A.A., Muhammad, M., Yem, N.I.Y.: RCM Analysis of Process Equipment: A Case Study on Heat Exchangers, (2011).
7. Uchida, S., Okada, H., Naitoh, M., Kojima, M., Kikura, H., Liste, D.H.: Improvement of plant reliability based on combination of prediction and inspection of wall thinning due to FAC. Nucl. Eng. Des. 337, 84–95 (2018). https://doi.org/10.1016/j.nucengdes.2018.06.007.
8. Yang, Z., Xu, B., Chen, F., Hao, Q., Zhu, X., Jia, Y.: A new failure mode and effects analysis model of CNC machine tool using fuzzy theory. 2010 IEEE Int. Conf. Inf. Autom. ICIA 2010. 582–587 (2010). https://doi.org/10.1109/ICINFA.2010.5512403.
9. Gupta, G., Mishra, R.P.: A Failure Mode Effect and Criticality Analysis of Conventional Milling Machine Using Fuzzy Logic: Case Study of RCM. Qual. Reliab. Eng. Int. 33, 347–356 (2017). https://doi.org/10.1002/qre.2011.
10. Ben-Daya, M., Raulf, A.: A revised failure mode and effects analysis model. Int. J. Qual. Reliab. Manag. 13, 43–47 (1996).
11. Braglia, M., Frosolini, M., Montanari, R.: Fuzzy TOPSIS Approach for Failure Mode, Effects and Criticality Analysis. Qual. Reliab. Eng. Int. 19, 425–443 (2003). https://doi.org/10.1002/qre.528.
12. Bowles, J.B., Peláez, C.E.: Fuzzy logic prioritization of failures in a system failure mode, effects and criticality analysis. Reliab. Eng. Syst. Saf. 50, 203–213 (1995). https://doi.org/10.1016/0951-8320(95)00068-D.
13. Adams, J., Parlikad, A., Amadi-Echendu, J.: A Bibliographic Review of Trends in the Application of 'Criticality' Towards the Management of Engineered Assets. In: Asset Intelligence through Integration and Interoperability and Contemporary Vibration Engineering Technologies, Lecture Notes in Mechanical Engineering. pp. 11–21 (2019).
14. Alvi, A., Labib, A.: Selecting next-generation manufacturing paradigms—an analytic hierarchy process based criticality analysis. In: Proc Inst Mech Eng: Part B: J Eng Manuf. pp. 1773–1786 (2001).
15. Saboya, G.L., Quelhas, O.L.G., Caiado, R.G.G., França, S.L.B., Meiriño, M.J.: Monte Carlo Simulation for Planning and Decisions Making in Transmission Project of Electricity. IEEE Lat. Am. Trans. 15, (2017). https://doi.org/10.1109/TLA.2017.7867172.
16. Ahmadi, A., Gupta, S., Karim, R., Kumar, U.: Selection of maintenance strategy for aircraft systems using multi-criteria decision making methodologies. Int J Reliab Qual Saf Eng. 17, 223–243 (2010).
17. Marriott, B., Garza-Reyes, J.A., Soriano-Meier, H., Antony, J.: An integrated methodology to prioritise improvement initiatives in low volume-high integrity product manufacturing organisations. . J Manuf Technol Manag. 24, 197–217 (2013).

18. Awad, M., As'ad, R.A.: Reliability centered maintenance actions prioritization using fuzzy inference systems. J. Qual. Maint. Eng. 22, 433–452 (2016). https://doi.org/10.1108/JQME-07-2015-0029.

19. Gupta, G., Mishra, R.P., Singhvi, P.: An Application of Reliability Centered Maintenance Using RPN Mean and Range on Conventional Lathe Machine. Int. J. Reliab. Qual. Saf. Eng. 23, (2016). https://doi.org/10.1142/S0218539316400106.

20. Marchiori, G., Formentin, F., Rampini, F.: Reliability-centered maintenance for ground-based large optical telescopes and radio antenna arrays. Ground-based Airborne Telesc. V. 9145, 91453M (2014). https://doi.org/10.1117/12.2057593.

21. Mishra, M., Keshavarzzadeh, V., Noshadravan, A.: Reliability-based lifecycle management for corroding pipelines. Struct. Saf. 76, 1–14 (2019). https://doi.org/10.1016/j.strusafe.2018.06.007.

22. Dawson, J.L., John, G., Oliver, K.: Management of corrosion in the oil and gas industry. Shreir's Corros. 3230–269 (2010). https://doi.org/10.1016/B978-044452787-5.00168-2.

23. Britton, C.F.: Corrosion monitoring and inspection. Shreir's Corros. 3117–3166 (2010). https://doi.org/10.1016/B978-044452787-5.00130-X.

24. Dawson, J.L.: Corrosion management overview. Shreir's Corros. 3001–3039 (2010). https://doi.org/10.1016/B978-044452787-5.00127-X.

25. Ceryno, P., Scavarda, L.F., Klingebiel, K. Supply chain risk: empirical research in the automotive industry. Journal of Risk Research (Print) 18, 1145–1164 (2015).

26. Ceryno, P., Scavarda, L.F., Klingebiel, K., Yuzbulec, G. Supply Chain Risk Management: A Content Analysis Approach. International Journal of Industrial Engineering and Management, 4, 141–150 (2013).

27. Clark, G.: A Review of Australian and New Zealand Investigations on Aeronautical Fatigue During the Period April 2005 to March 2007, (2017).

28. Ferreira, F.A.L., Scavarda, L.F., Ceryno, P.S., Leiras, A. Supply chain risk analysis: a ship-building industry case. International Journal of Logistics-Research and Applications, 21, 542–556 (2018).

29. Goossens, A., RJI, B., Van, D.L.: Exploring the use of the Analytic Hierarchy Process for maintenance policy selection. In: Safety, reliability and risk analysis: beyond the horizon—proceedings of the European safety and reliability conference, ESREL 2013. pp. 1027–1032. (2014).

30. Caiado, R.G.G., Scavarda, L.F., Gavião, L.O., Ivson, P., Nascimento, D.L. de M., Garza-Reyes, J.A.: A fuzzy rule-based industry 4.0 maturity model for operations and supply chain management. Int. J. Prod. Econ. 231, (2021). https://doi.org/10.1016/j.ijpe.2020.107883.

31. Saaty, T.L.: How to make a decision: the analytic hierarchy process. Eur. J. Oper. Res. 48, 9–26 (1990).

32. Marhavilas, P.K., Filippidis, M., Koulinas, G.K., Koulouriotis, D.E.: The integration of HAZOP study with risk-matrix and the analytical-hierarchy process for identifying critical control-points and prioritizing risks in industry – A case study. J. Loss Prev. Process Ind. 62, 103981 (2019). https://doi.org/10.1016/j.jlp.2019.103981.

33. Zadeh, L.A.: Fuzzy sets. Inf. Control. 8, 338–353 (1965).

34. Azevedo, B.D., Scavarda, L.F., Caiado, R.G.G., Fuss, M.: Improving urban household solid waste management in developing countries based on the German experience. Waste Manag. (2020). https://doi.org/10.1016/j.wasman.2020.11.001.

35. Caiado R.G.G., Neves H.F., Corseuil E.T., Bacoccoli L., de Castro A.R.P.: Development of Indicators for Monitoring the Regulatory Compliance of Static Equipment in Industrial Plants—an Empirical Study in the Oil and Gas Sector. IJCIEOM 2020. Springer Proceedings in Mathematics & Statistics, vol 337 (2020).

How Much Do the Objectives and KPIS of Maintenance Cover Sustainability Dimensions?

João Felipe Fraga Bianchi, Helder Gomes Costa ⓘ **,
and Leonardo Antonio Monteiro Pessôa** ⓘ

Abstract This research points out which of the Key Performance Indicators (KPIs) and objectives commonly applied to measure maintenance cover the three dimensions of triple bottom line (TBL). First, it was performed literature review focusing in maintenance KPIs that have been used to evaluate the performance of the maintenance and the maintenance objectives related to them, linking the results to the sustainability dimensions. In a second step, a survey was conducted to identify relationship among sustainability dimensions and maintenance objectives and KPIs. After all, the data collected were analyzed leading to the conclusion of the research. It was evidenced the unbalance of maintenance KPIs in covering the three dimensions of sustainability, once the KPIs and objectives analyzed mainly cover the economic dimension of sustainability which implies in a scarcity of such indicators in covering environmental and social aspects of maintenance operations. No weights were assigned to the maintenance KPIs in order to help managers in driving their efforts. It is suggested the development of maintenance KPIs, in order to cover aspects linked to companies' image, such as responsibility to customers. This research is inserted within an evident gap on the scientific literature: the scarcity of researches that link maintenance to sustainability. It contributes in fulfilling this gap, by mapping how much do the maintenance KPIs cover the sustainability dimensions. The results should be useful for managers and researchers, supporting them in reviewing maintenance indicators in order to produce more operations more sustainable.

Keywords Maintenance · Sustainability · Key performance indicators

J. F. F. Bianchi (✉) · H. G. Costa · L. A. M. Pessôa
Universidade Federal Fluminense, Rua Passo da Patria 156, Niterói, Rio de Janeiro, Brazil

L. A. M. Pessôa
e-mail: lampessoa@gmail.com.br

L. A. M. Pessôa
Centro de Análises de Sistemas Navais, Ed. 23 do, AMRJ – Ilha das Cobras, Rio de Janeiro, Rio de Janeiro, Brazil

167

A. M. Tavares Thomé et al. (eds.), *Industrial Engineering and Operations Management*,
Springer Proceedings in Mathematics & Statistics 367,
https://doi.org/10.1007/978-3-030-78570-3_13

1   Introduction

Maintenance can enhance productivity, quality, health and safety [1]. So, any framework designed to measure maintenance performance should encompass results in different areas [2–5]. In this scenario, sets of KPIs related to different aspects of maintenance have been developed recently in [6–9]. As sustainability objectives rise as strategical for organizations [10], companies should know how their maintenance process impact the TBL, as discussed in [11–16]. But, with a few exceptions such as [17, 18], sustainability is rarely mentioned in researches about maintenance. This paper contributes to fulfill his gab by highlight the relationship between TBL and the main KPIs mentioned in previous researches and also those used by companies. It discovered the KPIs most suitable to measure the accomplishment of each maintenance objective and which maintenance objective are related to each sustainability dimension.

2   Methodology

As the space for this paper is constrained, we focused in describing the main results instead of providing details about methodology. We highlight that literature review is a part of the methodology, once its results are used as input in the other research steps as it occurs in [19].

2.1   Literature Review

The steps for the literature review were inspired in [20–24]. First of all, it was done a search using the terms "maintenance" and "performance" and "indicators" or "KPI" or "metrics" into Scopus and Web of Science (WOS). The choice of these sources aims to reduce the probability of using "grey science" as mentioned in [25], or even paper published in predatory journals without a qualified peer review process. As the terms used for the query are not specific for the maintenance context, it was necessary to screen the articles removing those not related to industrial maintenance. The full-text of the remaining articles were read, searching for any KPIs used to measure performance of maintenance. A full list of KPIs used on these articles was built and from this step it was possible to perform quantitative analysis on the amount of KPIs used for each maintenance objective and sustainability dimension.

2.2 Exploratory Survey

Aside the literature review, it was explored the perception of maintenance professionals about the TBL. In order to elucidate the main goals and priorities of professional's companies, maintenance experts answered a survey about their main maintenance objectives, and which KPIs are used for each sustainability dimension. The invitation of maintenance professionals to answer the questionnaire occurred over a professional social network. A link to the questions was shared on several groups regarding maintenance. Sustainability groups were also chosen, because professionals within those groups may contribute with ideas and insights different than those related only to maintenance. The next step was the comparison between the answers from the survey and the results from the literature review. The questions about maintenance objectives and KPIs were open-ended, so it was necessary to refine and sort the answers in order to help the integration with the results from literature review.

2.3 Survey for Mapping Maintenance KPIs Linked to Sustainability

After selecting the main maintenance KPIs and objectives from the previous steps, a second survey was performed. This time the aim was to stablish what KPIs were adequate to measure de accomplishment of each objective, and assess which objectives were related to each sustainability dimension. It was announced on the same groups of the first survey, and an e-mail was sent to the previous respondents.

3 Results

3.1 Literature Review

The queries performed on the Scopus and Web of Science returned a initial list consisting of 1237 articles, as it appears in Fig. 1. After processing filtering steps, it resulted in a set composed by 77 articles for deeper analysis (see Fig. 2). As a result, some prominent topics were found and aggregated into categories. It follows some comments about the main finds in this step.

- The large number of papers modifying or creating KPIs, indicates that there is still a lack for the creation or improvement of existing metrics.
- The Overall Equipment Effectiveness (OEE) is the most mentioned indicator, as it is often modified.

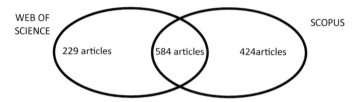

Fig. 1 Number of articles mapped in the scientific bases

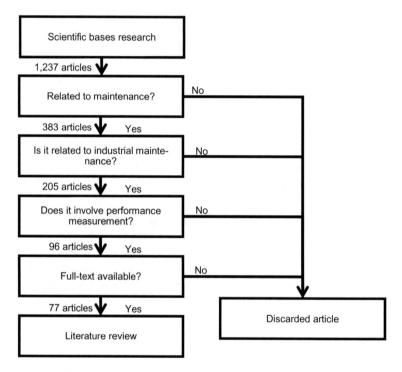

Fig. 2 Articles filtering process

- A rising topic was the comparison between different maintenance strategies to discover which of maintenance strategies performs better; or, even benchmark, among companies taking KPIs indicators for evaluation.
- In the case studies, KPIs were most applied to assess the results of actions designed to improve systems performance.
- The prediction or estimation of indicators were also an important topic within the literature, enabling the use of KPIs as inputs on conditions based maintenance systems.
- A few papers have explored the tradeoffs between different maintenance KPIs, such as the rate of unexpected failures and the remaining useful life of components.

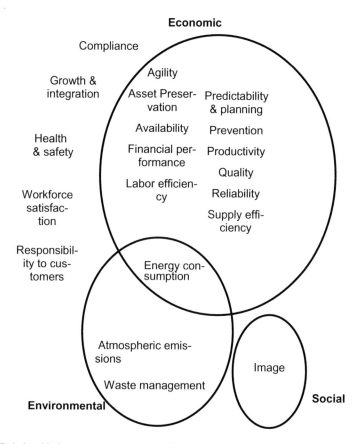

Fig. 3 Relationship between maintenance o objectives and sustainability dimensions

Some objectives and aspects are repeated on each dimension. It happens because such objectives increase overall maintenance performance, leading to better performance on three dimensions of sustainability. It should lead to a false impression of balance between the three dimensions of sustainability (Fig. 3).

3.2 Exploratory Survey

The previous section has risen a doubt about the importance attributed by maintenance to sustainability dimensions. In order to elucidate this questioning, a survey was applied to maintenance experts colleting their perceptions about how much maintenance KPIs cover sustainability dimensions. Such survey returned 92 answers, from people spread in 23 countries in Africa, Asia, Europe, Latin America, Middle East and North America.

The professionals were also asked if their companies where they have worked used indicators related to each sustainability dimension and which KPIs were used. The results demonstrate that 68.5% of the respondents could affirm that economic KPIs were used in their companies, while for the environmental and social dimensions the results were 42% and 30%, respectively.

Table 1 (See Appendix) shows the KPIs mentioned on the survey and those most found out in the literature, both categorized according to the maintenance objectives and dimensions. Despite the differences on objective selection, all the objectives from literature had at least one related KPI mentioned on the survey. But still there are some gaps on the literature and companies' KPIs. The objectives that would benefit the most for having more KPIs are asset preservation, atmospheric emissions, energy consumption and workforce satisfaction.

3.3   Survey for Mapping Maintenance KPIs Linked to Sustainability

A second questionnaire composed by two questions was applied to elucidate the links between sustainability dimensions, maintenance objectives and KPIs. The first question asked the respondents to check if there was a relationship between each objective-dimension pair. Table 2 demonstrates the percentage of positive answers, while Fig. 1 illustrates which objectives were considered significantly related to each sustainability dimension, assuming a level of significance of 5%.

The results in Fig. 1 show that most of the maintenance objectives relates exclusively to the economic dimension. Also, the only objected that have been considered for multiple dimensions were energy consumption, which relates to the economic and environmental dimensions at the same time. Besides, five objectives do not relate to any of the dimensions. Such results evidence that professionals are less prone to link objectives to dimensions. These results were unexpected and show that professionals tend to disregard the relation between objectives and dimensions when the impacts from achieving the objectives are indirect.

The second question of the survey aimed to assess the most adequate KPIs to measure the performance for each objective. After analyzing the answers, it resulted in 25 indicators that need to be taken into account when measuring the sustainability of the maintenance process: 22 of these KPIs are related exclusively to the economic dimension. 2 are related to the environmental dimensions and one is related to both of them. We noticed that there are no indicators for the social dimension. Table 3 presents the indicators and objectives considered for each dimension after the statistical analysis.

In general, the results in Table 2 demonstrate a higher level of prioritization of the economic dimension if compared to the results from the first survey or from the literature review. Part of this difference is due to the choice of objectives related to sustainability dimensions. All the five objectives that were not related to any of

Table 1 Main KPIs in literature and companies

Dimension	Objective	KPIs from literature	KPIs from companies
Economic	Agility	Mean time to repair; time to answer; average delay on jobs; % of jobs on hold; % of jobs completed on time; % of failures not corrected on the first contact	Mean time to repair, mean response time, backlog aging
Economic	Asset preservation	Maintenance cost/replacement value	Asset health index, maintenance cost per asset cost, average component lifespan
Environmental	Atmospheric emissions	Greenhouse gas emissions	Carbon footprint, plant pollutant emissions
Economic	Availability	Availability; OEE; unplanned downtime; downtime caused by spare parts delay	OEE, availability, total downtime
Environmental, economic and social	Compliance	Predicted requirements/Total requirements monitored by society; number of non-conformities	n° of environmental aspects and impacts
Environmental, economic	Energy consumption	Energy efficiency index	Water and electricity consumption per unit, fuel consumption, efficiency
Economic	Financial performance	Profit loss from failure; unexplored lifetime; Maintenance cost per unit; % of costs from corrective maintenance; % of costs from personnel; % of costs from spare parts; maintenance value; Indirect economy; total maintenance costs	Maintenance cost per unit, % of costs from workforce, % of costs from installations, % of cost from corrective maintenance, maintenance cost as % of Total Cost of Production, opportunity costs
Economic, Social	Growth and integration	Training hours per employee per year; improvement suggestions received in a year	Training hours, number of new initiatives

(continued)

Table 1 (continued)

Dimension	Objective	KPIs from literature	KPIs from companies
Environmental, economic and social	Health and safety	Mean time to respond emergencies; % of harmful residues; mean time to discover failures on security system; number of non-conformities on health and safety standards; potentially harmful failures; insurance costs; % of critically analyzed systems; loss of working hours due to accidents; n° of accidents/no of incidents	Criticality equipment analysis, pollution incidents, frequency of accidents
Environmental, economic and social	Image	Complaints from local communities; market share; awards	Number of social projects with involvement, frequency of community complaints
Economic and social	Labor efficiency	Manpower availability; % of extra hours; actual working hours/planned working hours; labor productivity; % of manpower from third-party solutions; actual working hours/available working hours	Man hours per order, % of extra-hours
Environmental, economic and social	Predictability and planning	Planned manpower/total available manpower; % of completed jobs	Cost adherence to planned, schedule completion, backlog size
Environmental, economic and social	Prevention	% of manpower allocated on corrective, preventive and proactive activities; preventive maintenance orders/total maintenance orders; % of failures on items that should have been inspected	% of reactive and preventive jobs

(continued)

Table 1 (continued)

Dimension	Objective	KPIs from literature	KPIs from companies
Economic	Productivity	OEE; produced volume; actual production/Maximum theoric production	OEE, total output, productivity loss
Economic	Quality	OEE; Success rate on work orders; rejected production/actual production; repetitive failures/total failures; % of work orders related to rework	OEE, % of orders from rework
Economic	Reliability	Mean time between failures or mean time to failure; failure frequency; % of equipment failing before the expected	MTBF, failure frequency, MTTR
Economic and social	Responsibility to customers	Recall events caused by maintenance; consumers complaint	Compliance with contract requirements, Stress level, satisfaction
Economic	Supply efficiency	Components unavailability; stock turnover; stock value/stock replacement value; supplier performance index; amount of emergency purchasing orders; utilization of items in contract	Stock turnover, stock value
Environmental, economic and social	Waste management	Scrap volume; % of recycled residues; % of incinerated residues; lubricant consumption	residues generation, recycling of wastes, savings from recycling initiatives
Economic and social	Workforce satisfaction	Absenteeism rate; employee satisfaction grades	Team stress level, satisfaction survey, turnover, absenteeism

sustainability dimensions were part of the social dimension on the initial analyses. Another noteworthy aspect about the objective selection is that most of objectives related to the economic dimension and a second or third dimension were then considered as exclusively regarded as an economic objective, the lack of KPIs related to the image objective has caused the final list of KPIs to ignore the social dimension completely. This scenario highlights the importance of developing indicators related to the image of companies in order to measure its impact on the social dimension.

Table 2 Relationship between objectives and dimensions

Objective	Code	Economic (%)	Environmental (%)	Social (%)
Agility	O1	86	7	24
Asset preservation	O2	86	34	14
Atmospheric emissions	O3	21	100	28
Availability	O4	79	14	24
Compliance	O5	52	55	41
Energy consumption	O6	72	69	21
Financial performance	O7	93	7	7
Growth and integration	O8	62	28	55
Health and safety	O9	62	55	52
Image	O10	38	28	72
Labor efficiency	O11	90	10	24
Predictability and planning	O12	97	28	17
Prevention	O13	83	52	34
Productivity	O14	90	21	17
Quality	O15	76	31	38
Reliability	O16	83	31	34
Responsibility to customers	O17	59	31	66
Supply efficiency	O18	83	31	17
Waste management	O19	24	100	28
Workforce satisfaction	O20	55	28	66

4 Conclusion

The results from the research demonstrated there is a lack of maintenance KPIs that covers the environmental and social dimensions of sustainability. Also, it was evidenced the unbalance of maintenance KPIs in covering the three dimensions of sustainability, once the KPIs and objectives analyzed mainly cover the economic dimension o sustainability implying in a scarcity of such indicators in covering environmental and social aspects of maintenance operations.

It is also noteworthy that maintenance professionals surveyed often disregarded the indirect impact that some objectives have on multiple sustainability dimensions. For this reason, objectives like health and safety, growth and integration, and compliance were not considered to impact any of the sustainability dimensions. Also, objectives like prevention and predictability and planning were considered related only to the economic dimension. A possible reason for this behavior is that achieving these objectives would lead to indirect impact on sustainability dimensions, rather than direct impact.

This research is into an evident gap: the scarcity of knowledge that link maintenance to sustainability. The results should support managers and researchers,

Table 3 List of maintenance KPIs linked sustainability

TBL Dimension	Objective	KPI
Economic	Agility	Average response time (beginning of repair)
	Reliability	Mean time between interventions
		Failure frequency
	Financial performance	Maintenance cost per unit
		Maintenance value
		Total maintenance costs
	Availability	Availability (Available time/Total time)
	Labor efficiency	Labor productivity
		Actual working hours/planned working hours
	Supply efficiency	Stock turnover
		Components unavailability
	Asset preservation	Actual lifespan/Projected lifespan
		Equipment lifespan/Warranty period
	Prevention	% of dedication to corrective, preventive and proactive maintenance
		% of failures on items that should have been inspected
	Predictability and planning	% of jobs completed
		Actual costs/Planned costs
		% of working hours used on unplanned orders
	Productivity	OEE
	Quality	% of work orders related to rework
		Rejected production/actual production
		Success rate on work orders
Environmental and Economic	Energy consumption	Energy efficiency
Environmental	Atmospheric emissions	Emission of poisonous pollutants
	Waste management	% of recycled residues
Social	Image	–

supporting them in reviewing maintenance indicators in order to produce operations more sustainable.

There is a lack of adequate maintenance indicators related to company image, which could imply in reducing organizations reputation. For this reason, it is recommended for future researches the development of KPIs able to measure the impacts of maintenance on companies' image. It may even be viable for this objective to incorporate others not related to sustainability dimensions, such as workforce satisfaction, responsibility to customers and Growth and integration.

The present work may serve also as a first step towards the practical application of the KPIs, in order to compare the performance of companies' maintenance systems. Other recommendations for future researches are the allocation of weights to the KPIs in order to identify which of KPIs are more relevant to companies.

Acknowledgements Conselho Nacional de Desenvolvimento Científico e Tecnológico Award Number: 314352/2018-0|Recipient: Helder Gomes Costa.

Coordenação de Aperfeiçoamento de Pessoal de Ensino Superior Award Number: 001|Recipient: Not applicable.

Appendix

See Table 1.

References

1. Parida A, Chattopadhyay G (2007) Development of a multi-criteria hierarchical framework for maintenance performance measurement (MPM). J Qual Maint Eng 13:241–258. https://doi.org/10.1108/13552510710780276
2. De Jong A, Smit K (2019) Collaboratives to improve industrial maintenance contract relationships. J Qual Maint Eng. https://doi.org/10.1108/JQME-07-2013-0050
3. Che-Ani AI, Ali R (2019) Facility management demand theory: Impact of proactive maintenance on corrective maintenance. J Facil Manag. https://doi.org/10.1108/JFM-09-2018-0057
4. Aju kumar VN, Gupta P, Gandhi OP (2019) Maintenance performance evaluation using an integrated approach of graph theory, ISM and matrix method. Int J Syst Assur Eng Manag. https://doi.org/10.1007/s13198-018-0753-6
5. Lai JHK, Man CS (2018) Performance indicators for facilities operation and maintenance (Part 2): Shortlisting through a focus group study. Facilities. https://doi.org/10.1108/F-08-2017-0076
6. Contri P, Kuzmina I, Elsing B (2012) Maintenance Optimization and Nuclear Power Plant Life Management—A Proposal for an Integrated Set of Maintenance Effectiveness Indicators. J Press Vessel Technol 134:031602. https://doi.org/10.1115/1.4005809
7. Muchiri P, Pintelon L, Gelders L, Martin H (2011) Development of maintenance function performance measurement framework and indicators. Int J Prod Econ 131:295–302. https://doi.org/10.1016/j.ijpe.2010.04.039

8. Raza T, Muhammad MB, Majid MAA (2016) A comprehensive framework and key performance indicators for maintenance performance measurement. ARPN J Eng Appl Sci 11:12146–12152

9. Van Horenbeek A, Pintelon L (2014) Development of a maintenance performance measurement framework-using the analytic network process (ANP) for maintenance performance indicator selection. Omega (United Kingdom) 42:33–46. https://doi.org/10.1016/j.omega.2013.02.006

10. Elkington J (2013) Enter the Triple Bottom Line. 23–38. https://doi.org/10.4324/978184977 3348-8

11. de Freitas JG, Costa HG (2017) Impacts of Lean Six Sigma over organizational sustainability. Int J Lean Six Sigma. https://doi.org/10.1108/ijlss-10-2015-0039

12. Amado dos Santos R, Méxas MP, Meiriño MJ, et al (2020) Criteria for assessing a sustainable hotel business. J Clean Prod 262. https://doi.org/10.1016/j.jclepro.2020.121347

13. Soares DAS da R, Oliva EC, Kubo EK de M, et al (2018) Organizational culture and sustainability in Brazilian electricity companies. RAUSP Manag J. https://doi.org/10.1108/RAUSP-07-2018-0038

14. Barata JFF, Quelhas OLG, Costa HG, et al (2014) Multi-criteria indicator for sustainability rating in suppliers of the oil and gas industries in Brazil. Sustain 6. https://doi.org/10.3390/su6 031107

15. Annarelli A, Battistella C, Nonino F (2020) A framework to evaluate the effects of organizational resilience on service quality. Sustain. https://doi.org/10.3390/su12030958

16. Ma J, Harstvedt JD, Jaradat R, Smith B (2020) Sustainability driven multi-criteria project portfolio selection under uncertain decision-making environment. Comput Ind Eng. https://doi.org/10.1016/j.cie.2019.106236

17. Sari E, Shaharoun AM, Ma'aram A, Mohd Yazid A (2015) Sustainable maintenance performance measures: A pilot survey in Malaysian automotive companies. Procedia CIRP 26:443–448. https://doi.org/10.1016/j.procir.2014.07.163

18. Sénéchal O (2017) Research directions for integrating the triple bottom line in maintenance dashboards. J Clean Prod 142:331–342. https://doi.org/10.1016/j.jclepro.2016.07.132

19. Ribeiro Paixão T, Pereira V, Costa HG, Fernandes Costa T (2017) Mapping perceptions about the influence of critical success factors in BPM initiatives. In: Sixth International Conference on Advances in Social Science, Management and Human Behaviour - SMHB 2017. Institute of Research Engineers and Doctors, LLC, Rome, pp 22–25

20. Pereira V, Costa HG (2015) A literature review on lot size with quantity discounts: 1995–2013. J. Model. Manag. 10:341–359

21. Méxas MP, Quelhas OLG, Costa HG (2012) Prioritization criteria for enterprise resource planning systems selection for civil construction companies: A multicriteria approach. Can. J. Civ. Eng.

22. De Carvalho Pereira F, Verocai HD, Cordeiro VR, et al (2015) Bibliometric analysis of information systems related to innovation. In: Procedia Computer Science

23. Méxas MP, Quelhas OLG, Costa HG, De Jesus Lameira V (2013) A set of criteria for selection of enterprise resource planning (ERP). Int J Enterp Inf Syst 9:44–69. https://doi.org/10.4018/jeis.2013040103

24. Moher D, Liberati A, Tetzlaff J, Altman DG (2009) Preferred Reporting Items for Systematic Reviews and Meta-Analyses: The PRISMA Statement. PLoS Med 6:e1000097. https://doi.org/10.1371/journal.pmed.1000097

25. Da Silva GB, Gomes Costa H (2015) Mapeamento de um núcleo de partida de referências em Data Mining a partir de periódicos publicados no Brasil. Gest e Prod 22:107–118. https://doi.org/10.1590/0104-530X792-13

Circular Economy and Supply Chain Management: Publications and Main Themes

Priscilla Cristina Cabral Ribeiro⬤, **Luiza Dantas Carpilovsky, and Carlos Francisco Simões Gomes**

Abstract This paper aims to analyze the publications about Circular Economy and Supply Chain Management. Supply Chain Management is a broad stream of research, and one of its themes is the Circular Economy. It becomes a relevant field of study in the last ten years, therefore some countries, journals, and authors, which have been exploring the Sustainability and SCM streams of research, are motivated to present their results in publications. Aiming to analyze these articles, a bibliometric research was used, firstly searching papers in Scopus and Web of Science databases. The collected data were analyzed and showed articles per country, per journal, per author, and year. The word clouds present the keywords and gaps prominent in the papers selected after some filters. Even though there are no concentration in the set of papers, the authors discovered that England was the country with more studies, as well as the Journal of Cleaner Production was the journal with the greatest number of publications. Similarly, in the authors' list, there is no concentration. When the authors compare the number of papers published in the period (2018–2020), seems that 2020 will overcome 2019. The main gaps are in Sustainability, Chains, and Research. The theme that is closer to the Circular Economy and Supply Chain, which also has many authors working at, is the Implementation of Circular Economy and Sustainability. Besides, in this subtheme the authors noticed that food sector and industry 4.0 received more attention, as sector and theme, respectively.

Keywords Circular economy · Supply chain management · Bibliometric studies

1 Introduction

Supply Chain Management (SCM) and Circular Economy (CE) represent two relevant concepts and, because of this, they are gaining a huge importance nowadays. The concept of SCM appears for the first time in literature in the mid-1980s. Thereby, [1] state that the interest in this concept is increasing since then, because the companies

P. C. C. Ribeiro (✉) · L. D. Carpilovsky · C. F. S. Gomes
Fluminense Federal University, Niterói, RJ 24210240, Brazil
e-mail: priscillaribeiro@id.uff.br

© The Author(s), under exclusive license to Springer Nature Switzerland AG 2021
A. M. Tavares Thomé et al. (eds.), *Industrial Engineering and Operations Management*,
Springer Proceedings in Mathematics & Statistics 367,
https://doi.org/10.1007/978-3-030-78570-3_14

noticed the benefits of the cooperative relationship inside and outside their organizations. Similarly, CE is one of the most discussed subjects, as it is responsible for impacting the supply chain in many good ways [2]. As an example, in food sector, the public perception of environmental impacts caused by processing and production often diverges from the facts revealed by scientific studies. The environmental management of supply chains can solve this problem by using necessary data and quality information supplied by research, not only as a basis for their decision-making processes, but also for marketing purposes [3].

Furthermore, a bibliometric analysis consists in a commonly method used to identify the development of a subject. Karanatsiou et al. [4] understand this concept as a statistical analysis, considering, for example, journals, articles, and authors. Albort-Morant et al. [5] state that through bibliometric it is possible to guide the research domain, once it gives shape, structure, and direction. Moreover, one of the advantages of the bibliometric is that it is possible to study a research area, based on the analysis of citations, geographical distribution, word frequency, and achieve decisive and useful conclusions [6].

As CE is receiving increasing attention over the years, many companies are aiming to improve their supply chain, through a sustainable approach [7]. In this view, it is crucial to analyze the implementation of CE and Sustainability. Barbaritano et al. [8] state that applying those concepts could bring economic, financial, and environmental improvements. Kayikci et al. [9] also affirm that this would be a solution to the waste problem within industries. Thereby, although there are some barriers that companies will have to face, implementing CE and sustainability will contribute in many significant ways.

The aim of this paper is to analyze the publications about Circular Economy and Supply Chain Management. The paper is organized as follows. Section 2 presents the Methodology followed by an analysis of the publications about the SCM and CE, a study of keywords and gaps word clouds, and an overview about the main theme related to the researchers interests and what was shown in the word clouds. The paper concludes in Sect. 4, with the theoretical and managerial implications.

2   Methodology

This paper was based on [10, 11] and aims to collect data for the analysis, Its research was done in two main databases: Scopus and Web of Science. As there was a large volume of data, specific filters were applied for each one of the databases, such as year, keywords, and categories. The articles were collected until July 3rd, 2020, in order that the ones that were published after this date are not in this analyze.

To begin with, the following words were searched at Scopus: "circular", "economy", "supply", "chain", and between them the connective "AND" was written, therefore, showing 577 results. After decided to narrow it by the years of 2018, 2019 and 2020, 432 documents were left. Another essential filter is the "Subject Area". In this way, the consecutives ones were chosen: "Environmental Science", "Engineering", "Decision Science" and "Economics, Econometrics and Finance", totaling 358 publications. As the sample was still large, it was chosen some keywords: "Sustainability", "Supply Chain", "Sustainable Supply Chain", "Environment", "Literature Review", "Green Supply Chain", "Brazil" and "European Union". Consequently, it was shown 143 papers. Finally, the option "Article" was selected from "Document Type", remaining, then, 89.

Secondly, it was done the research at Web of Science. Searching: "circular", "economy", "supply", "chain", each one in a different row, 647 documents were shown. Differently from Scopus, the first filter applied was "Document Type", narrowing the sample by articles, totaling 461. Afterward, it was chosen some categories that would be helpful to limit the results. The selected ones were: "Environmental Sciences", "Environmental Studies", "Engineering Manufacturing", "Engineering Industrial", "Management", "Business", "Economics" and "Engineering Multidisciplinary". As a result, 261 papers were left. Then, the word "Sustainability" was searched, aiming to refine the sample, thus, 122 articles were collected from Web of Science.

It is critical to emphasize the fact that Scopus and Web of Sciences had 42 articles in common. As the results from Scopus were analyzed first, the ones of Web of Science that were repeated, did not took part in the analysis. The results of these steps are in the next section.

3 Results

3.1 The Publications' Analysis

This subsection analyzes the chosen publications, based on criteria mentioned on the previous one. Thereby, presenting the countries, journals, and authors with the greatest number of publications, as well as the number of articles per year.

The 205 collected articles are distributed in 52 different countries. The number of articles published in a country is a good indicator to measure the degree of commitment with the field of study. The seven most productive countries, between 2018 and 2020, are illustrated by Table 1. These main countries published 182 articles, which represent 88.78% of the total publications. England is the number one on the list, with 36 articles (17.56%), followed by Italy, with 31 publications (15.12%) and India with 25 articles (12.20%). Then, France and United States appear with 23 articles each (11.22%). Finally, Brazil and China with 22 publications, representing 10.73%. It is necessary to highlight the fact that some authors have written the same article, but are from different countries. Therefore, the total number of articles is over 205.

It is also important to analyze the journals where the articles were published to observe the ones that are the most influent in the field of circular economy and supply chain. In this perspective, 56 journals are responsible for the publication of 205 papers. The top 12 represents 74.15% of this total (Table 2). The most prominent

Table 1 Articles per country

Countries	Articles per country	Percentage (%)
England	36	17.56
Italy	31	15.12
India	25	12.20
France	23	11.22
United States	23	11.22
Brazil	22	10.73
China	22	10.73

Table 2 Articles per journal

Journals	Articles per journal	Percentage (%)
Journal of Cleaner Production	41	20.00
Sustainability	26	12.68
Resources, Conservation and Recycling	23	11.22
International Journal of Production Research	16	7.80
Production Planning & Control	13	6.34
Management Decision	7	3.41
Science of the Total Environment	7	3.41
Business Strategy and the Environment	6	2.93
Journal of Manufacturing Technology Management	4	1.95
Energy	3	1.46
Journal of Environmental Management	3	1.46
Management of Environmental Quality	3	1.46

one is "Journal of Cleaner Production" with 41 publications (20%) and impact factor of 7.246 (2019).

Furthermore, it was observed the number of publications per author. Although there are 676 authors, only 22 are responsible for writing more than two articles. These ones represent 21.46% of the total, with 72 publications. However, some of those articles were written by one or more authors, therefore accounted more than once. In this view, to correct this flaw, the papers that appeared a few times, were only counted once. Thus, among the 205 articles in total, the 22 authors who had published the most are responsible for 44 of them. Due to this result, it is evident that there is no concentration of publications by authors, from 2018 to 2020, as the authors with the greatest number of articles, published five papers.

Finally, Fig. 1 shows the annual publication frequency between 2018 and 2020. As presented, the number of publications is increasing exponentially over the years, pointing out that the field of circular economy and supply chain has been receiving more attention. In 2018, 46 articles (22.44%) were published, representing less than

Fig. 1 Articles published
per year

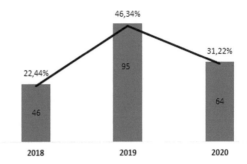

half of publications in 2019, which had 95 (46.34%) papers in total. Something has to be said, the data collected is referring until the first semester of 2020. The present year was responsible for publishing 64 articles (31.22%), showing a strong tendency to overcome the total of 2019.

The analysis of the 205 articles continues, aiming the better comprehension of the subject. To begin with, the years of 2019 and 2020 were selected, totaling 159 articles. This was done to observe which topics were explored the most and the ones that were not. Therefore, two word clouds were done, in order to have a better visualization.

The first word cloud was about the keywords. For its elaboration, all the ones present in the 159 articles were selected (Fig. 2). Based on its frequency, the word gains greater or lesser representation, which is reflected by its size. Thereby, analyzing the word cloud is a good way to observe the themes that were discussed more often. Some words have a clear connection with the theme, such as: "Sustainable" and "Environmental" and, for this reason, are expected to appear. Meanwhile,

Fig. 2 Keywords word
cloud

Fig. 3 Gaps word cloud

others are receiving increasing attention, for example: "Model", "Manufacturing" and "Food".

Secondly, it was done another word cloud, aiming to understand the gaps. The 159 articles were analyzed, to identify the gaps pointed out by them. After that, a table was developed with those information and, then, the word cloud was made (Fig. 3). The words "Sustainability", "Supply", "Chain", "Circular" and "Research" appeared with a higher frequency, as they are the main subject, in a generic way. However, other words that are also shown in the cloud, are sub-themes of those previously mentioned, and represent areas that were highlighted as necessary to be explored.

Finally, it is imperative to discuss the connection between both word clouds. The words that appeared prominently in the keywords cloud, must appear in reduced size, or not even appear in the cloud of gaps. This is due to the fact that if they have a greater representation in the keywords cloud, they correspond to themes and subjects that are most addressed in the articles, so they would not represent gaps. In addition, the words that appeared in the cloud of gaps are an important guide to observe the themes that need to be further explored.

The sample presented papers discussing Circular Economy, Sustainable Supply Chain and Sustainability. Based on the keywords/objectives, the main subthemes found in the 159 papers are: Circular supply chain and reverse logistic, Circular economy business model, Circular economy innovation and technologies, Implementation of circular economy and sustainability, and Transition to CE. Next section will discuss the main themes, Circular Economy and Supply Chain Management, and one of the subthemes that is closer to the main themes, Implementation of circular economy and sustainability.

3.2 Literature Review

Circular Economy. Since the creation of the Ellen MacArthur Foundation, in 2010, whose purpose consists in accelerate the transition from a linear economy to a Circular Economy (CE), this concept is becoming increasingly crucial and popular. In this view, according to [12], the CE is, currently, one of the most discussed terms among environmental economic scientists. As a matter of fact, CE emerges as a response to all the threats that the environment is facing, such as the amounts of waste, that are constantly increasing [13].

Jones and Comfort [14] believe that the concept of the CE is essential for building a more sustainable world and future. In this way, the analysis of this subject will begin with the contextualization of the linear economy. Garcés-Ayerbe et al. [15] affirm that the linear economy contributes to the deterioration of the environment, once that "goods are manufactured and then discarded as waste."

On the other hand, for [16], the CE is an economic system that replaces the "end of life", since it reduces, reuses in alternative ways, recycles and recovers materials in the production/distribution and consumption processes. Similarly, [17] believe that CE changes the perspective about the relationship between human society and nature.

As a reflection of its importance, [18] affirm that the largest nation on the planet, China, adopted the CE as its main structure for environmental changes and economic development over the next years. Therefore, [19] believe that CE is a promising concept, as it has been able to "attract the business community to sustainable development work".

Circular Economy and Supply Chain Management. Once the concept of Circular Economy (CE) was presented and the Supply Chain Management's is well known, it is also critical to analyze the link between them. Firstly, the discourse of circular economy is gaining space among companies, countries, and researchers [20]. In this way, [2] state that CE is being responsible for impacting the supply chain in many ways, transforming the production system, for example.

Gimenez and Sierra [21] affirm that one of the key challenges for companies is to develop strategies to achieve sustainability in the supply chain. Moreover, [22] point out that firms are increasingly trying to implement CE principles with the aim of gaining competitive advantage. As a consequence of the implementation of CE aspects in SCM, new concepts are emerging. In this context, Reverse Supply Chain Management has been developed as one of them.

Besides that, in the last decade, Closed-Loop Supply Chains (CLSC) has drawn attention in the subject of supply chain [23]. De Angelis et al. [24] explain that CSCL can be understood as the incorporation of CE principles in supply chains. In this regard, to create a CLSC system it is necessary to combine forward and reverse activities within supply chains [25]. In conclusion, all the concepts presented have a decisive relationship between them.

Implementation of Circular Economy and Sustainability. After grouping the main subjects into subthemes, it was possible to make a detailed analysis. Once all of them were observed, the chosen one was: "Implementation of Circular Economy and Sustainability". This is due to the fact that there was a great number of articles involving this theme, and it is responsible to connect two big areas of study. As it was shown in Fig. 3, those themes complement themselves, once "sustainability" and "circular" appeared prominently in the word cloud, while "implementation" do not have much visibility.

There were ten articles about the chosen subtheme. Once all of them were read, it was possible to observe some areas which have a greater importance, as it appeared frequently among the publications. In this way, in order to better explore them, the articles were divided by subjects.

Food Sector. Three out of the ten articles have approached the food sector, showing that this subject is gaining an increasingly importance.

Sehnem et al. [7] state that the food sector allows the implementation of many circular practices, such as using clean energy, water recycling and composting. And, according to the authors, those sustainable practices need to be incentivized because they directly affect the company's performance, as they are responsible to increase its competitive advantage.

One of the main problems of the food sector, highlighted by [9], is the food lost and waste. Its gravity is due to the fact that they cause social, environmental, and economic problems. In this view, the implementation of circular economy practices is a solution to the waste issue. As a practical example, it was observed, at a red meat sector company, that many byproducts, such as bones and skins, are discarded. However, this linear process can be converted into a circular one. Therefore, those wastes should be transformed into useful products that could be used as input for other sectors, and this would provide additional income for producers, for example. Finally, it is evident that applying circular economy concepts, with a sustainability approach, bring many advantages for companies.

Sharma et al. [26] also affirm the importance of circular economy implementation, as it helps companies to deliver products with higher quality and security in a most sustainable way. Furthermore, the authors point out that lack of technology and government policies are the main issues to adopt sustainability and circular economy. Firstly, advanced technology is essential to improve overall quality of the food supply chain and, also, to reduce food wastage. Secondly, governments have to be aware of their importance in this subject. This is because, they have to be supportive to organizations and farmers, through policies, in order to encourage the adoption of sustainable practices.

Industry 4.0. Another influential area among the ten articles was Industry 4.0. It is evident that this is a great ally in the implementation of Circular Economy and Sustainability.

According to [27], Industry 4.0 involves many technologies, as Internet of Things, Big Data and Artificial Intelligence, and all of them have a great importance to overcome barriers of Circular Economy implementation. This set of tools should be used to reduce waste and provide higher levels of productivity.

In this way, [28] also affirm that integrating technologies from Industry 4.0 and Circular Economy practices, could result in a solution of the waste problem, through the creation of a business model that reuses and recycles. Thereby, Industry 4.0 is an interesting path to set sustainable manufacturing. Finally, society will be impacted in the environmental, sociotechnical, and economic areas.

Other sectors. As the other four articles were about different sectors, they were grouped but still studied separately. Firstly, [29] analyzed the retail market and were able to explore the implementation of Circular Economy in this sector, as well as its barriers, opportunities, and challenges. Even though it is clear how necessary is applying Circular Economy concepts and Sustainability, many retail companies do not realize, and the ones that already implement, only do so for a small part of the firm. And this leads to the barriers, which make this implementation harder, such as: jobbers, plastic firms (products are wrapped in plastic film, consequently more difficult to recycle), secondary markets seen as threats. Finally, in order to explore Circular Economy, interesting options appear and should be applied, for example: remanufacturing, redesigning products and improving the reverse supply chains.

Secondly, [30] are responsible to study the manufacturing sector, especially among Small-Medium Enterprises (SME). To begin with, there are many companies that are implementing Circular Economy and encouraging their supply chain to do the same, though apply Circular Economy in SME is a slow process. In this way, three factors are associated with this implementation: financial advantage, material provision and resource reutilization. Thereby, to have a practical adoption, there is a requirement that is: have appropriate strategies, competence building and resource deployment.

Furthermore, [31] explored the cosmetics sector. They analyzed the Circular Economy practices of a company that is worried about the sustainability. This firm has already adopted recycling process and constantly is searching for actions that aim to minimize environmental degradation. It is worth to mention two significant aspects. The first one is that this company recognizes the importance of the innovation in this subject. Moreover, it is essential to invest in education along the value chain, in order to achieve better results in the cosmetics industry.

Finally, although Circular Economy concept is new within the luxury furniture sector, it is gaining increasingly attention and, for this reason, [8] studied this subject, aiming to investigate its implementation. The dimensions which encourage Circular Economy the most are environmental and economic. The first one is because adopting more sustainable practices will impact directly in the reduction of the environmental degradation. Besides this motivation, the economic dimension is essential, as the companies probably will have a great reduction of total costs, achieve growing sales, therefore, improving financial performances. In order to encourage companies to implement Circular Economy practices, Government should give fiscal and economic incentives. As it was already mentioned in other sectors, it is also recommended that schools and local institutions train the final market. This is due to the fact that once people value sustainable practices, they might prefer choosing companies with the same beliefs.

4   Conclusions

Once all the research was done, by analyzing the publications about SCM and CE, some conclusions appear. Firstly, it is observed that Europe is the continent which has the greatest number of articles, and this shows a big concern about CE and SCM subjects in this zone. Moreover, although emerging countries face many barriers to implement those concepts, there is a considerable number of publications by them.

Another interesting fact is that SCM and CE concepts are gaining increasingly importance. This can be noticed by the annual publication frequency. As it was analyzed the years of 2018, 2019 and 2020, it is clear that the number of articles published per year is rising significantly. Thereby, this is a promising field of research.

Afterwards, both word clouds provide essential information. The first one, about keywords. shows that besides the words which are expected to appear, for example: "Sustainable" and "Environmental", other words need to be highlighted as well. "Model" appears prominently as many companies are developing business model in basis of CE concepts. "Manufacturing" and "Food" are the main sectors within this theme. Secondly, the gaps word cloud points out areas which need to be explored. In this way, words such as: "implementation" and "technology" give crucial orientation about possible further researches.

Moreover, it was selected the subject of "Implementation of Circular Economy and Sustainability" to be explored better. This subtheme was chosen based on the word clouds results and its importance as connects two big areas of study. With this analysis, it was concluded that the Food Sector needs to be motivated to apply CE and sustainable concepts, because it is a great solution to the waste problem and the food loss. Thereby, it is essential that the companies and farmers are encouraged to adopt more sustainable approaches. Equally critical, the role of Industry 4.0 needs to be highlighted. Technologies from Industry 4.0 and CE practices should be integrated in order to help the transition from a linear economy to a circular one. So, companies need to invest in the technology sector, which include: Big Data, Internet of Things and Artificial Intelligence, for example.

In summary, the importance of CE and sustainability needs to be spread out, as those concepts provide benefits in many areas. Therefore, Governments should give economic and fiscal incentives, in order to encourage companies to adopt CE practices. Once those aspects are implemented and receive more attention among society, the environment will be much more sustainable, as it will have a reduction of its degradation.

References

1. Lummus, R.R., Vokurka, R.J.: Defining supply chain management: a historical perspective and practical guidelines. Industrial Management and Data Systems 99(1), 11–17 (1999).
2. Khan, S., Haleem, A., Khan, M.I.: Enablers to Implement Circular Initiatives in the Supply Chain: A Grey DEMATEL Method. Global Business Review (2020).
3. Ferreira, F.U., Robra, S., Ribeiro, P.C.C., Gomes, C.F.S., Almeida Neto, J.A., Rodrigues, L.B.: Towards a contribution to sustainable management of a dairy supply chain. Production 30, 1–13 (2020).
4. Karanatsiou, D., Misirlis, N., Vlachopoulou, M.: Bibliometrics and altmetrics literature review: Performance indicators and comparison analysis. Performance Measurement and Metrics 18(1), 16–27 (2017).
5. Albort-Morant, G., Henseler, J., Leal-Millán, A., Cepeda-Carrión, G.: Mapping the field: A bibliometric analysis of green innovation. Sustainability 9(6), 1011 (2017).
6. Liao, H., Tang, M., Luo, L., Li, C., Chiclana, F., Zeng, X.J.: A bibliometric analysis and visualization of medical big data research. Sustainability 10(1), 166 (2018).
7. Sehnem, S., Jabbour, C.J.C., Pereira, S.C.F., Jabbour, A.B.L.S.: Improving sustainable supply chains performance through operational excellence: circular economy approach. Resources, Conservation and Recycling 149, 236–248 (2019).
8. Barbaritano, M., Bravi, L., Savelli, E.: Sustainability and quality management in the Italian luxury furniture sector: A circular economy perspective. Sustainability 11(11), 3089 (2019).
9. Kayikci, Y., Ozbiltekin, M., Kazancoglu, Y.: Minimizing losses at red meat supply chain with circular and central slaughterhouse model. Journal of Enterprise Information Management 33(4), 791–816 (2019).
10. Gomes, C. F. S., Ribeiro, P. C. C., Freire, K. A. M.: Bibliometric research in Warehouse Management System from 2006 to 2016. In: 22nd World Multi-Conference on Systemics, Cybernetics and Informatics, pp. 200–204, WMSCI, Orlando (2018).
11. Oliveira, A. O., Oliveira, H. L. S.; Gomes, C. F. S.; Ribeiro, P. C. C.: Quantitative Analysis of RFID' Publications from 2006 to 2016. International Journal of Information Management 48, 185–192 (2019).
12. Geisendorf, S., Pietrulla, F.: The circular economy and circular economic concepts—a literature analysis and redefinition. Thunderbird International Business Review 60(5), 771–782 (2018).
13. Seroka-Stolka, O., Ociepa-Kubicka, A.: Green logistics and circular economy. Transportation Research Procedia 39, 471–479 (2019).
14. Jones, P., Comfort, D.: Towards the circular economy: A commentary on corporate approaches and challenges. Journal of Public Affairs 17(4), e1680 (2017).
15. Garcés-Ayerbe, C., Rivera-Torres, P., Suarez-Perales, I., Leyva-de la Hiz, D.I.: Is it possible to change from a linear to a circular economy? An overview of opportunities and barriers for european small and medium-sized enterprise companies. International Journal of Environmental Research and Public Health 16(5), 851 (2019).
16. Kirchherr, J., Reike, D., Hekkert, M.: Conceptualizing the circular economy: An analysis of 114 definitions. Resources, Conservation and Recycling 127, 221–232 (2017).
17. Prieto-Sandoval, V., Jaca, C., Ormazabal, M.: Towards a consensus on the circular economy. Journal of Cleaner Production 179, 605–615 (2018).
18. Murray, A., Skene, K., Haynes, K.: The circular economy: an interdisciplinary exploration of the concept and application in a global context. Journal of Business Ethics 140(3), 369–380 (2017).
19. Korhonen, J., Honkasalo, A., Seppälä, J.: Circular economy: the concept and its limitations. Ecological economics 143, 37–46 (2018).
20. Genovese, A., Acquaye, A.A., Figueroa, A., Koh, S.C.L.: Sustainable supply chain management and the transition towards a circular economy: Evidence and some applications. Omega 66, Part B 344–357 (2017).
21. Gimenez, C., Sierra, V.: Sustainable supply chains: Governance mechanisms to greening suppliers. Journal of business ethics 116(1), 189–203 (2013).

22. Fowler, S.J., Hope, C.: Incorporating sustainable business practices into company strategy. Business strategy and the Environment 16(1), 26–38 (2007).
23. Battini, D., Bogotaj, M., Choudhary, A.: Closed loop supply chain (CLSC): economics, modelling, management and control. International Journal of Production Economics 183, Part B 319–321 (2017).
24. De Angelis, R., Howard, M., Miemczyk, J.: Supply chain management and the circular economy: towards the circular supply chain. Production Planning & Control 29(6), 425–437 (2018).
25. Heydari, J., Govindan, K., Jafari, A.: Reverse and closed loop supply chain coordination by considering government role. Transportation Research Part D: Transport and Environment 52, Part A 379–398 (2017).
26. Sharma, Y.K., Mangla, S.K., Patil, P.P., Liu, S.: When challenges impede the process: for circular economy driven sustainability practices in food supply chain. Management Decision 57(4), 995–1017 (2019).
27. Pham, T.T., Kuo, T.C., Tseng, M.L., Tan, R.R., Tan, K., Ika, D.S., Lin, C.J.: Industry 4.0 to Accelerate the Circular Economy: A Case Study of Electric Scooter Sharing. Sustainability 11(23), 6661 (2019).
28. Nascimento, D.L.M., Alencastro, V., Quelhas, O.L.G., Caiado, R.G.G., Garza-Reyes, J.A., Lona, L.R., Tortorella, G.: Exploring Industry 4.0 technologies to enable circular economy practices in a manufacturing context. Journal of Manufacturing Technology Management 30(3), 607–627 (2019).
29. Frei, R., Jack, L., Krzyzaniak, S.A.: Sustainable reverse supply chains and circular economy in multichannel retail returns. Business Strategy and the Environment 26(5), 1925–1940 (2020).
30. Dey, P.K., Malesios, C., De, D., Budhwar, P., Chowdhury, S., Cheffi, W.: Circular economy to enhance sustainability of small and medium-sized enterprises. Business Strategy and the Environment 29(6), 2145–2169 (2020).
31. Sehnem, S., Pandolfi, A., Gomes, C.: Is sustainability a driver of the circular economy?. Social Responsibility Journal 16 (3), 329–346 (2019).

Framework for Implementation of Autonomous Maintenance with the HTO Approach

Paulo Cézar Loures, Luiz Felipe Scavarda, and Andréa Regina Nunes de Carvalho

Abstract Autonomous Maintenance (AM) is part of a maintenance strategy that focuses on the "man-machine" relationship to effectively carry out cleaning, lubrication, inspection and minor repairs. When properly implemented, AM can significantly improve productivity and quality, as well as reduce cost, therefore, it is an important area of operations management (OM). However, the industry has been challenged with numerous barriers to successfully implement AM and academia has done little to help industry in this regard. This paper addresses this current research-practice gap by offering as its main goal a framework for implementation of AM with the "Human, Technological and Organizational" (HTO) dimensions. It builds upon an action research conducted within a longitudinal study in a rolling mill production process. The adoption of the HTO approach by OM scholars has been successfully conducted in different cases and is well documented in the literature. However, this is the first research to use this approach within AM. The research findings indicate the adherence of the HTO dimensions to match the AM implementation challenges and reinforce the need for a holistic and combined perspective of these dimensions for business development. Practitioners can take stock of the lessons learnt within this action research and the offered framework to aid the implementation of AM in their operations.

Keywords Production systems · Total productive maintenance · Performance management · Action research · Rolling mill

P. C. Loures (✉) · L. F. Scavarda
Pontifical Catholic University of Rio de Janeiro (PUC-Rio), Marquês de São Vicente Street, 225 - Gávea, Rio de Janeiro, Rio de Janeiro 22451-900, Brazil
e-mail: paulo.loures@outlook.com

A. R. N. de Carvalho
Instituto Nacional de Tecnologia (INT), Av. Venezuela, 82 - Saúde, Rio de Janeiro, Rio de Janeiro 20081-312, Brazil

1 Introduction

Autonomous Maintenance (AM) is a concept that has been applied in production systems of several industries, being an important area of operations management (OM) [1]. AM's goal is to achieve an excellent degree of cleanliness, perfect lubrication and safe fastening to reduce machine breakdown and maintenance costs [2]. By practicing AM continuously, machines can operate at rated capacity, their breakdowns and interruptions occur rarely, which reduces capital expenditure on the machines [3]. Therefore, when properly implemented, AM can significantly improve the productivity, quality and cost reduction of a company's operations. However, companies have faced numerous obstacles to successfully implement AM [4, 5]. Moreover, academy has done little to help the industry in this direction as it has mainly focused on the implementation of the Total Productive Maintenance (TPM) [6]. This opens an important research-practice gap to be addressed by OM scholars.

Implementing AM is an organizational transformation that can affect its structures, management system and employee's responsibilities. It may increase employee motivation through a performance incentive system, enhance skills and result in a better information technology system [7]. Therefore, the "Human, Technological and Organizational" (HTO) dimensions should be addressed within an AM implementation process, as successfully applied by Carvalho et al. [8] within a similar context in the OM literature. The organizational dimension includes the vision of structures and processes as well as the actors roles and responsibilities within the corporation. The technological dimension contemplates both technologies and equipments of a productive system, including information systems, from local spreadsheets to integrated decision support systems that permeate several functional areas within the company [9]. A vision of organizational integration or a sophisticated information management system is of little use if the people involved have no vision of processes or knowledge of the numerous functionalities associated with information systems, showing the need for integration of the human dimension in the analysis. Therefore, the need for an adjustment and alignment between the HTO dimensions is essential, requiring a holistic perspective of the dimensions and not individual dimensions [10, 11].

This paper addresses the current research-practice gap by offering as its main goal a framework for implementation of AM with the HTO approach. The research findings build upon an action research conducted within a longitudinal study in a rolling mill production process. To achieve its goal, the paper is organized into five sections, being this first one the introduction. Sections 2 and 3 provide the theoretical foundations and research method, respectively. The results and discussions are presented in Sect. 4. The last section offers the authors conclusions and suggestions for future research.

2   Theoretical Foundation

This section provides the theoretical foundations for this research organized by first introducing AM and then the HTO approach.

2.1   Autonomous Maintenance (AM)

AM arose from the philosophy of TPM, which is a management model with a performance improvement approach to maintenance activities [12] that takes advantage of the skills and knowledge of people in an organization to accomplish maintenance effectiveness [13]. The aim of AM is to achieve an effective programme of cleaning, lubrication and equipment adjustments to inhibit deterioration and prevent machine breakdown [14]. AM is a strategy that focuses on the human-machine relationship to effectively carry out the following activities: cleaning, lubrication and fastening [2]. AM contemplates human capital development among operators and is supported by technicians and engineers to perform easy daily maintenance activities in addition to planned maintenance [15]. AM implementation ensures that each activity follow-up is critically analyzed before it is carried out and this reduces the chances and possibilities of ignoring important details in this process [16].

AM is not easy to implement due to the presence of many barriers [5]. Although, the reasons for the failure to carry out AM are discussed in research papers [4, 5], the literature does not demonstrate how to address such barriers neither mention alternatives to successfully achieve AM. Barriers within the business unit can cause a major setback to implementation [17]. For instance, there are people who see AM activities as additional work, as a threat to their work and even as an operational security regime [18].

Figure 1 presents an AM framework, adapted from Min et al. [15], which focuses on four improvements in the operational area: technical skills, attitude, culture and safety at work. The framework is developed as part of the operators' skills improvement programme providing a step-by-step guidance on how to train operators to maintain their own production equipment.

During Stage 1, the AM management and execution teams are formed. The management team is responsible for coordinating the work of the maintenance and production operators and focuses on process improvement regarding technology, automation and new techniques aimed at maintaining the equipment. The execution level team is responsible for equipment maintenance based on AM activities. During this stage, a maintenance schedule must be developed. Moreover, a pilot equipment for AM implementation is chosen. The criteria for this choice could be based on the frequency of deficiency or breakdown of the machine or equipment, according to Labib [19].

Stage 2 refers to safety training. Nordlöf et al. [20] highlights the importance of individual and collective responsibility for security. In this stage, enforcement-level

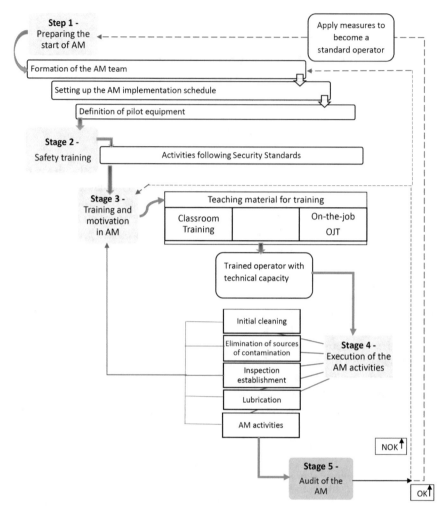

Fig. 1 AM framework (adapted from Min et al. [15])

personnel are aligned with the safety standards of the activities that will be undertaken within the AM. In the sequence, Stage 3 focuses on training and motivation in AM. Therefore, a skill training program is ministered in two phases. The first one is a classroom training with the participation of the coordination and execution levels. The second phase refers to the on the job training (OJT).

The execution of the AM occurs in Stage 4 and involves five main steps: (i) Initial cleaning (i.e., when the equipment stops working and the execution level staff—operators and technicians—carry out the general cleaning); (ii) Elimination of sources of contamination (i.e., identifying and correcting sources of contamination in hard-to-reach sites); (iii) Inspection establishment (i.e., monitoring and verifying the current condition of certain parts of the equipment); (iv) Setting standards and lubrication; and (v) AM activities (i.e., operation receives the daily maintenance plan and executes AM routine tasks).

Finally, Stage 5 refers to the auditing process where compliance with the cleaning, lubrication and inspection activities, carried out by the operator, is evaluated. The main objective of this stage is to enhance the practice of AM in search of continuous improvement. Top managers should be involved in this stage [15].

2.2 HTO Approach

The HTO concept originates from the safety work of the nuclear-air industry and refers to a multidisciplinary approach that can be applied in any business [21]. A holistic view of HTO is necessary for successful business development [8, 10]. The HTO approach has common points with the maintenance implementation process, as it involves all three dimensions. For instance, investments in prevention and strategic maintenance can lead to greater efficiency with fewer production stoppages, reduced costs, greater job satisfaction among employees, and reduced exposure to stress for employees [22].

Within the HTO concept, the "H" component focuses on the professional who aims to contribute to the business process [23]. This dimension can be described at a biological level (i.e., considering the human physiological system), at a knowledge level (i.e., when humans are considered information processing systems) and at a social level (i.e., individuals are members of social groups with different cultures, which will partly determine their values and habits) [24]. The technological aspect ("T") can be divided into two parts, one dedicated to maintaining the company's production capacity and another that consists of the information systems (hardware and software) that are used as decision support tools [23]. Ogén [22] defines that the organization's technical factors also include all technological factors existing in the organization's process, from input to output, including: tools, equipment and machines used in the production process; workflow design; technical procedures; and the technical knowledge of the organization's members. For information on main technologies focused for industry 4.0, please consult Caiado et al. [25]. Ogén [22] highlights that organizational aspects ("O") represent the formal elements of an organization that are developed to coordinate the behavior of employees and the different parts of the organization. These three components have been associated as enablers for different OM practices and applications [8, 10, 26, 27].

It is important to stress that these three dimensions do not work in isolation and that the interplay between them is important [10]. Berglund and Karltun [23] identify on the production scheduling process involving four companies that the technical ability to use IT systems, rapid thought processing and decision making are directly linked to human and technology dimensions. Moreover, Han et al. [28] discuss the adoption of mobile technology as a potential aid to support employees work and alert that the success of technology depends on how people use it in their work. Considering the "TO" interplay, Chandran [29] categorizes six major losses that

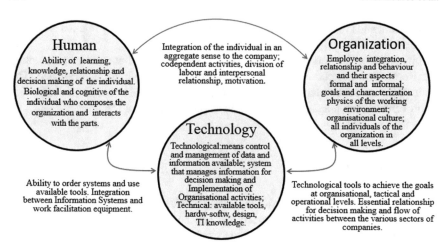

Fig. 2 HTO framework: dimensions and their interplay

can result from faulty equipment or operation and associates them with different costs for the organization. The author highlights that the elimination of these losses may effectively improve the Overall Equipment Efficiency (OEE) index. Kulkarni and Dabade [30], for instance, investigate knowledge sharing among maintenance operators and the impact of TPM on employees in terms of morale, job satisfaction and motivation, linking organizational and human dimensions. Carvalho et al. [8] offer several lessons that embrace the HTO dimensions within a real-life implementation of a simulator-based scheduling system. The authors consider the interplays between the production schedulers, the information systems and their integration and the simulator model itself.

Figure 2 presents a synthesis of the dimensions of the HTO with its implicit interplays in a framework.

3   Research Methodology

This research was conducted at a rolling mill. This study sample was chosen as it was facing huge challenges to implement AM in its operations. The paper builds upon the action research method, which consists of conducting actions, making them happen, and creating knowledge from their results [31]. What differs action research from consultancy practices, is the scientific contribution, the rigorous methodology and ensure that the intervention is informed and supported by theoretical considerations with no bias [32].

The research application follows Carvalho et al. [33, 34] and aims the solution impasses from the industry with the involvement of researchers and practitioners in a real live challenge towards understanding the process of change [35]. Therefore, one of the authors of this paper was involved as an "outside agent" participating in the changing process of the studied setting. He worked directly with the general director of the studied rolling mill company and with the production's, maintenance and operation's managers, as well as with their team composed by coordinators, analysts and technical staff. The research followed the steps offered in Coughlan and Coghlan [31] to gather, feedback and analyze data, and to plan, implement and evaluate action, preceded by a pre-step to comprehend the context and purpose, and a meta-step for monitoring. All the steps of the framework AM were conducted along a longitudinal study of 2 years within the rolling mill until the entire cycle was concluded.

4 Results and Discussions

This section offers the main findings of the action research, organized by problem characterization, implementation of the AM framework, and analysis from the HTO perspective.

4.1 Problem Characterization at the Rolling Mill

This research was conducted in a continuous rolling mill process (see Fig. 3) of an integrated steel production plant. This process transforms steel slabs into coils, by reducing the thickness of slabs in continuous casting, according to the dimensional and tolerance requirements of the end consumer. These coils, which are either inputs for internal processes or sold as finished products, are used in the automotive and white goods industries and other contexts. According to the company's historical data, the rolling mill process had a low utilization index due to a high maintenance stoppage rate. In this data analysis, the coilers were identified as the equipment

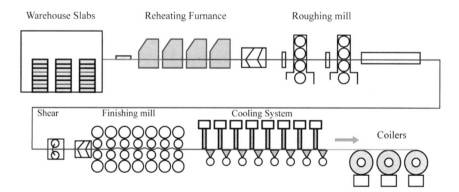

Fig. 3 Layout of the production flow

that most contributed to the inefficiency within this process. The causes for this shortcoming refer to factors such as premature breakage, leakage, sensor failure and poor lubrication.

Although the company has developed a maintenance strategy and implemented technology to address it, the shop floor operates within a cultural context which influences how employees handle maintenance tasks. In general, equipment suffers from chronic problems and operators usually ignore these problems to guarantee production outputs. This brings short-term consequences for the equipment and operators work becomes of a more stressed nature. Not to mention, the lack of preventive maintenance increases the probability of future breaks. Additionally, the preventive maintenance planning was considered inadequate as the round inspection operations still needed to be included within this planning.

Moreover, based on an exploratory research, two other factors contributed to decrease the overall effectiveness of the studied production process. The first one refers to the lack of training of the operation and maintenance staffs, especially for the new employees. The second issue concerns the out of date maintenance management technology used. The systems were not integrated which resulted in a faulty communication between operation and maintenance and a time-consuming maintenance scheduling process. For instance, in every work shift, technicians had to manually generate and print the inspection forms, conduct the inspection within the shop floor, manually input inspection data, generate the work orders for the abnormalities found and register adjacent abnormalities (i.e., abnormalities found outside the scheduled items).

4.2   Implementation of the AM Framework

This section discusses a number of issues related to the implementation case.

The teams involved.

The "outside agents" implementation team: The goal of the implementation team is to coordinate the systemic implementation of AM throughout the entire company. This team, which was designed by the plant's general director, is composed of a general coordinator, a programming technician and a support technician, both with skills in IT. The coordinator has the function of transmitting all the AM implementation regulations, guidelines, progress of the events and the implementation schedule. The programming technician is responsible for coding the IT system of each production sector using the integrated management system of planning and maintenance scheduling activities. The support technician is responsible for the implementation of this IT system and training of their users, including the hardware implementation of data collectors.

Management level team: The management level team is responsible for monitoring AM work progress, training, providing safety guidelines [36], implementing and coordinating the schedule and preparing inspection plans. In general, this team is composed of a sector coordinator of AM (i.e., an engineer who coordinates the maintenance plan of that sector), an engineer or specialist with knowledge of mechanics and fluids (e.g., oil, grease, water) and an electrical or automation engineer or specialist. The engineers or specialists are responsible for preparing course materials and training the execution team.

Execution level team: In general, the execution team of a given sector is composed of two technicians (i.e., for the areas of mechanics/fluids and electricity/automation) and all the production operators (i.e., in the rolling mill there were 4 workers and 1 supervisor per shift). The technicians support the operators in AM activities, coordinate inspection routines, planning and execution of inspection activities. On the other hand, after training, the operators perform AM activities, including cleaning, lubrication and maintenance inspection.

The information flow and technology adopted. To implement AM in the studied company, three IT system were basically involved. The framework is shown in Fig. 4. The first one is the organizations corporate system (ERP), which integrate all the organization's data and processes. The second one is the maintenance system, a departmental system that integrates the database of all maintenance planning and scheduling activities. Finally, the digital data collector, which was new to the company and implemented within this AM project, is a mobile handheld system that captures the entire schedule program of maintenance and maintenance operation. This collector has its own screen and keyboard and permits a high performance in terms of data collection and communication. It has eliminated paper printing and automated all steps in the inspection and programming system for AM.

The stages. The AM implementation process was organized in five stages. Table 1 summarizes these stages presenting the aims, main activities and participants involved in each one. The pilot equipment considered was the coiler, as mentioned above. Additionally, during stages 3 and 4, the general coordinator monitored the progress of the implementation schedule and the training activities, while the programming technician prepared the maintenance system with the new applications (i.e., round service and round collector) and the support technician performed

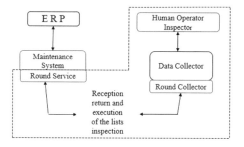

Data Collector- It is a logical unit of receipt, execution and return of the maintenance database inspection lists including the notes made in the areas.

Round Service - Application of the Data Collector installed in the server of the Maintenance Database for communication with the Collector.

Round Collector- Application installed on the Data Collector responsible for processing the data during the inspection in the area.

Maintenance System - Integrated system for managing maintenance planning and scheduling activities. (SIGAPPM)

Fig. 4 Flow of IT information from the data collector

Table 1 AM pilot equipment implementation stages

Stage	Aim	Activities	Participants
1	Preparing the start	Team formation (roles and responsibilities); Schedule definition; Definition of pilot equipment	All teams
2	Safety training	Basic training given by the Company's Safety Department	Execution level
		Specific training (on the pilot equipment) given by the Technical Department Engineer	Execution level
3	Training and motivation in AM	Opening held by the general coordinator, presentation of the AM program and its importance, responsibilities of each team member, schedule of the pilot project, objectives, expected results	All teams
		Theoretical technical training, ministered by the management level, on basic notions and functional techniques of each equipment	Execution level
		On the job training, ministered by the management level, where engineers perform activities with the trainees in the production area	Execution level
4	Execution of AM activities	Operators perform AM steps: initial cleaning, elimination of sources of contamination, inspection establishment, setting standards and lubrication and AM activities	Execution level
5	Audit of the AM	The main objective is to improve the AM practice in search of continuous improvement	Management and Execution level

the installation of the data collector and trained the execution level team. Before stage 5, the implementation team led an assisted operation phase during eight consecutive days after implementation. The team followed the operators during the inspection procedures, validating the performance of the IT system and the data collector.

Quantitative impacts registered. The quantitative impacts resulting from this work were measured during the first 8 days of inspection with the data collector. Within these days, the now trained AM operators inspected more than 1200 items, solved 18 urgent abnormalities, scheduled another 42 (considered not urgent) and identified 10 adjacent abnormalities. Additionally, a 75% reduction in the preparation time of inspection activities and a 35% reduction in the time spent in the AM inspection round were observed. Due to the implementation of AM in the coilers, the utilization rate of the rolling mill process went from 93.8 to 97.12% in the first 6 months. This represents a good recovery in terms of the resource capacity.

4.3 Analysis from the HTO Perspective

The successful implementation of AM in the studied context provided interesting lessons that embraced human, organizational and technological dimensions. These lessons are organized in three topics that highlight the interplays within these dimensions. The intention here is to reinforce the need for a holistic perspective of the HTO dimensions when conduction an AM implementation. Evidently these lessons refer to a particular situation and may not be generalizable to all business development processes.

- **HO Interplay**

AM implementation should be a top-down process. In this action research, the plant's director formally launched the AM implementation program. Rules were defined by him to guide this process. He designated an official implementation team linked directly to him which was responsible for conducting all AM projects within the company. This team carried out the AM guidelines to all the managers of the industrial areas. Through this strategy, the organization (O) was able to disseminate in a coordinated manner the goals of the project to all employees (H). This approach is in accordance with Min et al. [15] that highlight the importance of top-management involvement and support in a similar context.

Team training on equipment maintenance and operation should involve all machine operators and maintenance technicians. The knowledge sharing among these participants during the training events was crucial to address barriers and guarantee the success of the AM implementation. In these occasions, operators and maintenance technicians exchanged information and skills on maintenance and operation issues. For the organization, this knowledge apprehension and sharing made it easier to solve problems and improve the overall process performance. This is in accordance with Min et al. [15] that state that AM activities need to be implemented to increase knowledge and awareness among production operators and maintenance technicians.

Roles and responsibilities must be clearly defined among team members within an AM implementation process. Management and technical level teams were clearly defined and their respective functions were formalized. Not all operators are inspectors, but all new inspectors became operators. It is important to differentiate these operators with a new position as maintenance operators of AM. Min et al. [15] highlight that it is fundamental that the AM implementation starts with a well-defined AM team composition.

The organization should motivate AM teams by providing feedback and recognition. In the studied context, this motivational process began during the training events when operators were shown the importance of acquiring skills to accomplish results and achieve professional and organizational success. Moreover, the succeeded projects and activities, presented by the AM team, were formerly recognized by the organization. This recognition was disseminated throughout the entire shop floor through a visual management system and internal meetings. In short, an incentive system that appropriately recognizes team activities is essential for the success of an AM implementation project. Kulkarni and Dabade [30] state that knowledge sharing process with maintenance operators promotes a culture of ownership among them. This culture is related to increase in morale of employees and job satisfaction and motivation as well [37].

- **HT Interplay**

A customized IT solution for AM should be developed to guarantee adherence to the complexity of the industrial context. The availability of a new solution does not guarantee its acceptance. In this sense, it was essential to involve the technicians and operators of each operational area in the validation process of the IT solution (i.e., the data collector and the maintenance system) developed to support AM.

An easy-to-use system for data collection is essential when performing the inspection round. The inspections realized by the technical team (maintenance and operation) were carried out with the new technology by a data collector, which is user-friendly and easy to use on the factory floor. Han et al. [28] highlight the adoption of mobile technology to increase worker's productivity and performance.

The technological system may have a positive effect on the development of human capital. Particularly within this action research, maintenance operators acquired a technical capability to use IT system (e.g., when executing the inspection rounds with the data collectors) and to operate the automation systems of machines and equipment. Moreover, fast thought processing, decision making and problem solving were abilities that were developed and improved during the training events and the inspection rounds. Additionally, technicians acquired new skills to handle the now integrated structure of the maintenance system and its interfaces with the data collector and the ERP system.

- **TO Interplay**

After AM implementation, an assisted operation phase should be conducted. This was crucial to confirm if the AM implementation process was concluded. One may say that AM is completely implemented when the operator's inspection rounds (i.e., which include operation and maintenance items) are carried out 24 h a day with a 100% inspection of the equipment's critical items during that period. This is supported by Min et al. [15] that highlights that autonomous maintenance comprises the maintenance tasks that are performed daily by the operators themselves. To the extent of our knowledge, this is the first action research on AM that includes an assisted operation phase after AM implementation.

An efficient maintenance operation demands integrated technology systems. The main idea is that software applications should easily share information, through an integrated database. In the studied context, the ERP system was not explicitly designed to properly serve the needs of maintenance functions. Therefore, the proposed maintenance system connects the shop floor maintenance data to the ERP database, as it communicates both with the data collector and the ERP. This is in accordance with Carvalho et al. [8] that highlight the relevance of data acquisition and integration among systems in a similar context.

AM should contribute to aligning maintenance to the organization's goals. The goals for implementing AM are to reduce machine breakdown and reduce maintenance costs. AM implementation is supported by maintenance IT systems which capture machine stoppage rate and input data for maintenance cost calculation. This is essential for the overall business planning and for OEE index estimation.

Figure 5 presents the proposed HTO—AM framework which summarizes the interplays discussed within this section.

5 Conclusion

The implementation of AM involves many tasks and requires efforts, such as planning, execution and improvement in the maintenance bases. This paper addresses a research practice gap regarding the lack of studies in the literature that offer solutions to aid the industry in successfully implementing AM. The novelty of the research counts with the introduction of the HTO approach within an AM framework to offer a practical solution for practitioners with implications for OM scholars, what is enabled by an action research conducted in a rolling mill. The framework offered the company a systematic guidance to assist their management level in implementing

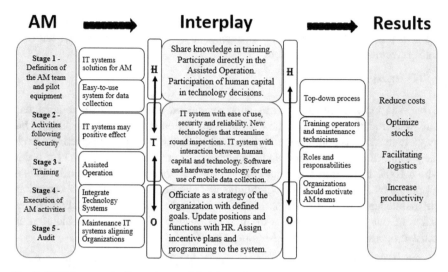

Fig. 5 HTO framework: dimensions and their interplay

AM practice. As a result, the company could obtain improvement in production performance, minimising machine breakdown and gaining in its utilization rate. The research findings indicate the adherence of the HTO dimensions to match the AM implementation challenges and reinforce the need for a holistic and combined perspective [38] of these dimensions for business development. Practitioners can take stock of the lessons learnt within this action research and the offered framework to aid the implementation of AM in their operations.

As research findings are limited to one single company, the authors suggest for additional studies embracing different companies towards generalization of the research findings [39]. Future research should also embrace other areas with a view on integrating AM with other existing maintenance techniques that are more technical, such as time-based maintenance, condition-based maintenance, reliability-centric maintenance and WCM (World Class Manufacturing).

Acknowledgements The following research agencies supported this work: Coordination for the Improvement of Higher Education Personnel (Coordenação de Aperfeiçoamento de Pessoal de Nível Superior—Brazil—CAPES) (Finance Code 001), Brazilian National Council for Scientific and Technological Development (CNPq) (311757/2018-9) and Research Support Foundation of the State of Rio de Janeiro (Fundação de Amparo à Pesquisa do Estado do Rio de Janeiro - FAPERJ) (APQ1 2019/ n.° 210.183/2019).

References

1. Lazim, H. M., and Ramayah, T. (2010). Maintenance strategy in Malaysian manufacturing companies: a total productive maintenance (TPM) approach. Business Strategy Series, 11 (6), 387–396.
2. Musmam, A.H, and Ahmad, R. (2018). Critical component identification and autonomous maintenance activities determination using fuzzy analytical hierarchy process method. International Journal of Industrial and Systems Engineering 28 (3), 360–378.
3. Mugwindiri, K. and Mbohwa, C. (2013) Availability performance improvement by using autonomous maintenance: the case of a developing country, Zimbabwe', World Congress on Engineering, July, England.
4. Ahuja, I.P.S., and Khamba, J.S. (2008). Strategies and success factors for overcoming challenges in TPM implementation in Indian manufacturing industry. Journal of Quality in Maintenance Engineering, 14 (2), 23–147.
5. Rajesh, A., Sandeep, G., Nikhil, D., and Deepak, Kumar. (2012). Analysis of barriers of Total Productive Maintenance (TPM). International Journal of Systems Assurance Engineering and Management, 4 (4), 365–377.
6. Alseari, A., and Farrel, P. (2020). Technical and operational barriers that affect the successful total productive maintenance (TPM) implementation: case studies of Abu Dhabi Power Industry. Ed. Springer International Publishing, 23 (2).
7. Zhongwei, W. and Qixin, C. (2010). Development of an autonumus in-pipe robot for offshore pipeline maintance. International Journal of Robotics and Automation, 37(2), 177–184.
8. Carvalho, A.N., Scavarda, L.F., and Lustosa, L. (2014). Implementing finite capacity production scheduling: lessons from a practical case. International Journal of Production Research, 52 (4), 1215–1230.
9. Lins, M. G., Zotes, L. P., and Caiado, R. (2019). Critical factors for lean and innovation in services: from a systematic review to an empirical investigation. Total Quality Management & Business Excellence, 1–26.
10. Gutierrez, D. M., Scavarda, L. F., Fiorencio, L., and Martins, R. A. (2015). Evolution of the performance measurement system in the Logistics Department of a broadcasting company: An action research. International Journal of Production Economics, 160, 1–12.
11. Kristensen, J., and Jonsson, P. (2018). Context-based Sales and Operations Planning (S&OP) research: A literature review and future agenda. International Journal of Physical Distribution & Logistics Management, 48 (1), 19–46.
12. Ahuja, I.P.S., and Kumar, P. (2009). A case study of total productive maintenance implementation at precision tube mills. Journal of Quality in Maintenance Engineering, 15, 241–258.
13. Relkar, A. S., and Nandurkar, K.N. (2012). Optimizing & Analysing Overall Equipment Effectiveness (OEE) Through Design of Experiments (DOE). International conference on modeling optimization and computing, Procedia Engineering, 38, 2973–2980.
14. Eti, M.C., Ogaji, S.O.T. and Probert, S.D. (2004). Implementing total productive maintenance in Nigerian manufacturing industries. Applied Energy (79), 385–401.
15. Min, C.S., Ahmad, R., Kamaruddin, S., and Azid, A.I. (2011). Development of autonomous maintenance implementation framework for semiconductor industries. International Journal of International Journal of Industrial and Systems Engineering 9 (3), 268–297.
16. Raheja, D., Llinas, J., Nagi, R., and Romanowski, C. (2006). Data fusion/data mining based on architecture for condition based maintenance. International Journal of Production Research, 44 (14), 2869–2887.
17. Swanson, L. L. (2001). Maintenance strategies to performance, Elsevier, International Journal of Production Economics, 70 (3), 237–244.
18. Endrenyi, J. (2001). The present status of maintenance strategies and the impact of maintenance on reliability, power Systems, IEEE Transactions, 16 (4), 638–646.

19. Labib, A. (1998). World-class maintenance using a computerized maintenance management system. Journal of Quality in Maintenance Engineering, 4 (1), 66–75.
20. Nordlöf H., Wiitavaara,B. Winblad,U., Wijk,K., and Westerling,R. (2014). Safety culture and reasons for risk-taking at a large steel-manufacturing company: Investigating the worker perspective. Safety Science, 73, 126–135.
21. Karltun, A. (2007). Forskarstött arbetet i själva verket – Att förbättra arbetssituationen för 15 000 brevbärare. Dissertation. n. 1122. Avdelningen för industriell arbetsvetenskap. Institutionen för ekonomisk och industriell utveckling. Linköpings Universität.
22. Ogén, O. (2011). Examensarbete inom Ergonomi och MTO, avancerad nivå, 15 hp KTH STH Campus Flemingsberg Examensrapport.
23. Berglund, M., and Karltun, J. (2007). Human, technological and organizational aspects influencing the production scheduling process. International Journal of Production Economics, 110, 160–174.
24. Daniellou, F. (2001). Epistemological issues about ergonomics and human factor. In International Encyclopedia of Ergonomics and Human Factors, Taylor & Francis, London and New York, (2), 43–46.
25. Caiado, R. C. G. G., Scavarda, L. F., Gavião, L. O., Ivson, P., Nascimento, D. L. M., Garza-Reyes, J. A. (2021). A fuzzy rule-based industry 4.0 maturity model for operations and supply chain management. International Journal of Production Economics, 231, 107883.
26. Hollmann, R. L., Scavarda, L.F., Thomé, A. M. T. (2015). Collaborative planning, forecasting and replenishment: a literature review. The International Journal of Productivity and Performance Management, 64, 971–993.
27. Matos, D. B. F., Scavarda, L. F., Caiado, R. G. G., Thomé, A. M. T. (2020). Supply Chain Management Practices in Small Enterprises: A Practical Implementation Guidance. In: Thomé A.M.T., Barbastefano R.G.; Scavarda L.F.; dos Reis J.C.G.; Amorim M.P.C. (eds). (Org.). Springer Proceedings in Mathematics & Statistics. 337ed.: Springer International Publishing, 141–153.
28. Han, S., P., Seppanen, M. and Kallio, M. (2004). Physicians' behavior intentions regarding the use of mobile technology: an exploratory study', AMCIS Pacific Asia Conference on Information Systems at AIS Electronic Library (AISeL), Shanghai, China.
29. Chandran, S. (2015) TPM implementation approach. Journal of Industrial Engineering and Management, publication 279059214.
30. Kulkarni, A., and Dabade, Dr. B. M. (2013). Investigation of Human Aspect in Total Productive Maintenance (TPM): Literature Review. International Journal of Engineering Research and Development, 5 (10), 27–36.
31. Coughlan, P., and Coghlan, D. (2002). Action research for operations management. International Journal of Operations & Production Management, 22 (2), 220–240.
32. O'Brein R. (1998). An Overview of the Action Research. Faculty of Information Studies, University of Toronto.
33. Carvalho, A.N., Oliveira, F., and Scavarda, L.F. (2015). Tactical capacity planning in a real-world ETO industry case: an action research. International Journal of Production Economics, 167, 187–203.
34. Carvalho, A.N., Oliveira, F., and Scavarda, L.F. (2016). Tactical capacity planning in a real-world ETO industry case: a robust optimization approach. International Journal of Production Economics, 180, 158–171.
35. Vizzon, J. S., Scavarda, L.F; Ceryno, P. S., and Fiorencio, L. (2020). Business process redesign: an action research. Gestão & Produção, 27 (2), e4305.
36. Muniz, M.V. P., Lima, G. B. A., Caiado, R. G. G., and Quelhas, O. L. G. (2018).Bow tie to improve risk management of natural gas pipelines. Process safety progress, 37(2), 169–175.
37. Caiado, R., Nascimento, D., Quelhas, O., Tortorella, G., and Rangel, L. (2018). Towards sustainability through Green, Lean and Six Sigma integration at service industry: review and framework. Technological and Economic Development of Economy, 24(4), 1659–1678.

38. Julianelli, V., Caiado, R. G. G., Scavarda, L. F., and Cruz, S. P. D. M. F. (2020). Interplay between reverse logistics and circular economy: critical success factors-based taxonomy and framework. Resources, Conservation and Recycling, 158, 104784.
39. Azevedo, B. D., Scavarda, L. F., Caiado, R. G. G., and Fuss, M. (2020). Improving urban household solid waste management in developing countries based on the German experience. Waste Management. Available in https://doi.org/10.1016/j.wasman.2020.11.001.

Definitions of Critical Risk Factors in an Effluent Drainage Network Project Using AHP: A Case of Study

J. C. Pereira and Y. Dorino

Abstract High population growth and water quality are directly related to some critical factors in treatment plants, and are often not perceived or ignored for prevalent economic reasons. The present study aims to present the risk factors that can influence the decision-making process in an effluent drainage network project. The available literature on risk analysis for this type of process is limited. Through questionnaires responses by specialists in the area in social networks. As a result, the most critical factor is (a) possibility of reusing treated effluents, odour generation, interaction with the neighbourhood and regional environmental legislation (b) know the local legislation as the first condition for an effluent treatment project, it is important to note that differences in legislation often make it impossible to implement a treatment project that is successful in one state to another. (c) study of interaction with the neighbourhood must take into account so that there are no problems with the existing Environmental Legislation in order to minimize the problems with people living near the effluent treatment plants. The conclusion is that the proposed method is an invaluable source for environment professionals and decision makers, in the sense that it augments their information on drainage network projects and help to identify critical risks factors and allow the implementation of actions to avoid project failure.

Keywords Treated effluents · Reuse · Water resources management · Affinity diagram · AHP

J. C. Pereira (✉)
National Laboratory for Scientific Computation (LNCC), Petropolis Catholic University, Petropolis, Rio de Janeiro, Brazil
e-mail: jpereira@lncc.br

Y. Dorino
Department of Master in Engineering, Petropolis School of Engineering, Petropolis Catholic University, Petropolis, Rio de Janeiro, Brazil
e-mail: yuri.dorinog@outlook.com

1   Introduction

The present study aims to present the risk factors that can influence the decision-making process in an effluent drainage network project. Much of the existing drainage system in Brazil is obsolete and outdated. These systems were designed for an old reality, where they were less urbanized and waterproofed. The projects were not created based on the changes that could occur in society and, consequently, the networks that were designed are shown to be insufficient and unable to drain this flow of sewage and garbage in general. In this sense, errors or inconsistencies in the drainage network can harm the project or cause this project to be achieved by compensating the externalities parameter values. After the decline of the Roman Empire, the population of countless cities in Europe and Asia decreases causing the abandonment of municipal services, and consequently, their deterioration [1]. In the following centuries, strategies related to the design of urban drainage and sanitation networks have not undergone any relevant progress. In sanitary terms, a certain regression is allowed during the Middle Ages, as personal hygiene and cleanliness are not at all a concern of the population [2]. Only from the eighteenth century "being clean" assumes a distinguishing status, being only available for some very specific social classes [2]. Available literature on risk analysis are for example the work of Fernández [3], the author states that Effluent is waste from human activities, such as industrial processes and the sewage network, which are released into the environment, in the form of liquids or gases. The process of drainage nets has been intensively studied in recent times. Another important study work is the one from Almeida and Monteiro [4], which states that from an economic point of view, it is also important to highlight the influence that the increase in treated volumes, due to excess risks, may represent in the management of treatment. Generally, the operating costs involved can turn out to be much higher than those expected values. Several citations about the subject from different authors are presented herein: Drainage channels are the lines along which river processes act to transport water and mineral material from a local region, allowing gravity processes on the slopes to continue to diminish landscapes. Drainage networks are the basis for defining drainage basins, an essential component in hydrological models and resource management plans [5]. Unless the terrain is rugged, the derived water channels tend to flow in parallel lines in the preferred directions generated by grid orientation sampling. In addition to the presence of noise that creates artificial wells [6], there are problems that prevent the successful design of fully connected drainage networks, with a single line width: the positioning of the ends of the drainage networks; and the allocation of drainage directions, in particular the allocation of drainage directions through flat areas and in closed depressions in the wells [7]. For Heller [8], the area of basic sanitation lacks an approximation with the public health perspective, visualizing its ends and not the means to achieve them and, thus, increasing the effectiveness of its actions. Brazil, from the 1950s, went through an accelerated urbanization process. In the mid-2000s, 138 million Brazilians lived in urban areas [9]. Urbanization was carried out disproportionately across the Brazilian territory, so that only nine metropolises concentrate about 40%

of the urban population in the country [10]. In the field of sanitation, the National Sanitation Plan (PLANASA) was created in the military regime in 1970. Another curious fact is that the municipalities assisted by FUNASA [11], have resources from the Ministry of Health. However, in 1993, it assisted 625 localities, that is, 6% of Brazilian municipalities, serving about only, 5 million people [12]. Recently, there has been a new attempt to mobilize public opinion in order to stimulate public power to intervene repressively against favelas, controlling their expansion, or even, as has been seen in some cases, proposing their removal so that problems, as basic sanitation were reduced [13]. PLANASA's main guidelines were the centralization of the sector around the State Water and Sewage Companies (CEAEs), which are properly concessionaires contracted or agreed with the municipalities for water distribution and sewage collection and final destination. However, favelas are regions of municipalities that, throughout Brazil's history, have not been prioritized by these public basic sanitation policies [14]. The convergent aspect of the situation of basic sanitation in the favelas refers to the presence of vectors or reservoirs of diseases, such as rats, cockroaches, among others. This finding suggests that the lack of integration between basic sanitation actions that involve all related services, namely, water supply, sanitation, public cleaning, rain drainage and vector control of communicable diseases, have implications from a practical point of view. No matter how much progress is made in addressing the deficits in one of these services, deficiencies in another can cause the persistence of the problems to be tackled, such as the risks to public health resulting from the unhealthy environment [14]. Another important work is the one from Collares [15], the author states that drainage system is an important indicator of the alterations occurred in the composition of the hydrographic basin environment, either due to changes in their structure and shape or due to channel loss and gain. The management entities responsible for the control, insertion and monitoring of the current drainage networks, have faced serious problems, associated with the occurrence of infiltrations, of an economic, structural and environmental condition [16]. For Coelho [16], currently, infiltrations into drainage networks play an important role in the dimensioning of drainage network, including specific regulations for estimating these volumes, in case there are no real registers. The affluences are of particular concern improper or not accounted for at the time of the project. To assess the quality of drainage networks, quantitative parameters of both networks and reference drainage networks are necessary for their evaluation, in addition to the overlap of both networks for visual analysis [3]. The drainage network of a geographic region defines the drainage paths for water from rain according to the relief of the region [17]. Within the scope of water resources management, there is a shortage of globally applicable systems that reference, indicate and identify, in a unique and efficient way, the spatial organization of hydrographic basins and respective drainage networks [18]. Sanitary use generates waste that is mostly separated into specific treatments. Wastewater contains liquid and solid human excreta, various cleaning products, food waste, disinfectant products and pesticides. Human excreta, mainly, derive microorganisms present in the waste. Characteristics of industrial effluents are inherent to the composition of raw materials, supply water and the industrial process. The concentration of pollutants in effluents is a function of losses in the process or water consumption [19]. The treatment processes to be adopted, their constructive forms and the materials to be used are considered based on the following factors: the regional environmental legislation, the climate, the local culture, the investment costs, the operating costs, the quality of the treated effluent, operational safety related to leaks of used chemicals or effluents, generation of odor, interaction with the neighborhood, possibility of reusing treated effluents [20]. The treatment systems must be used not only with the minimum objective of treating the effluents, but also to meet other premises. An important point to be noted is that unnecessary waste should not be generated by using the treatment [21]. The sensory characteristics of the effluents are very important and may be the object of the authorities' attention. The odor in industrial effluents may be due to the exhalation of organic or inorganic substances due to fermentation reactions resulting from mixing with sewage, aromas, solvents and ammonia from the leachate. The color of the effluents is another characteristic confusingly controlled by the legislation [20]. The color in the environment is the apparent color, composed of dissolved substances and turbidity [21]. According to Archela et al. [22], pollution can be defined as any form of alteration of natural properties, whether physical, chemical or biological, that may occur in the environment. Thus, it is possible to distinguish pollution from contamination, as this represents a potential risk to nature, and is therefore more harmful to the environment and human health. In several ways, water can affect human health: through direct ingestion, in food preparation; personal hygiene, agriculture, environmental hygiene, industrial processes or leisure activities [23]. The pathogenic bacteria found in water and/or food constitute one of the main sources of morbidity and mortality in our environment. They are responsible for numerous cases of enteritis, childhood diarrhea and epidemic diseases, which can result in lethal cases [23]. According to Mannarino et al. [24], a well-known example of public health disorders that may result from the lack of good effluent treatment is the cholera epidemic that hit London in the 19th century. In addition, pathogenic substances, such as sulfates, have laxative and toxic properties, other elements, such as zinc, arsenic and cyanides, can cause serious problems to human health [11]. None of the previous work listed above focused on the identification of the risk factors that can influence the decision-making process in an effluent drainage network project, so this study proposes to identify the critical factors to be considered in an effluent drainage network project.

2 Objectives

The purpose of the present study is first, to increase knowledge concerning the risks in effluent drainage network project. Second, to provide a method to assess the risk factors related to effluent drainage network project, based on an in-depth analysis of current literature about the process and knowledge from experts. The intention is to help professionals working in the process of treating of effluents in different sectors on how to prioritize process factors by providing knowledge about risks of critical factors that can influence the effluent treatment processes and the population and the environment around them. In order to attain the research objective, the following research questions are proposed: Research Question 1. What are the critical factors to be considered in an effluent drainage network project? Research Question 2: What can be done to mitigate the risks?

3 Methods

A literature search was done to identify the critical factors that could cause the effluent treatment process to be adopted and a field study was done to validate these factors with the professionals currently working with these processes. The research carried out for this study, carried out with Google Docs forms, was answered by experienced professionals in the field. This study aimed to obtain the degree of importance on the perspective of each professional on each of the externalities and the degree of importance varied between it was important, it has high importance, medium importance, low importance, being the "it is important" considered something that should not be left aside in any project and the other three (high importance, medium importance, low importance) expressed that it was important but not essential. As part of the elicitation process with specialists, one hundred questionnaires were answered and selected by different professionals (workers in different areas of the effluent treatment industry), these professionals were selected on the LinkedIn social network, considering the branch of the company they work or worked with, more than 10 years in the effluent treatment business and if they understood the treatment process.

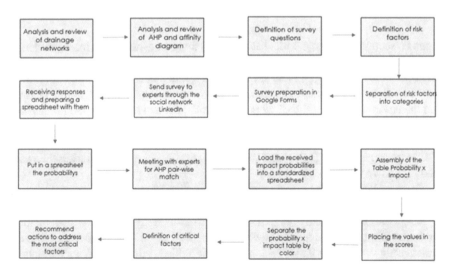

Fig. 1 Methodological flowchart. *Source* The author

As shown in Fig. 1, for the construction of the methodology of the present study, the research questions were defined in Chap. 1. The first step was to carry out a bibliographic search in databases to identify the state of the art of the literature on Drainage Networks and AHP in drainage networks. After this study, the research questions were then defined through gaps in the literature found. Then, the most important factors in a drainage network insertion process were defined and these factors were divided into categories according to the methodology of the affinity diagram, the factors were grouped by affinity, similarity and dependence by the conception of the authors. The research with the experts to define the probability of impact of the AHP was done with the Google Forms tool, the questions were equal combinations of each topic among themselves, and the answers were obtained through the LinkedIn network, the risk factors that lead to each risk within each category were identified by experts on the subject. The questionnaire was sent based on whether the specialist works or worked in the effluent drainage area, without considering working time or a specific area within the theme. Then, a meeting was held with experts to define the impacts of risk factors and the values found in a standardized matrix. With the data from the normalized matrix, the factors were prioritized, from the transformation of these values into scores from 1 to 5 and the values were colored according to the value found in this table. The questionnaire and meeting methodology were used so that the peer comparison was made with specialists from different nationalities, adhering to the presented spreadsheet. Thus, it was possible to determine the critical factors that most impact a process of inserting drainage networks and with that a methodology was created to mitigate the risks.

4 Results

It can be seen that the most critical activity and that can generate the most problems for any drainage network implantation is the insertion project. In the insertion project, great care must be taken, as there are countless errors that can affect this stage, such as: the types of sources they can generate, environmental and socio-economic factors and the environmental health of where a network will be inserted. In the process of inserting drainage networks, the following steps demonstrated in the mind map in Fig. 2 were defined, in which they can generate risks.

The management of the project management process generated during the case study is shown in Fig. 3.

The most important factors in a process were defined and classified into categories using the affinity diagram,

After defining the affinity diagram, as a result of this study the main critical factors were identified. The use of AHP evidenced the impact on the insertion of effluent drainage networks in the Paraty region, using the Table 1. After identifying the stages and analyzing each one, it was possible to find the risks of each one by eliciting the probabilities of specialists using the Google Forms tool and sharing the questionnaire with specialists on the LinkedIn website.

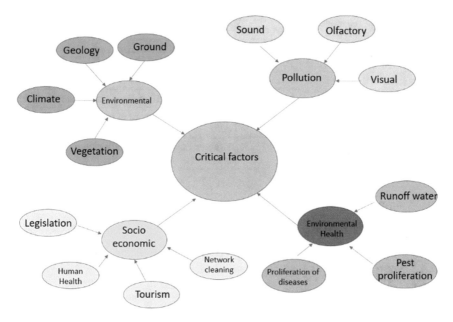

Fig. 2 Mental map for the process of insertion of drainage networks

Fig. 3 Map of the drainage network insert project

With this, the research followed the following steps: Choice of the target audience, Choice of specialists according to their time of experience in the area (chosen professionals working with water resources), Sending the questionnaire, Receiving the answers, Conference if the people who answered were the same ones in which the questionnaire was sent. Fifty experts answered the questionnaire. The impact factors were defined pairwise and placed in the AHP matrix, the results can be seen in Table 2 we see the normalized AHP values.

Table 1 Level of importance saaty [25]

Importance	Definition
1	Both elements are of equal importance
3	Moderate importance of one element over the other
5	Strong importance of one element over the other
7	Very strong importance of one element over the other
9	Extreme importance of one element over the other

Table 2 AHP

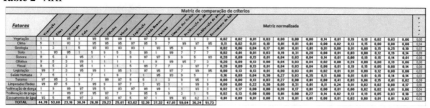

With the impact determined by the AHP, the score was assigned according to the Table 3.

After performing the multiplication and risk classification in each stage, the final values for the proposed model were finally found. These values are shown in Table 4 and are color coded, as per Table 5.

Based on the result of the study, the map of the initial process was revised to address the critical factors that affect the process, as shown in Fig. 5.

The risk factors that most impact the process of inserting effluent drainage networks are: Human Health, Legislation, Soil, Noise, Olfactory and Visual Pollution, Tourism and Network Cleaning. Regarding human health, basic sanitation incorporates the services of drinking water, sanitary sewage, garbage collection and urban drainage, widely associated with public health. A considerable number of people present illnesses or die due to precarious effluent drainage systems. A determining

Table 3 Methodologies scores

Probability level score			Impact level score		
Score	Probability level	Probability	Score	Impact level	Impact
5	Expected	More than 0.80	5	High	More than 0.16
4	Very probable	0.51–0.80	4	Elevated	0.12–0.16
3	Probable	0.31–0.50	3	Moderated	0.08–0.11
2	Improbable	0.11–0.30	2	Low	0.04–0.07
1	Almost no probability	Less than 0.10	1	Limited	Less than 0.04

Table 4 Table Probability x Impact

Factors	Probability	Impact	Score Prob.	Score Impact	Risks
Vegetation	0,61	0,05	4	2	8
Climate	0,49	0,04	3	2	6
Geology	0,51	0,07	4	2	8
Ground	0,57	0,09	4	3	12
Sound	0,61	0,08	4	3	12
Olfactory	0,60	0,08	4	3	12
Visual	0,57	0,08	4	3	12
Legislation	0,60	0,03	4	1	4
Human Health	0,52	0,13	4	4	16
Tourism	0,56	0,09	4	3	12
Network cleaning	0,56	0,10	4	3	12
Proliferation of diseases	0,56	0,04	4	2	8
Pest proliferation	0,47	0,08	3	3	9
Runoff water	0,59	0,03	4	1	4

Table 5 Classification of risk factors

			RISK				
			Limited	Low	Moderate	Elevated	High
			1	2	3	4	5
Probability	Almost no probability	1	1	2	3	4	5
	Unlikely	2	2	4	6	8	10
	Probable	3	3	6	9	12	15
	Very probable	4	4	8	12	16	20
	Expected	5	5	10	15	20	25
Risk Factor Classification							
1 -5	Insignificant	6 - 9	Tolerable	10 - 16	Undesireble	17 - 25	Intolerable

factor is planning, which can be done through the Municipal Basic Sanitation Plan (PMSB), which addresses the health needs of the municipalities and possible ways to achieve them. The population is one of the stakeholders in the drainage networks, however the executive branch is the one who should start the project, in this case the city hall. There are even ways to finance this action with government resources for the technical staff. Another factor is the investment that should largely come from the government, considering that most drainage companies are state or municipal. To ensure people's health, disinfection tunnels should be installed through all rain networks. Regarding the legislation, in January 97, Law No. 9,433 was created, known as the Water Law, which instituted the National Water Resources Policy and created the National Water Resources Management System. Considering that Brazil contains approximately 12% of all fresh water on the planet, the law emerged in a context in which water becomes increasingly scarce, with the concern that its distribution is equitable. Among the objectives of this law, the following can

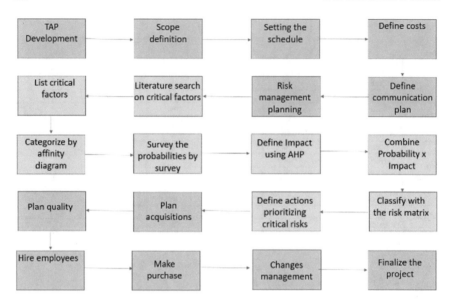

Fig. 5 Revised process map of the drainage network insert process

be highlighted: Define a water resources agenda, identifying priority management actions, programs, projects, works and investments, within a context that includes government agencies, civil society, users of water and the different institutions that participate in the management of water resources; Adapt the use, control and protection of water resources to social aspirations; Meet water demands with a focus on sustainable development; Balance water supply and demand, in order to ensure water availability in quantity, quality and reliability appropriate to different uses; Guide the use of water resources, considering variations in the hydrological cycle and development scenarios. Above soil, it is one of the most important components in an engineering work, as it is responsible for supporting the constructions. Therefore, for a good performance of the building, a good geotechnical project is fundamental. It is important to note that geotechnics studies the behavior of soils and rocks in relation to human actions. Its application is important in countless situations, such as preventing landslides, landslides, landslides and structural problems in buildings. To mitigate this risk, it is necessary to follow the following steps in the project to insert effluent drainage networks: Viability is responsible for expanding the level of detail and understanding of the soil, so that it is easier to predict the cost of the work and its completion time. This study allows professionals to identify what is technically possible and financially feasible, in addition to defining the expected schedule. It consists of preliminary recognition, inventory, pre-feasibility and preliminary design. The basic design stage is the stage in which the main building components are fixed and clarified, the descriptive memorial is created and the worksheets are assembled. In this phase, it is necessary to make structural calculations, drawings and technical specifications, in addition to gathering the necessary documents for the acquisition of equipment. Finally, the executive project stage consists of detailing and reviewing the basic project, without any changes to the pre-established parameters. Its main objective is to detail what was planned and will be implemented. Regarding noise pollution, sound at high and constant volumes should be considered as a potentially serious pollutant and as a serious threat to environmental health. The level of noise admitted in large urban centers by the World Health Organization (WHO), can reach up to 50 decibels, however, what is verified usually reaches 90 and 100 decibels. Therefore, any sound that exceeds 50 decibels, can already be considered harmful to health. It is considered an environmental crime and may result in a fine and imprisonment from 1 to 4 years. It is important to emphasize that the legislation on noise pollution is the responsibility of the municipalities, so it is the function of the prefecture of each Brazilian city, to create laws of silence and inspect them to be complied with. Among the federal laws, the Environmental Crimes Law, No. 9.605 of February 12, 1998, which provides for criminal and administrative sanctions for conduct and activities harmful to the environment. About pol olfactory use, the number of complaints related to olfactory pollution has increased considerably in recent years. Because it has a more subjective character than other forms of pollution, there is no specific national legislation that limits emissions of odorous substances. Regarding visual pollution, it was agreed to call visual pollution the excess of materials related to the degradation of public properties with the practice of graffiti, exposed garbage and buildings that lack maintenance. In order to improve and solve this "aesthetic problem" that has been affecting not only cities, but the human beings that inhabit it, Brazilian cities are betting on legislation that promotes the improvement of urban space. In addition to encouraging public policies that prioritize the reduction of numerous types of pollution in cities, promoting the quality of life of citizens, awareness of advertising companies and citizens themselves are essential to promote the improvement in the quality of life of the urban population. About tourism, these are the activities that people carry out during their travels and stay in places other than those they live in, for a period of time less than a consecutive year, for leisure, business and other purposes. As it is a city with greater financial activities arising from tourism, the sensory, visual and auditory part of any project must be viewed with great care so that it does not cause negative feelings for those who visit the city. It is important to even consider the place of insertion so that there are no problems in city traffic. To mitigate this risk, an impact study on the flow of cars and pedestrians in the city must be carried out. In addition, to have constant cleaning in the networks so that there are no problems with odor that can keep possible and constant visitors away. In addition, it must be inserted far from the historic center so that no visual pollution occurs. On cleaning networks, through Law 13.308/2016, the government will be in charge of cleaning and maintaining drains and drainage networks in cities, the goal is to prevent damage due to lack of maintenance of drains and networks, and thus prevent the occurrence of disasters such as floods. Cleaning actions must be carried out in places such as canals, manholes and wolf mouths. The measure must also be complemented by the recovery of public pavements and sidewalks, in addition to pruning, suppression and landscaping in various plants (trees, shrubs and palms), among other services to avoid collapsing the rain system with excess organic and inorganic materials.

5 Conclusion

The proposed method is important for several reasons. First, the risk factors that can influence the decision-making process in an effluent drainage network project were identified. In this study the methodology has been applied successfully in identifying the critical risk factors. Second, the proposed method is an invaluable source for environment professionals and decision makers, in the sense that it augments their information on drainage network projects and help to identify critical risks factors and allow the implementation of actions to avoid project failure. Third, the paper shows that the risk factors identified in this study are critical and must be controlled to avoid project failure. In answer to the first question, the model is ideal for the defined region. Showing itself effective in all the places where it was inserted. In response to the second question, the risk factors that most impact the process of inserting effluent drainage networks are: Human Health, Legislation, Soil, Noise, Olfactory and Visual Pollution, Tourism and Network Cleaning. Regarding human health, to ensure the health of people, disinfection tunnels should be installed through all rain networks. Regarding the legislation, follow what was established by the Water Law. Regarding noise and visual pollution, municipal laws must be observed. Regarding the cleaning of the networks, the measure should also be complemented by the recovery of public pavements and sidewalks, in addition to pruning, suppression and landscaping in various plants (trees, shrubs and palms), among other services to avoid collapsing the rain system with excess of materials organic and inorganic. On soil, it is important to note that geotechnics studies the behavior of soils and rocks in relation to human actions. Its application is important in countless situations, such as preventing landslides, landslides, landslides and structural problems in buildings. To mitigate this risk, it is necessary to follow the following steps in the project to insert effluent drainage networks: Viability, basic design and executive design. Regarding tourism, it is important to even consider the place of insertion so that there are no problems in the city's traffic. To mitigate this risk, an impact study on the flow of cars and pedestrians in the city must be carried out. In addition, to have constant cleaning in the networks so that there are no problems with odor that can keep possible and constant visitors away. In addition, it must be inserted far from the historic center so that no visual pollution occurs.

The main contribution to the scientific community is the identification of the most critical factors in the introduction of an effluent treatment plant. It allows the definition of actions prioritizing the most important factors. A future work would be a detailed study of each of the factors in different treatment plants.

References

1. Saldanha, J. (2003) Aspectos Históricos a Actuais da Evolução da Drenagem de Águas Residuais em Meio Urbano.
2. Burian, S.; Edwards, F. (2002) Historical Perspectives of Urban Drainage. Comunicação apresentada em 9th International Conference on Urban Drainage.
3. Fernández, D.; Valeriano, M..; Zani, H.; Andrade Filho, C. (2013) Extração Automática de Redes de Drenagem de Modelos Digitais de Elevação. Revista Brasileira de Cartografia, Vol. 64, No. 3.
4. Almeida, S.; Monteiro, P. (2004) Incidência de Caudais de Águas Pluviais em Redes de Drenagem de Águas Residuais - Dois casos de estudo em Municípios do Norte de Portugal. 7º Congresso da Água.
5. O' Callaghan, J.; Mark, D. (1984) The Extraction of Drainage Networks from Digital Elevation Data. Computer Vision, Graphics, and Image Processing. Vol. 28, pp. 323–344.
6. Fairfield, J.; Leymarie, P. (1991) Drainage networks from grid digital elevation models. Water Resourcer Research, Vol. 27, No. 5, pp. 709–717.
7. Tribe, A. (1992) Automated recognition of valley lines and drainage networks from grid digital elevation models: a review and a new method. Journal of Hydrology, Vol. 139, No. 1–4, pp. 263–293.
8. Heller, L. (1989) Esgotamento sanitário em zonas de urbanização precária. Belo Horizonte – MG. Master Dissertation, Universidade Federal de Minas Gerais.
9. IBGE. Instituto Brasileiro de Geografia e Estatística. Censo demográfico: dados da amostra. 2000.
10. IBGE. Instituto Brasileiro de Geografia e Estatística. (2001) Departamento de População e Indicadores Sociais. Tendências demográficas: uma análise dos resultados da sinopse preliminar do censo demográfico 2000. Rio de Janeiro, pp. 63.
11. Fundação Nacional de Saúde (2004) Manual de Saneamento. 3ª ed. Brasília: Fundação Nacional de Saúde, FUNASA, pp. 408.
12. Costa, A. (2003) Avaliação da Política Nacional de Saneamento. Rio de Janeiro – RJ. Tese de Doutorado, ENSP/FIOCRUZ, pp. 248.
13. Compans, R. (2007) A cidade contra a favela. A nova ameaça ambiental. Revista Brasileira de Estudos Urbanos e Regionais, Recife, Vol. 9, No. 1, pp. 83–99.
14. Gomes, U. (2009) Intervenções de saneamento básico em áreas de vilas e favelas: um estudo comparativo de duas experiências na região metropolitana de Belo Horizonte. Belo Horizonte – MG. Master Dissertation, Universidade Federal de Minas Gerais, 2009.
15. Collares, E. (2000) Avaliação de alterações em redes de drenagem de microbacias como subsídio ao zoneamento geoambiental de bacias hidrográficas: aplicação na bacia hidrográfica do Rio Capivari - SP. Doctoral Thesis, Escola de Engenharia de São Carlos, University of São Paulo, São Carlos.
16. Coelho, I. (2013) Variabilidade de Afluências às Redes de Drenagem de Águas residuais: Causas e Efeitos Versus Sustentabilidade Económica Um Caso de Estudo Master Dissertation, Faculdade de Engenharia Univescidade do Porto, Porto.
17. Santos, L. C.; Francisco, C. (2011) Avaliação dos Modelos Digitais de Elevação aplicados à extração automática de redes de drenagem. XV Simpósio Brasileiro de Sensoriamento Remoto - SBSR, Curitiba, PR, Brazil, pp. 1311.
18. Silva, N.; Ribeiro, C.; Barroso, W.; Ribeiro, P.; Soares, V.; Silva, E. (2008). Sistema de ottocodificação modificado para endereçamento de redes hidrográficas. Revista Árvore, Vol. 32, No.5, pp. 891–897.
19. Von Sperling, M. (1995) Princípios básicos do tratamento de esgotos. Belo Horizonte: Departamento de Engenharia Sanitária e Ambiental; UFMG, Princípios do Tratamento Biológico de Águas Residuárias, Vol. 3, pp. 196.
20. Giordano, G. (1999) Avaliação ambiental de um balneário e estudo de alternativa para controle da poluição utilizando o processo eletrolítico para o tratamento de esgotos. Niterói – RJ. Master Dissertation, Universidade Federal Fluminense, pp. 137.

21. Giordano, G. (2003) Análise e formulação de processos para tratamento dos chorumes gerados em aterros de resíduos sólidos urbanos. Rio de Janeiro – RJ. Doctor Tesis, PUC-Rio, pp. 257.
22. Archela, E.; Carraro, A.; Fernandes, F.; Barros, O.; Archela, R. (2003) Considerações sobre a geração de efluentes líquidos em centros urbanos. Geografia, Vol. 12, No. 1.
23. Fundação Nacional de Saúde (2006) Manual de Saneamento. 4ª ed. Brasília: Fundação Nacional de Saúde, FUNASA.
24. Mannarino, C.; Moreira, J.; Ferreira J.; Arias, A. (2013) Avaliação de impactos do efluente do tratamento combinado de lixiviado de aterro de resíduos sólidos urbanos e esgoto doméstico sobre a biota aquática. Ciência & Saúde Coletiva, Vol. 18, No. 11, pp. 3235–3243.
25. Saaty, T. L. (2009). Extending the Measurement of Tangibles to Intangibles. International Journal of Information Technology & Decision Making, Vol. 8, N. 1, p. 7–27. Available at SSRN: http://ssrn.com/abstract=1483438.

Decision-Making Process on Sustainability: A Systematic Literature Review

Renata Amaral Fonseca, Antônio Márcio Tavares Thomé, and Bruno Milanez

Abstract This paper addresses the relationships between sustainability and the decision-making process and investigates how the literature analyses this theme through a Systematic Literature Review (SLR). The Scopus and Web of Science databases were searched, resulting in 74 studies with full-text reading. The review presents some main categories of the studies and seeks to assess the decision-making breadth and depth. The findings reveal an increasing number of papers published on the subject in the last ten years. Among sustainability dimensions, environmental and economical are the most explored. Although the social dimension is relatively under-investigated, an increasing number of studies addressing the decision process's social themes are perceived, suggesting a growth trend. The use of resources, mainly energy consumption, and pollutant gas emissions are the sustainability problem most addressed. Regarding the area of activity, product development is prominent, highlighting the design phase. Further, the automotive industry is the more prevalent industry type in the studies. Finally, the review shows a concentration of articles dealing with decision making involving sustainability analytically, through a broader approach using different data types, including qualitative and subjective elements.

Keywords Sustainability · Decision-making process · Manufacturing

1 Introduction

This paper addresses two areas of research with academic and practical impact: decision-making and sustainability. Aiming to link them, it investigates the decision-making process involving sustainability issues. Sustainability themes are interpreted and addressed under different logics [e.g., 1–5], influencing how decisions are made.

R. A. Fonseca (✉) · A. M. T. Thomé
Pontifícia Universidade Católica do Rio de Janeiro, Rio de Janeiro, Rio de Janeiro, Brazil

B. Milanez
Universidade Federal de Juiz de Fora, Juiz de Fora, Minas Gerais, Brazil

© The Author(s), under exclusive license to Springer Nature Switzerland AG 2021
A. M. Tavares Thomé et al. (eds.), *Industrial Engineering and Operations Management*,
Springer Proceedings in Mathematics & Statistics 367,
https://doi.org/10.1007/978-3-030-78570-3_17

Different decision process' characteristics are drawn depending on how managers deal with the environmental, social and economic dimensions [4].

This problem gains relevance because the decision-making process matters, considering that variations on the process may result in different choices and different results [6]. Also, sustainable actions' effectiveness depends on the manner decisions are made. Being a sustainable company depends on the choices made by managers and "in the end, successful companies will have little option but to get involved in this rapidly emerging area" [7, p. 99] of sustainability.

This paper focuses on how the extant literature handles the decision-making process involving sustainability, investigating the main research topics and informing possible opportunities for future inquiry efforts. Through a Systematic Literature Review (SLR), it addresses two related research questions (RQs):

RQ1—What is the concentration of the extant research on sustainability and decision-making regarding geographical regions, sustainability dimensions and problems, areas of manufacturing activity and type of industry?

RQ2—How are the breadth and depth of the decision-making process involving sustainability issues covered by the extant research?

The remainder of the paper is composed of the theoretical background on the decision-making process and sustainability (Sect. 2), the research methodology (Sect. 3), content analysis and findings (Sect. 4) and conclusions (Sect. 5).

2   Decision-Making Process and Sustainability

This section describes the rationale and motivation to investigate the relationships between decision-making and sustainability. Economic theories of decision-making date centuries ago, in which decisions are deeply affected by the level of risk and uncertainty [8]. Sustainability, in turn, is a relatively more recent topic. According to Elkington [7], the concept of sustainable development gained prominence after the publication of "Our Common Future" report, in 1987, and with the United Nations Conference on Environment and Development in 1992.

A decision comprises "a specific commitment to action", and the decision process is "a set of actions and dynamic factors that begins with the identification of a stimulus for action and ends with the specific commitment to action" [9, p. 246]. Sustainability encompasses environmental, social and economic concerns [10], referring to the ability to meet the present generation's needs without compromising future generations' ability to meet their own needs [11].

Although the managers' cognition [4, 5] and practical experience [12] are influencing factors in the decision-making process, managers might tackle sustainability issues with a narrowly focused or ample view (breadth). Moreover, they might be more intuitive or prone to analytical decision-making (depth) according to different decision-making styles. Also, not only the decision-making process breadth and depth varies but also its effectiveness [6].

The decision process breadth relates to the more general or specialized form of information processing [4, 13, 14]. Depth, in this paper, considers the degree to which information processing grounds on intuitive methods or analytical procedures [6, 15], assessing the details of the information on sustainability issues [4]. The effectiveness relates to the measure that the action implemented meets the decision goals [6].

Regarding decision-making breadth, it can vary in terms of the diversity and focus of the information processed and their sources and the comprehensive or restricted way in which it considers environmental, social and economic sustainability issues [4]. Various factors may be perceived and considered for both the supply and demand for sustainable goods and services. For example, from the standing point of consumers, there is an ample spectrum of drivers behind decisions for sustainable purchasing as the consumer's involvement and concerns on sustainability issues, the trust on company and green product, the perceived informational utility of a green product [16], self-image [17] and positive attitudes toward sustainability issues [16, 17].

Managers carry different sustainability motivations [18] and behave distinctly in decision-making stages [4]. Regarding drivers for sustainability, the literature addresses issues such as firm image, market opportunities [19, 20], moral duty, business strategy, compliance with legislation, economic benefit, operational efficiency, employee satisfaction [20], human health and labour practices [21, 22]. From both the consumer and the producer perspectives, it seems that the decision-making process might include a large or narrow breadth of issues relating to sustainability and decision-making.

The breadth may relate to the type of decision (strategic, tactical or operational). In this sense, Zarte et al. [23] infer that integrating the three sustainability dimensions is geared to activities at the strategic planning level. At the operational level, decisions are influenced at most by the economic and environmental dimensions. Also, the breadth is related to the type of evaluation required. For example, analyses involving Life Cycle Assessment (LCA) approaches demand the decision-maker to consider a broad spectrum of data, going through the product life cycle stages and considering different environmental impacts [e.g., 24–33].

Regarding decision-making depth, it can be evaluated according to the use of intuitive or analytical methods, differentiating itself by major or slight structuring of processes, by nature of the processed information—quantitative and qualitative, as well by the level of objectivity and subjectivity allowed [15]. Different characteristics relating to the depth of general decision-making processes also apply to the specific case of sustainability [4, 12].

Although the decision-making process may have a different breadth and depth characteristics, the literature underlines a continuum formed by different decision-making rationales where the agents may be positioned between two poles [4, 15, 34]. For example, some decision support methods allow analytically processing a larger and varied range of information, addressing sustainability dimensions broadly and working with subjectivity and ambiguity situations.

Models using the Analytical Hierarchy Process (AHP) method [e.g., 25, 32, 35–41] allow individual or group subjective judgments, addressing different selection criteria and weights. However, the applied method breadth, including the way to address sustainability issues, may depend on the decision-makers' style. The fuzzy techniques apply to decision processes that involve dealing with the complexity of sustainability, conflicts about the nature of the objectives, subjective judgments, qualitative elements, and expressing ambiguous and vague situations [e.g., 35, 37, 39, 42–48].

The use of analytical methods in decision-making has also been applied to the relationships between sustainability and technology. Information technology appears in Shrouf and Miragliotta's [49] evaluation of how the Internet of Things could improve manufacturing facilities' energy efficiency. Zhang et al. [30] propose a framework based on Big Data Analytics applied to LCA management. Souza et al. [50] conduct an SLR looking at data mining and machine learning approaches to assist in sustainability issues.

Finally, different decision processes can lead to distinct choices and, therefore, variations in the decision's effectiveness. Thus, how the decision-making process takes place is relevant so that actions are consistent with the expected results [6].

3   Research Method

This paper employs the step-by-step approach proposed by Thomé et al. [51] for SLRs. The eight steps are: (i) formulating the problem, (ii) searching the literature, (iii) data gathering, (iv) quality evaluation, (v) data analysis and synthesis, (vi) interpretation, (vii) reporting, (viii) updating the review.

In the first step of the SLR, the reviewers' team was formed, the scope of the SLR and the research questions were defined, as well as the initial categories to classify the studies. The scope and focus of the literature review were carefully discussed and agreed-upon among reviewers.

Scopus and Web of Science (WoS) are the databases used for the literature search, in the second step of the SLR. The search was restricted to peer-reviewed articles and reviews published in English only, with no limitations on publication years and no constraints to indexed journals. The search did not include grey literature and snowball searches, which constitutes a limitation of the study. The keywords used for search, applied to title, abstract and keywords, are ("decision making process") AND (manufact* OR factory OR factories) AND ("sustainability" OR "sustainable" OR "green" OR "corporate social responsibility" OR "csr" OR "corporate sustainability" OR "eco-innovation" OR "green technology" OR renewable* OR "social responsibility" OR "environmental social responsibility" OR "social environmental management" OR "sustainab* development" OR "triple bottom line" OR eco-efficien* OR eco-effectiv* OR "sustainab* development indicator" OR sdi OR "ISO 14001").

The search returned 139 papers from Scopus and 125 from the Web of Science, resulting in a final set of 184 studies after excluding duplicates. The main papers included in the analysis are listed in the references. A full list of selected papers is available upon request to the first author. The following exclusion criteria were applied to papers (i) dealing with decisions on specific matters without exploring the link between sustainability and decision processes; (ii) with specific goals or applications with no focus on the relationship between sustainability and decision-making; (iii) introducing methods or models to assist in the decision-making process but not directed to sustainability matters; (iv) with surveys analyzing specific concerns, with no focus on decision-making and sustainability; (v) dealing with decisions regarding the location of facilities.

The exclusion process took three rounds. For each round, the researchers worked independently of one another and discussed the cases of disagreement. The inter-coders reliability was high, resulting in 96% of agreement, Scott's Pi and Cohen's Kappa of 0.912 and a 0.913 Krippendorff's Alfa. After the selection process, 74 papers were analyzed after full-text reading.

Steps 3 and 4 (data gathering and quality evaluation) consisted of populating a concept matrix in Excel, portraying each study in lines, and the studies' characteristics in columns. Papers were selected from peer-reviewed journals, thus subjected to a first quality screening of primary data. The next section covers steps 5, 6 and 7: data analysis, synthesis, interpretation and presentation of results. Step 8, the update of reviews, is beyond the scope of this paper.

4 Analysis and Findings

4.1 Classification of Papers

This section shows the papers' main classifications, expecting that observed results can contribute to the academic environment, showing the evolution of research, the main topics addressed, and possible gaps that can be filled through future inquiry efforts.

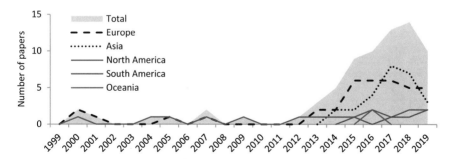

Fig. 1 Number of papers published over the years

Figure 1 exhibits the number of publications over the years, showing a step increase after 2011 and suggesting a greater concern in developing new proposals to support sustainability decisions. Another issue addressed herein reports to where the research efforts concentrate and how they evolve, portraying countries and institutions that are publishing most on the subject of decision-making and sustainability. This data can be useful for academics and practitioners, informing about places where new partnerships and collaboration can arise.

Regarding the number of publications per continent, Europe is leading (37 papers), and after comes Asia (26), North America (16), South America (5) and Oceania (2). A temporal aspect to observe is the evolution in the number of publications, mainly in Europe and Asia, suggesting a possible growth trend within the theme of sustainability decisions (Fig. 1). Analyzed papers involving Asia start recently in 2014, leveraged mainly by China, India, Iran and Malaysia. Studies from South America are also recent, starting in 2015 with the majority participation from Brazil.

A total of 30 countries and 110 different institutions were found in the database, characterizing a scattered distribution of publications. The USA is the most represented country in the number of papers (12). Italy is the second (11) and China comes after (9). It is important to note that a single article may have been written by authors from different countries, revealing different regions' collaborative work. While different countries were co-participants in developing the same study, other countries have conducted their research internally. For example, the USA, China, Italy, UK, and other countries launched collaborative efforts with other nations for publications. Brazil and Malaysia are some of the countries that produced papers without international partnership.

In addition to looking for the main actors in research on decision-making processes involving sustainability, this paper was also concerned with understanding how economic, environmental and social issues have been addressed. The results show that, in addition to the economic dimension, the environment is prevalent among the articles, constituting 93 and 96% of the analyzed sample, respectively. This result is not surprising due to the choice of keywords reported in Sect. 3—Research Method. Although data suggest some preference for environmental and economic concerns, it is observed that the social theme has been gaining space over time, despite being still under-investigated—it is present in half of the sample. Figure 2 shows the evolution in the number of publications involving the social theme, with strong growth after

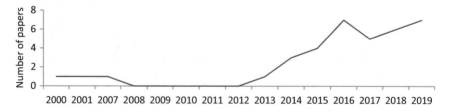

Fig. 2 Number of papers published involving the social theme of sustainability

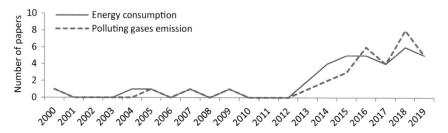

Fig. 3 Number of papers involving energy consumption and gas emissions problems

the year 2013, suggesting a positive growth trend of including the social dimension of sustainability in decision-making matters.

One relevant aspect to observe is how the researchers deal with sustainability problems regarding environmental and social dimensions. Results depicted in Fig. 3 show a predominance of environmental issues focused on resources consumption and atmospheric emissions. In addition to studies applying LCA approaches, energy consumption and gas emissions problems are the most prevalent. There is also an increase after 2013 in the number of papers considering energy consumption and gas emissions, suggesting a tendency for growth in studies involving these sustainability problems. Some of the main remaining problems encompass products End of Life (EOL) alternatives, highlighting the recycling, waste generation and social aspects related to workers' health and safety.

Another relevant issue refers to the company's activity related to the decision-making process. Results show that new product development appears in a prominent position, representing a quarter of the analyzed papers, mainly addressing EOL alternatives (e.g., design for recycling, design for remanufacturing) and LCA approaches (e.g., design for minimizing environmental impacts). Figure 4 depicts the most prevalent areas of activities regarding the decision-making process. The remaining areas (not listed in Fig. 4) encompass decisions regarding the manufacturing process design, waste management, ISO certification, use of environmental management systems, other factors to develop sustainable manufacturing. Other studies concerns are an understanding of enterprise and sustainable consumer behaviour.

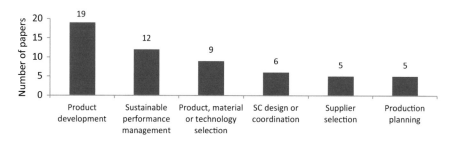

Fig. 4 Number of papers by decision-making area of activity

The articles in the sample did not show any growth trend in specific manufacturing areas of decision-making. Information regarding the area of activities is relevant because it helps readers think about which organizational departments, hierarchical levels and stakeholders could be involved in the decision-making. Depending on the area, strategic, tactical or operational decisions can be prevalent and different actors are involved in choosing alternatives.

Regarding industry sectors, nineteen categories were enlisted. The three most prevalent are the automotive, machinery/equipment/parts and electronic industries. The remaining encompasses metalworking, construction, food, digital and information technology, textile, chemical, energy/oil/gas, health and personal care, plastic, recycling, wood, maritime, infrastructure maintenance, agriculture, education and manufacturing in general. However, it is worth noting that it was not possible to classify the industry sector in 20% of the cases. Finally, selected papers deal mainly with focal companies rather than supply chains, which might be explained by the search strategy geared at papers addressing manufacturing decision-making.

4.2   Breadth and Depth of Decision-Making

This section discusses how sustainability-related decision-making can vary by the degree to which information is processed more broadly or more focused (breadth) and by the level of use of intuitive or analytical procedures (depth).

Firstly, it is worth mentioning that, due to the search strategy used, as detailed in the methodology section, it was expected that predominantly analytical articles would be retrieved—which was confirmed after data extraction. Regarding decisions breadth, it is noted that decision-makers can be indistinctively more restricted or broad while using analytical methods.

Among the 74 articles, 17 do not portray a decision model, guides or frameworks whose depth could be assessed. For example, they represent surveys and case study research investigating the motivations, barriers, and drivers for sustainable behaviour of companies, managers or consumers. However, they provide useful information on how comprehensive these agents are in adopting sustainable practices and may help companies develop sustainable strategies to attend consumer and organizational demands. Additionally, four decision support guides and frameworks allow for an ample range of data and analysis flexibility, depending on the manager's decision-making style. Thus, it was possible to evaluate the decision support models covered in 53 studies.

On the one hand, the first group of 19 papers uses analytical decision support models with structured data processing procedures and narrow information processing. The objective rather than subjective approaches are prevalent, and the processed data are predominantly quantitative, detailed and focused. We observe a preference in the use of optimization, simulation, forecasting and sensitivity analysis methods. Variables and parameters concerning environmental and economic impacts are prevalent; however, some focus on financial results and social themes are slightly explored.

On the other hand, the second group of 34 papers points to models or frameworks for decision support that allow analytically processing a wide and varied range of information. In this group, it is possible to verify an increasing inclusion of the social themes, addressing the dimensions of sustainability more comprehensively. Qualitative and subjective data are also observed in some papers, being analytically treated by some specific methods such as fuzzy techniques and AHP. Fuzzy techniques were widely applied in decision process encompassing qualitative data, vagueness, uncertainty, ambiguity and subjective judgments, as expected. Furthermore, many papers, in this group, support decisions aimed at new product development activities.

There is some tendency for the academy to seek alternatives that will support the decision-making process based on analytical procedures, but with broader approaches to information processing. Thus, this result aligns with the understanding of the existence of a continuum formed by different decision-making rationales [4, 15, 34].

5 Conclusion

This study reveals an increasing number of papers published in the last ten years, suggesting a rising interest by academics on the subject of sustainability decision-making. Europe is prominent in the number of published papers and seems to keep up with a growth trend, just as observed for Asia and slightly less for South and North America.

Looking at areas of activity requiring sustainable solutions helps in thinking about how sustainability is handled at the strategic, tactical or operational levels in the companies and the different stakeholders concerned with the decisions. The new product development activity, highlighting the design phase, is the most investigated topic, suggesting the attention for strategic rather than tactical or operational decisions.

Data also suggest an increasing effort to consider sustainability problems like energy consumption and polluting gas emissions on decision-making that seem to gain highlight after 2013. A gap and research opportunities were found for considering other matters such as social themes, waste generation, effluent emissions, and residual water, among others. Besides, the social dimension of sustainability is under-investigated compared to the environmental and economic ones. Finally, there is a concentration of articles dealing with decision-making involving sustainability analytically through a broader approach. This fact suggests an attempt to work with various data types, allowing the inclusion of qualitative and subjective elements in the analysis.

Acknowledgements The authors acknowledge the support of the following: Brazilian Coordination for the Improvement of Higher Education Personnel (CAPES)—Finance Code 001; National Council for Scientific and Technological Development (CNPq), Grant # 304931/2016–0,404682/2016–2 and 311862/2019-5, and the Research Support Foundation of the State of Rio de Janeiro (FAPERJ), Grant # E-26/203.252/2017; and the partner universities of Pontifical Catholic University of Rio de Janeiro (PUC-Rio) and Federal University of Juiz de Fora (UFJF).

References

1. Carroll, AB, Shabana, KM: The business case for corporate social responsibility: A review of concepts, research and practice. International Journal of Management Reviews 12(1), 85–105 (2010).
2. Smith, W., Lewis, M.: Toward a theory of paradox: A dynamic equilibrium model of organizing. Academy of Management Review 36(2), 381–403 (2011).
3. Gao, J., Bansal, P.: Instrumental and Integrative Logics in Business Sustainability. Journal of Business Ethics 112(2), 241–255 (2013).
4. Hahn, T., Preuss, L., Pinkse, J., Figge, F.: Cognitive frames in corporate sustainability: managerial sensemaking with paradoxical and business case frames. Academy of Management Review 39(4), 463–487 (2014).
5. Sharma, G., Jaiswal, AK: Unsustainability of Sustainability: Cognitive Frames and Tensions in Bottom of the Pyramid Projects. Journal of Business Ethics 148(2), 291–307 (2018).
6. Dean, JW, Sharfman, MP: Does decision process matter? A study of strategic decision-making effectiveness. Academy of Management Journal 39(2), 368–392 (1996).
7. Elkington, J.: Towards the Sustainable Corporation: Win-Win-Win Business Strategies for Sustainable Development. California Management Review 36(2), 90–100 (1994).
8. Edwards, W.: The theory of decision making. Psychological Bulletin 51(4), 380–417 (1954).
9. Mintzberg, H., Raisinghani, D., Theoret, A.: The Structure of "Unstructured" Decision Processes. Administrative Science Quarterly 21, 246–275 (1976).
10. Elkington, J.: Cannibals with forks: The triple bottom line of the 21st century business. Capstone, Oxford (1997).
11. WCED. Our Commom Future, https://sustainabledevelopment.un.org/content/documents/598 7our-common-future.pdf, last accessed 2020/07/29.
12. Sasse-Werhahn, LF, Bachmann, C., Habisch, A.: Managing Tensions in Corporate Sustainability Through a Practical Wisdom Lens. Journal of Business Ethics 163(1), 53–66 (2020).
13. Walsh, JP: Selectivity and selective perception: An investigation of managers' belief structures and information processing. Academy of Management Journal 31, 873–896 (1988).
14. Beyer, JM, Chattopadhyay, P., George, E. et al.: The selective perception of managers revisited. Academy of Management Journal 40(3), 716–737 (1997).
15. Simon, HA: Making management decisions: the role of intuition and emotion. Academy of Management Perspectives 1(1), 57–64 (1987).
16. Wei, CF, Chiang, CT, Kou, TC, Lee, BCY: Toward Sustainable Livelihoods: Investigating the Drivers of Purchase Behavior for Green Products. Business Strategy and the Environment 26, 626–639 (2017).
17. Xu, L., Prybutok, V., Blankson, C.: An environmental awareness purchasing intention model. Industrial Management and Data Systems 119(2), 367–381 (2019).
18. Bansal, P., Roth, K.: Why companies go green: A model of ecological responsiveness. Academy of Management Journal 3(4), 717–736 (2000).
19. Braglia, M., Petroni, A.: Stakeholders' influence and internal championing of product stewardship in the Italian food packaging industry. Journal of Industrial Ecology 4(1), 75–92 (2000).

20. Dey, PK, Petridis, NE, Petridis, K. et al.: Environmental management and corporate social responsibility practices of small and medium-sized enterprises. Journal of Cleaner Production 195, 687–702 (2018).

21. Martinsson, C., Lohela-Karlsson, M., Kwak, L. et al.: What incentives influence employers to engage in workplace health interventions?. BMC Public Health 16(1), 854 (2016).

22. Kaur, A., Sharma, PC: Social sustainability in supply chain decisions: Indian manufacturers. Environment, Development and Sustainability 20(4), 1707–1721 (2018).

23. Zarte, M., Pechmann, A., Nunes, I.L.: Decision support systems for sustainable Manufacturing surrounding the product and production life cycle – A literature review. Journal of Cleaner Production 219, 336–349 (2019).

24. Ribeiro, C., Ferreira, JV, Partidário, P.: Life cycle assessment of a multi-material car component. International Journal of Life Cycle Assessment 12(5), 336–345 (2007).

25. Choi, JK, Ramani, K.: An Integrated Decision Analysis for the Sustainable Product Design. In: Proceedings of the ASME 2008 International Manufacturing Science and Engineering Conference, vol. 1, pp. 311–318. ASME, Evanston (2009).

26. Du, G., Karoumi, R.: Life cycle assessment of a railway bridge: comparison of two superstructure designs. Structure and Infrastructure Engineering 9(11), 1149–1160 (2013).

27. Narayanan, A., Witherell, P., Lee, JH et al.: Identifying the Material Information Requirements for Sustainable Decision Making. In: Proceedings of the ASME 2013 International Design Engineering Technical Conferences and Computers and Information in Engineering Conference, 33rd Computers and Information in Engineering Conference, vol 2B. ASME, Portland (2014).

28. Eastwood, MD, Haapala, KR: A unit process model based methodology to assist product sustainability assessment during design for Manufacturing. Journal of Cleaner Production 108, 54–64 (2015).

29. Resta, B., Gaiardelli, P., Pinto, R., Dotti, S.: Enhancing environmental management in the textile sector: An Organisational-Life Cycle Assessment approach. Journal of Cleaner Production 135, 620–632 (2016).

30. Zhang, Y., Ren, S., Liu, Y. et al.: A framework for Big Data driven product lifecycle management. Journal of Cleaner Production 159, 229–240 (2017).

31. Favi, C, Campi, F., Germani, M., Manieri, S.: Using design information to create a data framework and tool for life cycle analysis of complex maritime vessels. Journal of Cleaner Production 192, 887–905 (2018).

32. Kolotzek, C., Helbig, C., Thorenz, A. et al.: A company-oriented model for the assessment of raw material supply risks, environmental impact and social implications. Journal of Cleaner Production 176, 566–580 (2018).

33. Vozzola, E., Overcash, M., Griffing, E.: Environmental considerations in the selection of isolation gowns: A life cycle assessment of reusable and disposable alternatives. American Journal of Infection Control 46(8), 881–886 (2018).

34. Kirton, M.: Adaptors and Innovators: a description and measure. Journal of Applied Psychology 61(5), 622 – 629 (1976).

35. Chunhua, F., Shi, H., Guozhen B.: A group decision making method for sustainable design using intuitionistic fuzzy preference relations in the conceptual design stage. Journal of Cleaner Production 243, 118640 (2020).

36. Huang, S., Zhang, H.: Green supply chain management of automotive manufacturing industry considering multiperspective indices. IEEJ Transactions on Electrical and Electronic Engineering 14(12), 1787–1795 (2019).

37. Solgi, E., Moattar Husseini, SM, Ahmadi, A., Gitinavard, H.: A hybrid hierarchical soft computing approach for the technology selection problem in brick industry considering environmental competencies: A case study. Journal of Environmental Management 248, 109219 (2019).

38. Taborga, CP, Lusa, A., Coves, AM: A proposal for a green supply chain strategy. Journal of Industrial Engineering and Management 11(3), 445–465 (2018).

39. Govindan, K.., Darbari, JD, Agarwal, V., Jha, PC: Fuzzy multi-objective approach for optimal selection of suppliers and transportation decisions in an eco-efficient closed loop supply chain network. Journal of Cleaner Production 165, 1598–1619 (2017).
40. Khatri, J., Srivastava, M.: Technology selection for sustainable supply chains. International Journal of Technology Management and Sustainable Development 15(3), 275–289 (2016).
41. Jagadish, Ray A.: Green Cutting Fluid Selection using Multi-attribute Decision Making Approach. Journal of The Institution of Engineers (India): Series C 96(1), 35–39 (2015).
42. Mohammadi, H., Farahni, FV, Noroozi, M., Lashgari, A.: Green supplier selection by developing a new group decision-making method under type 2 fuzzy uncertainty. International Journal of Advanced Manufacturing Technology 93, 1443–1462 (2017).
43. Grandhi, S., Wibowo, S.: Sustainability Performance Evaluation of Automotive Manufacturing Companies. In: 11th Conference on Industrial Electronics and Applications (ICIEA), pp. 1725–1730. IEEE, Hefei (2016).
44. Shahryari Nia, A, Olfat, L., Esmaeili, A. et al.: Using fuzzy Choquet Integral operator for supplier selection with environmental considerations. Journal of Business Economics and Management 17(4), 503–526 (2016).
45. Singh, S., Olugu, EU, Musa, SN: Development of sustainable manufacturing performance evaluation expert system for small and medium enterprises. Procedia CIRP 40, 608–613 (2016).
46. Wu, C., Barnes, D.: Partner selection for reverse logistics centres in green supply chains: a fuzzy artificial immune optimization approach. Production Planning and Control 27(16), 1356–1372 (2016).
47. Beng, LG, Omar, B.: Integrating axiomatic design principles into sustainable product development. International Journal of Precision Engineering and Manufacturing - Green Technology 1(2), 107–117 (2014).
48. Remery, M., Mascle, C., Agard, B.: A new method for evaluating the best product end-of-life strategy during the early design phase. Journal of Engineering Design 23(6), 419–441 (2012).
49. Shrouf, F., Miragliotta, G.: Energy management based on Internet of Things: Practices and framework for adoption in production management. Journal of Cleaner Production 100, 235–246 (2015).
50. Souza, J., Francisco, A., Piekarski, C. et al.: Data mining and machine learning in the context of sustainable evaluation: a literature review. IEEE Latin American Transactions 17(3), 372–382 (2019).
51. Thomé, AMT, Scavarda, LF, Scavarda, AJ: Conducting systematic literature review in operations management. Production Planning & Control 27(5), 408–420 (2016).

Business Simulation Games: A Comparative Analysis for Teaching of Production Engineering

Sofia Emi Maia Pinto Ishihara, Annibal Affonso Neto, and Clóvis Neumann

Abstract Business simulation games have been applied in several fields of college education as important motivational elements and learning tools, a way of bringing the content taught in the classroom closer to real-world problems. There are few comparative studies of the main features of simulation games available to support decision making in selecting the best options for different contexts. The aim of the present study is to analyse the largest number of business simulation games suitable for Production Engineering courses. The methodology used in the development of the study was qualitative research, especially the content analysis and document analysis. Sixteen business simulation games were analysed and classified in two dimensions using document and content analysis: general aspects and operations aspects; in 12 decision variables. The main contribution of the present study was to propose an analysis model and to make possible the joint analyse of simulation games by 13 different companies from 6 different countries.

Keywords Business simulation · Games classification · Comparative analysis · Production engineer

1 Introduction

COVID-19 pandemic precipitated important changes in education system, in particular in the teaching learning process. Because of the need to replace face-to-face classes with classes mediated by technology, synchronous or asynchronous ones, there was the necessity to rethink everything from hours, contents, methodologies, didactics and even the reorganization of the academic semester at universities. This

S. E. M. P. Ishihara (✉) · A. A. Neto · C. Neumann
University of Brasília, Brasilia, Federal District 70910-900, Brazil

A. A. Neto
e-mail: annibal@terra.com.br

C. Neumann
e-mail: clovisneumann@unb.br

intensified the use of remote activities in order to ensure students' interest and motivation in order to avoid dropout.

Three years ago, the Production Engineering Program at the University of Brasília began to implement the Special Topics in Business Simulation (STBS) course in order to develop skills that could facilitate students' employability at the end of the undergraduate major. The course has among its goals to improve students' links with companies and support developments in learning, especially in project-based learning (Problem Based Learning—PBL).

Business simulation games feature the reproduction of specific contexts of companies, covering the areas of finance, marketing, production and association between functions [1].

STBS course was developed in order to provide participants the experience in business and industrial management and business challenges through the application of business game simulation, in order to reproduce, in the most reliable way possible, the professional challenges that graduates will face in managerial decision making, involving the areas of marketing, human resources, finance, production, accounting and strategic administration.

STBS course was planned in order to align with the contents covered in an interdisciplinary set: design of production systems, factory design and layout, strategic management, financial management, management accounting, marketing management, production management and planning and control of production.

A fundamental phase was the analysis and comparison of simulation games. In this stage, some initial requirements were identified and analysed to classify the simulation games, so that it was possible to simulate decisions in an industrial environment and to get as close as possible to market situations: be comprehensive so that the student is in control of the data; provide detailed and updated reports; present performance indicators (Key Performance Indicator—KPI); simulate sudden changes in the market in which students could be rewarded for responding to these, in a time variable and random interventions in the real world, in order to simulate various horizons of business planning during the school term.

2   Business Simulation Games

From an educational perspective, games are learning tools and important motivational elements, a way of bringing content taught in the classroom closer to real-world problems, a "learn by doing" or a "practical" approach of the learning process. For this reason, business simulations become increasingly popular teaching methods in undergraduate courses [2, 3].

Schafranski and Tubino [1] affirm "for both individuals and organizations, learning to adapt to new situations is becoming as important as having a good performance in old situations". Although applied to the competitive context among companies, this statement portrays a trend for all modern situations, marked by high dynamism. In the scenario of pandemic and social distance, the adoption of distance classes proved to be the solution for educational institutions. Adapting to this new trend is a challenge that involves the entire traditional learning process.

In contrast to traditional teaching, simulation bridges the gap between the reality of the classroom and the business world through experiential learning experiences in which students analyze, design, implement and control business strategies.

Business simulations are used as learning platforms that stimulate students' interest in the "game", in addition to providing a structured learning environment. Introduced in 1956 by the American Management Association [4], simulations have grown considerably in popularity within business schools in recent years. The research showed that students perceive the simulations as: involving; useful; effective learning tools; and effective in promoting teamwork [5].

There are several ways to conceptualize what a simulation is. "A simulation is an exercise involving the reality of function in an artificial environment, a case study but with the participants inside" [6].

Simulation are usually company or industry business games in which players learn by managing a simulated company, most often with an industry or competitive environment. Students usually make decisions, but not exclusively, allocating resources [7]. The simulations can focus on the internal mechanisms/dynamics of the company, its interaction with the environment or both. In some games, decisions focus on a specific function or business analysis or cover many functional areas.

Simulation can be applied in different contexts, obtaining different functions: as a technique to help in decision making by providing operating information, performance, capacity, amount of manpower, viewing future problems and possible changes in preventive and predictive way for the creation of new systems; it represents a basis for budgets [8]; it instigates students' entrepreneurial sense [9, 10]; it helps the student learning [1, 11–14]; it is used in training, enabling employees to understand the implication of their duties in the supply chain [1]; it allows managers to test scenarios and experiment with decisions that are not subject to testing in practice [15].

Neumann et al. [12, 13] shows that with a school in which students have an experience with real cases, such as project-based learning in (PBL), the maturity level of training of the undergraduate students is incremented. Thus, the student's preparation for entering the job market is linked to the traditional teaching format, in which the gap between teaching programs and industrial reality is evident. Neto et al. [16] complement this idea by stating that if the combination of knowledge, skills and attitude in the classroom results in the student's competence, then traditional teaching, based only on the dissemination of content, does not contribute to the preparation of students to the challenges of the modern world.

Nonaka and Takeuchi [17] developed the theory of the spiral of learning. In this theory, knowledge retention increases according to the stimuli in the four stages of knowledge creation (Socialization, Externalization, Combination, Internalization) that are interconnected by the transformation of tacit and explicit knowledge. In other words, for the spiral to work, dialogues, field building, learning by doing and correlation with explicit knowledge are needed. This theory is shown to be completely in line with the business simulation learning proposal, also highlighting the gap in traditional education.

A business game is a complex game designed to provide students with the opportunity to learn by doing, involving them in a simulated real-world experience, in order to involve them in a management situation as close to reality as possible.

The business game involves the participants in the management process, developing their decision making capacity and compelling them to adopt a managerial point of view. The games are realistic, with the aim of simulating the real environment of the business world. Participants are divided into teams (companies) and immersed in companies that are to produce and sell products. Teams operate in simulated markets similar to real markets.

In industrial or production engineering, games have been used in business resource planning. Chwif and Barretto [18] developed a Didactic Operations Simulation Model (DOSM), which uses discrete event simulation as the main mechanism time advance. Leger [19] described the application of a simulation game using various SAP R/3 modules to teach ERP concepts to undergraduate and graduate students.

Adelsberger et al. [20] developed a simulation game "MASH Bikes", whose processes and model are derived from a bicycle factory. The game information processing is modelled on the SAP R/3 system and the production processes are modelled with the ARENA simulation system. The focus of the game is to help students to understand the complete logistics chain and the business processes involved in it.

Cox and Walker [10], considering that the concepts of balancing, planning and controlling the production line are difficult to understand for students who have no experience in a manufacturing company, developed the non-computerized game Socratic Dice & Penny to help students to understand the impact of restrictions and non-restrictions on production, line quality and efficiency and traditional programming, Kanban and drum buffer rope.

Bringelson et al. [21] developed the computer simulation game "NCTB" to help students to make decisions in an interdisciplinary group. The focus was on teaching four functional areas: purchasing, production planning and control, quality control and marketing.

Learning process with business games occurs at several levels. Students learn from the information contained in the dynamics of the game, in the risk assessment and risk-taking process, and by analysing the results of decisions made through management reports and performance indicators (KPI).

3   Methodology

The methodology used in the development of the study was qualitative research, in particular the content analysis and document analysis. Content analysis is "a research methodology that uses a set of procedures to produce valid inferences from a text. These inferences are about senders, the message itself, or the audience of the message [22].

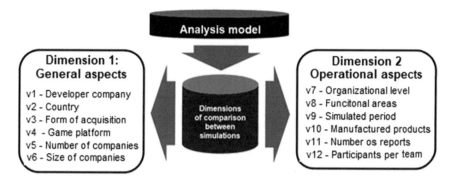

Fig. 1 Analysis model

Content analysis is "a research technique for making replicable and valid inferences from texts (or other significant issues) for the contexts of their use" [23].

With the company game manuals in hand, the material was read in four moments: exploratory, selective, analytical and interpretive reading [24]. In the exploratory reading, several simulation game manuals were examined in full in order to seek a relationship between their content and the analysis variables.

In this phase, the selection of documents to be analyzed was carried out. Through the documentary analysis, the game manuals of the simulation companies from 16 (sixteen) suppliers were available on the websites of the simulators developers: Bernard [25]; Management [26]; General Management [27]; Cesim Firm [28]; Production Management [29]; Production Systems Simulation Laboratory [30]; Moto Cycle [31]; Shoe Maker [31]; Simulare [32]; Industrial Strategic Simulator [33]; Industrial Simulator [34]; Production Planning Simulation [35]; Supply Chain and Channel Management [26]; Topaz Management Simulation [36]; Virtonomics Entrepreneur [37]; Virtual Business Retailing [38]. Data were collected between July and September 2020.

After the selection of simulation games, the analytical measurements were performed for the grouping of the information available. Using documentary and content analysis, including the analysis variables, and the analysis model (see Fig. 1), 16 (sixteen) selected simulation games were analyzed by 12 (twelve) variables classified in 2 (two) dimensions, each with 6 (six) variables.

4   Results

Dimension 1 gathers the variables Developer Company, Country, Form of acquisition, Game platform, Number of companies and Size of companies, called general aspects (see Table 1).

In dimension 1, the first variable analyzed (variable 1) is in relation to the companies that developed the games. 16 simulation games analyzed were developed by 13

Table 1 Dimension 1: general aspects

Simulation Game	Developer company (v1)	Country (v2)	Form of acquisition (v3)	Game platform (v4)	Number of companies (v5)	Size of companies (v6)
Bernard	Bernard Simulação Gerencial	Brazil	Paid	Online	10	S - M - L
Business Management	Marketplace Simulations	USA	Paid	Online	–	L
General Management	TOPSIM GmbH	Germany	Paid	Download or online	3 a 10 teams	S - M - L
Cesim Firm	Cesim Business Simulation Games	Finland	Paid	Online	2 a 12 teams	L
Production Management—PM	UFSC	Brazil	Free	Download	Single player	S - M - L
Production Systems Simulation Laboratory—PSSL	UFSC	Brazil	Free	Download	Single player	S - M - L
Moto Cycle	Simformer Business Simulation	USA	Paid	Online	Multiplayer	–
Shoe Maker	Simformer Business Simulation	USA	Paid	Online	Multiplayer	–
Simulare	Simulare Jogos Empresariais	Brazil	Paid	Online	–	S - M - L
Industrial Strategic Simulator	OGG Simulação Empresarial	Brazil	Paid	Online	–	L
Industrial Simulator	TINO Empresarial	Brazil	Paid	Online	8	S
Production Planning Simulation—PPS	LDP Jogos de Empresas	Brazil	Paid	Online	49 participants	–
Supply Chain and Channel Management	Marketplace Simulations	USA	Paid	Online	–	L
Topaz Management Simulation	SDG Simuladores e Modelos de Gestao	Portugal	Paid	Online	–	L

(continued)

Table 1 (continued)

Simulation Game	Developer company (v1)	Country (v2)	Form of acquisition (v3)	Game platform (v4)	Number of companies (v5)	Size of companies (v6)
Virtonomics Entreprenuer	Virtonomics Business Simulation Games	Cyprus	Paid	Online	Multiplayer	S - M - L
Virtual Business Retailing	Knowledge Matters	USA	Paid	Online	Single player	S - M - L

different companies. Only 3 of these companies had more than one game analyzed, UFSC (2), Marketplace Simulations (2) and Simformer Business Simulation (2), while for all the others only one game was analyzed. Regarding the country of origin of the companies that developed the games (variable 2), the companies are located in 6 countries. The countries with the most simulation games analyzed were Brazil (7) and the USA (5), but games from Germany, Finland, Portugal and Cyprus were also analyzed.

Regarding the Acquisition Form (variable 3), only SSPL and PM games, both developed at the UFSC's Production Systems Simulation Laboratory, are free, while all the other ones are paid. In terms of the platform used in the game (variable 4) most simulation games operate online, and it is only necessary to download the file for SSPL and PM games from the UFSC Production Systems Simulation Laboratory, while the game TOPSIM's General Management uses online or download platform.

Regarding the Number of Companies for each simulation game (variable 5), it can be noted that Virtonomics Entrepreneur, Motor Cycle and Shoe Maker simulators allow to simulate "Multiplayer online", that SPP also has no limit on companies. CESIM simulator allows 2 to 12 companies, Bernard simulator allows up to 10 companies, TINO simulator allows up to 8 companies, General Management simulator allows to simulate 3 to 10 companies. In SSPL and PM simulation games, companies compete for the highest PI (Performance Index). It was not possible to obtain information about the other simulation games.

As for the Size of the Simulated Companies (variable 6), 6 (six) simulation games (Business Management, Cesim Firm, Industrial Strategic Simulator, Supply Chain and Channel Management, Topaz Management Simulation and General Management) are configured to operate only in large-sized company; and Industrial Simulator from Tino Empresarial for small-sized company only.

Another 6 (six) simulation games were set up to operate in large, medium and small sized companies: Bernard, PM, SSPL, Simulare, Virtual Business Retailing and Virtonomics Entrepreneur. It was not possible to obtain this information about the other simulation games.

Dimension 2 displays the variables: Simulated games, Organizational level, Functional areas, Simulated period, Manufactured product, Number of reports e Number of participants recommended per team, called operational aspects (see Table 2).

Table 2 Dimension 2: general aspects

Simulation Game	Organizational level (v7)	Functional areas (v8)	Simulated period (v9)	Manufactured product (v10)	Number of reports (v11)	Participants per. team (v12)
Bernard	Strategic and Operational	Production management, Costs, Financial management, Sales, Human resources, Marketing	Quarterly	Phsyical goods	5	Free
Business Management	Strategic	Marketing, Product development, Manufacturing, Sales, Accounting, Financial management	Quarterly	Microcomputer business	5	–
General Management	Strategic and Operational	Marketing, Sales, Product development, Purchasing, Manufacturing, Human resources, Financial management	–	Graphic design industry	–	3 to 5
Cesim Firm	Strategic	Marketing, Sales, Manufacturing, Quality, Supply chain, Human resources, Financial management	Quarterly	Pharmaceutical industry	5	1 to 8

(continued)

Table 2 (continued)

Simulation Game	Organizational level (v7)	Functional areas (v8)	Simulated period (v9)	Manufactured product (v10)	Number of reports (v11)	Participants per team (v12)
Production Management—PM	Tactical and Operational	Production management, Manufacturing, Purchasing, Financial management	Monthly or weekly	Furniture industry	3	Up to 5
Production Systems Simulation Laboratory—PSSL	Tactical and Operational	Production management, Manufacturing, Purchasing, Financial management	Monthly or weekly	Knitting industry	3	Up to 5
Moto Cycle	Strategic	Production management, Human resources, Supply chain, Financial management, Marketing	Weekly	Motorcycle industry	–	1 to 5
Shoe Maker	Strategic	Production management, Human resources, Supply chain, Financial management, Marketing	Weekly	Shoes industry	–	1 to 5

(continued)

Table 2 (continued)

Simulation Game	Organizational level (v7)	Functional areas (v8)	Simulated period (v9)	Manufactured product (v10)	Number of reports (v11)	Participants per team (v12)
Simulare	Strategic and Operational	Supply chain, Financial management, Production process management, Human resources	Hours to months	6 options	25+	3 to 5
Industrial Strategic Simulator	Strategic	Manufacturing, Financial management, Marketing, Human resources	Hours to months	Phsyical goods	8+	–
Industrial Simulator	Operational	Sales, Production management, Human resources, Financial management, Manufacturing	Quarterly	Phsyical goods	3	Up to 4
Production Planning Simulation—PPS	Operational	Manufacturing, Human resources, Financial management	Quarterly	Fictional products	6	Up to 4
Supply Chain and Channel Management	Strategic	Marketing, Product development, Sales, Manufacturing, Human resources, Accounting, Financial Management, Supply chain	Quarterly	Microcomputer business	5	–

(continued)

Table 2 (continued)

Simulation Game	Organizational level (v7)	Functional areas (v8)	Simulated period (v9)	Manufactured product (v10)	Number of reports (v11)	Participants per team (v12)
Topaz Management Simulation	Strategic	Marketing, Financial management, Human resources, Manufacturing	Quarterly	3 products	11	3 to 5
Virtonomics Entreprenuer	Strategic and Operational	Manufacturing, Marketing, Supply chain, Financial management, Human resources	–	Clothes	2	–
Virtual Business Retailing	Strategic and Operational	Production management, Marketing, Purchasing, Financial management	–	Grocery/convenience store	7	–

Regarding the simulated Organizational Level (variable 7), 7 (seven) of the games operate focused on the strategic level: Business Management, Cesim Firm, Motor Cycle, Shoe Maker, Industrial Strategic Simulator, Supply Chain and Channel Management, Topaz Management Simulation. Another 4 (four) of the games operate focused on ranging from the strategic to the operational level: Bernard, General Management, Simulare and Virtronomics Entreprenuer.

Production Management (PM) and Production Systems Simulation Laboratory (PSSL) games from UFSC operate mainly between the tactical and operational levels, while Industrial Simulator and Production Planning Simulation (PPS) simulators operate mainly at the operational level.

Regarding the simulated Functional Areas (variable 8), all 16 (sixteen) games selected simulate Financial Management area, 12 (twelve) of them simulate Human Resources (HR) area, 11 (eleven) Manufacturing area and also 11 (eleven) Marketing area. The least simulated functional areas are Production Process Management (1), Quality (1), Costs (1), Accounting (2) and Product development (3).

The simulation games that encompass more functional areas are: Supply Chain and Channel Management (8), General Management (7) and Cesim Firm (7), Bernard (6) and Business Management (6). While the simulation games that include less functional areas are: Production Planning Simulation (PPS) (3), Production Management (PM) and Production Systems Simulation Laboratory (PSSL) from UFSC. Simulare, Industrial Strategic Simulator, Topaz Management Simulation and Virtual Business Retailing with 4 (four) each.

Regarding the simulated Period (variable 9), 7 (seven) of the simulators simulate quarterly periods: Business Management, Bernard, Cesim Firm, Industrial Simulator, Production Planning Simulation (PPS), Supply Chain and Channel Management and Topaz Management Simulation.

Simulators Production Management (PM) and Production Systems Simulation Laboratory (PSSL) from UFSC simulate monthly to weekly periods. Moto Cycle and Shoe Maker simulators simulate weekly periods and Industrial Strategic Simulator simulate a period of hours to months. It was not possible to obtain information from the simulators General Management, Virtronomics Entreprenuer and Virtual Business Retailing.

Regarding the Manufactured Product (variable 10), 10 (ten) games simulate the manufacture of specific products: Business Management, Supply Chain and Channel Management simulate the production of computers, TOPSIM's General Management simulates the functioning of a graphic design industry, Cesim Firm simulates the manufacture of medicines, Production Management (PM) and Production Systems Simulation Laboratory (PSSL) simulate the production of a furniture industry and a knitting industry, respectively. Simulators Moto Cycle and Shoe Maker, both from Simformer Business Simulation, simulate, as the name suggests, motorcycle manufacturing and shoe manufacturing. Virtual Business Retailing simulates a grocery/convenience store.

The other games do not simulate the manufacture of specific goods. Bernard simulator simulates the manufacture of durable consumer goods. Simulare presents 6 product options, while Production Planning Simulation (PPS) game simulates the manufacture of a fictitious product and Topaz Management Simulation simulates the manufacture of three types of products.

As for variable 11, the simulators that creates the largest number of reports are: Simulare generates more than 25 (twenty-five) different reports, Topaz Management Simulation 11 (eleven), Industrial Strategic Simulator emits more than 8 (eight) reports and Virtual Business Retailing 7 (seven). The simulators that generate less reports are: Virtronomics Entrepreneur 2 (two), Production Management (PM), Production Systems Simulation Laboratory (PSSL) and Industrial Simulator with 3 (three) reports each. It was not possible to obtain this information regarding the General Management, Moto Cycle and Shoe Maker simulators.

Regarding the recommended Number of Participants per team (variable 12), Bernard simulator does not limit the number of participants and up to 8 participants can participate 8 in Cesim Firm simulator. Most simulators recommend up to 5 (five) participants per team, such as General Management, Production Management (PM), Production Systems Simulation Laboratory (PSSL), Moto Cycle and Shoe Maker, Topaz Management Simulation and Simulare.

TINO Empresarial's Industrial Simulator recommends up to 4 (four) participants per team, as well as Production Planning Simulation (PPS) game. It was not possible to obtain this information regarding the following simulators: Business Management, Industrial Strategic Simulator, Supply Chain and Channel Management, Virtonomics Entrepreneur and Virtual Business Retailing.

5   Conclusion

Because of the COVID-19 pandemic the use of business simulation games became even more important, as it provides several benefits for students' learning. This article aimed to comparatively analyze the business simulation games for Production Engineering courses and to present an analysis model with dimensions, helping to choose the best option for each case.

As a recommendation for future studies, new research could be carried out to signal the improvement of the students' learning experience and to highlight other factors intrinsic to the learning process.

References

1. Schafranski LE, Tubino DF (2013) Simulação Empresarial em Gestão da Produção, 1st edn. Editora Atlas, São Paulo
2. Faria AJ (1998) Business simulation games: Current usage levels—An update. Simul Gaming 29:295–308
3. Keeffe MJ, Dyson DA, Edwards RR (1993) Strategic management simulations. A current assessment. Simul Gaming 24:363–368

4. Cohen K, Rhenman E (1961) The role of management games in education and research. Manag Sci 7(2):131166
5. Lainema T, Lainema K (2007) Advancing acquisition of business know-how: Critical learning elements. J Res Technol Educ 40(2):183–198
6. Thavikulwat P (2004) The architecture of computerized business gaming simulations. Simul Gaming 35(2):242269
7. Summers GJ (2004) Today's business simulation industry. Simul Gaming 35(2):208–241
8. Oliveira JB (2007) Simulação Computacional: Análise de um sistema de manufatura em fase de desenvolvimento. Repositório Universidade Federal de Itajubá, Itajubá
9. Yen WCY, Lin HH (2020) Investigating the effect of flow experience on learning performance and entrepreneurial self-efficacy in a business simulation systems context. In: Interactive Learning Environments, pp 1–16. Published online
10. Cox JF, Walker ED (2004) Using a Socratic game to introduce basic line design and planning and control concepts. Decis Sci J Innov Educ 2:77–82
11. Igidio SS, De Paula DDFF, Rodrigues AC, Gontijo TS, Braga LBM (2017) A contribuição de jogos de simulação na aprendizagem de alunos de engenharia de produção. In: Revista Espacios, vol 38, no 30, pp 23–31
12. Neumann C, Affonso Neto A, Aquere AL, Carvalho MTM (2018) The contribution of simulation games for the learning process of lean education in production engineering undergraduate course. In: 5th ELEC, pp. 185 – 192. Department of Production and Systems of School of Engineering of the University of Minho, Braga
13. Neumann C, Affonso Neto A, Rodrigues MMVOC (2018) Project-based learning (PBL) applied to developing a production systems project in conjunction with factory design and layout (FDL) methodology. In: 10th PAEE and 15th ALE, pp 352–359. PAEE association, Brasília
14. Miranda IG, Heras AD (2020) Uso de videojuegos de simulación empresarial como complemento de aprendizaje em el área de ingeniería de organización. Dirección y organización, vol 70, pp. 19–27. Revista Dyo, Espana
15. Alessa B (2010) Uma análise de jogos de empresas na área de planejamento da produção e uso integrado de sistema de informações. Repositório Universidade Estadual Paulista "Júlio de Mesquita Filho". Bauru, SP, Brazil
16. Affonso Neto A, Neumann C, Rodrigues MMVOC (2018) The adoption of business simulation as a pedagogical alternative within a problem-based learning approach (PBL) in engineering education. In: 10th PAEE and 15th ALE, pp 352–359. PAEE Association, Brasília
17. Nonaka IN, Takeuchi H (1995) The knowledge creating company – How Japanese companies create the dynamics of innovation, 1st edn. Oxford University Press, Oxford
18. Chwif L, Barretto MRP (2003) Simulation models as an aid for the teaching and learning process in operations management. In: Proceedings of the 2003 Winter Simulation Conference, vol 2, 1994–2000, December 2003, pp 7–10
19. Leger P-M (2006) Using a simulation game approach to teach ERP concepts. HEC Montreal, Groupe de recherche en systemes d'information, Montreal
20. Adelsberger HH, Bick MH, Kraus UF, Pawlowski JM (1999) A simulation game approach for efficient education in enterprise resource planning systems. In: Proceedings of ESM 99— Modeling & Simulation: A Tool for the Next Millennium, Warsaw
21. Bringelson LS, Lyth DM, Reck RL, Landeros R (1995) Training industrial engineers with an interfunctional computer simulation game. Comput Ind Eng 29:89-92
22. Weber RP (1985) Basic content analysis. Sage, Beverly Hills, CA
23. Krippendorff K (2004) Content analysis: An introduction to its methodology, 2nd edn. Sage, Thousand Oaks, CA
24. Gil AC (2010) Como elaborar projetos de pesquisa, 5. edn. Atlas, São Paulo
25. Bernard industrial page, https://bernard.com.br/simuladores/industrial-sind/. Last accessed 2020/08/26
26. Marketplace simulation homepage, https://www.marketplace-simulation.com/simulations. Last accessed 2020/09/05
27. TOPSIM homepage, https://topsim.com/en/education/. Last accessed 2020/09/04

28. Cesim homepage, https://www.cesim.com/br/compare-as-simulacoes. Last accessed 2020/09/02
29. LSSP – GP homepage, http://lssp.deps.ufsc.br/index_arquivos/GP.htm. Last accessed 2020/08/26
30. LSSP – PCP homepage. http://lssp.deps.ufsc.br/index_arquivos/LSSP_PCP.htm. Last accessed 2020/08/26
31. Simformer business simulation homepage, https://simformer.com/marketplace/. Last accessed 2020/09/05
32. Simulare homepage, https://simulare.com.br/jogo_de_empresas_sobre/. Last accessed 2020/08/23
33. OGG homepage, https://ogg.com.br/simuladores/. Last accessed 2020/09/04
34. Tino homepage, http://tinoempresarial.com.br/. Last accessed 2020/08/26
35. LDP homepage, http://www.ldp.com.br/?sessao=publico. Last accessed 2020/08/29
36. SDG homepage, http://www.sdg.pt/public_html/00_sdg/index.html. Last accessed 2020/09/02
37. Virtonomics homepage, https://virtonomics.com/pt/entrepreneur/. Last accessed 2020/09/04
38. Knowledge matters homepage, https://knowledgematters.com/highschool/retailing/. Last accessed 2020/09/04
39. Dod J (1999) Effective Substances. In: The dictionary of substances and their effects. Royal Society of Chemistry. Available via DIALOG. http://www.rsc.org/dose/title of subordinate document. Cited 15 Jan 1999
40. Slifka MK, Whitton JL (2000) Clinical implications of dysregulated cytokine production. J Mol Med. https://doi.org/10.1007/s001090000086
41. Smith J, Jones M Jr, Houghton L et al (1999) Future of health insurance. N Engl J Med 965:325–329
42. South J, Blass B (2001) The future of modern genomics. Blackwell, London

Descriptive Bibliometric Analysis on Vaccine Supply Chain Management for COVID-19

Paulo Henrique Amorim Santos⬤ and Roberto Antonio Martins⬤

Abstract Inefficient supply chain management could jeopardize or limit an effective vaccination campaign as well as the immunization programs. As we face a COVID-19 pandemic outbreak and population immunization is the best solution, it is critical to map the vaccine supply chain management to identify risks and opportunities for the Industrial Engineering and Operations Management researchers and practitioners. This paper aims to map the field, applying descriptive bibliometric analysis. The application of a string search in the science index Web of Science resulted in a sample of 808 documents. We used the Bibliometrix R for data processing. The scientific production of the domain has been rising since 2006. The few high productivity authors focus on Health Sciences and Immunology studies even though the results unfold a growing interest by academics from Engineering, Business and Economics, Operations Research, and Management Sciences domains. Vaccine and other health sciences journals are the most productive and present a higher impact. The authors' keywords dynamics analysis shows a resettling of technical terms as "stabilization" and "formulation" to supply chain management keywords as "logistics". Evidence points out the need for more Industrial Engineering and Operations Management research to increase vaccine distribution effectiveness.

Keywords Vaccine · Supply chain management · Cold chain · COVID-19 · Bibliometrics

1 Introduction

Vaccination is one of the most cost-effective methods to prevent and control infectious diseases like COVID-19. Immunization saves 2 to 3 million lives each year and a global vaccination coverage improvement could prevent an additional 1.5 million deaths [1]. The past decade witnessed several new vaccines' development

P. H. A. Santos (✉) · R. A. Martins
Federal University of Sao Carlos, Sao Carlos, São Paulo 13565-905, Brazil

© The Author(s), under exclusive license to Springer Nature Switzerland AG 2021
A. M. Tavares Thomé et al. (eds.), *Industrial Engineering and Operations Management*,
Springer Proceedings in Mathematics & Statistics 367,
https://doi.org/10.1007/978-3-030-78570-3_19

253

and approval, contributing to protecting bacterial and viral infections [2]. Besides, older vaccines were upgraded [2].

Besides preventive vaccination, reactive vaccination can occur during an outbreak of an infectious disease [3]. Researchers are currently attempting to develop a vaccine for Coronavirus disease 2019 (COVID-19), an infectious disease caused by the severe acute respiratory syndrome. It has caused socio-economic turmoil and more than a million deaths worldwide [4]. Healthcare resource allocation is scarce in a public health crisis and successful vaccination campaigns depend on the logistics operations' effectiveness [3]. Hence, the vaccine supply chain's strategy development is critical to address logistics and distribution obstacles before the vaccine is ready for application [5].

Generally, vaccine availability depends on managing a fragile supply chain that deals with specific equipment and procedures to maintain the vaccine quality [6]. The vaccines' instability often emerges as a critical problem during clinical development and commercial distribution [2]. Vaccines may become useless when exposed to temperatures outside the appropriate range. Recently, numerous vaccine-related adverse events have occurred globally, especially in developing countries, due to vaccines spoilage by inappropriate temperatures exposition during their transportation [7]. Therefore, such a supply chain's design is challenging and requires technical and managerial expertise as requirements are generally applicable to all vaccine supply chains [8]. Frequently, the combination of formulation approaches and efficient supply chain management can stabilize the vaccines [2].

The necessity of developing, producing, and distributing new vaccines to address medical requirements remains a high priority for improving public health [2]. Moreover, many developing countries do not have sufficient cold chain capacity and effective vaccine regulation policies and penalties [9]. Logistical aspects of vaccination have been attracting the interest of the Operations Management and Industrial Engineering researchers. Overviewing current literature on vaccine logistics and supply chain helps identifying cooperation opportunities for the management sciences [3]. Therefore, this article describes the vaccine supply chain management field, offering opportunities for developing research.

2 Research Design

Bibliometric methods are the most appropriate research method to describe a scientific domain regarding scientific production and its impact. The results could help explain the domain's boundaries and pose new insights. The analysis focuses on the document's metadata as authors, journals, authors' keywords, authors' affiliations, and countries [10, 11].

A bibliometric analysis follows the steps proposed by Zupic and Cater [10]: (1) Research design: research question definition, choose of adequate methods for data analysis; (2) Data compilation: scientific index selection, capture, and treatment of data applying filters; (3) Data analysis: bibliometric software selection, data cleaning, and statistical analysis; (4) Visualization: definition of preferred visualization methods and selection of appropriate software; and (5) Interpretation: describe and interpret the results.

The scientific index considered for document collection was the Web of Science, preferable for DOI reference linkage. We defined the following criteria to ensure consistency in document eligibility: (a) Only articles, early access articles, and reviews published in journals with a peer-review process; (b) No time filter; and (c) only documents in English. We applied the following search string in the Web of Science scientific index: TS = (("vaccine*" AND "supply chain*") OR ("vaccine*" AND "cold chain*")). The * symbol means that the suffix may vary. The feature allows considering the term derivations. Consequently, it improves search results. We processed the sample registers using the "Bibliometrix" R package, version 3.0.0 [11], running in the RStudio integrated development environment, version 1.2.5042. We also elaborated customized graphs using Excel spreadsheets and the R package "ggplot2" [12].

3 Results

We carried out the document search on 05/10/2020, resulting in a sample of 808 documents (681 articles, six early access articles, and 121 reviews), published in 324 journals. Figure 1 illustrates the scientific production field and its impact (citations) from 1977 to 2020. The annual scientific production (green line) has been rapidly increasing since 2006. Two spikes occurred in 2017 (95) and 2018 (91). Regarding the impact (purple line), two peaks of impact occurred in the early years, namely in 1991 (2.8) and 1995 (7.8). After 2006, the documents have also granted the field a

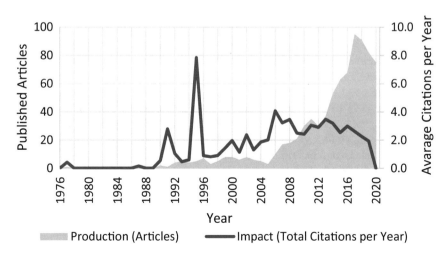

Fig. 1 Annual scientific production and impact

Table 1 Most globally cited articles

Paper	Total citations	TC per year	Ranking TC per year
Fine [13]	933	35,9	1
Giudice and Campbell [14]	304	20,3	2
Nochi et al. [15]	223	15,9	4
Le, TP; Coonan, KM; Hedstrom, RC; Charoenvit, Y; Sedegah, M; Epstein, JE; Kumar, S; Wang, RB; Doolan, DL; Maguire, JD; Parker, SE; Hobart, P; Norman, J; Hoffman, SL, 2000, Vaccine	178	8,5	11
Matthias et al. [6]	135	9,6	7
Kumru et al. [2]	121	17,3	3
Dal-Mas, M; Giarola, S; Zamboni, A; Bezzo, F, 2011, Biomass Bioenerg	118	11,8	5
Palatnik-De-Sousa, CB, 2008, Vaccine	118	9,1	8
Amorij et al. [17]	114	8,8	9
Chen et al. (2011)	107	10,7	6
Chen and Kristensen [18]	104	8,7	10
Makidon, PE; Bielinska, AU; Nigavekar, SS; Janczak, KW; Knowlton, J; Scott, AJ; Mank, N; Cao, ZY; Rathinavelu, S; Beer, MR; Wilkinson, JE; Blanco, LP; Landers, JJ; Baker, JR, 2008, Plos One	101	7,8	12

high impact profile (4.1). As expected, the curve has declined in recent years because the documents published recently earn fewer citations.

Table 1 complements Fig. 1, presenting the highest impact documents regarding total citations and total citations per year of publication. Fine [13] is the most cited article (933 citations and 35.9 citations per year), followed by Giudice and Campbell [14] (304 citations and 20.3 citations per year) and Nochi et al. [15] (223 citations and 15.9 citations per year). Fine deal with BCG vaccine stability; Giudice and Campbell focus on needle-free vaccine delivery; and Nochi et al. [15] propose a rice-based mucosal vaccine as a needle-free delivery. Fine [13] elucidates the peak on impact in 1995 (Fig. 1). Sutter et al. [16] that deal with a poliomyelitis outbreak explain the spike in 1991. Finally, Kumru et al. [2] exhibit a significant citation per year (17.3) coping with vaccine instability along the supply chain. All outlets, which published articles listed in Table 1, are from health and biological sciences domains.

Table 2 complements the previous information, focusing on the citations by sample documents (local citations). Matthias et al. [6] is the most influential paper concerning local citations. The article offers a systematic literature review of freezing temperatures effects in the vaccine supply chain. Following, Zaffran et al. [19], Kumru et al. [2], and Techathawat et al. [20] focus on vaccine stability. Almost all documents in Table 2 earned more than 50% of their citation from the sample documents. The

Table 2 Most locally cited articles (within the sample)

References	Local citations	Global citations	Local/total (%)
Matthias et al. [6]	84	135	62
Zaffran et al. [19]	38	68	56
Kumru et al. [2]	36	121	30
Techathawat et al. [20]	35	43	81
Kaufmann, JR; Miller, R; Cheyne, J, 2011, Health Affair	35	49	71
Thakker, Y; Woods, S, 1992, Brit. Med. J	30	46	65
Bell et al. [21]	28	58	48
Amorij et al. [17]	28	114	25
Lee et al. [23]	28	47	60
Nelson, C; Froes, P; Van Dyck, AM; Chavarria, J; Boda, E; Coca, A; Crespo, G; Lima, H, 2007, Vaccine	27	35	77
Otto, BF; Suarnawa, IM; Stewart, T; Nelson, C; Ruff, TA; Widjaya, A; Maynard, JE, 1999, Vaccine	26	40	65
Wang et al. [22]	26	57	46
Murhekar, MV; Dutta, S; Kapoor, AN; Bitragunta, S; Dodum, R; Ghosh, P; Swamy, KK; Mukhopadhyay, K; Ningombam, S; Parmar, K; Ravishankar, D; Singh, B; Singh, V; Sisodiya, R; Subramanian, R; Takum, T, 2013, B. World Health Organ	26	28	93

exceptions are Kumru et al. [2], Bell et al. [21], Amorij et al. [17], and Wang et al. [22]. That set of papers influenced intensively other sample papers.

Figure 2 illustrates the prominent field authors regarding their scientific production and impact. The bubble size represents the number of articles published in the year, and the color intensity is proportional to the total number of citations per year. The most productive authors, Lee BY (30 articles since 2011) and Brown ST (28 papers since 2011) co-authored many papers and focused on the thermostability and vaccine supply chain's freezing capacity and thermal performance. Connor DL (19), Norman BA (18), Rajgopal J (17), Wasteska AR (15), Haidari LA (13), and Chen SI (12) also co-authored several papers with Lee BY and Brown ST. They have significantly contributed to the field since 2011. The authors cited previously high productivity explain the growing scientific field production in Fig. 1. However, those authors' articles do not figure among the most cited (Tables 1 and 2). The exceptions are Kristensen and Chen (Table 1, [18]), Kristensen (Table 2, [19]), Lee et al. (Table 2, [23]).

Table 3 details the authors' influence inside the sample offering information on the most cited by sample authors. Lee BY (227 citations) is the most productive author and the most influential author inside the sample. Following him, Chen DX, Matthias DM, Kartoglu U, Amorij JP, and Lydon P also had considerable influence on sample

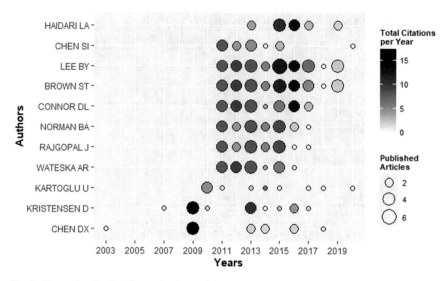

Fig. 2 Top authors' scientific production and impact

Table 3 Most cited authors (within the sample)

Author	Citations
World Health Organization	507
WHO	483
Lee B. Y.	227
Chen D. X.	87
Matthias D. M.	84
UNICEF	65
PATH	64
Kartoglu U.	61
Amorij J. P.	60
Lydon P.	60

authors. It is worthy of noting the impact of the World Health Organization (WHO) and UNICEF. Table 3 also helps explain the previous findings from Tables 1 and 2 and Fig. 2 concerning high global citations, high rates of local per global citations, and high productivity. The authors cited previously are co-authors of many articles cited by sample authors.

Figure 3 offers a different perspective on the most relevant authors by applying Lotka's Law. The law predicts that scientific productivity follows an inverse square law and only a small number of authors are likely to publish many articles in a domain [24]. According to Lotka's Law, the sample studied has a higher frequency than expected for "occasional" authors, i.e., those who have published only one article. Although there are a few high productivity authors in the field, there are also

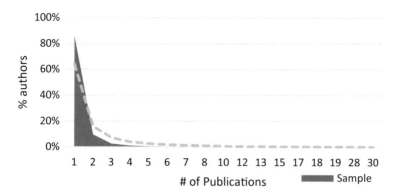

Fig. 3 Lotka's Law applied to sample documents

Table 4 Top journals of the field

Source	Articles	First publication year	Total citations	H index
Vaccine	199	1992	3276	175
Expert Review of Vaccines	20	2009	421	81
Plos One	20	2008	379	300
Human Vaccines & Immunotherapeutics	19	2012	116	50
Bulletin of the World Health Organization	17	1990	539	158
BMC Public Health	15	2005	146	130
International Journal of Pharmaceutics	15	2010	162	204
Journal of Infectious Diseases	14	2000	86	241
European Journal of Pharmaceutics And Biopharmaceutics	10	2011	107	150
Journal of Controlled Release	9	2002	387	256
Scientific Reports	9	2014	32	179
Biologicals	8	2003	157	53
American Journal of Tropical Medicine and Hygiene	7	2003	126	144

many "occasional" authors. This finding points out a lack of central authors in the domain.

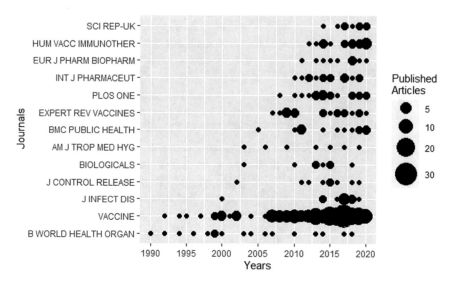

Fig. 4 Journals dynamics

Table 4 offers the sample leading journals based on the number of articles published, the total number of citations, and the journal's h index. Most outlets have started publishing articles in the 2000s, except for Vaccine and Bulletin of the World Health Organization. Vaccine dominance is outstanding (199 articles) regarding published articles and impact (citation and h index). Plos One, International Journal of Pharmaceutics, Journal of Infectious Diseases, and Journal of Controlled Release also pose a high h index. It is relevant to highlight the lack of Industrial Engineering, Operations Management, and Management Science journals.

Figure 4 complements Table 4, offering a dynamic perspective of the outlets. B. World. Health. Organ. and Vaccine are pioneers in the field. Figure 4 reinforces Vaccine dominance in recent years. Vaccine has also been publishing steadily since 2006, and Expert Rev. Vaccines has also been publishing steadily since 2014. The findings strengthen the health sciences journals' presence in the field.

Figure 5 sheds light on the authors' research area dynamics regarding scientific production and impact. The findings help understand the weakness of Industrial Engineering, Operations Management, and Management Science authors, papers, and journals in the vaccine supply chain field. We prefer to use the research area which the authors' identified instead of WoS research areas. Health Sciences are much longer present in the domain concerning scientific production and impact. Biological Sciences as Microbiology, Biotechnology, and Biochemistry have joined up in the last two decades, showing significant productivity. In the last ten years, the field has been growing the interest of research areas such as Engineering, Operations Research and Management Sciences, and Business and Economics, but it is not representative.

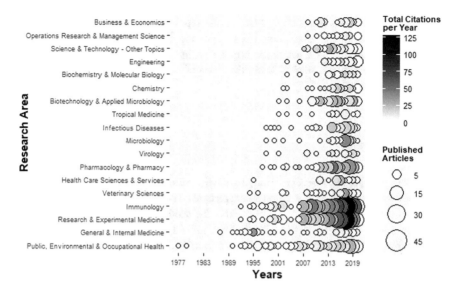

Fig. 5 Research area dynamics

Moving to research themes in the field and using the authors' keywords as proxies, Fig. 6 presents the dynamics of the 23 most used authors' keywords. We joined similar terms as "cold chain" and "cold-chain", "vaccine" and "vaccines", and "immuniza-tion" and "immunisation". As expected, the terms "cold chain" and "vaccine" shows high frequency because they are like field proxies. Following them, the most cited

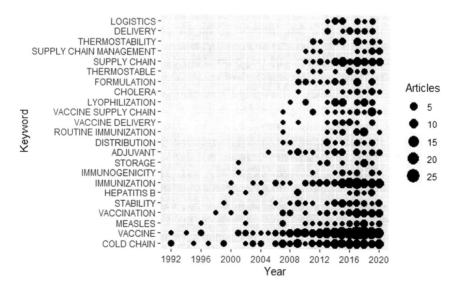

Fig. 6 Author's keywords dynamics

keywords relate to stabilization and formulation approaches. Subsequently, "immunization" and "delivery" also show high incidence. The most recent keywords are supply chain management terms as "supply chain" and "logistics". Finally, "thermostability", "storage", and "distribution" stand out as some challenges of this specific supply chain.

4 Conclusions

This paper provides a step towards closing the literature gap by offering a bibliometric study on the vaccine supply chain management. The annual scientific production has been rapidly increasing since 2006 and the production curve shape shows the scholars' rising interest (Fig. 1). The findings unfold two interesting points.

The first point is a highly productive group of authors in the field (Fig. 2). They co-authored some articles and presented a significant impact over the sample authors regarding the high local citation and the high rate between local and global citations (Tables 2 and 3). It is relevant to highlight that we cannot check if those authors' high citation results from self-citation. Although they are very productive, there is still no central set of authors who produce consistently over time in the field (Fig. 3). Despite that, those top productivity authors require attention from the domain's scholars. They focus on technical issues rather than management matters. That links to the next point.

The second and most crucial point is the dominance of technical issues over management matters. The bibliometric analysis on most cited papers (global and local), the most productive authors, the most relevant journals, and the most cited research areas and keywords by sample authors reveals technical issues' preeminence (Health and Biological Sciences). The weakness of scientific production and the impact of Industrial Engineering, Operations Management, and Management Science scholars on the field is apparent (Fig. 5). That is troublesome because the cold chain (the most cited keyword—Fig. 6) depends on technical (lyophilization, thermostable, thermostability—Fig. 6) and management (supply chain management, logistics, delivery—Fig. 6) issues to deliver the quality vaccines to immunize the population against diseases like Covid-19. Especially regarding the current pandemics, logistics operations will play a key role in delivering millions of vaccine doses quickly to people worldwide. Haghani et al. [25] demonstrate similar concerns regarding COVID-19. More specifically, Wang et al. [26] warm about the challenges in COVID-19 distribution regarding the vaccine stabilization. Rowan and Laffey [27] highlight the items' shortage (medication, personal, and protective equipment) at the beginning of the current pandemics. Therefore, there are many risks and opportunities in improving the vaccine supply chain management that requires more attention and research efforts from Industrial Engineering, Operations Management, and Management Science scholars.

Acknowledgements The authors acknowledge the financial support by the Coordenação de Aperfeiçoamento de Pessoal de Nível Superior—Brasil (CAPES)—Financing Code 001.

References

1. World Health Organisation. Immunization. https://www.who.int/news-room/facts-in-pictures/detail/immunization (2019). Accessed 07 October 2020
2. Kumru, O. S., Joshi, S. B., Smith, D. E., Middaugh, C. R., Prusik, T., Volkin, D. B.: Vaccine instability in the cold chain: mechanisms, analysis and formulation strategies. Biologicals (2007). https://doi.org/10.1016/j.biologicals.2014.05.007
3. Duijzer, L. E., van Jaarsveld, W., Dekker, R.: Literature review: The vaccine supply chain. Eur. J. Oper. Res. (2018). https://doi.org/10.1016/j.ejor.2018.01.015
4. Worldometer. COVID-19 Coronavirus Pademic. https://www.worldometers.info/coronavirus (2020). Accessed 07 October 2020
5. Deo, S., Manurkar, S., Krishnan, S., Franz, C.: COVID-19 Vaccine: Development, Access and Distribution in the Indian Context. ORF Issue Brief. https://www.orfonline.org/research/covid19-vaccine-development-access-and-distribution-in-the-indian-context-69538/ (2020). Accessed 07 October 2020
6. Matthias, D. M., Robertson, J., Garrison, M. M., Newland, S., Nelson, C.: Freezing temperatures in the vaccine cold chain: a systematic literature review. Vaccine (2007). https://doi.org/10.1016/j.vaccine.2007.02.052
7. Lin, Q., Zhao, Q., Lev, B.: Cold chain transportation decision in the vaccine supply chain. Eur. J. Oper. Res. (2020). https://doi.org/10.1016/j.ejor.2019.11.005
8. Lemmens, S., Decouttere, C., Vandaele, N., Bernuzzi, M.: A review of integrated supply chain network design models: Key issues for vaccine supply chains. Chem. Eng. Res. Design (2016). https://doi.org/10.1016/j.cherd.2016.02.015
9. Ashok, A., Brison, M., LeTallec, Y.: Improving cold chain systems: Challenges and solutions. Vaccine (2017). https://doi.org/10.1016/j.vaccine.2016.08.045
10. Zupic, I., Čater, T.: Bibliometric methods in management and organization. Organ. Res. Methods (2015). https://doi.org/10.1177/1094428114562629
11. Aria, M., Cuccurullo, C.: bibliometrix: An R-tool for comprehensive science mapping analysis. J. Inf. (2017). https://doi.org/10.1016/j.joi.2017.08.007
12. Wickham, H.: ggplot2: elegant graphics for data analysis. Springer, New York (2016).
13. Fine, P. E.: Variation in protection by BCG: implications of and for heterologous immunity. The Lancet (1995). https://doi.org/10.1016/S0140-6736(95)92348-9
14. Giudice, E. L., Campbell, J. D.: Needle-free vaccine delivery. Adv. Drug. Deliv. Ver. (2006). https://doi.org/10.1016/j.addr.2005.12.003
15. Nochi, T., Takagi, H., Yuki, Y., Yang, L., Masumura, T., Mejima, M., Nakanishi, U., Matsumura, A., Uozumi, A., Hiroi, T., Morita, S., Tanaka, K., Takaiwa, F., Kiyono, H.: Rice-based mucosal vaccine as a global strategy for cold-chain-and needle-free vaccination. Proc. Nation Acad. Sci. (2007). https://doi.org/10.1073/pnas.0703766104
16. Sutter, R. W., Patriarca, P. A., Cochi, S. L., Pallansch, M. A., Kew, O. M., Hall, D. B., Suleiman, A. J. M., EI-Bualy, M. S., Bass, A. G., Alexander, J. P.: Outbreak of paralytic poliomyelitis in Oman: evidence for widespread transmission among fully vaccinated children. The Lancet (1991). https://doi.org/10.1016/0140-6736(91)91442-W
17. Amorij, J.P., Huckriede, A., Wischut, J., Frijlink, H.W., Hinrichs, W.L.J.: Development of Stable Influenza Vaccine Powder Formulations: Challenges and Possibilities. Pharm. Res.-Dordr (2008). https://doi.org/10.1007/s11095-008-9559-6
18. Chen, D., Kristensen, D.: Opportunities and challenges of developing thermostable vaccines. Expert Rev. Vaccines, (2009). https://doi.org/10.1586/erv.09.20

19. Zaffran, M., Vandelaer, J., Kristensen, D., Melgaard, B., Yadav, P., Antwi-Agyei, K. O., Lasher, H.: The imperative for stronger vaccine supply and logistics systems. Vaccine (2013). https://doi.org/10.1016/j.vaccine.2012.11.036

20. Techathawat, S., Varinsathien, P., Rasdjarmrearnsook, A., Tharmaphornpilas, P.: Exposure to heat and freezing in the vaccine cold chain in Thailand. Vaccine (2007). https://doi.org/10.1016/j.vaccine.2006.09.092

21. Bell, K.N., Hogue, C.J.R., Manning, C. Kendal, A.P.: Risk factors for improper vaccine storage and handling in private provider offices. Pediatrics (2001). https://doi.org/10.1542/peds.107.6.e100.

22. Wang, L. X., Li, J. H., Chen, H. P., Li, F. J., Armstrong, G. L., Nelson, C., Ze, W. Y., Shapiro, C. N.: Hepatitis B vaccination of newborn infants in rural China: evaluation of a village-based, out-of-cold-chain delivery strategy. B. World Health Organ. (2007). https://doi.org/10.2471/blt.06.037002

23. Lee, B. Y., Cakouros, B. E., Assi, T. M., Connor, D. L., Welling, J., Kone, S., Djibo, A., Wateska, A. R., Pierre, L., Brown, S. T.: The impact of making vaccines thermostable in Niger's vaccine supply chain. Vaccine (2012). https://doi.org/10.1016/j.vaccine.2012.06.087

24. Leimkuhler, F., Chen, Y.: "A Relationship between Lotka' s Law, Bradford' s Law, and Zipf' s Law". J. Am. Soc. Inf. Sci. (1986). https://doi.org/10.1002/(SICI)1097-4571(198609)37:5<307::AID-ASI5>3.0.CO;2-8

25. Haghani, M., Bliemer, M. C. J., Goerlandt, F., Li, J.: The scientific literature on Coronaviruses, COVID-19 and its associated safety-related research dimensions: A scientometric analysis and scoping review. Safety Sci. (2020). https://doi.org/10.1016/j.ssci.2020.104806

26. Wang, J. L., Peng, Y., Xu, H. Y., Cui, Z. R. Williams, R. O.: The COVID-19 Vaccine Race: Challenges and Opportunities in Vaccine Formulation. AAPS PharmSciTech (2020). https://doi.org/10.1208/s12249-020-01744-7

27. Rowan, N. J., Laffey, J. G.: Challenges and solutions for addressing critical shortage of supply chain for personal and protective equipment (PPE) arising from Coronavirus disease (COVID19) pandemic - Case study from the Republic of Ireland. Sci. Total. Env. (2020). https://doi.org/10.1016/j.scitotenv.2020.138532

Disaster Influencing Migratory Movements: A System Dynamics Analysis

Luiza Ribeiro Alves Cunha, **Joaquim Rocha dos Santos**,
and **Adriana Leiras**

Abstract The number of migrants, refugees, and asylum seekers have dramatically increased in the past few years as a result of instability, socio-economic crises, and natural and human-made disasters. This growth of migratory movements has attracted the attention of academics interested in social-impact oriented research. In this context, this paper proposes a causal loop diagram to analyze the influence of disasters in migratory movements. Based on a systematic literature review (SLR) encompassing the system dynamics method, we identify variables that relate disasters and migratory movements. Thus, we contribute to the literature by proposing a taxonomy (list of variables) and a framework (a causal loop diagram) to support decision-makers in operations management resolutions.

Keywords Disaster · Systematic literature review · System thinking

1 Introduction

The global migration crisis requires that researchers and policymakers comprehend the magnitude of the consequences of these migrations and the influencing factors. Society as a whole needs to discuss the problem and propose solutions to the fact that millions of people leave their homes daily and deal with a lack of respect for human and fundamental rights every year. Due to inhuman situations, researches show that forced migration brings post-traumatic stress disorder, major depression, or psychotic illnesses in these people [1–3].

L. R. A. Cunha (✉) · A. Leiras
Pontifical Catholic University of Rio de Janeiro, Rio de Janeiro, Brazil

A. Leiras
e-mail: adrianaleiras@puc-rio.br

J. R. dos Santos
São Paulo University, São Paulo, Brazil
e-mail: jrsantos@usp.br

© The Author(s), under exclusive license to Springer Nature Switzerland AG 2021
A. M. Tavares Thomé et al. (eds.), *Industrial Engineering and Operations Management*,
Springer Proceedings in Mathematics & Statistics 367,
https://doi.org/10.1007/978-3-030-78570-3_20

Researches have discussed migrants, refugees, and asylum seekers challenges but, although disasters have increasingly impacted society, few papers encompass the influence of disasters on migratory movements [4, 5]. Between 2000 and 2016, 11,374 natural disasters occurred, including the 2004 tsunami in Indonesia, Hurricane Katrina in 2005, and the Haiti earthquake in 2010, leading to more than 1 million dollars in economic damage, more than 1 million casualties, and more than 3 billion people affected [6]. The 2010 Haiti earthquake triggered massive population movement such that less than twenty days after the earthquake, migratory movements had caused the population of the capital Port-au-Prince to decrease by an estimated 23% [7], and led thousands of Haitians to flee the country in search of safety and survival [8]. In light of the ongoing political crisis in Venezuela, massive migration to other countries has also marked the Venezuelan most significant population mobilization, with more than 4 million Venezuelans forced to leave their homeland [9].

As the impact of natural (such as hurricanes, floods, earthquakes, droughts) or human-made (such as refugee crises, political crises, terrorist attacks) disasters—disruptions that physically affect a system and threatens its priorities and goals—has increased, the importance of studying different ways to mitigate, prepare, respond, and recovery to them has also gained importance [10–13]. In this context, to reduce the impact of disasters worldwide by helping decision-makers to make better decisions in disaster operations, academics and practitioners have developed diverse methods and tools [14, 15]. One of the most commonly recognized techniques in operation management is the simulation, as it provides a subsidy to perform experiments in the real world, opportunity to participate in model development, gain a more detailed understanding about the problems they face, and help prepare aid workers beyond what theory and knowledge can offer [6].

System dynamics (SD) has been used to study a variety of subjects, including project management [16], finance [17], logistics [18, 19], city management [20], economic development [19], water resources management, and urban water systems modeling [21, 22]. Also, some researchers have used SD to study problems related to hazards and disasters, for example, vehicle fleet management [23], road-rush-repairs after earthquakes [24], trade-offs between a provision of relief assistance and capacity building in humanitarian organizations [25], and investments in disaster management capabilities and pre-positioning of inventory [26]. More recently, studies applying SD to study refugee crises have also gained attention [4, 5].

In this context, the present research aims to analyze the influence of disasters in migratory movements to answer the following research question: what are the central relationships that decision-makers should focus on for managing migratory movements due to disasters? We deliver a taxonomy, a framework, and a research agenda, as proposed by Torraco [27] as results of a systematic literature review (SLR). The contribution of the present study relies on gathering a list of variables relating disasters to migratory movements (taxonomy) and on developing a causal loop diagram (CLD) connecting these variables (framework). Up to our knowledge, this is the first research attempt to apply a causal loop diagram to study the influence of disasters in migratory movements using variables found in academic literature searches.

The remainder of this paper is structured as follows: Sect. 2 outlines the research methodology. Section 3 addresses the results. Section 4 presents the causal loop diagram based on the papers retrieved from SLR. Finally, Sect. 5 provides the concluding remarks.

2 Research Methodology

SD method consists of breaking down a complex problem into a set of variables and parameters and assigning mathematical rules to determine the interactions between them [28]. SD modeling starts with a problem statement, represented in terms of the evolution of key variables over time. Such key variables provide clues about significant "stocks" (or levels) that describe the state of the system and "flows" (or rates) that change such stocks over time [25]. Feedback processes (and other elements such as delays and nonlinearities) are incorporated to capture the interconnection among different parts determining the dynamics of the system [29]. Once the problem is caught and represented in a causal loop diagram (CLD), it is possible to translate them into a mathematical simulation model, using a stock and flow diagram, making it achievable for managers to approach the model to study the consequences of interactions among variables, test the side effects of decisions, and systematically explore new strategies. Thus, CLDs are an essential tool for representing and communicating the feedback structure of a system and consists of variables connected by arrows denoting the causal influences among the variables [29].

To define the list of variables to be connected in the CLD, we use the SLR methodology. The present research adopts the eight-step proposed by Thomé et al. [30] to conduct the SLR: (i) research problem formulation; (ii) literature search; (iii) data collection; (iv) quality assessment; (v) data analysis and synthesis; (vi) interpretation; (vii) presentation of results; (viii) updating of the review.

We described the research problem formulation in the introductory section. In the accomplishment of the literature search, we selected the Scopus and Web of Science databases since they have the most extensive catalogue of indexed peer-reviewed journals [31]. Peer-reviewed papers were considered as the primary source for the literature review due to its academic relevance. However, conference papers were also added in the SLR to avoid bias risks. The set of keywords was defined to cover the topic broadly enough to prevent artificial limitation of the documents obtained at the same time, providing limits to exclude undesirable results [32]. The search considered two groups of keywords (one regarding disaster and other regarding SD), and the search used the title, abstract, and keyword topic: TITLE-ABS-KEY (disaster) OR TITLE-ABS-KEY(*migra*) OR TITLE-ABS-KEY (refugee) OR TITLE-ABS-KEY ("asylum seeker") AND TITLE-ABS-KEY("system dynamics").

The definition of eligible articles for this study considers articles related to disasters or migratory movements using SD. Equally important, the exclusion criteria in the abstract and full reading steps include papers discussing animal migrations, planetary transition, and urban movements (people leaving the countryside and going to big cities due to industries and other jobs).

Still, in the second stage, two Ph.D. specialists with more than four years of experience in disaster research were responsible for reviewing the titles and abstracts of the database documents, and the reliability of the process has been verified. Reliability, then, was verified by agreement between the reviewers (independent) and by the Krippendorff's alpha agreement index. The results show that the reliability indexes of the alpha Krippendorff reached 0.872. Alphas greater than 0.8 are considered acceptable [33].

3 Results

The majority of papers cover natural disasters such as flood, earthquake, hurricane, typhoon, volcano eruption, and drought [34–46]. Still, there are also papers surrounding human-made disasters as terrorism, and nuclear disaster [47–49]. Besides the papers covering natural and human-made disasters, 8 studies embrace migratory movements [4, 5, 50–55], and some papers do not specify the type of disaster [56–61].

Table 1 summarizes the variables related to disasters and migratory movements found in the sample of 69 articles selected in the SLR.

Table 1 Variables related to disasters and migratory movements

Variables	References
Disaster prevention	[34, 57]
Disaster preparedness	[26, 35, 38, 44, 58]
Awareness level (AL)	[35, 37, 44]
Damage and Loss	[34, 35, 43, 62, 63]
Affected population (AP)	[42, 43, 46, 47, 63–65]
Unaffected population (UAP)	[42, 46]
Relief operation	[28, 36, 46, 47, 61]
Recovery operations	[39, 40, 45, 64, 66, 67]
GAP between restorations and necessary restorations	[39, 40]
GAP between people requiring and receiving relief	[46, 47]
Mass migration out (MMO)	[4, 5, 49]
Press interest	[35, 48, 59]
Pressure to provide relief/recovery	[46]
International mobilization	[26, 35, 65]

(continued)

Table 1 (continued)

Variables	References
Humanitarian organizations mobilization	[36, 46, 59, 61]
Government mobilization	[39, 47, 65, 68]
Volunteers	[46, 63, 69]
Workers	[46, 47, 62, 63, 67, 69, 70]
Supplies	[28, 41, 42, 59, 64, 71, 72]
Donations	[59, 60, 73]
Black market, fraud and corruption	[73]
Resettled Iimmigrant	[54]
Illegal immigrant	[50, 54]
Socio-economic crises	[4]
Civil conflicts	[4]
Migration risks	[4]
Information of other immigrants	[50]
Desire for foreign labor	[52]
Politics regarding immigrants	[4, 53–55]
Pre-immigration and resettlement stressors	[53]
Desired foreign labor	[52]
Public–private partnership regarding immigrants resettlement	[4]
Impacts on countries receiving immigrants	[4]
Willingness to receive immigrants	[4]
Economic Impacts	[4]
Social security/safety impact	[4]
Income	[4]
Unemployment rate	[4]
Population resistance	[4]

These variables become even clear as they relate to each other. Thus, Sect. 3 presents the prosed CLD to detail the relations between variables.

4 Framework

After analyzing the papers, we propose a framework (CLD), using Vensim PRO software (Ventana System), to summarize the literature.

When a disaster occurs, the population is divided into the affected and unaffected population. Figure 1 shows that the affected ones can either became unaffected or influence the mass migration out; that is, a flow of people leaving the affected place.

The awareness level (AL) stock variable—the awareness of being at risk—is influenced by: (i) Natural AL, means that each person living in a disaster-prone area has a certain degree of disaster awareness [35] and accepts it up to a certain level; (ii) affected population (AP), and; (iii) Unaffected population (UAP). Thus, the increase of UAP, decreases the awareness level while the increase of AP, increases the awareness level, causing the pressure to disaster prevention and preparedness.

The awareness level, in turn, influences the prevention/ preparedness to new disasters, that will affect the damage that a disaster can cause. In the CLD, we use the auxiliary variable "Damage and Loss" in a general manner but, papers address besides physical injuries [42, 46], some infrastructure damages such as closed/restricted roads [40, 45], ports [56, 66], hospitals [58], industrial facilities [39], houses [64, 67], public buildings [40] and communication systems [28].

The mass migration rate (MMO) is influenced by the affected population [42, 73] and also has a natural influx (natural MMO). The awareness level still influences the mass migration rate, as the effect of the awareness can increase the number of affected population migration out. Lastly, the mass migration rate is affected by disaster prevention/ preparedness as the more disaster prevention actions; fewer people desire to leave their houses. Regarding the prevention/ preparedness to new disasters, the literature also addresses the adoption of early warning systems [37, 57].

Figure 1 also presents the government, humanitarian organizations (HO), international mobilization, and volunteers. The mobilization of these actors will influence the available labor force and supplies. Donations also affect the available supplies—here, we can mention the unsolicited donations that pose a challenge during disasters [59, 60]. Black market donation sales, fraud, and corruption also negatively affect the supplies [73]. Regarding international mobilization, custom clearance is also an essential factor influencing the arrival of supplies [26, 65].

Humanitarian workers and supplies are essential for relief operations. The difference between the people that require relief (the affected population) and the people that received relief; call media attention (press interest), consequently pressuring more the actors to provide relief [48]. Similarly, the recovery operations—such as the debris management [45]—present a difference between the damage restorations and the damages to restore that call the media interest, also creating more pressure to provide recovery on the involved actors.

Figure 2 represents the complete CLD, where we describe the flow of people. Of those who migrate and became immigrants, some manage to settle in another country, becoming a resettled immigrant and, some of them became illegal immigrants due to border crossing or overstaying a visa [50]. Figure 2 also shows the flow of those illegal immigrants that are repatriated, and the flow of illegal immigrants that become resettled immigrants.

Figure 2 also brings to the CLD new important variables. The variable 'Factors influencing migratory movements' represents other regular factors found in the literature, beyond disasters, that usually influence the mass migration out rate, those being: socio-economic crises [4], civil conflicts, migration risks [4], information of other immigrants—the resettled ones [50], politics regarding immigrants [4, 54], and the desire for foreign labor [52]. The variable 'Factors influencing resettle rate', in

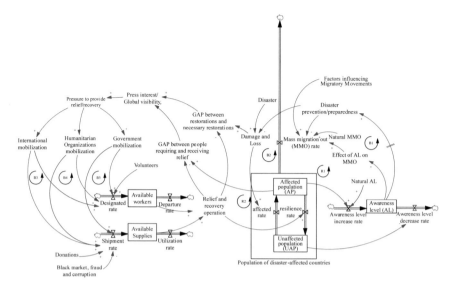

Fig. 1 Disaster causal loop diagram

turn, represents the variables influencing the resettle rate, those being: the willingness to receive immigrants, the impacts on the countries receiving immigrants and, the public–private partnership in the process of immigrants resettlement [4].

Two new stock variables can also be observed in Fig. 2: 'Resident population' and 'Available Labor Force'. The gap between the resident population and the target labor population makes the foreign labor force more or less desired, therefore influencing the mass migration out rate. Regarding the workforce (those immigrants that can work and are looking forward to this), the pre-immigration and resettlement stressors are represented as influencing the stock of workers available. Those stressors are related to health care access, to adverse working conditions, to social exclusions, food insecurity due to low wages, and can be minimized with politics protecting those immigrant workers [53].

By adding the available workforce, the income and the unemployment rate of the country that receive the immigrants, are affected. By affecting the income, the willingness to receive immigrants and the impacts on the country that receive immigrants are also affected, influencing the resettle rate (the rate of immigrants that effectively are accepted in another country).

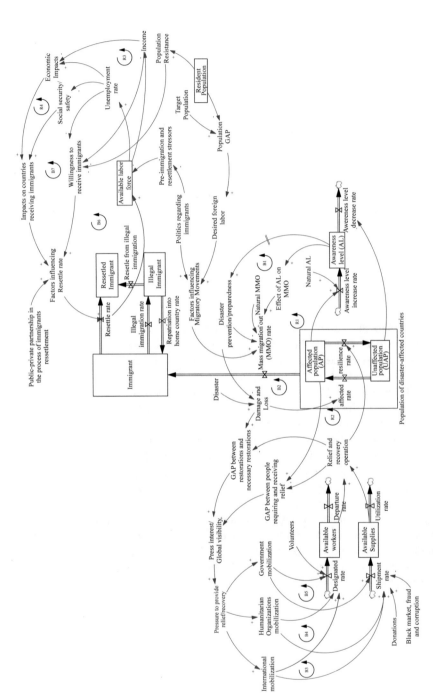

Fig. 2 Migratory movement and disaster causal loop diagram

5 Conclusion

This paper analyzes the disaster influence in migratory movements based on a SLR and a CLD. We identify the essential variables to be used in the diagram through a SLR and deliver a taxonomy list, where the variables and the references are represented. After the identification of the variables related to disasters and migratory movements literature, the developed CLD helped to answer the research question: How disaster influence migratory movements from the perspective of system dynamics? Our findings indicate that disaster influence the mass migration out rate. Thus, in addition to answering the proposed research question, the present study identified other regular factors that usually influence migratory movements: socio-economic crises, civil conflicts, migration risks, information of other immigrants, politics regarding asylum seekers, and the desire for foreign labor.

The proposed framework (CLD) structures and integrates different perspectives regarding disasters and migratory movements, and forms a basis of the academy view, allowing new studies to validate, add or contrast what already exists. Furthermore, the CLD is a new result for the academy as well as a tool so that practitioners can refresh decision-making skills.

Based on the gaps identified in the academic literature, Future studies should analyze and aggregate to the CLD other stakeholder's participation, such as humanitarian organizations, government, international mobilization, volunteers, and media. Corroborating with other researchers [28, 34], we also believe that a real case—practical implementation—should be used to validate the CLD here presented, to understand the most robust interactions in different circumstances, and to add other variables not covered by the academic literature to the diagram.

Besides, the interactions between the stock variables, auxiliary variables, and rates of the proposed CLD also deserve attention. To do that, we suggest the development of stock and flow diagrams to simulate real cases and validate the relationships here demonstrated.

Acknowledgements Coordination for the Improvement of Higher Education Personnel (CAPES) [Finance Code 001].

References

1. Hollifield, M., Warner, T.D., Lian, N., Krakow, B., Jenkins, J.H., Kesler, J., Westermeyer, J.: Measuring trauma and health status in refugees: a critical review. Jama 288(5), 611–621 (2002).
2. Fazel, M., Wheeler, J., Danesh, J.: Prevalence of serious mental disorder in 7000 refugees resettled in western countries: a systematic review. The Lancet 365(9467), 1309–1314 (2005).
3. Steel, Z., Chey, T., Silove, D., Marnane, C., Bryant, R.A., Van Ommeren, M.: Association of torture and other potentially traumatic events with mental health outcomes among populations exposed to mass conflict and displacement: a systematic review and meta-analysis. Jama 302(5), 537–549 (2009).

4. Yang, J., Dong, H.: A prediction based migration route evaluation method for refugees. In 10th International Congress on Image and Signal Processing, 1–5 (2017).
5. Dolezal, O., Tomaskova, H.: System dynamics in migration modeling. Proceedings of the 31st International Business Information Management Association Conference (IBIMA), 5056–5060 (2018).
6. Mishra, D., Kumar, S., Hassini, E.: Current trends in disaster management simulation modelling research. Annals of Operations Research, 1–25 (2018).
7. Lu, X., Bengtsson, L., Holme, P.: Predictability of population displacement after the 2010 Haiti earthquake. Proceedings of the National Academy of Sciences 109(29), 11576–11581 (2012).
8. Thomaz, D.: Post-disaster Haitian migration. Forced Migration Review, 43 (2013).
9. UNHCR. Refugees and migrants from Venezuela to 4 million: UNHCR and IOM 2019. https://www.unhcr.org/news/press/2019/6/5cfa2a4a4/refugees-migrants-venezuela-top-4-million-unhcr-iom.html, last accessed 2020/08/27.
10. Van Wassenhove, L.N.: Humanitarian aid logistics: supply chain management in high gear. Journal of the Operational Research Society 57(5), 475–489 (2006).
11. Leiras, A., de Brito Jr, I., Peres, E.Q., Bertazzo, T.R., Yoshizaki, H.T.Y: Literature review of humanitarian logistics research: trends and challenges. Journal of Humanitarian Logistics and Supply Chain Management 4(1), 95–130 (2014).
12. Condeixa, L.D., Leiras, A., Oliveira, F., de Brito Jr, I.: Disaster relief supply pre-positioning optimization: A risk analysis via shortage mitigation. International journal of disaster risk reduction 25, 238–247 (2017).
13. Bravo, R.Z.B., Leiras, A., Cyrino Oliveira, F.L.: The Use of UAV s in Humanitarian Relief: An Application of POMDP-Based Methodology for Finding Victims. Production and Operations Management 28(2), 421–440 (2019).
14. Coles, J.B., Zhang, J., Zhuang, J.: Bridging the research-practice gap in disaster relief: Using the IFRC Code of Conduct to develop an aid model. Annals of Operations Research (2017).
15. Pyakurel, U., and Dhamala, T.N.: Continuous dynamic contraflow approach for evacuation planning. Annals of Operations Research 253(1), 573–598 (2017).
16. Cheng, J.K.; Tahar, R.M., Ang, C.L.: Understanding the complexity of container terminal operation through the development of system dynamics model. International Journal of Shipping and Transport Logistics 2(4), 429–443 (2010).
17. Melse, E.: The financial accounting model from a system dynamics' perspective. Social Science Electronic Publishing, 7624 (2006).
18. Campuzano, F., Mula, J., and Peidro, D.: Fuzzy estimations and system dynamics for improving supply chains. Fuzzy Sets and Systems 161(11), 1530–1542 (2010).
19. Barber, P. López-Valcárcel, B.G.: Forecasting the need for medical specialists in Spain: Application of a system dynamics model. Human Resources for Health 8(1), 24 (2010).
20. Haase, D. and Schwarz, N.: Simulation models on Human Nature interactions in urban landscapes: A review including spatial economics, system dynamics, cellular automata and agent-based approaches. Living Reviews in Landscape Research 3(2), 1–45 (2009).
21. Gastélum J.R., Valdés J.B., Stewart S.: A decision support system to improve water resources management in the Conchos basin. Water Resource Management 23, 1519–1548 (2009).
22. Winz, I., Brierley, G., Trowsdale, S.: The use of system dynamics simulation in water resources management. Water Resource Management 23, 1301–1323, (2009).
23. Besiou, M., Stapleton, O., Van Wassenhove, L.N.: System dynamics for humanitarian operations. Journal of Humanitarian Logistics and Supply Chain Management (2011).
24. Zhang T.Z., Lu Y.M.: Study on simulation and optimization of the road rush-repair model after disaster. Applied Mechanics and Materials 50(2), 298–303 (2011).
25. Gonçalves, P.: Balancing provision of relief and recovery with capacity building in humanitarian operations. Operations Management Research (2011).
26. Kunz, N., Reiner, G., Gold, S.: Investing in disaster management capabilities versus pre-positioning inventory: a new approach to disaster preparedness. International Journal of Production Economics 157, 261–272, (2014).

27. Torraco, R.J.: Writing Integrative Literature Reviews: Guidelines e Examples. Human Resource Development Review 4(3), 356–367 (2005).
28. Diedrichs, D.R., Phelps, K., Isihara, P.A.: Quantifying communication effects in disaster response logistics: A multiple network system dynamics model. Journal of Humanitarian Logistics and Supply Chain Management 6(1), 24–45 (2016).
29. Sterman, J.D. Business Dynamics: Systems Thinking and Modeling for a Complex World, McGraw-Hill, Irwin, PA (2000).
30. Thomé, A.M.T., Scarvada, L.F., Scarvada, A.J.: Conducting Systematic Literature Review in Operations Management. Production Planning & Control 27(5), 408–420 (2016).
31. Mongeon, P., Paul-Hus, A.: The journal coverage of Web of Science and Scopus: a comparative analysis, Scientometrics 106, 213–228 (2016).
32. Cooper, H.: Research synthesis and meta-analysis: A step-by-step approach, Applied Social Research Methods Series 2, Sage Publications, Thousand Oaks, CA, USA (2010).
33. Krippendorff, K.: Content analysis: An introduction to its methodology. Pennsylvania: USA: SAGE Publications (2012).
34. Song, K., You, S., Chon, J.: Simulation modeling for a resilience improvement plan for natural disasters in a coastal area. Environmental pollution 242, 1970–1980 (2018).
35. Powell, J.H., Mustafee, N., Chen, A. S., Hammond, M.: System-focused risk identification and assessment for disaster preparedness: Dynamic threat analysis. European Journal of Operational Research 254(2), 550–564 (2016).
36. Cohen, J., Quilenderino, J., Bubulka, J., Paulo, E. P.: Linking a throughput simulation to a systems dynamics simulation to assess the utility of a US Navy foreign humanitarian aid mission. Defense & Security Analysis 29(2), 141–155 (2013).
37. Simonovic S.P., Ahmad S.: Computer-based model for flood evacuation emergency planning. Natural Hazards 34(1), 25–51 (2005).
38. Ahmad, S., Simonovic, S. P.: Modeling human behavior for evacuation planning: A system dynamics approach. In Bridging the Gap: Meeting the World's Water and Environmental Resources Challenges, 1–10 (2001).
39. Hwang, S., Park, M., Lee, H.S., Lee, S., Kim, H.: Postdisaster interdependent built environment recovery efforts and the effects of governmental plans: Case analysis using system dynamics. Journal of construction engineering and management 141(3), 04014081 (2014).
40. Hwang, S., Park, M., Lee, H. S., Lee, S.: Hybrid simulation framework for immediate facility restoration planning after a catastrophic disaster. Journal of Construction Engineering and Management 142(8), 04016026 (2016).
41. Peng, M., Peng, Y., Chen, H.: Post-seismic supply chain risk management: A system dynamics disruption analysis approach for inventory and logistics planning. Computers & Operations Research 42, 14–24 (2014).
42. Peng, M., Chen, H., Zhou, M.: Modelling and simulating the dynamic environmental factors in post-seismic relief operation. Journal of Simulation 8(2), 164–178 (2014).
43. Kuwata, Y., Takada, S.: Effective emergency transportation for saving human lives. Natural Hazards 33(1), 23–46 (2004).
44. Gillespie, D. F., Robards, K. J., Cho, S.: Designing safe systems: Using system dynamics to understand complexity. Natural Hazards Review 5(2), 82–88 (2004).
45. Kim, J., Deshmukh, A., Hastak, M.: A framework for assessing the resilience of a disaster debris management system. International Journal of Disaster Risk Reduction 28, 674–687 (2018).
46. Sopha, B.M., Asih, A.M.S.: Human resource allocation for humanitarian organizations: a systemic perspective. In MATEC Web of Conferences 154, 01048 (2018).
47. Yu, W., Lv, Y., Hu, C., Liu, X., Chen, H., Xue, C., Zhang, L.: Research of an emergency medical system for mass casualty incidents in Shanghai, China: a system dynamics model. Patient preference and adherence 12, 207 (2018).
48. Gunawan, I., Gorod, A., Hallo, L., Nguyen, T.: Developing a system of systems management framework for the Fukushima Daiichi Nuclear disaster recovery. In 2017 International Conference on System Science and Engineering (ICSSE), 563–568 (2017).

49. Ager, A. K., Lembani, M., Mohammed, A., Ashir, G. M., Abdulwahab, A., de Pinho, H., Zarowsky, C.: Health service resilience in Yobe state, Nigeria in the context of the Boko Haram insurgency: a systems dynamics analysis using group model building. Conflict and health 9(1), 30 (2015).

50. Shearer, N., Khasawneh, M., Zhang, J., Bowling, S., Rabadi, G.: Sensitivity analysis of a large-scale system dynamics immigration model. In 2010 IEEE Systems and Information Engineering Design Symposium, 78–81 (2010).

51. Pedamallu, C.S., Ozdamar, L., Akar, H., Weber, G.W., Özsoy, A.: Investigating academic performance of migrant students: A system dynamics perspective with an application to Turkey. International Journal of Production Economics 139(2), 422–430 (2012).

52. Ansah, J. P., Riley, C. M., Thompson, J. P., Matchar, D. B.: The impact of population dynamics and foreign labour policy on dependency: the case of Singapore. Journal of Population Research 32(2), 115–138 (2015).

53. Sonmez, S., Apostolopoulos, Y., Lemke, M. K., Hsieh, Y.C.J., Karwowski, W.: Complexity of occupational health in the hospitality industry: Dynamic simulation modeling to advance immigrant worker health. International Journal of Hospitality Management 67, 95–105 (2017).

54. Vernon-Bido, D., Frydenlund, E., Padilla, J.J., Earnest, D.C.: Durable solutions and potential protraction: The syrian refugee case. In Proceedings of the 50th Annual Simulation Symposium, 19 (2017).

55. Palmer, E.: Beyond proximity: Consequentialist ethics and system dynamics. Systems Engineering Applied to Evaluate Social Systems: Analyzing systemic challenges to the Norwegian welfare state (2017).

56. Kwesi-Buor, J., Menachof, D.A., Talas, R.: Scenario analysis and disaster preparedness for port and maritime logistics risk management. Accident Analysis & Prevention (2017).

57. Pujadi, T.: Early warning systems using dynamics system for social empowerment society environment. In 2017 International Conference on Information Management and Technology (ICIMTech), 304–309 (2016).

58. Voyer, J., Dean, M. D., Pickles, C. B.: Hospital evacuation in disasters: uncovering the systemic leverage using system dynamics. International Journal of Emergency Management 12(2), 152–167 (2016).

59. Besiou, M., Pedraza-Martinez, A.J., Van Wassenhove, L.N.: Vehicle supply chains in humanitarian operations: Decentralization, operational mix, and earmarked funding. Production and Operations Management 23(11), 1950–1965 (2014).

60. Ozpolat, K., Rilling, J., Altay, N., Chavez, E.: Engaging donors in smart compassion: USAID CIDI's Greatest Good Donation Calculator. Journal of Humanitarian Logistics and Supply Chain Management 5(1), 95–112 (2015).

61. Heaslip, G., Sharif, A.M., Althonayan, A.: Employing a systems-based perspective to the identification of inter-relationships within humanitarian logistics. International Journal of Production Economics 139(2), 377–392 (2012).

62. Chow, V. W., Loosemore, M., McDonnell, G.: Modelling the impact of extreme weather events on hospital facilities management using a system dynamics approach, (2012).

63. Xu, J., Xie, H., Dai, J., Rao, R.: Post-seismic allocation of medical staff in the Longmen Shan fault area: case study of the Lushan Earthquake. Environmental Hazards 14(4), 289–311 (2015).

64. Diaz, R., Kumar, S., Behr, J.: Housing recovery in the aftermath of a catastrophe: Material resources Perspective. Computers & Industrial Engineering 81, 130–139 (2015).

65. Turner, R.: Barriers to customs entry at the time of disaster in developing countries: Mitigating the delay of life-saving materials. World Customs Journal 9(1), 3–14 (2015).

66. Cho, H., Park, H. Constructing resilience model of port infrastructure based on system dynamics. Disaster Management 7(3), 245–254 (2017).

67. Kumar, S., Diaz, R., Behr, J.G., Toba, A.L. Modeling the effects of labor on housing reconstruction: A system perspective. International Journal of Disaster Risk Reduction 12, 154–162 (2015).

68. Zhong, Z., Yang, S., Duan, Y.: Sustainable Development of Typhoon Prone Coastal Areas Based on SD Model. Journal of Coastal Research 83(1), 754–769 (2018).

69. Xie, H., Rao, R.: A system dynamics model for medical staff allocation in post-Wenchuan earthquake relief. In 2014 IEEE Workshop on Advanced Research and Technology in Industry Applications (WARTIA), 307–311 (2014).
70. Armenia, S., Tsaples, G., Carlini, C.: Critical Events and Critical Infrastructures: A System Dynamics Approach. In International Conference on Decision Support System Technology, 55–66 (2018).
71. Min, P., Hong, C.: System dynamics analysis for the impact of dynamic transport and information delay to disaster relief supplies. In 2011 International Conference on Management Science & Engineering, 93–98 (2011).
72. Zhao, S.: Research on Information Feedback Mechanism of Emergency Supplies Based on System Dynamics. In ICLEM 2010: Logistics For Sustained Economic Development: Infrastructure, Information, Integration, 2235–2240 (2010).
73. Xu, L., Meng, X., Xu, X.: Natural hazard chain research in China: A review. Natural Hazards 70(2), 1631–1659 (2014).

The Deprivation Cost in Humanitarian Logistics: A Systematic Review

Maria Angélica Silva⊙ and **Adriana Leiras**⊙

Abstract The main objective of humanitarian logistics is to save lives and alleviate human suffering before and after a disaster onset. However, the available resources may not be enough to meet the beneficiaries' demand. Thus, the deprivation cost is a way of quantifying human suffering, given an unmet demand. In this paper, we carried out a systematic review of the literature covering deprivation cost and unmet demand to analyze the state of the art of modelling in humanitarian logistics. With the results obtained, we deliver a taxonomy to synthesize the models used. Finally, we suggest directions for future studies.

Keywords Deprivation cost · Unmet demand · Systematic literature review

1 Introduction

In 2019, disasters affected approximately 95 million people and claimed 11,755 fatalities, in addition to causing about 130 billion dollars of damage worldwide [1]. A disaster can be defined as "a severe disruption in the functioning of a community or society at any scale, due to extreme events interacting with exposure conditions, vulnerability and adaptability, leading to impacts such as human, material, economic or environmental losses" [2].

During and after disasters, some commodities (such as water) are essential for human survival, considering that the shortage can have a significant impact. Besides, one of the bottlenecks of Humanitarian Logistics (HL) are supplies. The shortage of supplies may occur either in the demand forecasting process, in the acquisition process or the distribution to the population [3]. For this reason, Holguín-Veras et al.

M. A. Silva (✉) · A. Leiras
Pontifical Catholic University of Rio de Janeiro—PUC-Rio, Rua Marquês de São Vicente 255, Brazil

A. Leiras
e-mail: adrianaleiras@puc-rio.br

[4] proposed an innovative way of quantifying the suffering of people affected by disasters. This suffering, when quantifiable, is considered as deprivation cost.

This concept has gained space in the literature, and some authors have included in the solution of several problems. Holguín-Veras et al. [5] developed different approaches to consider human suffering and its implications. One of these approaches was used by Biswal et al. [6] to verify the applicability of RFID in the humanitarian supply chain. The insertion of this cost is essential since when considering a distribution of supplies, it is necessary to know the priority of the population, that is, to know how much or to whom should be offered.

Shao et al. [7] analyzed the development of research on deprivation costs and conclude the need to develop new models in HL, which unite the state of the art and practice. Besides, they warn that human suffering or indicators of suffering are commonly neglected, and indirect measures are generally used, which make it challenging to use in practice.

Other authors use a penalty cost for the demand not satisfied, in other words, the unmet demand. The objective is to minimize the unmet demand. According to Turkes et al. [8], the penalty costs for unmet demand are applied to circumvent the simple minimization of costs that would result in no provision of services. Because of that, in the objective function, the minimization of the unmet demand is included [9–11]. Similarly, Bravo et al. [12] propose a methodology using unmanned aerial vehicles (UAVs) to find disaster victims using higher priorities in places where it is more likely to find victims.

In this context, this paper aims to bring state of the art and propose a taxonomy on the models used to represent the deprivation cost or/and unmet demand in HL. For this, we conducted a systematic literature review of 70 articles.

2 Methodology

We conduct a Systematic Literature Review (SLR) to answer the following research question: what the state of the art of deprivation cost models is. In our SLR, we use the eight steps of Thomé et al. [13]. These steps are: (i) planning and formulation of the research problem; (ii) literature research; (iii) data collect; (iv) quality assessment; (v) data analysis and synthesis; (vi) interpretation of results; (vii) results presentation; and (viii) review update. In the first step, to direct and structure the research, we define the scope of the SLR, additionally presents the objectives of the review. Thus, we applied this method to identify state of the art about unmet demands and deprivation costs in the humanitarian context.

According to Mongeon and Paul-Hus [14], the Scopus and Web of Science (WoS) databases have the broadest coverage of the literature in the engineering area. Therefore, we chose these bases for the second stage, which consists of the selection of databases and keywords, review of abstracts, application of exclusion criteria, review of complete texts and snowball technique.

We chose the combination of keywords so that the topic was broad enough to avoid limitations and unwanted results. The search considered the following structure in the titles, abstracts, and keywords: ("deprivation cost*" OR "unmet demand*") AND ("disaster" OR "relief" OR "humanitaria* logistic*" OR "humanitaria*" OR "humanitaria* operation*"). Our search was on September 3, 2020 and resulted in 69 documents in the Scopus base and 52 in the WoS. Figure 1 shows the results of our search, based on the Preferred Reporting Items for Systematic Reviews and Meta-Analyses (PRISMA) flow diagram [15].

After that, we analyzed these documents according to the following exclusion criterion: Criterion 1: Exclusion of conference papers and book chapters; Criterion 2: Articles that do not have deprivation cost models; Criterion 3: Articles out of humanitarian logistics area.

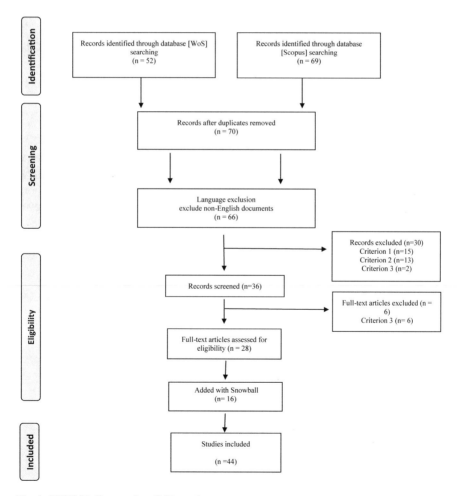

Fig. 1 PRISMA diagram from RSL results

Subsequently, we read the selected articles complete and use the same exclusion criteria applied in the abstract. After that, we apply the snowball technique, which is characterized by searches backwards and forwards of the citations presented in the selected articles. This technique is essential because it allows an increase in the coverage of selected documents related to the theme [16]. Snowball inclusion criteria were shortage cost, unsatisfied demand, and delay cost. After applying the exclusion criteria and applying the snowball technique, it ended with 44 (forty-four) articles.

Henceforth, we move on to the third step (data collection), which we carried out by identifying and informing the articles selected in the previous step. We created a standard data collection model to allow full traceability, and we use a matrix format. We filled out this matrix in spreadsheet with the data content of the bibliographic search as title, authors, year of publication, type of disaster, description of the models and Methodology used. In fourth step, we can ensure the quality assessment by the use and selection of peer-reviewed articles indexed in the Scopus and WoS databases. In the next section, we show a taxonomic analysis that results from the fifth, sixth and seventh steps, with the papers selects. The eighth step, that is an update of the SLR, is out of the scope of this paper.

3 Results

In the context of humanitarian logistics, the selected forty-four articles, are published in the period from 2012 to 2020. As highlighted by Shao et al. [7], humanitarian logistics as part of operations management is growing and gaining more attention. This trend can be confirmed by Fig. 2 that shows the distribution of publications over time.

Table 1 presents the proposed taxonomy. As can be seen from Table 1, despite the introduction of the deprivation cost in the literature, some articles continue to use

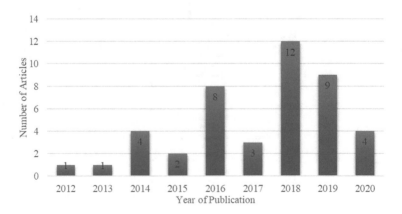

Fig. 2 The number of articles per year of publication

Table 1 Taxonomy

Article	Year	Types of disaster	Case of study	Modeling approach	Goods/service	Phase of disaster	Unmet demand or deprivation cost
Yushimito et al. [17]	2012	Hurricane	Louisiana and Mississippi, USA	The deprivation cost function is sensitive to the type and time interval without the commodity	–	Preparedness	Deprivation cost
Holguin-Veras et al. [5]	2013	General	Numerical experiments	Deprivation cost is a function of deprivation time	Water	Response and Recovery	Deprivation cost
Rezaei-Malek and Tavakkoli- Moghaddam [24]	2014	Earthquake	Seattle area, USA	Total p enalty cost of unsatisfied demand that considers the priority ofeachdemandp oint oor each supply implicitly by value penalty values	Medical supplies	Response	Unmet demand
Kelle et al. [25]	2014	Hurricane	Louisiana Gulf Coast, USA	Deprivation cost is a simple constant shortage penalty model	Water, canned meal, cot, tarp, and generators	Preparedness and response	Deprivation cost
Das and Hananka [21]	2014	Earthquake	Great East Japan	The deprivation cost increases with the late delivery	–	Response	Deprivation cost

(continued)

Table 1 (continued)

Article	Year	Types of disaster	Case of study	Modeling approach	Goods/service	Phase of disaster	Unmet demand or deprivation cost
Huang et al. [22]	2015	Earthquake	Sichuan, China	Deprivation Cost is modeled as a a linear functionofdelay costiwhich is measuredby the time of delay and the consequence of delay	–	Response	Deprivation cost
Khayal et al [26]	2015	General	Numerical analysis	Deprivation cost is a function of the delayedunits ofresourcesi andthe mitt delay penatty cost. The deprivation cost increase with delays in time requiredtosarisfy the demand	–	Response	Deprivation cost

(continued)

Table 1 (continued)

Article	Year	Types of disaster	Case of study	Modeling approach	Goods/service	Phase of disaster	Unmet demand or deprivation cost
Diedrichs et al. [27]	2016	General/hurricane	Numerical simulations/southeastern coast of the USA	Deprivation Cost is modeled as a probability functionfordeatha giventime step ma process similar to building an actuarial life table. The probability ofdeathby deprivationofa commodity at a time follows a sigmoid function	High priority commodities: water, food/low priority commodities: medical supplies, shelter, blankets	Response	Deprivation cost
Pradhananga et al. [28]	2016	General	Numerical analysis	Deprivation cost increases exponentially with the deprivation time	–	Prep aredness and response	Deprivation cost
Serrato-Garcia et al. [23]	2016	Flooding	Acambay de Ruiz Castaneda, M exico	Deprivation cost is a linear function of deprivation time	Food and beverage, water and sanitation, medicines and medical supplies, shelter and p ersonal needs	Response	Deprivation cost

(continued)

Table 1 (continued)

Article	Year	Types of disaster	Case of study	Modeling approach	Goods/service	Phase of disaster	Unmet demand or deprivation cost
Perez-Rodriguez and Holguin- Veras [29]	2016	General	Numerical experiments	Deprivation cost is a generic function that only depends on deprivation time and rely on the p riority of the commodities	Commodities	Response and Recovery	Deprivation cost
Holguin-Veras et al. [30]	2016	Flooding	Colombian Caribbean Region	Deprivation cost functions are nonlinear functions of the deprivation time. In most cases i exponential functions	Water	Response M itigation,	Deprivation cost
Rezaei-Malek et al. [10]	2016	Earthquake	Tehran, Iran	The penalty cost of each unmet demand is determined based on its priority in dem lmdlpoint following scenario	Medical commodities	Preparedness, Response and Recovery	Unmet demand
Cavdur et al. [11]	2016	Earthquake	Bursa, Turkey	Cost (penalty) of each unit of unmet demand multiplied by the unmet demand of commodity in the neighborhood	Water, food and medical kit	Response	Unmet demand

(continued)

Table 1 (continued)

Article	Year	Types of disaster	Case of study	Modeling approach	Goods/service	Phase of disaster	Unmet demand or deprivation cost
Pradhananga et al. [31]	2016	General	Numerical analysis	Deprivation cost is represented by a shortage cost multiplied by demand of affected areas minus the quantity ordered	–	Prep aredness and response	Deprivation cost
Rivera-Royero et al. [32]	2016	Flooding	Departmento del Atlantico, Colombia	Deprivation cost function considers the waiting time, type of relief pallet and the particular characteristics of each demand point	Non-perishable food kit, basic personal care products	Response	Deprivation cost
Lodree et al. [33]	2017	General	Numerical simulations	Deprivation cost consider a holding cost that is linearly proportional to the expected number of customers remaining in the system	Medical service	Response	Deprivation cost

(continued)

Table 1 (continued)

Article	Year	Types of disaster	Case of study	Modeling approach	Goods/service	Phase of disaster	Unmet demand or deprivation cost
Condeixa et al. [34]	2017	Flooding	Vale do Paraiba region in Sao Paulo, Brazil	A fixed penalty multiplied by the unmet demand. The penalty is calculated through the accumulated deprivation cost of 72 h of the unmet demand	–	Preparedness and response	Deprivation cost
Ni et al. [35]	2017	Earthquake	Yushu in Qinghai Province, PR China	Deprivation cost is a generic faction that depends on deprivation time and the number of individuals	–	Mitigation and Preparedness	Deprivation cost
Biswal et al. [6]	2018	General	India	Deprivation cost is a linear function of deprivation time	Food	Mitigation and Preparedness	Deprivation cost

(continued)

Table 1 (continued)

Article	Year	Types of disaster	Case of study	Modeling approach	Goods/service	Phase of disaster	Unmet demand or deprivation cost
Moreno et al. [36]	2018	Flooding	Serrana region of Rio de Janeiro state, Brazil	The two proposed deprivation cost functions depends on the priority of the commodities. The function depends on the deprivation time, the number of people affected and the maximum deprivation cost per person per commodity	Food, domestic hygiene kits, personal hygiene kits, medical kits and water	Preparedness and response	Deprivation cost
Loree and Aros-Vera [37]	2018	General	Numerical Experiments	Nonlinear deprivation cost function	–	Response	Deprivation cost
Yu et al. [38]	2018	General	Numerical Experiments	Deprivation cost grows exponentially with the deprivation time. Proposes a variant of the deprivation costs, starting state-based deprivation cost (SSDC)	Water	Response	Deprivation cost

(continued)

Table 1 (continued)

Article	Year	Types of disaster	Case of study	Modeling approach	Goods/service	Phase of disaster	Unmet demand or deprivation cost
Chap man and Mitchell [39]	2018	Hurricane	New Orleans, EUA	Deprivation cost is a convex monotonic increasing function of the distance traveled	–	Preparedness	Deprivation cost
Gutjahr and Fischer [40]	2018	Earthquake	Kathmandu, Nepal	The deprivation intensity grows according to a quadratic function. The term deprivation cost is for the cumulative deprivation intensities	Perishable food or medical help	Response	Deprivation cost
Cantillo et al. [41]	2018	Floods and Earthquakes	Colombia	The deprivation cost function is convex and nonlinear on the deprivation time. The mathematical modeling allows the variations on the willingness to pay	Basic food kit	Response	Deprivation cost
Gralla and Goentzel [42]	2018	General	Simulation training	Deprivation cost is a function that increases non-linearly as the deprivation time increases	Shelter, health, water and sanitation, food	Response	Deprivation cost

(continued)

Table 1 (continued)

Article	Year	Types of disaster	Case of study	Modeling approach	Goods/service	Phase of disaster	Unmet demand or deprivation cost
Chakravarty [18]	2018	Flooding	New Orleans, EUA	Deprivation cost is a convex increasing function and depends on the deprivation time. This function defines the cost of saving a life	–	Mitigation, preparation, and response	Deprivation cost
Zhu et al. [43]	2019	Flooding	Houston Flood of Texas, USA	Deprivation cost is a three-stage absolute function. First, the absolute deprivation cost increases exponentially with the length of waiting time. Then, it shows a linear reduction. Finally, a new exponential increase on the absolute deprivation cost appears when victims are transported to the medical center	Medical service	Response	Deprivation cost
Cantillo et al. [19]	2019	General	Numerical experiments	Deprivation cost is modeled as a generic function that depends only on the deprivation time	Water	Preparedness, response,	Deprivation cost

(continued)

Table 1 (continued)

Article	Year	Types of disaster	Case of study	Modeling approach	Goods/service	Phase of disaster	Unmet demand or deprivation cost
Yu et al. [44]	2019	General	Numerical experiments	Deprivation Cost is represented by the State-based Deprivation Cost (SSDC)—the cumulative deprivation cost based on the starting state of each stage. The deprivation function is a exponential fuction	Water	Response (first 3 days)	Deprivation cost
Hu and Dong [45]	2019	Hurricane	The southeastern United States	Deprivation is a penalty cost that increases each time when lead time interval increases	Water	Preparedness and response	Deprivation cost
Fard et al. [46]	2019	General	Numerical experiments, (Iraq, Kenya, Liberia, Syria and Sudan)	Deprivation cost is approximated by an exponential function with respect to the deprivation time	Vehicle transportation cprviPPS services	Mitigation and Preparedness	Deprivation cost

(continued)

Table 1 (continued)

Article	Year	Types of disaster	Case of study	Modeling approach	Goods/service	Phase of disaster	Unmet demand or deprivation cost
Wang et al. [47]	2019	Earthquake	Ya'an, Sichuan Province, China	The deprivation level function that increases with the deprivation time, and are a weakly monotonic, strictly convex at a first and strictly concave thereafter	Food	Mitigation and Preparedness	Deprivation cost
Turkes et al. [8]	2019	General	Numerical experiments Seattle area, the United States	Unmet demand is models as unit undersupply (unmet demand) penalty cost of commodity multiplied by undersupply of commodity at a vertex in a scenario. They use the shortage amount of each product in each zone multiplied by a penalty and the	–	Mitigation and Preparedness	Unmet demand

(continued)

Table 1 (continued)

Article	Year	Types of disaster	Case of study	Modeling approach	Goods/service	Phase of disaster	Unmet demand or deprivation cost
Roshan et al. [9]	2019	Earthquake		importance of each product according to the area. They also use a second objective function to minimize the maximum unmet demand for each product type	Pharmaceutical Materials	Response	Unmet demand
Cotes and Cantilho [48]	2019	Flooding	Caribbean region, Colombia	Deprivation cost is a linear function of deprivation time	Basic food kit	Preparedness	Deprivation cost
Paul and Wang [49]	2019	Earthquake	Los Angeles, USA	Deprivation functions consider three levels of severity to adjusted survivability time using a piece-wise linear function with four subfunctions	Medical Supplies	Preparedness and response	Deprivation cost

(continued)

Table 1 (continued)

Article	Year	Types of disaster	Case of study	Modeling approach	Goods/service	Phase of disaster	Unmet demand or deprivation cost
Paul and Zhang [50]	2019	Hurricane	Numerical experiments	Deprivation costs was modeled as three functions due to the delay in treating three levels of severity of victims, without taking into account permanent disability	Medical service	Preparedness and response	Deprivation cost
Wang et al. [20]	2020	Hurricane	Hypothetical study region	Deprivation cost function is estimated according to FEMA severity, dividing each level into four subfunctions according to the deprivation time. The function is a monotonically increasing nonlinear convex function concerning deprivation time	Water	Preparedness	Deprivation cost

(continued)

Table 1 (continued)

Article	Year	Types of disaster	Case of study	Modeling approach	Goods/service	Phase of disaster	Unmet demand or deprivation cost
Sakiani et al	2020	General	North West of Iran	Deprivation ratio is calculated by multiplying the deprivation cost per person, the population under risk of deprivation and the remaining number of periods in the planning horizon	Durable goods and consumable goods	Response and recovery	Deprivation cost
Rezaei et al. [52]	2020	Earthquake	Numerical experiments	A penalty is considered for unsatisfied and/or lost demands	Fuel	Response and Recovery	Unmet demand
Rivera-Royero et al. [53]	2020	Flooding	Departmento del Atlantico, Colombia	The deprivation cost function is an exponential function dependent on the deprivation time	Non-perishable food, hygiene	Preparedness and products response	Deprivation cost

the unmet demand variation. Although the objective is the same (i.e., to minimize human suffering), among the selected papers, most of them explore the deprivation cost when compared to unmet demand (in thirty-eight out of forty-four articles).

About the modelling approach, for the deprivation cost, most articles use a function that depends on the deprivation time [5, 17–20]. However, in some cases, there is another parameter. Das and Hananka [21], for example, incorporate an urgency index, to introduce an agent-based model framework for integrating human interests. Huang et al. [22] use a linear deprivation cost function in their integrated multi-objective optimization model for allocation and distribution of emergency supplies. Serrato-Garcia et al. [23] also use a linear function to build a multi-objective optimization model and an information system based on mobile technology, that aims to support decision-makers in the HL operations.

Whereas some papers consider a linear function [6, 33, 49], other papers use an exponential function with different coefficients [30, 37, 38, 42, 44, 46, 53].

Diedrichs et al. [27], Pérez-Rodriguez and Holguín-Veras [29], and Moreno et al. [36] use different functions depending on the priority of the commodities. Zhu et al. [43] apply three types of the deprivation cost functions to develop two models differentiated by considerations on the identical and diverse injured degrees, and they propose the deprivation cost as one of the decision-making objectives. Paul and Zhang [50] also consider three functions of deprivation costs to develop a two-stage stochastic model that optimizes first the locations of points of distribution, medical supply level and transportation capacity, and second the transportation decisions or path.

Another deprivation function presented in the literature depends on the level of deprivation. Wang et al. [47] develop a performance measurement model using this type of function that aims to measure the effectiveness, efficiency, and cost-effectiveness of an HL operation. Moreover, Chakravarty [18] employ the deprivation cost function to define the cost of saving lives in a three-decision making framework that aims to study how relief providers can provide investment in leave capacity for flooding integrating with supply procurements and rapid response.

Concerning the papers based on the unmet demand approach, the most used function is a penalty cost multiplied by the total unmet demand [8, 11, 24, 52]. There are other variations such as including the priority of the product based on the demand point [9, 10], but the main idea is the same. Moreover, some articles use the concept of deprivation cost, but the function seems the unmet demand [25, 31, 34, 45]. For example, Kelle et al. [25] consider a shortage penalty model to suggest a decision criterion for disaster preparedness and apply in the supply of commodities before a hurricane. Sakiani et al. [51] present a similar function and propose a model for distributing and redistributing relief items. Since in some circumstances some locations have things leftover and other places are missing, they calculated their function by multiplying the deprivation cost per person, the population under risk and the remaining time.

Earthquakes are the most studied disaster type [9–11, 21, 22, 24, 35, 40, 47, 49, 52]. Ni, Shu and Song [35] propose a min–max robust model to capture the uncertainties in demand and other parameters and minimize the deprivation cost for the case of an earthquake in China. Gutjahr and Fischer [40] discuss the equity criterion in the case of an earthquake and propose a new deprivation cost function.

The idea of an appropriate level of service is also used by Chapman and Mitchell [39] to build a formulation to choose a set of distribution points to be opened from a list of available locations and to assign population to a distribution point.

Another point is the different applications of deprivation cost or unmet demand. Some studies use the location-allocation model, such as Khayal et al. [26]. The authors proposed a location model for dynamic selection of temporary distribution facilities and the allocation of resources for emergency response planning. Prad-hananga et al. [28] also studied the same problem and proposes a three-echelon network model that aims to integrate emergency preparedness and response planning on the distribution of emergency supplies.

About the phase of the disaster (mitigation, preparedness, response, or recovery), the most studied is the response phase [9, 11, 21–24, 26, 27, 30, 32, 33, 37, 38, 40–44]. Rivera-Royero et al. [32] develop a dynamic model to serve demand, that aims to help in the distribution of the relief supplies after the disaster, according to the level of urgency. Cantillo et al. [41] also study the response phase, they use the theory of discrete choices to assess deprivation costs due to the deprivation time of a basic food kit. They concluded that this approach is useful to estimate the social costs of HL operations.

Some papers include the same idea of the social cost, that is the logics cost plus deprivation cost. Holguín-Veras et al. [5], for example, suggest the use of social costs, in the objective functions of post-disaster humanitarian logistics models. In the same direction, Cotes and Cantilho [48] propose a facility location model for pre-positioning the supplies in the preparedness phase and minimize the global social costs.

4 Conclusions

This study presented a systematic review of the literature focusing on deprivation cost in humanitarian logistics. The findings prove the growing interest in this theme of HL operations. However, few articles have an applied case study; most are numer-ical examples. Thus, studies with practical applications are necessary to investigate whether the deprivation costs modeling describe the reality and whether it is possible to generalize the modeling, since in many cases the costs used are specific for a type of disaster or good/service.

We propose a taxonomy summarizing what has been done in the theme of depri-vation cost, the case studies used, and which phase of the disaster was studied. In this context, much has been done, but there is still room for research. The 44 selected papers show different ways of modelling the deprivation cost in various phases of a disaster lifecycle, nevertheless, the recovery and mitigation phases are the less covered. Still, many goods and services have not been addressed, for example the energy that is interrupted in some disasters. Another gap is the type of disaster.

We have found no application in human-made disasters such as terrorism and war. Therefore, research on levels of deprivation and deprivation costs can be interesting to analyze how they behave, whether they should be used at the same time or exclude each other, as both are methods of quantifying suffering human.

Acknowledgements This study was financed in part by the Coordenação de Aperfeiçoamento de Pessoal de Nível Superior—Brasil (CAPES)—Finance Code 001. Also thank FAPERJ for the research funding.

References

1. Centre of Research for The Epidemiology of DisasterS (CRED). International Disaster Database 2019, http://www.emdat.be/, last accessed 2020/08/25.
2. United Nations Office for Disaster Risk Reduction (UNISDR), https://www.undrr.org/, last access: 2020/10/08.
3. Balcik, B.; Beamon, B.M.; Smilowitz, K.R. Last mile distribution in humanitarian relief. Journal of Intelligent Transportation Systems, v.12, n.2, p.51–63, (2008).
4. Holguín-Veras, J., Jaller, M., Van Wassenhove, L. N., Pérez, N., & Wachtendorf, T. On the unique features of post-disaster humanitarian logistics. Journal of Operations Management, 30(7-8), 494–506, (2012).
5. Holguín-Veras, J., Pérez, N., Jaller, M., Van Wassenhove, L. N., & Aros-Vera, F. On the appropriate objective function for post-disaster humanitarian logistics models. Journal of Operations Management, 31(5), 262–280, (2013).
6. Biswal, A. K., Jenamani, M., & Kumar, S. K. Warehouse efficiency improvement using RFID in a humanitarian supply chain: Implications for Indian food security system. Transportation Research Part E: Logistics and Transportation Review, 109, 205–224, (2018).
7. Shao, J., Wang, X., Liang, C., & Holguín-Veras, J. Research progress on deprivation costs in humanitarian logistics. International Journal of Disaster Risk Reduction, 101343, (2019).
8. Turkeš, Renata; Cuervo, Daniel Palhazi; Sörensen, Kenneth. Pre-positioning of emergency supplies: does putting a price on human life help to save lives?. Annals of Operations Research, v. 283, n. 1, p. 865–895, (2019).
9. Roshan, Mohaddeseh; Tavakkoli-Moghaddam, Reza; Rahimi, Yaser. A two-stage approach to agile pharmaceutical supply chain management with product substitutability in crises. Computers & Chemical Engineering, v. 127, p. 200–217, (2019).
10. Rezaei-Malek, M., Tavakkoli-Moghaddam, R., Zahiri, B., & Bozorgi-Amiri, A. An interactive approach for designing a robust disaster relief logistics network with perishable commodities. Computers & Industrial Engineering, 94, 201–215, (2016).
11. Cavdur, Fatih; Kose-Kucuk, Merve; Sebatli, Asli. Allocation of temporary disaster response facilities under demand uncertainty: An earthquake case study. International Journal of Disaster Risk Reduction, v. 19, p. 159–166, (2016).
12. Bravo, R.Z.B., Leiras, A. and Cyrino Oliveira, F.L. The Use of UAV s in Humanitarian Relief: An Application of POMDP-Based Methodology for Finding Victims. Production and Operations Management, v. 28, n. 2, p. 421–440, (2019).
13. Thomé, A. M. T., Scavarda, L. F., & Scavarda, A. J. Conducting systematic literature review in operations management. Production Planning & Control, 27(5), 408–420, (2016).
14. Mongeon, P., & Paul-Hus, A. The journal coverage of Web of Science and Scopus: a comparative analysis. Scientometrics, 106(1), 213–228, (2016).
15. Moher, D., Shamseer, L., Clarke, M., Ghersi, D., Liberati, A., Petticrew, M., ... & Stewart, L. A. Preferred reporting items for systematic review and meta-analysis protocols (PRISMA-P) 2015 statement. Systematic reviews, 4(1), 1, (2015).

16. Jalali, Samireh; Wohlin, Claes. Systematic literature studies: database searches vs. backward snowballing. In: Proceedings of the 2012 ACM-IEEE international symposium on empirical software engineering and measurement. IEEE, p. 29–38 (2012).

17. Yushimito, Wilfredo F.; Jaller, Miguel; Ukkusuri, Satish. A Voronoi-based heuristic algorithm for locating distribution centers in disasters. Networks and Spatial Economics, v. 12, n. 1, p. 21–39, (2012).

18. Chakravarty, Amiya K. Humanitarian response to hurricane disasters: Coordinating flood-risk mitigation with fundraising and relief operations. Naval Research Logistics (NRL), v. 65, n. 3, p. 275–288, (2018).

19. Cantillo, Victor; Macea, Luis F.; Jaller, Miguel. Assessing vulnerability of transportation networks for disaster response operations. Networks and Spatial Economics, v. 19, n. 1, p. 243–273, (2019).

20. Wang, Xinfang Jocelyn; Paul, Jomon A. Robust optimization for hurricane preparedness. International Journal of Production Economics, v. 221, p. 107464, (2020).

21. Das, Rubel; Hanaoka, Shinya. An agent-based model for resource allocation during relief distribution. Journal of Humanitarian Logistics and Supply Chain Management, (2014).

22. Huang, K., Jiang, Y., Yuan, Y. and Zhao, L.Modeling multiple humanitarian objectives in emergency response to large-scale disasters. Transportation Research Part E: Logistics and Transportation Review, v. 75, p. 1–17, (2015).

23. Serrato-Garcia, M.A., Mora-Vargas, J. and Murillo, R.T. Multi-objective optimization for humanitarian logistics operations through the use of mobile technologies. Journal of Humanitarian Logistics and Supply Chain Management, (2016).

24. Rezaei-Malek, M. and Tavakkoli-Moghaddam, R. Robust humanitarian relief logistics network planning. Uncertain Supply Chain Management, v. 2, n. 2, p. 73–96, (2014).

25. Kelle, Peter; Schneider, Helmut; Yi, Huizhi. Decision alternatives between expected cost minimization and worst case scenario in emergency supply–Second revision. International Journal of Production Economics, v. 157, p. 250–260, (2014).

26. Khayal, Danya, Rojee Pradhananga, Shaligram Pokharel, and Fatih Mutlu. A model for planning locations of temporary distribution facilities for emergency response. Socio-Economic Planning Sciences, v. 52, p. 22–30, (2015).

27. Diedrichs, D.R., Phelps, K. and Isihara, P.A. Quantifying communication effects in disaster response logistics. Journal of Humanitarian Logistics and Supply Chain Management, (2016).

28. Pradhananga, Rojee, Fatih Mutlu, Shaligram Pokharel, José Holguín-Veras, and Dinesh Seth. An integrated resource allocation and distribution model for pre-disaster planning. Computers & Industrial Engineering, v. 91, p. 229–238, (2016).

29. Pérez-Rodríguez, Noel; Holguín-Veras, José. Inventory-allocation distribution models for post-disaster humanitarian logistics with explicit consideration of deprivation costs. Transportation Science, v. 50, n. 4, p. 1261–1285, (2016).

30. Holguín-Veras, J., Amaya-Leal, J., Cantillo, V., Van Wassenhove, L.N., Aros-Vera, F. and Jaller, M. Econometric estimation of deprivation cost functions: A contingent valuation experiment. Journal of Operations Management, v. 45, p. 44–56, (2016).

31. Nikkhoo, Fatemeh; Bozorgi-Amiri, Ali; Heydari, Jafar. Coordination of relief items procurement in humanitarian logistic based on quantity flexibility contract. International Journal of Disaster Risk Reduction, v. 31, p. 331–340, (2018).

32. Rivera-Royero, D., Galindo, G. and Yie-Pinedo, R. A dynamic model for disaster response considering prioritized demand points. Socio-economic planning sciences, v. 55, p. 59–75, (2016).

33. Lodree, E.J., Altay, N. and Cook, R.A. Staff assignment policies for a mass casualty event queuing network. Annals of Operations Research, v. 283, n. 1, p. 411–442, (2019).

34. Condeixa, L.D., Leiras, A., Oliveira, F. and de Brito Jr, I. Disaster relief supply pre-positioning optimization: A risk analysis via shortage mitigation. International journal of disaster risk reduction, v. 25, p. 238–247, (2017).

35. Ni, Wenjun; Shu, Jia; Song, Miao. Location and emergency inventory pre-positioning for disaster response operations: Min-max robust model and a case study of Yushu earthquake. Production and Operations Management, v. 27, n. 1, p. 160–183, (2018).

36. Moreno, A., Alem, D., Ferreira, D. and Clark, A. An effective two-stage stochastic multi-trip location-transportation model with social concerns in relief supply chains. European Journal of Operational Research, v. 269, n. 3, p. 1050–1071, (2018).

37. Loree, N., & Aros-Vera, F. Points of distribution location and inventory management model for Post-Disaster Humanitarian Logistics. Transportation Research Part E: Logistics and Transportation Review, 116, 1–24, (2018).

38. Yu, L., Zhang, C., Yang, H. and Miao, L. Novel methods for resource allocation in humanitarian logistics considering human suffering. Computers & Industrial Engineering, v. 119, p. 1–20, (2018).

39. Chapman, Amy Givler; Mitchell, John E. A fair division approach to humanitarian logistics inspired by conditional value-at-risk. Annals of Operations Research, v. 262, n. 1, p. 133–151, (2018).

40. Gutjahr, Walter J.; Fischer, Sophie. Equity and deprivation costs in humanitarian logistics. European Journal of Operational Research, v. 270, n. 1, p. 185–197, (2018).

41. Cantillo, V., Serrano, I., Macea, L.F. and Holguín-Veras, J..Discrete choice approach for assessing deprivation cost in humanitarian relief operations. Socio-Economic Planning Sciences, v. 63, p. 33–46, (2018).

42. Gralla, Erica; Goentzel, Jarrod. Humanitarian transportation planning: Evaluation of practice-based heuristics and recommendations for improvement. European Journal of Operational Research, v. 269, n. 2, p. 436–450, (2018).

43. Zhu, L., Gong, Y., Xu, Y. and Gu, J.. Emergency relief routing models for injured victims considering equity and priority. Annals of Operations Research, v. 283, n. 1–2, p. 1573–1606, (2019).

44. Yu, L., Yang, H., Miao, L. and Zhang, C. Rollout algorithms for resource allocation in humanitarian logistics. IISE Transactions, v. 51, n. 8, p. 887–909, (2019).

45. Hu, Shaolong; Dong, Zhijie Sasha. Supplier selection and pre-positioning strategy in humanitarian relief. Omega, v. 83, p. 287–298, (2019).

46. Keshvari Fard, M., Eftekhar, M. and Papier, F. An approach for managing operating assets for humanitarian development programs. Production and Operations Management, v. 28, n. 8, p. 2132–2151, (2019).

47. Wang, X., Fan, Y., Liang, L., De Vries, H., and Van Wassenhove, L.N. Augmenting fixed framework agreements in humanitarian logistics with a bonus contract. Production and Operations Management, v. 28, n. 8, p. 1921–1938, (2019).

48. Cotes, Nathalie; Cantillo, Victor. Including deprivation costs in facility location models for humanitarian relief logistics. Socio-Economic Planning Sciences, v. 65, p. 89–100, (2019).

49. Paul, Jomon A.; Wang, Xinfang Jocelyn. Robust location-allocation network design for earthquake preparedness. Transportation research part B: methodological, v. 119, p. 139–155, (2019).

50. Paul, Jomon A.; Zhang, Minjiao. Supply location and transportation planning for hurricanes: A two-stage stochastic programming framework. European Journal of Operational Research, v. 274, n. 1, p. 108–125, (2019).

51. Sakiani, Reza; Seifi, Abbas; Khorshiddoust, Reza Ramazani. Inventory routing and dynamic redistribution of relief goods in post-disaster operations. Computers & Industrial Engineering, v. 140, p. 106219, (2020).

52. Rezaei, M., Afsahi, M., Shafiee, M. and Patriksson, M. A bi-objective optimization framework for designing an efficient fuel supply chain network in post-earthquakes. Computers & Industrial Engineering, v. 147, p. 106654, (2020).

53. Rivera-Royero, Daniel; Galindo, Gina; Yie-Pinedo, Ruben. Planning the delivery of relief supplies upon the occurrence of a natural disaster while considering the assembly process of the relief kits. Socio-Economic Planning Sciences, v. 69, p. 100682, (2020).

Measuring Universities' Strategic Commitment to the Sustainable Development Goals

Carolina Grano◉ and Vanderli Correia Prieto◉

Abstract Higher Education Institutions (HEIs) have been called to merge the pursue of sustainability through the commitment to the sustainable development goals (SDGs). In order to be more successful in their efforts, this commitment must be present into the strategic level of the planning and then deployed into actions. Despite the importance of the theme there is a lack of studies about the SDGs in the strategic level of the universities. To contribute to fulfill this gap this study adapted and applied a set of qualitative sustainability indicators based on the metrics from the 2020 edition of Times Higher Education Impact Ranking to measure HEIs' strategic level commitment to the SDGs. The study was conducted in three Brazilian Universities and the content analysis method and descriptive statistics were applied. Results show that the studied universities included activities related to the SDGs, what confirms that HEIs play a fundamental role helping society to achieve sustainable development.

Keywords Sustainability · Sustainability assessment · Higher education institutions · Universities · Sustainability in higher education · Sustainable development goals · Strategic planning · Content analysis

1 Introduction

Sustainability is one of the greatest challenges to be faced by humanity in twenty-first century. In that context, Higher Education Institutions (HEIs) play a key role contributing to the achievement of sustainable development through education, research, management and operations, and community outreach [1] and many of them have embraced the values of sustainability, such as the enhancement and preservation of the territory, the improvement of community welfare, the economic development based on knowledge, social equity and the capacity of the subjects involved to work

C. Grano (✉) · V. C. Prieto
Federal University of ABC, São Bernardo do Campo, São Paulo, Brazil

V. C. Prieto
e-mail: vanderli.prieto@ufabc.edu.br

© The Author(s), under exclusive license to Springer Nature Switzerland AG 2021
A. M. Tavares Thomé et al. (eds.), *Industrial Engineering and Operations Management*,
Springer Proceedings in Mathematics & Statistics 367,
https://doi.org/10.1007/978-3-030-78570-3_22

for the common good [2], what means that sustainable development at university level is connected to all dimensions of sustainability and the establishment of a sustainable development university policy is rooted to the commitment to make a positive difference in the world [3].

Research into sustainability in higher education has shown that there is a predominance of studies focused on reporting, assessing and analyzing HEIs experiences and results towards sustainability and the SDGs [4]. Although it has also shown the importance of planning for the achievement of sustainability results, even if it is a gap in the literature and there are fewer works dedicated to study sustainability in universities' strategic plans. This gap has already been identified by other researchers, which pointed out that there was a lack of empirical research on the connections between strategic planning and sustainability in higher education [5].

In response this gap, the present work aims to adapt and apply a set of qualitative sustainability indicators based on the metrics from the 2020 edition of Times Higher Education Impact Ranking to measure HEIs' strategic level commitment to the SDGs.

This paper is divided in six sections and it is organized as follows. Section 2 provides insights from the literature about the relationship between HEIs and sustainable development. Section 3 describes the method for the empirical study. The main outputs from the empirical study are shown in Sect. 4. Section 5 discusses the results presented in the previous section. Section 6 concludes the article.

2   Background: Sustainability in Higher Education

During the last decades, sustainability has become a global concern related to complex and unprecedent survival, social and political issues [6]. In 1972, the United Nations Conference on the Human Environment (Stockholm Conference) [7] was the first global event to address the sustainability issue. In the following decade, the World Commission on Environment and Development (WCED) was established and, in 1987, presented the report "Our Common Future", that says "humanity has the ability to make development sustainable to ensure that it meets the needs of the present without compromising the ability of future generations to meet their own needs" [8].

From 1987 to the twenty-first century, the global discussions about sustainable development grew bigger and in 2015 the United Nations presented the 2030 Agenda, which is "a plan of action for people, planet and prosperity" [9]. That means the Agenda 2030 understands sustainable development as a concept based on three pillars: social, environmental and economic. It is made of 17 Sustainable Development Goals (SDGs) and 169 targets, all of them being integrated and indivisible, and balancing the three dimensions of sustainable development [9]. The SDGs provide a platform in which organizations, society and governments can contribute to solving world's sustainable development challenges [10].

Sustainability is a goal for today, the SDGs are a compass, and universities have a critical role to play as change agents [11], once since the beginning of the concerns about sustainability, HEIs have been recognized as a fundamental part to achieve it, because they can equip people with skills, knowledge and understanding to address sustainability challenges and perform research that advances the sustainable development agenda [10]. Also, HEIs can act as test beds for new ideas, leading by example and sharing and implementing findings and technology with society [12], what means there is a variety of functions being incorporated by HEIs, such as engaging communities and promoting regional and local advancement that equally results in sustainable practices [13].

Not only HEIs have been recognized as organizations that are capable to help society to achieve sustainable development, as they are expected to be committed to the sustainable transformation, what can be observed by the existence of a variety of sustainability assessment tools applied to HEIs and research about them, as shown in Table 1, which presents some of those tools and studies where they have been mentioned.

Besides assessing and analyzing sustainable experiences, research has also shown that planning can be an extremely useful tool in supporting HEIs initiatives to integrate the three dimensions of sustainable development in their operations and to ensure the principles of sustainability are realized and put into practice [40]. Also, the lack of planning is a barrier to be eliminated for the successful implementation of long-term sustainability initiatives by the universities [1, 40], since HEIs with sustainable development policies are more likely to have initiatives as green campus procedures, sustainability teaching and local sustainable development activities, when compared to those who do not [3].

3 Method

3.1 Tool Selection and Adaptation

Considering that ideal methodologies to assess sustainability in HEIs should be holistic and equally reflect the different dimensions of sustainability [3], and that the 2030 Agenda with the SDGs is the current trend for sustainability efforts, the most suitable tools, among those listed in Table 1, were STARS and THEIR. The main difference between them is that STARS' indicators are linked to one or more SDGs and THEIR's indicators were made from the SDGs and its targets. In addition, THEIR publishes its results yearly [36], what makes the strategic planning analyses done with a THEIR based tool easily comparable to the ranking results. Because of these reasons, THEIR stood out as the most suitable choice for this study.

THEIR is one of the products from Times Higher Education, an United Kingdom based company that offers university rankings and specialized solutions for research HEIs [41]. Any HEI offering undergraduate programs can apply for THEIR [42]. As other university rankings, it classifies HEIs according to their normalized scores, which are obtained from numeric, evidence-based, and continuous data answers [42].

Table 1 Sustainability assessment tools applied to HEIs

Assessment tool	Studies	Description	Year	SDGs
Auditing instrument for sustainability in higher education (AISHE)	[14–19]	Audit method based on a quality management model to verify how sustainability is addressed in each area of a HEI. It is focused on sustainability in education, with less emphasis in research and operations [20]	2001	No
Campus ecology	[15–17]	Created by the Center for Sustainable Systems from Michigan University, it is a model to monitor and report HEIs' environmental impacts, based on a set of quantitative metrics [21]	1999	No
Campus sustaina-bility assessment framework (CSAF)	[14–19]	Model built from other existing assessment tools, designed to evaluate campus sustainability in Canadian universities through 175 indicators [22]	2003	No
College sustainability report card (CSRC)/green report card	[14, 15, 17, 19, 23–25]	Tool that evaluates environmental sustainability in educational institutions from Canada and the United States. It is based on grades for 52 indicators divided in 9 categories. It was discontinued in 2012 [26]	2007	No
Conference of rectors of universities (CRUE)	[14, 15]	Model to benchmark and evaluate sustainability activities and strategies from Spanish universities [27]	2007	No

(continued)

Table 1 (continued)

Assessment tool	Studies	Description	Year	SDGs
Graphical assessment of sustainability in universities (GASU)	[14–17, 19, 28]	Tool that graphically analyses the sustainability efforts from universities, considering economic, social and environmental dimensions, based on guidelines from the Global Report Initiative for Sustainability [29]	2007	No
Sustainability assessment for higher education (SAHTE)	[16]	Brazilian model that enables to compare the sustainability performance of service operations between different individual institutions by using a common methodology [16]	2016	No
Sustainability assessment questionnaire (SAQ)	[14–16]	Qualitative survey planned to evaluate sustainability in HEIs in order to foster the debate about the role of sustainability in higher education [30]	2001	No
Sustainability tracking, assessment and rating (STARS)	[14–16, 18, 19, 23–25, 31, 32]	Model to support the development of reports to help HEIs and follow their progress in sustainability. It is possible to create reports with or without grades. The model was updated and nowadays it has indications of the SDGs in its indicators [33]	2006	Yes
Times higher education impact ranking (THEIR)	[34]	University ranking that evaluates universities with metrics based on the 17 SDGs [35, 36]	2019	Yes

(continued)

Table 1 (continued)

Assessment tool	Studies	Description	Year	SDGs
UI green metric world university rankings	[14, 16–19]	It was created by the Universitas Indonesia Association with help from specialists in university rankings. Its goal is to rank universities worldwide based on their sustainability actions [37]	2010	No
Unit-based sustai-nability assessment tool (USAT)	[14, 15, 18]	Tool created to give sustainability instruments to African and Asian universities [38, 39]	2009	No

THEIR scores are obtained from a set of indicators based on the SDGs [43]. There are between 3 and 6 indicators for each SDG and they were adapted to a simplified qualitative version of themselves, to make them appliable to the analysis of the strategic plans.

3.2 Content Analysis

The content analysis technique can be divided in three phases: pre analysis, material exploration, and results treatment and interpretation [44]. In this research, it was performed a qualitative analysis, considering the presence or absence of policies or actions related to the SDGs in each HEI strategic plan, with support of the adapted indicators.

Three Brazilian universities were selected from the 2020 edition of THEIR [36]: Federal University of Sao Paulo (UNIFESP), Federal University of ABC (UFABC) and Federal University of Ceara (UFC), and their current strategic plans (in 2020) were collected from their websites.

The material exploration happened in two steps: first, the interpretative reading of documents and second, the material rearrangement to match the analysis criteria. The interpretative reading had an exploratory function, as its purpose was to identify excerpts from the original texts describing the university's policies and actions towards sustainable development that could be linked to one or more indicators. The material rearrangement consisted in building a three-columns support table for each HEI with one line for each indicator. The indicators were listed in the first column of the table, and the selected excerpts in the second column right next to the matching indicators.

The process of identification of the excerpts and their connection to the indicators followed some inclusion and exclusion criteria: (a) ideally, full sentences, or even full paragraphs should be selected as an excerpt, in order to preserve the context; (b) in case of longer parts of the original text dedicated to the same subject, the most relevant sentences should be selected to summarize it and to form the excerpt; (c) parts of charts and tables could also be transformed in excerpts; (d) if one excerpt matched more than one indicator, it was placed next to all related indicators; (e) the excerpts should describe policies and actions taken by the institution; (f) current and planned policies and actions should be selected to the excerpts; (g) if a policy or action related to one or more indicators was mentioned multiple times in different parts of the original text, multiple excerpts about it should be produced; (h) if a sentence had specific terms that could be related to the SDGs, but it was not deep enough to present a policy or action, it could not be selected; (i) if the sentence was deep enough, but it presented a general definition, a standard or a regulation, it could not be selected; (j) it was possible to have excerpts with a subtle connection to the indicators; (k) the researcher should be careful to not assume something that was not written.

The third phase of the content analysis was treatment and interpretation. The treatment was a quantification of the results obtained from the previous phase to transform the support chart into something measurable and comparable. First, for each document, binary grades (1/0) were attributed in the third column of the support table. In this grading system, 1 means the presence of one or more excerpts related to the indicator and 0 means the absence of any excerpt related to the indicator. Then, for each SDG, the indicators grades were added, resulting in integer scores for the SDGs. Since the number of indicators were not the same for all the SDGs and could vary between 3 and 6, the document's final score for each SDG was given by the integer score divided by the number of indicators in that SDG, what resulted in a real number between 0 and 1. The final scores for each document were summarized in another table, what allowed the results interpretation and discussion, with the identification of the main SDGs in each document, and eventually the comparison to the published THEIR results.

4 Results

The process led to two outputs: a set of qualitative indicators based on THEIR's metrics [43] related to the SDGs (Table 2), and the result of the content analysis, represented by a set of scores that indicates the SDGs found in the strategic plans (Table 3).

Table 2 List of indicators

Indicators
SDG 1. No poverty: (1) Research into poverty; (2) Financial aid to low-income students; (3) Anti-poverty programs; (4) Programs for the community (support to local businesses, financial aids to the community, courses, and trainings, etc.)
SDG 2. Zero hunger: (5) Research into hunger; (6) Initiatives to minimize food waste on campus; (7) Student hunger (providing affordable and healthy food options, providing sustainable food choices, including vegetarian and vegan food, etc.); (8) Agricultural education including sustainable aspects; (9) Partnerships with community for sustainable agriculture (trainings, support to family agriculture and local producers, providing access to university facilities, etc.)
SDG 3. Good health and well-being: (10) Research into health; (11) Health professions education; (12) Health impact (community outreach programs promoting health and well-being, collaborations with local entities, providing access to university facilities, mental health support for students and staff, smoke free policy for the university, etc.)
SDG 4. Quality education: (13) Research into pedagogy; (14) Primary school teaching education; (15) Providing other learning opportunities (providing access to educational resources for those not studying at the university, hosting events thar are open to the general public, promoting activities that are accessible to all); (16) Policies for 1st generation students
SDG 5. Gender equality: (17) Research developed by women or related to gender issues; (18) Policies for 1st generation female students; (19) Policies to encourage applications by women and to promote their access to the higher education, especially in subjects where they are underrepresented; (20) Policies to provide leadership/senior positions to women; (21) Policies to adjust the proportion of women in all courses; (22) Progress measures (policies of non-discrimination against women and transgender people, support to parents, women's mentoring schemes, etc.)
SDG 6. Clean water and sanitation: (23) Research into water; (24) Control of the volume of water used per person on campus; (25) Water usage and care; (26) Water reuse; (27) Initiatives for the community (providing educational opportunities, promoting support to water conservation off campus, cooperate with governments on water security)
SDG 7. Affordable and clean energy: (28) Research into energy; (29) University measures (promoting energy efficiency by the construction of new buildings or the renovation of existing ones, energy efficiency plan, etc.); (30) Energy use density (energy used per m^2 floor space of the university buildings); (31) Energy and the community (educational actions, promoting renewable energy, partnerships with local industry, providing assistance for start-ups that foster and support a low-carbon economy/technology)
SDG 8. Decent work and economic growth: (32) Research (cite score or papers per staff); (33) Employment practice (pay all staff and faculty at least the living wage, recognizing unions, policies against discrimination on workplace, policy against modern slavery, policy on elimination of gender pay gaps); (34) Economic impact (university income per number of employees); (35) Policies to help students to find permanent job positions; (36) Employment security
SDG 9. Industry, innovation and infrastructure: (37) Research; (38) Patents related to research developed in the university; (39) Spin-offs (support to entrepreneurship); (40) Research related to industry demands

(continued)

Table 2 (continued)

Indicators

SDG 10. Reduced inequalities: (41) Research into reducing inequalities; (42) Policies for 1st generation students; (43) Support to foreign students from developing nations; (44) Inclusion of students with disabilities; (45) Inclusion of employees with disabilities; (46) Measures against discrimination (policies against discrimination, measures for admitting underrepresented groups, etc.)

SDG 11. Sustainable cities and communities: (47) Research into sustainable cities; (48) Arts and heritage (providing access to libraries, buildings and/or monuments of cultural significance, museums, and green spaces; contribution to the local arts and cultural preservation); (49) Spend on local arts and heritage; (50) Sustainable practices (providing access to affordable housing to students and staff, prioritizing pedestrian access on campus, constructing new buildings to sustainable standards, partnership with local authorities to develop affordable housing to the local residents)

SDG 12. Responsible consumption and production: (51) Research into responsible consumption and production; (52) Operations (policies on ethical sourcing of food and supplies, waste disposal, use minimization, and extension of those policies to the suppliers); (53) Waste recycling; (54) Publication of sustainability reports

SDG 13. Climate action: (55) Research into climate change; (56) Low carbon energy use; (57) Environmental education including disaster planning (educational programs, climate action plan, partnerships with government and local community groups); (58) Commitment to carbon neutral university

SDG 14. Life below water: (59) Research into aquatic ecosystems; (60) Supporting aquatic ecosystems through education; (61) Supporting aquatic ecosystems through action (supporting or organizing events that promote sustainable use of aquatic resources, developing technologies to minimize or prevent damage to aquatic ecosystems, working directly to maintain and extend ecosystems under threat); (62) Water sensitive waste disposal (standards and guidelines for water discharges, action plan to reduce plastic waste on campus, policy on preventing and reducing marine pollution); (63) Maintaining a local ecosystem (plan to minimize physical, chemical and/or biological alterations of related aquatic ecosystems, monitoring the health of aquatic ecosystems, collaborating with the local community in efforts to maintain aquatic ecosystems)

SDG 15. Life on land: (64) Research into land ecosystems; (65) Supporting land ecosystems through education; ((66) Supporting land ecosystems through action (policies to ensure the conservation, restoration and sustainable use of terrestrial ecosystems associated with the university, and to identify, monitor and protect threatened species); (67) Land sensitive waste disposal (standards and guidelines for water discharges, action plan to reduce plastic waste on campus, policy on preventing and reducing marine pollution)

SDG 16. Peace, justice and strong institutions: (68) Research into law and international relations; (69) Governance (having elected representation on the university's highest governing body, recognizing students' unions, policies and procedures to identify local stakeholders external to the university and engage with them, publishing university's principles and commitments, policy on supporting academic freedom, publishing university financial data); (70) Participation in local, regional and national government (and others); (71) Law and Social Sciences related subjects Education

SDG 17. Partnership for the goals: (72) Research into SDG with international partnerships; (73) Relationships with NGOs, regional and national government; (74) Publishing outputs across all SDGs; (75) Education for the SDGs

Table 3 Scores

University	Score per SDG																	Total
	1	2	3	4	5	6	7	8	9	10	11	12	13	14	15	16	17	
UNIFESP	0,5	0,4	1,0	1,0	0,3	0,0	0,0	0,6	0,8	0,8	0,8	0,0	0,0	0,6	0,0	0,8	0,3	7,8
UFABC	0,8	0,2	0,7	0,8	0,2	0,0	0,3	0,6	0,8	0,5	0,8	0,5	0,3	0,0	0,0	0,8	0,3	7,1
UFC	0,5	0,2	0,3	0,5	0,0	0,4	0,3	0,6	0,8	0,5	0,8	0,5	0,5	0,2	0,5	0,5	0,3	7,2

5 Discussion

Results show that the metrics from THEIR [43] could be successfully adapted into a set of qualitative indicators appliable to the sustainability analysis of HEIs' strategic plans. They also show the feasibility of the method, which can be employed to broader studies. With help of the indicators and the content analysis method described in Sect. 3, it was possible to measure the commitment of the studied universities to sustainable development in their strategic plans.

Although the strategic plan is a general document and HEIs are not required to mention sustainable development in it, Table 3 shows that, even in a small sample like this, for each SDG there was at least one example of actions or policies carried on by the universities. Table 3 also reveals that 10 of the 17 SDGs were covered by all documents of the sample, and as expected, the three analyzed plans achieved high scores for SDG 4 (quality education). This result emphasizes the importance of HEIs and confirms their role as institutions that are engaged to local, national and global development, and that they are able to lead and help society to pursue sustainable development.

Table 3 also provides the opportunity to identify the SDGs with the biggest scores for each university, what can be understood as the SDGs that had most participation in the strategic plans: 3 and 4 for UNIFESP, 1, 4, 9, 11 and 16 for UFABC, and 9 and 11 for UFC. This can be compared to the results of the 2020 edition of THEIR [36]. The ranking shows the following SDGs as the ones with the highest scores: 5, 8, 4 and 17 for UNIFESP, 7, 6, 11 and 17 for UFABC, and 8, 4, 2 and 17 for UFC.

This difference can be explained by the generality of the analyzed documents. As the universities must approach a wide set of subjects in their strategic plans, so it is possible that they have fostered sustainability-oriented actions that were not described in their strategic plans but were described in THEIR application form. It is also possible that actions and policies that were described in the strategic plans have accomplished low scores in THEIR evaluation.

In both cases, a low score in the strategic plan analysis, does not mean that the plan lacks quality or that the institution does not have sustainability policies. It just means that the university's policies and actions towards sustainable development were not formalized and prioritized in their strategic plans.

6 Conclusion

This work brough a different approach to research into the employment of HEIs sustainability assessment tools. In general, studies describe tools used to measure the institution's sustainability performance or to support the production of sustainability reports. In this paper we proposed a combination of an adapted version of an existing sustainability assessment tool and the content analysis technique to evaluate the commitment of universities towards the SDGs. This approach also highlighted the importance of strategic planning to sustainability-oriented policies and actions, which could receive more attention in the strategic plan writing process.

The research showed the feasibility of the use of THEIR's adapted indicators and content analysis to evaluate HEIs' commitment to sustainability in their strategic plans. Besides evaluation, the presented method can also support HEIs' strategic planning activities.

This paper also brought a social contribution by showing that Brazilian universities are actually committed to sustainable development and that, besides offering higher education, they engage activities capable of bringing development locally and nationally, because promoting development is part of their *raison d'être*. That is why we were capable to find policies and actions related to the 17 SDGs even though the analyzed documents did not explicitly mention the SDGs.

For future studies, the method presented in this paper can be applied to a bigger set of universities and even be associated to other content analysis techniques in order to create a full chart of HEIs' actions and policies towards the SDGs. Other possibilities are improvement of the indicators, the construction of a new tool exclusively thought to evaluate and support the inclusion of sustainable development aspects in HEIs strategic plans, and the study of HEIs' actions to fight COVID-19 in comparison to their strategic plans and the SDGs.

Acknowledgements The authors thank CAPES for the financial support for the publication of this work.

References

1. Leal Filho, W., Skanavis, C., Kounani, A., Brandli, L.L., Shiel, C., Paço, A. do, Pace, P., Mifsud, M., Beynaghi, A., Price, E., Salvia, A.L., Will, M., Shula, K.: The role of planning in implementing sustainable development in a higher education context. J. Clean. Prod. 235, 678–687 (2019). https://doi.org/10.1016/j.jclepro.2019.06.322.
2. Paletta, A., Bonoli, A.: Governing the university in the perspective of the United Nations 2030 Agenda: The case of the University of Bologna. Int. J. Sustain. High. Educ. 20, 500–514 (2019). https://doi.org/10.1108/IJSHE-02-2019-0083.
3. Leal Filho, W., Brandli, L.L., Becker, D., Skanavis, C., Kounani, A., Sardi, C., Papaioannidou, D., Paço, A., Azeiteiro, U., de Sousa, L.O., Raath, S., Pretorius, R.W., Shiel, C., Vargas, V., Trencher, G., Marans, R.W.: Sustainable development policies as indicators and pre-conditions

for sustainability efforts at universities. Int. J. Sustain. High. Educ. 19, 85–113 (2018). https://doi.org/10.1108/IJSHE-01-2017-0002.
4. Grano, C., Prieto, V.C.: Sustainable Development Goals in Higher Education. In: 26th IJCIEOM – International Joint Conference on Industrial Engineering and Operations Management. , Rio de Janeiro (2020).
5. Bieler, A., McKenzie, M.: Strategic planning for sustainability in Canadian higher education. Sustain. 9, 1–22 (2017). https://doi.org/10.3390/su9020161.
6. Avila, L.V., Beuron, T.A., Brandli, L.L., Damke, L.I., Pereira, R.S., Klein, L.L.: Barriers to innovation and sustainability in universities: an international comparison. Int. J. Sustain. High. Educ. 20, 805–821 (2019). https://doi.org/10.1108/IJSHE-02-2019-0067.
7. United Nations: Report of The United Nations Conference on the Human Environment, Stockholm (1972).
8. WCED: Our Common Future, Oslo (1987).
9. United Nations: Transforming our world: the 2030 Agenda for Sustainable Development. In: United Nations 70th General Assembly. UN, New York (2015).
10. Mori Junior, R., Fien, J., Horne, R.: Implementing the UN SDGs in Universities: Challenges, Opportunities, and Lessons Learned. Sustain. J. Rec. 12, 129–133 (2019). https://doi.org/10.1089/sus.2019.0004.
11. Purcell, W.M., Henriksen, H., Spengler, J.D.: Universities as the engine of transformational sustainability toward delivering the sustainable development goals: "Living labs" for sustainability. Int. J. Sustain. High. Educ. 20, 1343–1357 (2019). https://doi.org/10.1108/IJSHE-02-2019-0103.
12. Shawe, R., Horan, W., Moles, R., O'Regan, B.: Mapping of sustainability policies and initiatives in higher education institutes. Environ. Sci. Policy. 99, 80–88 (2019). https://doi.org/10.1016/j.envsci.2019.04.015.
13. Symaco, L.P., Tee, M.Y.: Social responsibility and engagement in higher education: Case of the ASEAN. Int. J. Educ. Dev. 66, 184–192 (2019). https://doi.org/10.1016/j.ijedudev.2018.10.001.
14. Alba-Hidalgo, D., Benayas del Álamo, J., Gutiérrez-Pérez, J.: Towards a Definition of Environmental Sustainability Evaluation in Higher Education. High. Educ. Policy. 31, 447–470 (2018). https://doi.org/10.1057/s41307-018-0106-8.
15. Berzosa, A., Bernaldo, M.O., Fernández-Sanchez, G.: Sustainability assessment tools for higher education: An empirical comparative analysis. J. Clean. Prod. 161, 812–820 (2017). https://doi.org/10.1016/j.jclepro.2017.05.194.
16. Drahein, A.D.: Proposta de Avaliação de Práticas Sustentáveis nas Operações de Serviço em Instituições de Ensino Superior da Rede Federal de Educação Profissional, Científica e Tecnológica, (2016).
17. Drahein, A.D., De Lima, E.P., Da Costa, S.E.G.: Sustainability assessment of the service operations at seven higher education institutions in Brazil. J. Clean. Prod. 212, 527–536 (2019). https://doi.org/10.1016/j.jclepro.2018.11.293.
18. Góes, H.C. de A., Magrini, A.: Higher education institution sustainability assessment tools. Int. J. Sustain. High. Educ. 17, 322–341 (2016). https://doi.org/10.1108/IJSHE-09-2014-0132.
19. Parvez, N., Agrawal, A.: Assessment of sustainable development in technical higher education institutes of India. J. Clean. Prod. 214, 975–994 (2019). https://doi.org/10.1016/j.jclepro.2018.12.305.
20. Roorda, N.: AISHE Auditing Instrument for Sustainability in Higher Education, (2001).
21. Center for Sustainable Systems: Campus Ecology: Environmental Performance Assessment and Reporting Economicology 1.0, 1.5, 2.0, http://css.umich.edu/project/campus-ecology-env ironmental-performance-assessment-and-reporting-economicology-10-15-20, last accessed 2020/06/28.
22. Cole, L.: Assessing Sustainability on Canadian University Campuses: Development of a Campus Sustainability Assessment Framework. (2003).
23. Bullock, G., Wilder, N.: The comprehensiveness of competing higher education sustainability assessments. Int. J. Sustain. High. Educ. 17, 282–304 (2016). https://doi.org/10.1108/IJSHE-05-2014-0078.

24. Sepasi, S., Braendle, U., Rahdari, A.H.: Comprehensive sustainability reporting in higher education institutions. Soc. Responsib. J. 15, 155–170 (2019). https://doi.org/10.1108/SRJ-01-2018-0009.

25. Shi, H., Lai, E.: An alternative university sustainability rating framework with a structured criteria tree. J. Clean. Prod. 61, 59–69 (2013). https://doi.org/10.1016/j.jclepro.2013.09.006.

26. Sustainable Endowments Institute: The College Sustainability Report Card, http://www.greenreportcard.org/, last accessed 2020/06/28.

27. The Platform for Sustainability Performance in Education: Conference of Rectors of Universities, CRUE – Spain, http://www.eauc.org.uk/theplatform/uam_crue_caped_spain, last accessed 2020/06/28.

28. Lozano, R.: The state of sustainability reporting in universities. Int. J. Sustain. High. Educ. 12, 67–78 (2011). https://doi.org/10.1108/14676371111098311.

29. Higher Education and Research for Sustainable Development: Graphical Assessment of Sustainability in Universities (GASU), https://www.iau-hesd.net/en/actions/2228/graphical-assessment-sustainability-universities-gasu, last accessed 2020/06/28.

30. ULSF: Sustainability Assessment Questionnaire, http://ulsf.org/sustainability-assessment-questionnaire/, last accessed 2020/06/28.

31. Maragakis, A., Van den Dobbelsteen, A.: Sustainability in Higher Education: Analysis and Selection of Assessment Systems. J. Sustain. Dev. 8, 1–9 (2015). https://doi.org/10.5539/jsd.v8n3p1.

32. Urbanski, M., Leal Filho, W.: Measuring sustainability at universities by means of the Sustainability Tracking, Assessment and Rating System (STARS): early findings from STARS data. Environ. Dev. Sustain. 17, 209–220 (2015). https://doi.org/10.1007/s10668-014-9564-3.

33. AASHE: STARS Technical Manual version 2.2, (2019).

34. Torabian, J.: Revisiting Global University Rankings and Their Indicators in the Age of Sustainable Development. Sustain. (United States). 12, 167–172 (2019). https://doi.org/10.1089/sus.2018.0037.

35. Ross, D.: We're including all 17 SDGs in the 2020 University Impact Rankings, https://www.timeshighereducation.com/blog/were-including-all-17-sdgs-2020-university-impact-rankings, last accessed 2020/06/28.

36. THE: Impact Rankings 2020, https://www.timeshighereducation.com/rankings/impact/2020/overall#!/page/0/length/25/sort_by/rank/sort_order/asc/cols/undefined, last accessed 2020/06/28.

37. UI Green Metric: Welcome to UI GreenMetric, http://greenmetric.ui.ac.id/what-is-greenmetric/, last accessed 2020/06/28.

38. Rhodes University: Unit-Based Sustainability Assessment Tool, https://www.ru.ac.za/elrc/publicationsandresources/unit-basedsustainabilityassessmenttoolusattool/, last accessed 2020/06/28.

39. The Platform for Sustainability Performance in Education: USAT (Unit-Based Sustainability Assessment Tool), http://www.eauc.org.uk/theplatform/usat_unit-based_sustainability_assessment_tool, last accessed 2020/06/28.

40. Leal Filho, W., Pallant, E., Enete, A., Richter, B., Brandli, L.L.: Planning and implementing sustainability in higher education institutions: an overview of the difficulties and potentials. Int. J. Sustain. Dev. World Ecol. 25, 712–720 (2018). https://doi.org/10.1080/13504509.2018.1461707.

41. THE: Times Higher Education: Helping their world's universities to achieve excellence, https://www.timeshighereducation.com/about-us, last accessed 2020/06/28.

42. THE: Impact Rankings: FAQs, https://www.timeshighereducation.com/world-university-rankings/impact-rankings-faqs, last accessed 2020/06/28.

43. THE: THE University Impact Rankings 2020 metrics, https://www.timeshighereducation.com/sites/default/files/breaking_news_files/sdg_poster_p1_2020_s_updated_latest.pdf, last accessed 2020/06/28.

44. Bardin, L.: Análise de Conteúdo. Edições 70/Almedina Brasil, Sao Paulo (2011).

The Adaptation of the Engineer to the Service Sector and the Maintenance of Economic-Social-Environmental Sustainability

Cesar B. Z. de Oliveira⬤, Mellany Juhlly Machado⬤, Dayse Mendes⬤, and Kellen Santos⬤

Abstract In parallel with the technological revolution there was a decrease in industrial activities, diminishing the necessity for labor in the sector, on the other hand, the demand for service providing professionals grew. Besides, the Sars-coV-19 pandemic brought a new set of concerns for society in all dimensions of sustainability: economic, social, and environmental. Further this scenario of growth in the importance of the third sector for World GDP, and the necessary changes concerning to the social, environmental and economical questions, a concern for capacitating specified labor for the resolution of problems in services is overlooked. Therefore, this article has the objective investigate the capacity of engineers to act in the service sector. To accomplish this goal a methodological procedure was structured and an investigation that can be characterized as a descriptive, a survey type design research was launched. Quantitative data regarding the existence of undergraduate and graduate degrees in the field of Service Engineering was collected. Although, it is considered that even though a few initiatives in this direction, they are still incipient. It was found, as final considerations, the importance of a service engineering professional, justified by the global demand for labor capable of solving the daily problems faced in the third sector, as well as the relevance in a graduation specific to the services sector.

Keywords Triple bottom line · Third sector · Engineering courses

1 Introduction

At the beginning of the twentieth century, the great problems of Western society were found within the factories. Progressivism and the Industrial Revolution opened space for the insertion of new technologies in people's routines and the big factories took

C. B. Z. de Oliveira (✉) · M. J. Machado · D. Mendes · K. Santos
Centro Universitário Internacional—Uninter, Av. Luiz Xavier, no. 103, Centro, Curitiba, Paraná, Brazil

© The Author(s), under exclusive license to Springer Nature Switzerland AG 2021
A. M. Tavares Thomé et al. (eds.), *Industrial Engineering and Operations Management*,
Springer Proceedings in Mathematics & Statistics 367,
https://doi.org/10.1007/978-3-030-78570-3_23

over the urban scene of that time. People's lives in cities have come to be governed by this industrial context. Thus, it is natural that tools, models and techniques had to be developed in order to improve the performance of factories and other organizations in the industrial sector. The economic sustainability of society depended on the effectiveness of deliveries from these factories and, consequently, on the resolution of their problems. In this scenario, the engineer rises, a professional who masters the art of applying scientific knowledge in the solution of practical problems. Not that the engineering profession did not exist previously, but the engineer was perfectly suited to meet the technical demands of the society that ended up associating this professional specifically for the industry. However, is it up to the engineer to work only in the factories? Considering the information so far, it is believed that a more comprehensive observation is needed on other possibilities, such as the service sector, which in the course of this article will show its strong growth and its considerable attention to the criteria of economic-environmental-social sustainability.

2 Sustainability, Services, and Engineering

Sustainability concerns the development of companies in a complex economic environment, in which they are required to be sustainable in the economic, social, and environmental spheres. These three pillars of sustainability are called the Triple Bottom Line. The concept was born with the publication of Elkington's work [1], Cannibals with Forks: Triple Bottom Line of twenty-first Century Business, published in 1997, in which the author mentions that implementing sustainability policies should be part of the routine of organizational managers of the twenty-first century.

This occurs because, in addition to these policies making it possible to reduce the imbalances caused by the current social economic model, they could also bring competitive advantages to organizations as they would have actions, as well as their image, more appropriate to a new form of interaction between companies and the society. For Elkington, the concept of sustainability is already part of the business world, even with some resistance on the part of some organizations and some countries, and it must be treated in several dimensions, such as technological, management, as well as values, intentions, and organizational behaviors.

The environmental dimension is related to the understanding that the resources available in the environment are scarce, showing a limit on their consumption. The challenge is to rethink the current model of economic development, based on this understanding that there are no inexhaustible sources of matter and energy. It is worth mentioning that the organizations that most generate environmental impact are industrial as opposed to service organizations, which have a reduced impact intensity. The social dimension considerer the concern with the eradication of basic social problems such as extreme poverty, hunger, unemployment, in short, meeting basic human needs. Finally, the economic dimension is related to sustainability not only in strict terms of organizational survival, but also due to its potential impact on other dimensions. Capra [2] tells us that "economic systems are constantly changing and evolving, depending on the equally changing ecological and social systems in which they are implanted. To understand them, we need a conceptual framework that is also capable of changing and adapting to new situations".

In times of pandemic, it is possible to observe all this intertwining between the dimensions of sustainability and the need to think and rethink the concept of what it means to be sustainable is explicit. After some time of the initial events of Sars-coV-19, it is possible to observe that the economic and social instability that has been installed in the world will demand from organizations, and society, changes in the way of acting and thinking about the organizational impacts in each of the three dimensions of the triple bottom line. This is a new and important problem that presents itself to society: how to be environmentally and socially sustainable while rebuilding and reorganizing the world economy. Figure 1 shows the slow recovery of the world economic scenarios.

Projections of GDP growth in the pre-pandemic period showed a discreet growth, however, significant when considering the GDP projection in recent years. It appears that despite the fall in 2020 due to the social isolation proposed by the World Health Organization (WHO) and ratified by countries around the world, an improvement in 2021 is expected due to expectations for the launch of an immunization vaccine

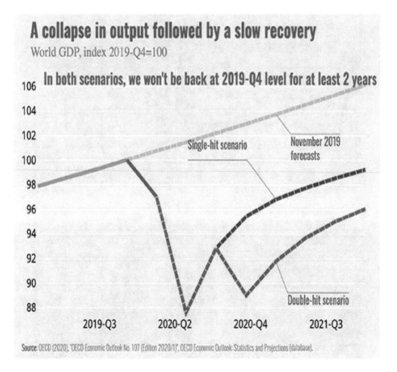

Fig. 1 World GPD projection [3]

in relation to SARS-coV-19 [4]. Also considering projections of a trend for GDP in the 2020/2021 Biennium [5], countries most impacted by Sars-coV-19 due to the number of deaths have a negative GDP trend for the 2020/2021 biennium. On the other hand, countries less affected by the Sars-coV-19 pandemic and those that were able to control the spread of the virus faster showed a growth trend, as shown Fig. 2.

In this context of the pandemic, it was also clear that services are a fundamental economic sector for contemporary society. Even before the pandemic, it was possible to observe a slowdown in the world Gross Domestic Product (GDP), a slowdown due to a series of commercial tensions that weakened manufacturing activity and investment in this sector. On the other hand, in contrast to the weakness of the

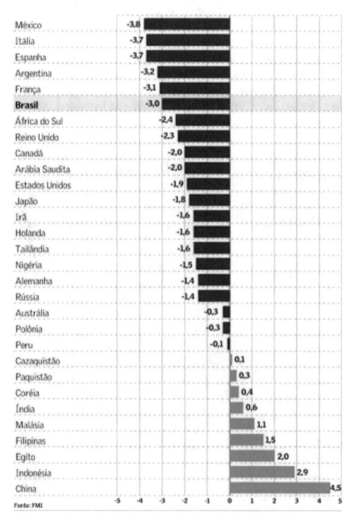

Fig. 2 Performance predicted for the 30 largest economies in the world [5]

Fig. 3 Leading economic sector of the countries [6]

industry, the services sector was steady or even growing in almost the whole world, keeping the labor market warm and showing itself to be the most representative activity in terms of national GDPs. The post-pandemic economic recovery will be gradual and will bring new opportunities for organizations that are able to observe the new market niches, especially those that are able to meet the demands for services, whether in person or at a distance. Given its economic relevance, it is essential a better understand this sectoral activity. Services can be defined as a set of activities that are carried out to meet a customer's needs. Thus, it is a product, but a product that brings greater difficulty in analysis, planning and production, since it is immaterial, intangible, and its consumption is happening simultaneously with its production. These two characteristics, intangibility, and simultaneity distinguish the production of a service from the production of a good. So, it is important to understand these two concepts. Figure 3 shows the economic sector by countries.

Concerning simultaneity, this characteristic can be explained by noting that services cannot be stored, as they are produced and consumed at the same time. The closest thing you get to the idea of stock in services is the customer waiting, that is, the queue. That is, in a service, the operation takes place as an open production system, in which all the impact of variations in demand suffers. As for intangibility, it concerns the fact that the consumer cannot touch the purchased product, as the service is immaterial. Other characteristics that can be mentioned regarding services are heterogeneity, that is, two services will never be the same since each consumer will have a different experience, even if the service, in theory, is the same; perishability, since, since a service cannot be stored, it must be consumed at the time of production; and the non-ownership of the service by the consumer, because when you purchase a service, you have only the right to receive it [7].

According to Krick [8], modern engineering can be understood as a profession essentially dedicated to the application of a certain set of knowledge of certain qualifications and a certain attitude in relation to the creation of devices, structures and processes used to convert resources to forms appropriate to the meeting human needs. In this way, the engineer uses scientific knowledge to apply it in the creation of an element valued by society. In this sense, modern engineering is perfectly suited to resolving issues related to services. Although in the popular imagination, engineering is still tied only to industry, the growth of the economic services sector and, consequently, the performance of the engineer in this sector is significant. In fact, it should be noted that the view that society has about the importance of services for the economy has been changing significantly, as services account for the largest share of the Gross Domestic Product—GDP of several nations [9]. Thus, there is a new field of action opening for engineers, so that they can work, from the perspective of sustainability, solving problems in the third sector.

It remains to be seen whether engineers are being trained for this new opportunity that arises, once again, to demonstrate its usefulness to society from the perspective of services and sustainability. In this sense, in terms of methodological procedures, an investigation was carried out that can be characterized as descriptive research with a survey-type design, insofar as we made systematic observations about the chosen problem [10], that is, whether there is validity in training engineers to performance in service. The method that characterizes this initial study of the theme is bibliographic research combined with documentary research. The data collected for this study originated from secondary sources, among them records and documents from private or official sources, such as pedagogical course projects, curricular structures, history of Engineering courses around the world; data on the third sector and its impact on GDP; Google Trends data; among others. The collection took place between the months of May to September 2020. The treatment of the data was carried out through techniques appropriate to qualitative data such as the categorization of data; as well as simple statistical techniques, suitable for studies with quantitative data, such as measures of central tendency.

2.1 Qualification of the Engineer to Work in Services

A simple search for the term "service engineering" in Google Trends for the period from Oct/2019 to Oct/2020 has shown itself, over that time, to be continuous and, with few exceptions, always close to the peak of interest in the observed period. This means that, over the past year, there was a demand for the subject and that demand has been maintained over the months. Although this is a quite simple indication, it gives us a first indication that there is a society's interest in the topic. Figure 4 shows the trends to offered vacancies in the *services engineering area*.

Another simple survey that proves interest in Service Engineering was carried out on a professional social network between 09/28/2020 and 05/10/2020. Vacancies

Fig. 4 Service engineering in the google trends [11]

Table 1 Service engineering courses

Continent	Undergraduate	Graduate
South America	1	2
North America	1	1
Asia	1	0
Europe	6	8

were identified in all continents over the course of a week, it was possible to count more than 17,000 jobs being offered in the *services engineering area.*

With these numbers as initial indicators of the importance of the area, a systemic search for courses related to engineering and services, spread around the world, was carried out. And, at the time of this publication, a total of 20 service engineering courses have been identified, at the undergraduate and graduate levels. The collected data can be seen in Table 1.

Analyzing the profiles of graduates made available by educational institutions, it is possible to perceive a great diversity among the understandings of what a Service Engineering would be and how it could contribute to society. It is observed that half of the courses (50%) are related to assignments related to Civil Engineering services and their consequences. 15% are related to Mechanical Engineering and product design. Another 15% is assigned to services in Electrical Engineering and Telecommunications. Finally, 20% of courses are related to service management more generally. In other words, only 4 (four) of the courses raised in the research have in their profile the aim of training students to train professional engineers who can work properly in the third sector.

As for the association with sustainability, it was possible to observe the concept proposed in four courses, from the twenty profiles of graduates analyzed. Specifically considering each of the dimensions of the triple bottom line, there are again four of the twenty courses proposing a graduate with an understanding of environmental and economic issues. However, only one of the twenty courses cites the social issue.

3   Final Considerations

With the technological revolution, pertinent to the advent of a post-industrial society, there was a decrease in industrial activities, reducing the need for labor in this sector, on the other hand, there was an increase in the demand for professionals in services. In addition, the pandemic will impose greater concern on society in terms of sustainability, in its three dimensions: environmental, social, and economic. Thus, it is relevant that educational institutions address the issue of sustainability in their courses, as natural resources are increasingly scarce, the world population increasingly larger and future professionals must be trained to use the resources that remain in a sustainable way, obtaining the needs of the population and still guaranteeing them in the future.

Observing the current events, it is possible to cognize the great impact in all dimensions of sustainability that the way of life of the current society has caused, as well as to plan the necessary changes to revert the situation in which we find ourselves. If these changes do not happen, and there is a continuity in activities harmful to the environment, the levels of quality of life started to decrease around 2030 [12], but with healthy techniques and conscious use of resources it will be possible to restore a good part of the diversity and still maintain a high standard of quality of life.

Despite the importance of the sustainability theme and the relevant growth in the third sector, it was observed in this research that there is still no representative quantitative concern in adequately training specific workers to solve problems in services, especially regarding Engineering, let alone in relation to the issue of sustainability. Thousands of engineering courses are offered by hundreds of educational institutions around the world and out of these thousands, we reached the tens of thousands in terms of service-oriented engineering and the units' house when it comes to engineering, services, and sustainability. It is understood that the initiative to offer courses aimed at Service Engineering is still incipient.

When fulfilling the general objective proposed for this article, the necessary opening of courses to train professionals with an undergraduate degree in Service Engineering is observed. It was noticed that there is a greater emphasis on traditional areas of engineering such as Civil Engineering, but it is observed that General Services Engineering, in turn, begins to appear. In this way, the importance of a Service Engineering professional in front of the world demand for professionals to solve the various problems faced in the daily life of the third sector is evidenced, as well as the relevance in a specific training for this performance. As potential future research, we understand that it is necessary to further investigate the reasons for the lack of concern with a specific professional for the service sector, identify and analyze the demands of the third sector in relation to the professional competences of interest, map the competences, skills and attitudes professionals expected for the Service Engineering professional, as well as designing a curriculum proposal for the general service engineer.

Please note that the first paragraph of a section or subsection is not indented. The first paragraphs that follows a table, figure, equation etc. does not have an indent, either.

Subsequent paragraphs, however, are indented.

References

1. Elkington, J., Cannibals with forks: the triple bottom line of 21st century business. Capstone, Oxford (1997).
2. Capra, F. O ponto de mutação: a ciência, a sociedade e a cultura emergente. 24nd edn. Editora Pensamento, São Paulo (2003).
3. The Economist, Intelligence Unit Homepage, https://www.oecd.org/economic-outlook/june-2020, last accessed 2020/10/02.
4. Júnior, J. R. de C. S.; Marco A. F. H.; Levy, P. M.; De Carvalho, L. M.; De Moraes, M. L.; Garcia, P. M.. Atividade econômica: revisão das previsões de crescimento 2020/2021.
5. Carta de Conjuntura, n. 47, 2° trimestre de 2020. IPEA, Brasília (2020). 5. Conceição, A., https://valorinveste.globo.com/mercados/brasil-epolitica/noticia/2020/07/06/brasil-tera-o-6o-pior-desempenho-entre-30-maioreseconomias-do-mundo.ghtml, last accessed 2020/10/05.
6. Statistics Times, http://statisticstimes.com/economy/countries-by-gdp-sectorcomposition.php, last accessed 2020/09/30.
7. Fitzsimmons, James A.; Fitzsimmons, MJ. Administração de serviços: operações, estratégia e tecnologia de informação. 7nd edn. AMGH, Porto Alegre (2014).
8. Krick, E. V. Introdução à Engenharia. 2. ed.. Livros Técnicos e Científicos, Rio de Janeiro (1979).
9. IBGE. Pesquisa Anual de Serviços 2017, https://biblioteca.ibge.gov.br/visualizacao/periodicos/150/pas_2017_v19_informativo.pdf, last accessed 2020/10/30.
10. Dos Santos, A. R., Metodologia científica: a construção do conhecimento. DP&A Editora, Rio de Janeiro (1999).
11. Google Trends, https://trends.google.com/trends/explore?q=service%20engineering, last accessed 2020/10/06.
12. Iberdrola Homepage, https://www.iberdrola.com/meio-ambiente/superexploracao-dosrecursos-naturais, last accessed 2020/10/05.

Analysis of Biosafety Actions in Supermarkets: The Customer View

Kidja Maria Ramalho Frazão, João Vitor da Costa Santos, Aldaiza Rayssa Santos, Pedro Henrique do Nascimento Silva, and Lucas Ambrósio Bezerra de Oliveira⬛

Abstract This article aims to analyze the perception and importance of biosafety actions in supermarkets. Besides, we want to understand how such actions can be classified under the perception of customers. For this, we established seven variables for the evaluation of biosafety. This research has an exploratory character because it analyzes biosafety actions and their effects under the perception of quality and customer satisfaction in a pandemic scenario. The results suggest that customers rate the actions as extremely important, but the perception of quality in carrying them out is reasonable. Six of the seven biosecurity measures were classified as one-dimensional attributes, indicating that customer satisfaction is proportional to performance. Item/action 6, which deals with the use of facial masks, was the action that obtained the best evaluation in matters of perception and importance. The Kano Model indicated that item/action 5, which deals with demonstrating the concern for customers' health, is the one that most satisfies them.

Keywords Biosafety · Servperf · Kano model · Supermarket · COVID-19

1 Introduction

One of SARS-Cov 2 (COVID-19) impacts on organizations is the requirement for biosafety protocols to function during the pandemic [1, 2]. Even essential sectors, such as supermarkets, need to implement actions that promote the reduction or elimination of health risks for workers and customers [3]. It is perceived that such actions can influence the perception of the quality of the services provided. Therefore, it is important to understand the perception and importance that customers have concerning biosafety actions, and seek to understand how the actions can be classified from the perspective of supermarket customers.

K. M. R. Frazão · J. V. da Costa Santos · A. R. Santos · P. H. do Nascimento Silva · L. A. B. de Oliveira (✉)
Federal University of the Semi-Arid Region, 59625-900, Mossoró, Brazil
e-mail: lucasambro@ufersa.edu.br

© The Author(s), under exclusive license to Springer Nature Switzerland AG 2021
A. M. Tavares Thomé et al. (eds.), *Industrial Engineering and Operations Management*,
Springer Proceedings in Mathematics & Statistics 367,
https://doi.org/10.1007/978-3-030-78570-3_24

Therefore, this article aims to analyze the perception and importance of actions that compose biosafety actions in supermarkets, using the logic of attribute evaluation proposed in the SERVPERF model. Besides, we want to understand how the variables can be classified under the perception of customers and, for this, the attractive quality Kano Model will be used.

This work is relevant and justified because supermarkets are considered essential activities [3, 4]. Therefore, biosafety actions need to be systematically analyzed to understand the effects on the perception of service quality and customer satisfaction and because there are still few studies on the effects of COVID-19 on customer behavior in supermarkets [5]. Thus, in addition to contributing to the improvement of such services' services, this work will make it possible to understand the perception and classification of actions and attributes by integrating two essential tools: SERVPERF and Kano Model.

2 Background

This section will present the context of the sector analyzed in this article, the tools SERVPERF and Kano Model, and a notion about biosafety actions. This background will provide support to carry out the necessary analyzes for the paper.

2.1 COVID-19 and Impact on Services Provided by Supermarkets

One of the impacts of COVID-19 on the supermarket sector is economical. As bars and restaurants stopped serving in person, part of the population started to eat at home, increasing the supermarket segment's demand. At the height of the pandemic, there was a growth of 16.5% in the period accumulated between March and August 2020 [6]. Along with demand, there was an increase in the number of customers shopping at these establishments, needing biosafety actions.

Another impact observed is the search for an understanding of the effect that biosafety actions can have on the perception of the quality of services provided and supermarket customers' satisfaction. Thus, we seek to understand the perception, importance, and classification of possible attributes for this new factor from the customers' perspective.

2.2 Biosafety

Biosafety is *"a set of actions designed to prevent, control, mitigate or eliminate risks inherent in activities that may interfere or compromise the quality of life, human health and the environment"* [7]. In this way, biosafety can be characterized as strategic and essential for supermarkets, especially in pandemic times.

In Brazil, supermarkets have performed several biosafety actions, such as [8]: cleaning of shopping carts and baskets, availability of 70% INPM alcohol, safe distance between people, checking body temperature of customers and employees, constant check-out cleaning and physical environment, control of the number of customers inside the establishment, requirement for the use of a facial mask.

2.3 SERVPERF and Kano Model

Understanding the perceived quality and classification of attributes under the customer's view can be done through the SERVPERF and Kano Model.

O SERVPERF [9] is a model for measuring quality and is based on the paradigm that quality assessment can be done based on the customer's perception of the item evaluated. Therefore, it is not necessary to know the customer's expectations about the service and, due to this characteristic, it differs from SERVQUAL. In the light of SERVPERF dynamics, the model's logic can be used to evaluate any items related to the provision of services that, in the case of this research, involve biosafety actions.

The Kano Model [10] measures satisfaction and classifies customer preferences into five attributes: must-be (M), one-dimensional (O), attractive (A), indifferent (I), and reverse (R). Must-be (M) attributes, if present and sufficient, do not generate satisfaction, and if they are not present or insufficient, they generate dissatisfaction. One-dimensional attribute (O) suggests that satisfaction is equivalent to performance level: if high, greater satisfaction and, if low, less satisfaction or dissatisfaction. In an attractive attribute (A), an item's low performance or non-existence does not generate dissatisfaction and generates satisfaction if present. However, if present, it will bring satisfaction proportional to the level of service. The indifferent attribute (I) does not influence satisfaction, and a reverse attribute (R) does not have the same meaning for customers: for some, it can generate dissatisfaction if it has a specific attribute, and for others, it can generate satisfaction. In addition to these, there are questionable attributes (Q) that suggest some inconsistency in the customer's responsibility.

An attribute is classified by comparing a client's responses concerning two questions/statements: one with functional form (positive sense) and another with dysfunctional form (negative sense). Figure 1 shows the procedure and classification matrix.

In addition to the classification, the Kano Model allows calculating a coefficient of satisfaction (SC—better) and dissatisfaction (IC—worse) [12, 13], as presented in Eqs. (1) and (2), respectively.

Statements n (Action)		Dysfunctional Form				
		1 I like it that way	2 I am expecting it to be that way	3 I am neutral	4 I can accept it to be that way	5 I dislike it that way
Functional Form	1 I like it that way	Q	A	A	A	O
	2 I am expecting it to be that way	R	I	I	I	M
	3 I am neutral	R	I	I	I	M
	4 I can accept it to be that way	R	I	I	I	M
	5 I dislike it that way	R	R	R	R	Q

Statements	Grade
N. 1	A
N. 2	...
N. 3	...
N. 4	...
N. 5	...
...	...
N. n	...

Fig. 1 Classification procedure and matrix. *Source* Adaptation from the Madzík [11]

$$SC\% \ (better) = \frac{A\% + U\%}{A\% + U\% + O\% + N\%} \tag{1}$$

$$IC\% \ (worse) = \frac{(U\% + O\%) * (-1)}{A\% + U\% + O\% + N\%} \tag{2}$$

The satisfaction coefficient indicates increased satisfaction with the analyzed attribute's presence; the dissatisfaction coefficient indicates the proportion of decreased satisfaction if the analyzed attribute does not exist. The closer to the limit (1 and –1), the greater the customer's satisfaction or dissatisfaction, respectively. Based on these coefficients, it is possible to create a classification matrix for the "Better-Worse" attributes.

Thus, the integration of the SERVPERF and KANO tools makes it possible to understand the perception and classification of the new attribute in the supermarket sector, which is relevant in times of the pandemic, biosafety.

3 Methodology

This work is of an applied nature, with exploratory and descriptive research objectives, using the quantitative and survey-type research methods [14].

The research was conducted in a city that is an economic hub in the interior of the State of Rio Grande do Norte, selected because it is strategic for the study's region in August 2020. The research instrument was divided into four parts. The first part aimed to identify the participant's social and economic profile (gender, age, frequency of purchases, and always bought at the same supermarket).

In addition to these questions, seven items were proposed, shown in Table 1. Each one was used to measure the perception, importance, and classification of attributes in an adapted way, corresponding to the second, third, and fourth parts of the questionnaire.

Table 1 Research variables

Item	Variable
Item 1	The supermarket sanitizes shopping carts and baskets before use by customers
Item 2	Gel alcohol is available in visible and easily accessible locations
Item 3	The supermarket makes use of marking to keep the distance in the queues
Item 4	Cashier operators follow safety regulations against COVID-19
Item 5	The supermarket, in this pandemic period, shows concern for the health of its customers
Item 6	The use of a mask is mandatory at the supermarket
Item 7	The supermarket offers online shopping (or by phone) and delivery

The research participants evaluated the statements about perception using a 5-point Likert scale (1 strongly disagree to 5 completely agree); to answer the statements related to importance, the same type of scale was used, but varying the meaning of the scale (1 Not at all critical up to 5 Extremely important); and to classify the attributes (Kano Model) the participants used the scale shown in Fig. 1.

Data analysis consisted of three stages. The first was to verify the internal reliability of the responses of primary interest variables, using Cronbach's Alpha. Therefore, the index of 0.60 was defined as the cutoff point as the minimum acceptable value for this exploratory study. The analysis of perception and importance was performed through descriptive statistical analysis and quartile analysis. The analysis of the data referring to the Kano model was performed according to the procedure related to the method [12, 13], besides calculating the coefficient better (SC) and worse (IC).

4   Results

4.1   Customer Profile and Reliability of Responses

The research sample consisted of 304 people from the research's target city and its surroundings, 71.1% women and 28.9% men. As for the interviewees' age, the average was 29.65 (standard deviation = 10.89), ranging from 18 to 73 years.

It was identified that 50.3% of respondents always buy at the same supermarket, while 49.7% buy at different establishments. 31.9% indicated that they make purchases once a week, while 21.7% stated that they make purchases 2–3 times a week; a minority goes shopping in supermarkets every day.

It was found that the Cronbach's alpha of the question groups (perception, importance, satisfaction, and dissatisfaction coefficient) reached the specified cutoff point, as shown in Table 2.

Thus, such alpha values suggest a good fit of the data.

Table 2 Alpha values

Question group	Alpha values
Perception	0.687
Importance	0.819
Satisfaction coefficient	0.989
Dissatisfaction coefficient	0.930

Table 3 Perception analysis

Item	Average	Standard deviation	Modo
Item 1	3.72	1.39	5
Item 2	3.96	1.34	5
Item 3	4.02	1.34	5
Item 4	4.41	0.91	5
Item 5	3.9	1.12	5
Item 6	4.85	0.58	5
Item 7	3.07	1.57	5

4.2 Analysis of the Perception of Biosafety Actions

The assessment of supermarket customers' perception of these establishments' biosafety actions was based on the SERVPERF model; therefore, based on users' observations. Table 3 presents the data for this analysis.

As the questions assess customers' perception, it can be seen that in items 4 and 6, respectively, cashier operators follow safety rules to combat COVID-19, and the use of a mask has been mandatory in the supermarket. The common opinion among users causes a positive perception that presents a standard deviation of less than one and an average greater than 4.

The items that showed less evaluation in the perception of the services provided are items 1, 5, and 7, suggesting that either the establishment is not complying or that users do not feel confident in executing such actions.

4.3 Analysis of the Importance of Biosafety Actions

To complement the analyses carried out in Sect. 4.2, it was asked about the perceived importance of actions to confront COVID-19. As shown in Table 4, users perceive all the actions analyzed as of great importance, with high averages and low standard deviations (SD). However, item 7, which deals with the offer of online purchases (or by phone) and home delivery, was the issue that presented the lowest average and the highest standard deviation.

Table 4 Analysis of the importance of actions

Item	Average	Standard deviation	Modo	Priority*
Item 1	4.73	0.84	5	High
Item 2	4.81	0.64	5	Moderate
Item 3	4.84	0.58	5	Low
Item 4	4.87	0.48	5	Low
Item 5	4.79	0.62	5	Moderate
Item 6	4.89	0.48	5	Low
Item 7	4.38	1.05	5	Critical

*Calculated based on quartile

The use of a mask (Item 6) has been frequent since the beginning of the pandemic, in which it is the item most used by society, which, being mandatory, creates a barrier against the virus. The importance of other equipment such as cleaning with gel alcohol (Item 2) has become a habit in new people's routine, forming responsibility and respect.

Based on the average values, it was possible to estimate the prioritization to improve combat actions. Item 7, being the item with the lowest average, obtained a high priority. Therefore, it must be improved with greater urgency; it is followed by item 1 (with high priority), items 2 and 5 (with moderate priority), and items 3, 4, and 6 (with low priority).

4.4 Analysis of the Importance of Biosafety Actions

Kano's attractiveness model helps to classify organizational actions from the customers' point of view. With the model, it is possible to classify them according to their impact on user satisfaction. Based on the Kano model algorithm, it was possible to obtain the "Best-Worse" matrix presented in Fig. 2.

The hygiene of the carts (1), the alcohol gel available (2), the distance between the lines (3), the cashier operators follow safety rules (4), the supermarket shows concern for the health of customers (5) and the mandatory use of masks (6) these were obtained as one-dimensional, therefore, the greater the performance in these items the greater the satisfaction of customers.

For this reason, it is reported the importance of improving these items, mainly items 1, 2 and 5, in order to seek to guarantee the satisfaction of its customers, having great advantages when compared to its competitors.

It is noteworthy that item 5 obtained a higher satisfaction coefficient (0.86), suggesting that customers are satisfied with the concern and care that the supermarket is taking with the pandemic of COVID-19.

Concerning item 7, which deals with the offer of online or telephone purchases and home delivery, it was observed that it was classified as an attractive attribute,

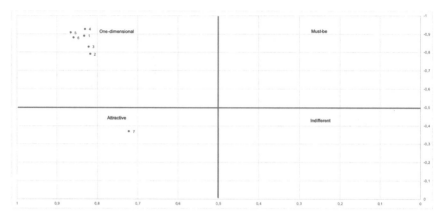

Fig. 2 Diagram better-worse

that is, it indicates that if present, it can bring satisfaction and be a differential, but if it does not exist it may not generate dissatisfaction for the customer. As a critical point, this factor must be considered and put into practice if possible, as it exceeds customers' expectations, ensuring their loyalty.

5 Conclusions

SARS-Cov 2 (COVID-19) inflicted several changes in organizational dynamics. One is the adaptation to the context of biosafety. This set of actions can influence the perception of quality and customer satisfaction.

Thus, the research carried out in the supermarket sector established seven biosafety actions and identified that customers' perception concerning them is reasonable, considering the average values around grade 4. Action 6, which deals with masks' use, was the action that obtained the best assessment of perception, indicating that customers expect to use them.

Regarding the importance of biosafety actions, the results suggest that they are critical and that organizations should pay attention to issues such as the hygiene of shopping carts and shopping baskets. Once again, item 6 was the most important.

Based on the Kano Model, it was also observed that item 7 (online or telephone purchase offer) was classified as an attractive attribute and all the others as one-dimensional attributes. Therefore, attention should be paid to such items as the one-dimensional ones because they indicate an improvement in satisfaction when they present high performance. The Kano Model indicated that item 5, which deals with showing concern for the health of customers, is the item with the most excellent satisfaction for customers.

Thus, even if exploratory, this work contributes to managerial practice in supermarkets, providing insights for the sector and assisting in developing the literature on determinants of perceived quality and satisfaction in the service segment.

As an indication for future work, it is suggested to carry out broad research involving the traditional dimensions of SERVPERF and the biosafety items used in this research to verify, in a model of structural equations or linear regression, whether the biosafety determinant explains the perceived quality. Among the research limitations, it stands out the fact that there is still little literature to support arguments and hypotheses that biosafety influences perceived quality and satisfaction.

References

1. SEBRAE (2020) Protocolos sanitários para seu minimercado, supermercado ou mercearia | Sebrae. https://www.sebrae.com.br/sites/PortalSebrae/ufs/al/artigos/protocolo-sanitarios-para-seu-minimercado-supermercado-ou-mercearia,b37b70abe35b3710VgnVCM1000004c 00210aRCRD. Accessed 16 Oct 2020
2. SINCOVAGA (2020) Supermercados adotam soluções para reduzir os riscos de contágio - Coronavírus - Sincovaga. https://www.sincovaga.com.br/supermercados-adotam-solucoes-para-reduzir-os-riscos-de-contagio-coronavirus/. Accessed 16 Oct 2020
3. Hakovirta M, Denuwara N (2020) How COVID-19 Redefines the Concept of Sustainability. Sustainability 12:3727. https://doi.org/10.3390/su12093727
4. BRASIL SG (2017) Decreto no 9.127, de 16 de agosto de 2017.
5. Anastasiadou E, Chrissos Anestis M, Karantza I, Vlachakis S (2020) The coronavirus' effects on consumer behavior and supermarket activities: insights from Greece and Sweden. Int J Sociol Soc Policy ahead-of-print: https://doi.org/10.1108/IJSSP-07-2020-0275
6. CIELO SA (2020) Coronavírus: Impacto do Covid-19 no varejo. In: Cielo. https://www.cielo.com.br. Accessed 16 Oct 2020
7. Lessa D (2014) Biossegurança, o que é? In: Fiocruz. https://portal.fiocruz.br/noticia/biosseguranca-o-que-e. Accessed 16 Oct 2020
8. ABRAS (2020) Site ABRAS. In: ABRAS - Assoc. Bras. Supermerc. https://www.abras.com.br/. Accessed 16 Oct 2020
9. Cronin JJ, Taylor SA (1992) Measuring Service Quality: A Reexamination and Extension. J Mark 56:55–68. https://doi.org/10.2307/1252296
10. Kano N (1984) Attractive quality and must-be quality. Hinshitsu Qual J Jpn Soc Qual Control 14:39–48
11. Madzík P (2018) Increasing accuracy of the Kano model – a case study. Total Qual Manag Bus Excell 29:387–409. https://doi.org/10.1080/14783363.2016.1194197
12. Berger C (1993) Kano's methods for understanding customer-defined quality. Cent Qual Manag J 2:3–36
13. Löfgren M, Witell L (2005) Kano's Theory of Attractive Quality and Packaging. Qual Manag J 12:7–20. https://doi.org/10.1080/10686967.2005.11919257
14. Turrioni JB, Mello CHP (2012) Metodologia de pesquisa em engenharia de produção. Programa Pós-grad Em Eng Produção Universidade Fed Itajubá Itajubá UNIFEI

Framework for the Development of Competencies Using a Matrix for Decision Making: A Case Study

Rafaela Plantes Pavloski, Mario Luis Nawcki, Edson Pinheiro de Lima, Fernando Deschamps, and Sergio Eduardo Gouvea da Costa

Abstract Strategic operations management is essential for business survival in an increasingly dynamic world, in which an increasing amount of data is generated all the time. Resource management, including competencies, requires solutions that are agile and based on information that can assist management teams in choosing professionals, who are able to provide the best solutions to old problems, in addition to facing the new scenarios. This article aims to present a procedural structure that helps managers in choosing competences to be developed, based on the analysis of contextual information and operational data. The proposed structure, built on the basis of studies by authors who evolved in the approaches to the proposed theme, was validated using data from an organization that uses the World Class Manufacturing (WCM) methodology, detailing the Cost Deployment concepts (CD) and its application in a real context. The method used was the creation of a framework and its application in a case study. As a result, it was possible to show an alternative for the choice of competences based on operational data and information, and not on the decision maker's intuition.

Keywords Strategic performance management · Skills · WCM · Cost deployment

1 Introduction

Due to the emergence of new technologies and constant evolutions in the demands of consumers, changes in the products and services presented by companies become

R. P. Pavloski (✉) · M. L. Nawcki · E. P. de Lima · F. Deschamps · S. E. G. da Costa
Pontificia Universidade Catolica Do Parana - PUCPR - Industrial and Systems Engineering, Curitiba, Brazil

E. P. de Lima · S. E. G. da Costa
Universidade Tecnologica Federal Do Parana - UTFPR - Industrial and Systems Engineering, Pato Branco, Brazil

F. Deschamps
Universidade Federal Do Parana - UFPR - Mechanical Engineering, Curitiba, Brazil

© The Author(s), under exclusive license to Springer Nature Switzerland AG 2021
A. M. Tavares Thomé et al. (eds.), *Industrial Engineering and Operations Management*,
Springer Proceedings in Mathematics & Statistics 367,
https://doi.org/10.1007/978-3-030-78570-3_25

necessary, which affects the way in which organizations structure their activities and define their priorities. One of the needs resulting from this process is the training of workers, so that they can deal with new demands related to knowledge about methods and tools, such as the use of automation [1] and the need for reading and analyzing data. Companies from different branches must invest more and more in the training of their employees in the coming years, aiming at the creation of distinctive competences, which according to [2] are organizational skills, not possessed by competitors, and that meet the needs of customers, thus being a source of advantage. Barney [3] states that the discovery of strenghts and weaknesses, as well as opportunities generated by the environment and neutralization of threats, enables organizations to obtain a competitive advantage.

According to [4], "resource-based management" points out that some companies can obtain sustainable competitive advantage by focusing on the role of resources that are largely internal to the company's operations. Simply put, "above average" performance is more likely to be the result of key capabilities (or competencies) inherent in a company's resources, than its competitive position in its industry.

The decision on which competencies should be developed is usually made by managers, who are based on their own experience, or on external trends that influence the company's operation. In a context where changes occur more and more quickly, the evolution of analytical methods aimed at developing skills can be an important way of aligning with the company's strategy, as seen in Mills, Platts and Bourne [5, 6], studies focused on the vision of competence as an organizational resource. The proposal of this work is to suggest a structural framework of the competence identification process, which was validated based on the analysis of the internal processes of an automotive company, which uses the principles of World Class Manufacturing. The proposed structure aims at greater assertiveness in actions, through alignment with the strategy and the context of the organization that uses it.

Another issue considered in the construction of the model was the approach used, which could consider the qualitative or quantitative dimensions of the observed situation. According to [7], the qualitative approach deals with the construction of theories based on the interpretation of experiences, whereas the quantitative approach starts from the confirmation of a hypothesis. Among the possibilities for using the qualitative dimension, there is the direct observation of the experience of workers with the development of skills, while the quantitative dimension demands the use of data generated by the operation, previously determined. An alternative to this second use is the use of financial criteria to choose training that meets the most urgent operational needs, in view of the losses observed by the company, and which must be mitigated as soon as possible. Cost Deployment, which according to [8] is the fundamental pillar in the WCM program that scientifically detects waste and losses, leading to the choice of improvement projects within the organization, is used in this work as a method to assist in the identification of gaps related to the competencies that must be developed more urgently.

The work was structured according to the following sections:

- The first stage of the study required a literature review on topics related to the problem presented. As it is a process that brings long-term results, and which aims to obtain competitive advantage, the researched themes were analyzed from the strategic point of view. The concepts of competence and Cost Deployment already used by the authors were also presented, as they are specific characteristics of both the study itself and the context found in the analyzed company. The objective of this step is to present past contributions that can bring relevant insights to the study.
- Then, a framework for the analysis of the research problem is proposed, based on the bibliographic references in the literature review, with the addition of other references of prior knowledge of the authors. This structure was subsequently evaluated using data provided by an automotive company located in the southern region of Brazil, in the city of Curitiba.
- The last part of the study presents the results obtained, as well as the conclusions of the authors, accompanied by suggestions for future analysis.

2   Theoretical Background

The theoretical background stage started with a literature review, as described in Fig. 1, a process that was carried out aiming at identifying the current state of publications on the topic addressed in this research. This allows not only research opportunities to be identified, but also useful references and discoveries made previously.

Fig. 1 Literature review process performed by the authors

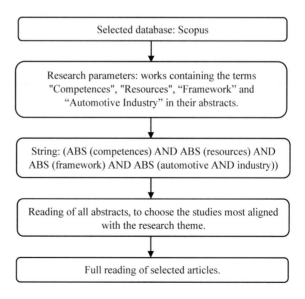

Selected database: Scopus

Research parameters: works containing the terms "Competences", "Resources", "Framework" and "Automotive Industry" in their abstracts.

String: (ABS (competences) AND ABS (resources) AND ABS (framework) AND ABS (automotive AND industry))

Reading of all abstracts, to choose the studies most aligned with the research theme.

Full reading of selected articles.

Table 1 Literature review results portfolio

Document title	Authors	Year
Types of green innovations: Ways of implementation in a non-green industry	Calza, F., Parmentola, A., Tutore, I.	2017
Requirements communication and balancing in large-scale software-intensive product development	Pernstål, J., Gorschek, T., Feldt, R., Florén, D.	2015
Principles of ISO management system integration/transition (ISO 9001:2000/ISO 14,001:1996)	Surette, E.	2005
Resource and operations analysis for agile manufacturing system development	Jin, K., Palaniappan, A.	2004

The database chosen for the search was Scopus, which has been used previously by the authors. Regarding the research parameters, after carrying out some tests that did not return a satisfactory number of results, it was defined that the papers should contain the terms "competences", "resources", "framework" and "automotive" in their abstracts, bearing in mind that the purpose of this work is the creation of a framework that contemplates the development of skills in the automotive industry. The search brought six results, one of which was a book of conference proceedings, not being possible to clearly identify the chapter that brought the result mentioned, which made it necessary to eliminate it from the portfolio. The abstracts of the other five results were read, and of these, the four that demonstrated alignment with the research topic were read in full by the authors, so that their contributions could be identified.

Table 1 shows the four articles that were fully read, which were published after the 2000s, what may indicate a recent interest. The applications and parallel themes addressed in these works also refer to recent challenges, such as sustainability, technology, and agile manufacturing.

The work of [9] addresses one of the challenges derived from the application of green innovations, which is the implementation of this type of technology in the automotive industry, one that is known for not being historically and widely adept at using it. This creates the need for the development of new work practices and organizational skills. The authors use The Green Innovation Landscape Map, a matrix proposed by [10] to classify the efforts of four companies in the automotive industry as routine, disruptive, radical or architectural, depending on the necessary changes in their business models and technical competences. The authors recommend the establishment of partnerships, that can facilitate access to sustainable innovations. Pernstål et al. [11] bring in their work the analysis of the application of the BRASS framework, which contains four dimensions related to communication, seeking improvements in the requirements engineering process in a Swedish automotive company, something motivated by the growth of the creation of products heavily software-based. The document published by [12] deals with the rules of the International Organization for Standardization, citing the tools necessary for a "controlled, systematic and transparent integration" of the ISO 9001:2000 and the ISO 14,001:1996 management systems, acompanied by actual case studies. Finally, in the extended abstract published by [13], it is addressed that companies that seek the implementation of agile manufacturing must include the supply chain in their efforts, which brings the need to align information and activities. These contributions demonstrate that the development of skills must be analyzed with the use of well-structured tools, which contemplates the challenges encountered, that are broad and include requirements that are not always controlled by the organization.

In order to build the proposed framework, the search for a definition of the main themes was carried out, as well as its association with the development of skills. Magalhães [14] states that the development of intangible assets is an important source of competitive advantage, due to the possibility of developing a new relationship with customers, with the market and within the organization itself. In the latter case, one way of making this type of change is by mobilizing the skills and motivation of employees to make improvements in processes, which helps in the development of new skills. Mills et al. [15] delimited in a study the five categories of competences, which are: central competences, which help in the survival of the company; distinctive competences, which are recognized by customers and distinct from those of competitors; organizational competencies, which are characteristic of a business unit; support skills, which assist in the execution of other activities; and finally, dynamic capability, which is the adaptive ability of an organization in the face of its competencies.

The framework proposed by [5] contains three categories, which represent the necessary set for organizations: competencies for resource development and coordination, as well as competencies that are perceived by the consumer. This classification demonstrates that not only the internal aspects must be observed, as well as the impact on deliveries made to those interested in the results generated by the work performed. The same authors proposed [6] an approach to the analysis of resources, divided into five stages, as described in Fig. 2.

3 Research Design

The next stage of the study is the proposal of a framework model for data collection and analysis, that makes it possible to identify the necessary competencies, which are prioritized through cost management criteria, defined by the company. Zangiski, Pinheiro de Lima and Gouvea da Costa [16], presented in their work a map that lists several variables related to competencies and organizational learning, namely: organizational competences and capabilities, organizational learning, organizational knowledge (strategic resource), organizational beliefs system, processes, activities, resources, dissemination and sharing of information and knowledge, results sharing and discussion, skills mobilization and integration, and operational strategy. This shows the importance of observing data and organizational structures, which influence the development of competencies.

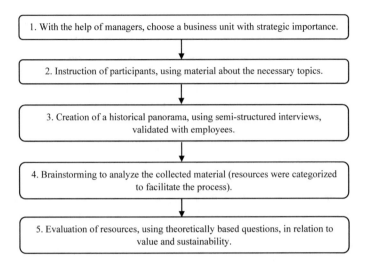

Fig. 2 Action plan for resource analysis, as described by Mills, Platts and Bourne [6]

Based on the work of Mills, Platts and Bourne [5, 6] in which the authors develop a structured framework for the identification of resources as competencies within companies, in addition to the need to choose the competences to be developed by the management team, an initial model is proposed for the identification of these elements using the Cost Deployment methodology, as described in Table 2.

The objective is to observe the instrumentalization of the proposed framework, based on the analysis of its application in the quality area of a manufacturing operation. There are four stages, the first of which deals with an environmental analysis, the second is the collection and analysis of the data by the controllership sector, the third is the identification of the necessary skills through problems that recur over longer periods, and finally the choice of skills through a functionally mixed team, which has HR, controllership, and engineering. There is a similarity to that proposed by the

Table 2 Framework model proposed by the authors

Step	Description
Contextualization	Environmental analysis, to survey elements that have an impact on the development of skills
Collection and analysis of operational data	Analysis performed by responsible sector/function, results in a reference on the current state
Gaps identification	Organization and description of gaps in skills development based on the information generated
Choice of actions	Use of cost deployment matrices to choose and prioritize actions

Cost Deployment methodology, from Yamashina and Kubo [17], which scientifically and systematically establishes a cost reduction program.

The proposed model has the function of operationalizing the choice of competences, based on the main losses of the Operation, which, as Son [18] points out, is reliable information related to costs, and which is useful for carrying out economic analysis. The model also identifies cost elements that should be included in the analysis of advanced manufacturing systems, and proposes a way to estimate them.

4 Results

The case study has been conducted at one production plant at a Brazilian automotive manufacturer. In total, 11,000 engines and 13,000 transmissions were produced by them in 2019. The focus of this case study is the internal Quality that supports machining and the assembly line. For reasons of data privacy protection, the company will be called Alpha. Since 2007 Alpha has been using World Class Manufacturing (WCM) approach, and it is possible to say that there is a high level of maturity on the methodology application.

First step performed was the cost deployment, that has fundamental importance to achieve the desired results. Data collection at the root cause level—the measurement of losses (physical units such as time, grams, cubic meters per hour, etc.) were performed in the most detailed way possible, and at the root cause level of the loss. The various systems implemented by the factory can assist in data collection and in Quality there is a system that has the information needed for our study. According to [18], data can be collected manually or automatically, and in an automated and connected factory, measurement is easier and more reliable. Quality concentrates the problems pointed out as Operation in the Quality Assurance Matrix, which has failure modes described in the lines, meanwhile its columns describes the operations and stations. It receives the data that are entered in the Quality system, whenever non-conformities occur in the operation.

Since all non-conformities are grouped in the system and organized into a matrix (Quality Assurance), the controller analyses the data according to the frequency that a particular failure mode occurs, the amount being computed, as well as the financial impacts that it causes. Son [18] argues that the cost of non-conformity is the loss due to the failure of finished products to meet the quality standards set by a company and customers, that combines the conventional costs of internal failure and external failure.

Costs of non-conformities are estimated at different levels: easy repair performed in the same station that occurrence happened, medium repair performed in an adjustment area in the final of assembly line, and hard repair made in the vehicle. For scrap there is a similar scale, and it is up to Quality to assess at which level the non-conformity is classified.

Through the Quality Assurance Matrix, which is processed in a frequency of three times a year, called waves, improvement projects can be launched to the support areas to eliminate that non-conformity. It is possible to understand, with an assessment among several waves, which are the problems that are repeated and are not properly solved. Through an approach that considers longer periods of time, it is possible to identify the lack of competence to deal with certain NCs, and the employees that solve the same problem several times, not attacking the root cause or the phenomena.

Based on data collected, it was possible to identify NCs that are not properly solved, and one of the achieved objectives is an understanding of the lack of competence identified. Following for the next step, scenario analysis is necessary. To perform this analysis, some elements of the organization are required, such as: Human Resources, Engineering and Controllership. HR has the function of mediating and directing discussions, that bring out the differences between the vision focused on costs in some sectors, with that of engineering, which points out technical issues.

5   Discussion

The application of the methodology is based on the costs of the Operation, more specifically internal Quality, having the cost structure described by Son [18] as the beginning of the proposed approach. The work is focused on costs per failure, or non-conformities, that is internally defined as: easy repair performed in the same station that occurrence happened, medium repair performed in an adjustment area in the final of assembly line, hard repair made in the vehicle; for scrap there is a similar scale. According to Son [18], all the costs related to Quality are called "Relatively ill-structured cost".

The occurrences of non-conformities in the process are collected manually by their insertion in the Quality system, with information such as: part number affected, causative station and detecting station. It also contains data related to the type of adjustment and its respective cost, rejection, and inspection costs. It was possible to realize that manual data collection is not ideal, since that a significative percentual of NCs do not go to the system.

After collection, the data are summarized in a matrix called QA (Quality Assurance Matrix), with the lines containing the failure mode, and the columns containing the stations/operations in the area in question. It is possible to see the biggest impacts by cost, since the occurrences are posted by type of adjustment, each containing a cost and quantities. The costs, once mapped, are sent to the controllership, which manages all the losses of the operation with the cost matrices.

With the data grouped and summarized in matrices for long periods, usually more than two years, it is possible to identify losses that were well treated and eliminated, and those that occurred again. This step in the framework is called "loss evolution", and seeks to show whether a problem has been addressed with the appropriate resources and skills. Another analysis is from the scenario, that shows the market conditions, new technologies and trends that will influence the portfolio of products offered to the consumer.

Through the data history, evolution of losses and analysis of future scenarios, the framework is finally concluded, with the analysis of necessary resources and skills based on facts and data. The management team can evaluate the problems that affect the operation, and provide resources in a timely manner, whether through training, searching for professionals in the market, or even with the help of external experts and consultants. The decision of which path to follow is not based solely on the manager's experiences, values and beliefs.

Figure 3 has a mapping of the steps performed, based on the identification, analysis, and classification of costs.

The use of the described model validates the initial proposal and shows the application for the management of resources and competencies, which goes beyond the elimination of losses and waste. An advantage in the case study is that the data and matrices already existed, and the framework used the data to help identifying the pointed-out losses that were not eliminated. To exemplify the findings, the analysis of the root cause was highlighted by the demonstration that the analyses and treatment made before did not understand the phenomenon, meaning that it treats the effects, not the causes. This is a typical situation where the professional is a specialist that solve the same problem several times, but does not attack root cause.

6 Conclusion

In this paper, an approach to manage resources and competences is proposed, considering that a lot of times there is a process to it, but the decision of what competences to develop is made by experience and intuition of the managers. Based on the Cost deployment (CD) approach, a foundation to map and quantify losses and wastes is build, and through these data it is possible to recognize repeated problems that the organization does not have competence neither resources to attack and solve. Another point that is essential in changing times is the analysis of context, and an understanding based not only on the past and present, but evaluating new scenarios, products, digitalization and more recently, a world dived in a pandemic.

The presented study had as objective the construction of a framework of the process of identification and choice of competences to be developed, using the Cost Deployment method as a decision criterion. A very relevant issue is that organizations need to have a structured way of collect data for wastes and losses, and a way to work with them, enabling to see where organization does not have competence to solve its problems. A long period of time is needed to understand what problems are reoccurring, and if the reason of that is the lack of competence.

The proposed approach can be a complement for organizations that already use CD as a methodology to attack wastes and losses, using it to guide and elect where to put energy regarding resources and competences. For the organizations that does not use CD, this work can be an inspiration and a starting point, showing that is possible to use a quantitative approach, and not only the qualitative ones.

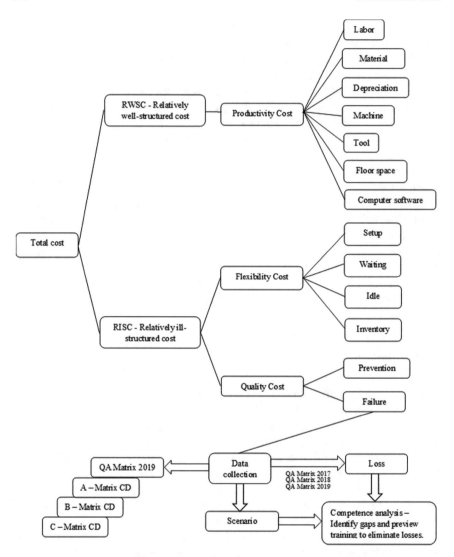

Fig. 3 Mapping the cost breakdown process

One limitation in this work is that the framework application is closed to the context of the automotive industry analyzed, but this can be an opportunity for future analysis being carried out in other organizations, with a view to consolidating the proposed structure.

References

1. Brogårdh, T.: Present and future robot control development—An industrial perspective. Annual Reviews in Control 31(1), 69–79 (2007).
2. Selznick, P.: Leadership in Administration: A Sociological Interpretation. Harper & Row, New York (1957).
3. Barney, J.: Firm resources and sustained competitive advantage. Journal of management 17(1), 99–120 (1991).
4. Lewis, M., Slack, N.: Operations strategy. Pearson Education (2014).
5. Mills, J., Platts, K., Bourne, M.: Competence and resource architectures. International Journal of Operations & Production Management 23, 977–994 (2003a).
6. Mills, J., Platts, K., Bourne, M.: Applying resource-based theory. International Journal of Operations & Production Management 23, (2), 148–166 (2003b).
7. Newman, I., Benz, C.R., Ridenour, C.S.: Qualitative-quantitative research methodology: Exploring the interactive continuum. SIU Press, Illinois (1998).
8. Giovando, G., Crovini, C., Venturini, S.: Evolutions in manufacturing cost deployment. Global Business and Economics Review 22(1–2), 41–52 (2020).
9. Calza, F., Parmentola, A., Tutore, I.: Types of green innovations: Ways of implementation in a non-green industry. Sustainability 9(8), 1301 (2017).
10. Pisano, G.P.: You Need an Innovation Strategy. Harvard Business Review 93, 44–54 (2015).
11. Pernståhl, J., Gorschek, T., Feldt, R., Florén, D.: Requirements communication and balancing in large-scale software-intensive product development. Information and Software Technology 67, 44–64 (2015).
12. Surette, E.: Principles of ISO Management System Integration/Transition (ISO 9001: 2000/ISO 14001: 1996) (No. 2005–01–0536). SAE Technical Paper (2005).
13. Jin, K., Palaniappan, A.: Resource and Operations Analysis for Agile Manufacturing System Development. In: IIE Annual Conference Proceedings (p. 1). Institute of Industrial and Systems Engineers (IISE) (2004).
14. Magalhaes, R.: A knowing-in-practice framework for the corporate governance of information systems/technology. International Journal of Business Information Systems 3(1), 40–62 (2008).
15. Mills, J., Platts, K., Bourne, M., Richards, H.: Strategy and Performance: Competing through Competences. Cambridge University Press, Cambridge (2002).
16. Zangiski, M.A.S.G., Pinheiro de Lima, E., Gouvea da Costa, S.E.: Organizational competence building and development: Contributions to operations management. International Journal of Production Economics 144(1), 76–89 (2013).
17. Yamashina, H., Kubo, T.: Manufacturing cost deployment. International Journal of Production Research 40(16), 4077–4091 (2002).
18. Son, Y.K.: A cost estimation model for advanced manufacturing systems. The International Journal of Production Research 29(3), 441–452 (1991).

Advanced Manufacturing or Industry 4.0 Scholarly Works: Are They Relevant to Technology Development?

Izaskun Alvarez-Meaza⬭, Jon Borregan-Alvarado⬭,
Ernesto Cilleruelo-Carrasco⬭, and Rosa Maria Rio-Belver⬭

Abstract Advanced Manufacturing and Industry 4.0 technologies represent the new strategic paths of the manufacturing industry to retain the globalized market competitiveness. However, few studies have attempted to analyze the impact of scholarly works related to those technologies in results obtained from R&D developments, such as patents. Therefore, the aim of this work is to determine which technological patterns are followed by those patents that make use of scholarly works related to Advanced Manufacturing or Industry 4.0 as Non Patent Literature. The results obtained show that the main beneficiaries of the patents are companies that develop medical devices and embedded software, generally collaborating in small isolated research groups, and the technological development carried out is mainly directed towards semiconductor devices and medical devices and pumps. In addition, it should be noted that despite being a highly developed topic in the scientific field (16,118 documents), very few works (113 documents) are being cited as Non Patent Literature in patents (288 patents).

Keywords Industry 4.0 · Advanced manufacturing · Non-patent literature

1 Introduction

The overview change on the industrial sector, now focusing on the real added value created, directs manufacturing strategies towards Industry 4.0 (I 4.0), also known as the fourth industrial revolution [1, 2]. The introduction of I 4.0 technologies, such as Cyber-Physical Systems (CPS), Internet of Things (IoT), Big data, Cybersecurity, Augmented Reality, Additive Manufacturing, etc. has opened the way to adopt new business strategies, strongly linked to production systems based on Lean

I. Alvarez-Meaza (✉) · E. Cilleruelo-Carrasco · R. M. Rio-Belver
Technology, Foresight and Management Research Group, Department of Industrial Engineering and Management Engineering, University of the Basque Country UPV/EHU, Bilbao, Spain
e-mail: izaskun.alvarez@ehu.eus

J. Borregan-Alvarado
Unilever Foods Industrial España, Leioa, Spain

Management [3, 4]. The new concepts that are being introduced into manufacturing are specifically related to Advanced Manufacturing and Industry 4.0. Both represent two forms of expression that have the same purpose related to manufacturing systems, they are complex and defined in many ways, which makes it seem like an ever-evolving concept.

A review of the literature allows us to identify articles that develop different classifications related to these new approaches in productive systems. Esmaelian et al. [5] classify such productive systems, reviewing and grouping them by tangible and intangible elements based on technical and engineering aspects of manufacturing systems. Alcácer and Cruz-Machado [2] and Pagliosa et al. [3] conduct a literature review and define the main technologies associated with Industry 4.0, such as, Cloud computing, Big Data, Simulation, Augmented reality, Additive manufacturing, Autonomous robots, Cybersecurity, CPS and Horizontal and Vertical System Integration, among others. In addition, Wagire et al. [6] define the research landscape of Industry 4.0 through latent semantic analysis, where the research themes turn out to be Interoperability, Advanced Manufacturing Systems, IoTs; Smart city, Big data analytics, Simulation and Cloud manufacturing.

According to the classical theory of innovation [7, 8], scientific publications are the prelude to future technological development or patents. Hence, the importance of scientific research to justify the development of a patent, also known as non-patent literature in the patent. According to Nagaoka [9], there is a positive correlation between the use of scientific literature in patents and the quality of the patent. Furthermore, Og et al. [10] categorize the backward references to non-patent literature as an "*ex ante* indicator" of the value of a patent; and their measurement allows them to conclude that together with forward citations they are the most important indicators of a patent value. Patents provide an exclusively detailed source of information of inventive activity [11] and are one of the most important indicators to evaluate the performance of industry research and development (R&D) [12]. However, a review of the scientific literature has shown that there are no research papers that analyze the non-patent literature (NPL) that is used to justify the patents or the impact of science on technology, i.e. on patents. Therefore, the aim of this paper is to identify the performance of the NPL related to advanced manufacturing used in patents and the trending of technological developments, in order to assess a scientific research community output indicator.

2 Methodology

The research methodology is carried out in three steps. The first is based on *retrieving data*, especially scholarly works related to Advanced Manufacturing and I 4.0. For that, a complete open source research platform called *Lens.org* is used, which contains a global patent database and scholarly works database (Microsoft Academic, PubMed and Crossref). The allocation was accomplished testing different terms and queries iteratively by gathering data from the scholarly works database. With a view to build

Fig. 1 Scheme followed in defining the search query

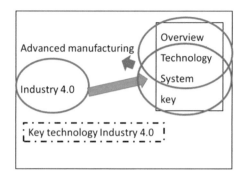

the main query terminology, the particular terms that occur in previous Advanced Manufacturing and Industry 4.0 related work were established. The search query was built using "additive manufacturing or Industry 4.0" and "technology or systems or overview or key" (see Fig. 1) as Title, Abstract, Keywords and Field of Study fields. The main query used in the Lens database retrieved a total of 16,118 scholarly works, of which only 113 have been cited by patents.

The second step is to *clean up the refined database,* using text-mining tools. The scholarly works database and the related patents database were imported into Vantage Point^R (VP) software, text-mining software that helps us to clean data and analyze them through a combination of statistics.

In the last step, the scholarly work and the patent *performance profile* is generated and a *network analysis* is carried out. The networks are generated and visualized through the *Gephi* software [13].

3 Results

3.1 Scholarly Work Performance Profile

The evolution of scholarly works cited in patents over the last twenty years is represented in Fig. 2. However, it should be noted that publication timespan of the 113 works ranges from 1979 to 2019. In addition, the last ten years have been the most productive, and the main types of document are Journal articles and Conference Proceedings.

In terms of academic performance, the main producing countries are the USA (28 documents) and China (13 documents), followed by countries from the European continent such as the UK, Germany and Italy. According to Institutions, the most active are the University of California-Berkeley and Katholike Universiteit Leuden, each with 4 documents. However, among the most active institutions are large companies such as Mentor Graphics (Siemens group), Micron Technology and the Ford Motor Company, all of them from USA (see Fig. 3). As far as the publisher

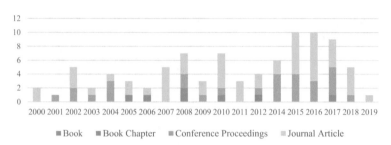

Fig. 2 Publication trend from 2000–2019

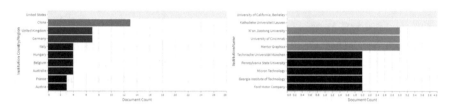

Fig. 3 The most active countries and institutions

is concerned, the IEEE is the main one (25 documents) and its publications are associated with a very technical field.

The identification of the main research topics has been carried out through network analysis of scholarly works' fields of study, based on a co-occurrence matrix. The co-occurrence network of fields of study was plotted using Gephi and we considered the fields of study that have co-occurred at least 9 times. Figure 4 shows the co-occurrence network, with 5 clusters being identified: the three most important of which have close proximity and, therefore, affinity, such as the research fields of Advanced Manufacturing (Manufacturing engineering, 3D printing, Nanotechnology…), Engineering (MES, ICT, CPS, I 4.0, Robotics, Systems engineering, Automation…) and Computer Science (software, Scheduling…). The fourth and fifth are more specific to their field of research, and are Material Science-composite material and Physics-Optics.

3.2 Patent Performance Profile

Once the trend of the articles cited in the patents has been defined, the profile of these patents is analyzed. For this purpose, we answer these questions; When was the invention published? Who are the main beneficiaries and their partners? In which countries is it being protected? And which are the main fields of technological application that these patents cover? As shown in Fig. 5, the publication of patents takes

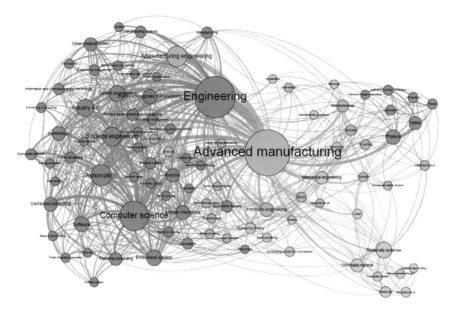

Fig. 4 Fields of study co-occurrence network

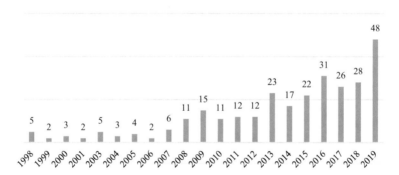

Fig. 5 Patent publication trend from 1998 to 2019

place during the period 1998–2019, however, there is a greater tendency to publish patents citing NPL related to advanced manufacturing or I 4.0 in the last 10 years.

According to inventive performance, the top applicants or beneficiaries of the patents are developers of medical devices, such as, TC1 LLC, Thoratec LLC and Styker Corporation (19, 13 and 7 patents, respectively). In addition, other applicants are IBM, Mentor Graphics and Siemens, as computer and embedded software developers, iROBOT Corporation specializing in domotics and artificial intelligence, Penn State Research Foundation as research centers, Tau Metrix Inc developing semiconductor and metrology technologies and General Electric as an electric devices developer (See Fig. 6). This underscores that all top applicants are from the USA.

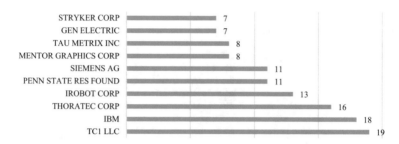

Fig. 6 Main applicants

As far as technological cooperation between applicants is concerned, the applicant's collaboration network, as shown in Fig. 7, identifies the main collaborations. It should be noted that not all of the main applicants collaborate, nonetheless, among the top ones, the Penn State Research Foundation works with the Thoratec Corporation to form one of the most intense groups in terms of collaboration, being located in the center of the network. To a lesser extent, other applicants also collaborate, such as IBM, Gen Electric, iRobot and Tau Metric.

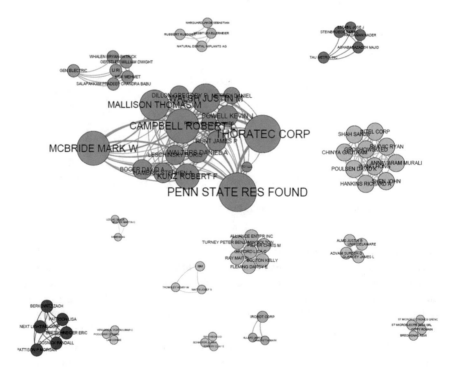

Fig. 7 Applicant collaboration networks

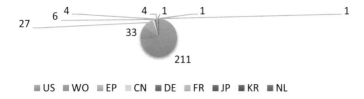

Fig. 8 Main jurisdiction countries

As far as the countries where these patents are being protected or the market allocation strategy of companies are concerned, USA is the main jurisdiction with 211 patent families protected (See Fig. 8).

According to the main technological fields that those patents cover, the Cooperative Patent Classification (CPC) was analyzed. Table 1 presents the main CPCs, allowing us to identify the most important research fields of the defined patents. Three main research fields are identified, Semiconductor devices (H01L), Medical

Table 1 Main CPCs

Number of PFs	CPC	Definition
20	H01L2924/00	Electricity-Basic Electric Elements-Semiconductor Devices; Electric Solid State Devices Not Otherwise Provided For-Indexing scheme for arrangements or methods for connecting or disconnecting semiconductor or solid-state bodies as covered by H01L 24/00
19	H01L2924/0002	Electricity- Basic Electric Elements- Semiconductor Devices; Electric Solid State Devices Not Otherwise Provided For-Indexing scheme for arrangements or methods for connecting or disconnecting semiconductor or solid-state bodies as covered by H01L 24/00
18	A61M1/101	Human Necessities-Medical Or Veterinary Science; Hygiene-Devices For Introducing Media Into, or onto, The Body-Suction or pumping devices for medical purposes; Devices for carrying-off, for treatment of, or for carrying-over, body-liquids
18	A61M1/1034	
18	A61M1/122	
18	A61M1/125	
16	A61M1/1024	
12	A61M1/1012	
12	G01R31/318511	Physics-Measuring; Testing- Measuring Electric Variables; Measuring Magnetic Variables- Arrangements for testing electric properties; Arrangements for locating electric faults; Arrangements for electrical testing characterized by what is being tested not provided for elsewhere
10	A61M1/1008	Human Necessities-Medical Or Veterinary Science; Hygiene-Devices For Introducing Media Into, Or onto, The Body-Suction or pumping devices for medical purposes; Devices for carrying-off, for treatment of, or for carrying-over, body-liquids

Devices for introducing media into/onto the body (A61M) and Physics: measuring electric variables (G01R).

Furthermore, a network analysis based on the co-occurrence of CPCs has been performed in order to analyze the relationship between the main fields of development. Different isolated fields of action have been identified, and the main group of technological development (with the largest number of clusters: blue-purple-light green) is directed towards the development of Semiconductor Devices, in collaboration with technology linked to Physics (static stores and measuring electric variable), among others. A second field of technological development is mainly related to Medical Devices (green) and linked to the field of mechanical engineering through Pumps or machines for liquids. A third group is focused on technological development related to performing operations (manipulators) (B25J) and Physics: Controlling/Regulating systems in general (G05B) and Climate change mitigation technology in production (Y02P). Finally, the fourth important group identified is linked to chemical processes (B01J) and combinatorial chemistry (C40B) (Fig. 9).

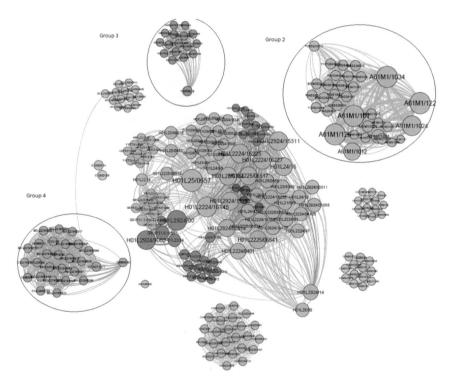

Fig. 9 Co-occurrence network of CPCs

4 Conclusions

The research work carried out in this paper allows us to conclude that few scholarly works related to advanced manufacturing or Industry 4.0 are used as NPL in the justification of patents. Despite scholarly works being developed all over the tri-polar world (North America-Europe-Asia), the USA stands out, highlighting the scientific involvement of its industrial sector. The fields of research covered are strongly linked to the areas of engineering, production and computer science; and a large number of scholarly works have been published with technical publishers. Regarding the use of these scholarly works in technological development or patents, it can be concluded that the main beneficiaries of the developed technology are companies related to medical device developers, software developers, robotics and electric device developers. These applicants usually collaborate in small groups and in an insulated way; and the patents are aimed at their particular field of industrial development (mainly medicine, electricity, mechanical engineering). As several studies indicate [9, 10], the patent value is positively correlated with the number of NPL citations; in this case, this study has been limited to study the characteristics of the patents that quote scientific literature related to Industry 4.0 or Advanced Manufacturing. Future research will lead us to identify and analyze more thoroughly the scientific literature applied in these patents, the number of NPLs in these patents (not only related to I 4.0 or AM, and so, identify the fields of research that are related in these patents), and another important indicator of the patents such as forward citations, in order to further analyze quantitatively the value of these patents.

References

1. L. Barreto, A. Amaral, and T. Pereira, "Industry 4.0 implications in logistics: an overview," *Procedia Manuf.*, vol. 13, pp. 1245–1252, 2017.
2. V. Alcácer and V. Cruz-Machado, "Scanning the Industry 4.0: A Literature Review on Technologies for Manufacturing Systems," *Eng. Sci. Technol. an Int. J.*, vol. 22, no. 3, pp. 899–919, 2019.
3. M. Pagliosa, G. Tortorella, and J. C. E. Ferreira, "Industry 4.0 and Lean Manufacturing: A systematic literature review and future research directions," *J. Manuf. Technol. Manag.*, 2019.
4. D. L. M. Nascimento *et al.*, "Exploring Industry 4.0 technologies to enable circular economy practices in a manufacturing context: A business model proposal," *J. Manuf. Technol. Manag.*, vol. 30, no. 3, pp. 607–627, 2019.
5. B. Esmaeilian, S. Behdad, and B. Wang, "The evolution and future of manufacturing: A review," *J. Manuf. Syst.*, vol. 39, pp. 79–100, 2016.
6. A. A. Wagire, A. P. S. Rathore, and R. Jain, "Analysis and synthesis of Industry 4.0 research landscape: Using latent semantic analysis approach," *J. Manuf. Technol. Manag.*, vol. 31, no. 1, pp. 31–51, 2019.
7. E. M. Roggers, *Diffusion of innovations*, 4th editio. Free Press, 1995.
8. J. M. Utterback and W. J. Abernathy, "A dynamic model of process and product innovation," *Omega*, vol. 3, no. 6, pp. 639–656, Dec. 1975.
9. S. Nagaoka, "Assessing the R&D management of a firm in terms of speed and science linkage: Evidence from the US patents," *J. Econ. Manag. Strateg.*, vol. 16, no. 1, pp. 129–156, 2007.

10. J. Y. Og, K. Pawelec, B. K. Kim, R. Paprocki, and E. S. Jeong, "Measuring patent value indicators with patent renewal information," *J. Open Innov. Technol. Mark. Complex.*, vol. 6, no. 1, 2020.
11. Organisation for Economic Co-operation and Development., *OECD patent statistics manual.* OECD, 2009.
12. Z. Griliches, "Patent Statistics as Economic Indicators: A Survey," *J. Econ. Lit.*, vol. 28, pp. 1661–1707, 1990.
13. M. Bastian, S. Heymann, and M. Jacomy, "Gephi: An open source software for exploring and manipulating networks. BT - International AAAI Conference on Weblogs and Social," *Int. AAAI Conf. Weblogs Soc. Media*, pp. 361–362, 2009.

Assessment of CO$_2$ Emission in the Soil–Cement Brick Industry: A Case Study in Southwest Paraná

Rogério Expedito Restelli, Atilas Ferreira de Paiva, Diego Audrey de Lima Lamezon, Edson Pinheiro de Lima, and Fernando José Avancini Schenatto

Abstract This article discusses a case study about the analysis of greenhouse gases (GHGs) emissions into a soil–cement brick industry in southwest Parana. A literature review was carried out on the relevant themes, carbon footprint calculation methods, technologies to reduce emissions and perceived benefits for the sustainability of the planet. The case study analyzed the whole process of manufacturing and the supply chain, collecting data for the assessment of GHG emissions. It used the GHG Protocol (Greenhouse Gas Protocol) tool to know how much would be the total of CO$_2$ emitted by the brick industry studied. It also addressed technological measures that can be adopted to contribute to the reduction of these emissions. In response, specifically to the production method of the evaluated industry and considering the current geographical conditions, the process generates a total of 2.16 tons of carbon dioxide (CO$_2$) per month. Based on the study details, it was evident the importance of the geographical location of the property in relation to suppliers of raw materials to reduce the impact that is currently caused.

Keywords Carbon footprint · Green production · Sustainability · Ecological brick · Soil–cement brick

1 Introduction

To reduce environmental impact, sustainable attitudes are seen as a priority. Less aggressive products derived from green production gain relevance. Green production is the application of environmental and socially sensitive practices to reduce the negative impact of manufacturing activities and at the same time, harmonize the pursuit of economic benefits [1].

R. E. Restelli (✉) · A. F. de Paiva · E. P. de Lima · F. J. A. Schenatto
Federal University of Technology, Pato Branco, Paraná, Brazil
e-mail: ppgeps-pb@utfpr.edu.br

D. A. de Lima Lamezon · E. P. de Lima
Pontifical Catholic University of Paraná, Curitiba, Paraná, Brazil

The parameters for measuring emissions of greenhouse gases are given from the carbon footprint measurement. The carbon footprint is a measure of the exclusive total amount of carbon dioxide which is directly or indirectly caused by an activity, or emissions driven through the stages of a product's life [2].

According to The Ministry of Environment [3], the Brazilian government is committed to reducing emissions of gases into the atmosphere when formulating its Intended Nationally Determined Contribution (INDC) in the Paris Agreement in 2016. On that occasion, it proposed to reduce gas emissions in the country by 43% by the year 2030.

This study fits as a contribution focused on sustainability and the environment. Sustainability means meeting our own needs without compromising the ability of future generations to meet their needs [4].

In India, was observed that the soil–cement brick is the most efficient material of the alternative for walls, consuming only a quarter of the energy of burnt clay brick [5]. The production of ceramic bricks in the conventional method of production with the burning, consumes 1.4 m^3 of wood to curing a thousand bricks units with temperatures in the range of 750 to 900 °C during 18 h [6]. The production process of soil–cement bricks does not require burning in furnaces; therefore, it does not burn firewood and consequently reduces the emission of gases into the atmosphere [7].

In Brazil, the soil–cement brick is popularly known as ecological brick, which is produced by compacting a mixture of sandy soil (material which is above the level of water sources and that does not degrade springs and adjacent regions) and 12.5% of cement [7]. These bricks are produced by pressing and do not require subsequent burning, therefore they are called "green bricks" [8].

In this scenario, the purpose of this article is to analyze the total emission of carbon dioxide in the manufacturing process of soil–cement bricks. So, how much would be the total CO_2 emitted by a soil–cement brick industry? Based on the results, what technological measures can be adopted to contribute to the reduction of these emissions?

To deepen the theme and better support this analysis, a methodological procedure of content analysis was adopted by identifying and using relevant tools with parameters already certified by the literature.

2   Methodological Procedures

2.1   Procedures for Selecting Articles from the Bibliographic Portfolio

The articles in the references of this research were selected according to the procedure and the methodological rigor of the Fast Systematic Literature Review (FSLR). The FSLR aims to optimize the traditional Systematic Literature Review (SLR) by simplifying tasks necessary for the implementation of the research, allowing better focus on the content without losing quality considering a shorter period of time [9].

The search was restricted to assess only articles because they are classified by the authors as more reliable information materials. All documents specified in the English language and published between 2009 and 2019, only indexed in the Scopus database. The search keywords included the Operations Management knowledge area, with a bias in the use of technologies to reduce CO$_2$ emissions.

At first, in the FSLR four search terms were developed: Carbon Footprint Methods and Sustainable Operations, Carbon Footprint Methods and Sustainability, Carbon Footprint Methods and Operations Management, and finally, Carbon Footprint Methods and Green Production. Thus, a set of 486 articles were formed with these combinations.

The 486 items were entered into the repository Mendeley, however, as the search was considered an extensive to analysis, a sample was defined by statistical calculation, considering margin of 95% confidence with an error margin of 5%, resulting in 215 articles to adhesion test. With the titles whether or not aligned to the theme, the next filter considered for the adhesion test was reading the summaries. Selecting the articles, another portfolio was created in sequence with the agreement of the keywords. After the adherence test, it came to the number of 37 items perfectly aligned to the theme. Finally, it was verified the documents that made the contents available in full to enable qualitative analyzes. Two of them were not located in full, ending with the number of 35 articles in that study.

3 Content Analysis

With the content analysis, the selected articles undergo approaches, called lenses. Specific lenses were adapted on the subject, to filter out only those considered relevant by the authors, meeting the objectives of this research. To the extent that the information obtained it is confronted with existing information, you can reach broad generalizations, which makes the content analysis one of the most important tools for the analysis of mass communications [10].

Content analysis is the set of communications analysis techniques aimed at obtaining, by systematic and objective procedures for describing the content of messages, indicators (quantitative or not) that allow the inference of knowledge relating to production/reception conditions (inferred variables) of these messages. As shown in Fig. 1, it is possible to visualize the sequence of the procedures adopted

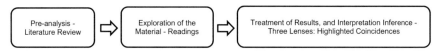

Fig. 1 Content analysis steps—prepared by the authors (adapted from Bardin, 2016)

for the result in three lenses with coincident connections [11].

3.1 Lens 01: Carbon Footprint Calculation Methods

Some tools are already considered relevant among the various methods of calculating the carbon footprint. In the research, many articles were noticed using The Compound method based on Financial Accounts (MC3) developed around the year 2000 by Juan Luis Doménech, who initially designed a tool to assess the ecological and carbon footprint of organizations. Currently, the MC3 is recognized by the Spanish Observatory for Sustainability (SOS) as a valid methodology to assess and reduce GHG emissions [12].

In addition to ISO 14,064–1/2006 established by the International Organization for Standardization (ISO), the approach to GHG Protocol also stands out with its methodology. The GHG Protocol is the most respectable guidance for assessments of greenhouse gases [13]. This tool was developed by the World Business Council for Sustainable Development (WBCSD) and the World Resources Institute (WRI) in 2004. Noting several inferences, the authors have chosen to use the spreadsheet method GHG Protocol in evaluating the case study of this article.

3.2 Lens 02: Technologies for Emissions Reduction

Clean technology should be understood as a tool for eliminating pollution through lower impact production, which involves reducing pollution at its source by replacing inputs, in addition to recycling in the process or creating radically new production processes [14].

Metrics for sustainability refines to collect data covering the environmental, economic, and social dimensions of industries. As in a two-way street, the industry also gains by identifying improvements to more sustainable management. New techniques, technologies, and new products emerge. As an example, the photovoltaic power generation expanding. In the same direction, the SUB-RAW index aims to compare materials performance for the replacement of raw materials by reusable materials, focusing on more viable solutions due to the depletion of finite resources [15].

3.3 Lens 03: Perceived Benefits for Sustainability

The carbon footprint is one of the indicators that contribute to the quantification metrics, in order to assess the damage by CO_2, reduce or eliminate them. Along the same lines, there is the preservation of the chemical footprint, water footprint, soil, biodiversity, social footprints, economic, ecological footprint, among many others [16].

In common among the various articles, we could see the highlight for the Life Cycle Assessment (LCA) of the products. LCA is a way to quantify the environmental impacts of a process or product by examining all stages of production, distribution, use, and recycling destination. The results can be used to improve the designs of these products that have less impact on the environment [17]. LCA is conventionally characterized by the approach "cradle-to-grave" [16].

The intentions found in the articles analyzed are in alignment with the theme of this study, with good works for the common good of humanity and the entire ecosystem. The use of renewable and fewer corrosive sources is combined with the desire for an ever deeper awareness of preservation.

4 Case Study

4.1 Analysis of the Production Process of Soil–Cement Bricks

Company and Product

The organization object of the study is a soil–cement brick industry. The company has its own headquarters and a factory area of approximately 350m^2, situated in the city of Palmas, in the state of Parana. They started in the brick production industry in 2005 with only family members. The industry was not satisfied with the conventional method of brick production, so they developed some machines according to their needs to achieve better quality.

One-third of the gas that causes global warming is associated with construction [18]. This type of sustainable bricks assists in preserving the environment by not being burned and generate less waste in the constructions [19]. It also argues, the walls are not broken for electrical and plumbing installations, do not use wood to build columns and dispense the coating and mass for laying the bricks.

Production Characteristics

The main features of the soil–cement brick production system are linked to high volume with high repeatability and low diversity, providing well-defined parts, standardized and low unit cost.

Production Physical Arrangement

In the case studied, it was found a production physical arrangement by product, also known as in-line production, where the soil and cement pass through the sequence of processes in which the machinery and equipment were arranged physically. As shown in Fig. 2, it is a production line with two employees performing activities in the mixing and compaction sector.

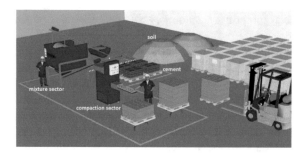

Fig. 2 Layout of production—prepared by the authors (adapted from Alroma, 2019)

For purposes of the calculation of CO_2 emissions, the two employees reside in a neighborhood in the host city, at a distance of 4.7 km of the company and they commute to work every day on their motorcycles.

Economical Order Quantity—Soil and Cement

At the Economical Order Quantity (EOQ) inventory levels of raw materials are based on production capacity [20]. The data are analyzed and the number of raw materials to be requested at the point of lowest stock level is calculated. In the study of a related case, the production line with an automatic compression press has nine seconds uninterrupted cycles, which provides an average production of 3,000 units per day on a working day of eight hours, already discounting the cleaning times of the equipment. Currently, the company produces the model measures the $250 \times 125 \times 70$ cm (L × W × H) with a weight of 3.2 kg each in a proportion of eight parts of soil to one of cement (12.5%).

According to the company, preferably working with materials recently extracted impacts directly in the quality and strength of the final product. Therefore, special attention should be paid on the purchase of the soil, as shown in Fig. 3, with weekly supplies in a constant replenishment of raw material. With the data obtained, it can measure the monthly amount of raw material consumed and units produced. When considering a production of 40 h a week, in a month of 20 days, it obtained a

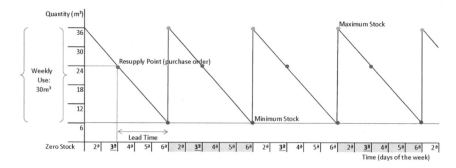

Fig. 3 Economical order quantity (EOQ) of soil (by the authors)

production of 60,000 units monthly. The delivery time of the raw material (soil) for each charge it is 3 days.

The sandy soil is from a mine extraction of sand from the city of União da Vitória - Paraná, whose supplier is 132 km from the headquarters of the brick factory. The sandy soil is transported by the extraction company, which owns a dump truck. This moves exclusively 30m³ loaded with material to Palmas-PR, returning empty by the same path.

The cement used by the company is of the Portland CPV-ARI type, it has the characteristic of high initial resistance and reaches 66% of its strength on the third day [21]. According to the company, even though the cement is valid for ninety days, the more recent its manufacturing date, the better the results obtained in the final product. Therefore, it is suggested lots of purchases in moderate quantities and carried out more frequently. For cement EOQ, the delivery time of product is three workdays.

In this case, it was not considered the amount of consumption as a metric reference to the minimum inventory, but the division of consumption to fortnightly purchases, according to Fig. 4. All this, taking into account the empirical knowledge of the company studied on the advantage of using newly produced cement. The EOQ's graphics allow analyzing and better understand the dynamics of the company to purchase supply, therefore facilitate the measurement of transport costs and its GHG emissions.

Two hundred and forty bags of cement are transported by truck from the city of the supplier Curitiba-PR to Palmas-PR with a total distance of 401 km. This service is performed via third-party freight in the freight-return mode (does not return), therefore, it is not considered double the mileage for the calculation of CO$_2$ emission.

Production Process

Its production process fits in the production batches, where the inputs go through the same sequence of machines and equipment resulting in standardized parts.

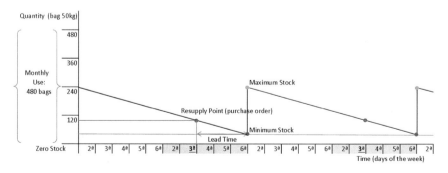

Fig. 4 Purchase economic Lot Cement (by the authors)

Fig. 5 Sequence of the production—prepared by the authors (adapted from Alroma, 2019)

In the process, the preparation of the raw material starts with a machine called multiprocessor. Figure 5 shows the process sequence which will be approached point-to-point. The 50 kg of cement is accommodated in the compartment called the dosing shell (1). Then the soil, in the natural conditions of extraction, is allocated over the cement in the dosing shell. This compartment has the size of a dosage of 400 kg of soil where the edge defines the level. A Clark 2 T forklift with a loader bucket attached works to fill the dosing shell (1) for one hour a day. Its operation generates a consumption of 10 L of gasoline per hour.

A geared device drives the dosing shell through rails to the top of the equipment to load the mixing cylinder (2), shown in Fig. 6, which is positioned on the top of the equipment. When the cylinder is filled, it starts to rotate slowly, while inside, a rotor axis rotates in high rotation, crushing and mixing the soil quickly with the cement.

The transfer of material from the mixing cylinder to the screening stage (3) takes place by gravity, taking advantage of the space below the cylinder through the sieve.

With vibratory and back-and-forth motions, the material reaches the storage silo (4) that is fully insulated from the wind, preventing evaporation of moisture established. The conveyor belt (5) drives the material to the press machine (6), which compresses the material with twelve tons by the hydraulic system as illustrated in Fig. 7.

According to the company studied, compact parts with high humidity makes it possible to achieve better quality levels of the final product. As the soil of the deposit comes naturally with a considerably high degree of moisture, the ideal humidity settings for the pressing vary with the addition of 1 to 10 L of water per batch, only when necessary. The company has no record of how much water it consumes because it uses only water collected from rainfall and natural sources.

An air compressor model 10 ft/100 L and it is required to drive the press and a command for cleaning the bricks when removed from the press. The bricks are then

Fig. 6 Mixing cylinder
(ALROMA, 2019)

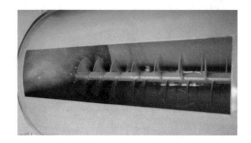

Fig. 7 Automatic press
(ALROMA, 2019)

conducted on the pallet and positioned on a rotating table (7) with the appropriate ergonomic working height. The finished pallet is packed and accommodated in a large space (8) for healing. They rest for three days in the shadow. At that stage, the forklift is used for an average of one hour per day for accommodation and handling pallets in stock. According to the company studied, the high degree of moisture in the mixture is retained on the packaged pallet, which ensures sufficient moisture to cure the cement. Cement takes 28 days for full cure, however the third day the bricks are already sufficient strength for transport [21].

It was recorded the need for 30 min of the forklift to load 10 pallets (5,000 bricks) in the transport truck. This freight for the delivery of bricks, it is always hired by the client of the city/destination. Therefore, it was found that the product cycle for the industry encloses with the forklift loading onto the truck.

Since the production does not make dust, does not generate noise, waste or effluent, this type of plant does not require the state environmental license, being sufficient to Exemption from State Environmental License (abbreviation in Portuguese—DLAE) a legitimately accepted document in Paraná under Resolution No 51/2009/SEMA [22].

The process has a constant quality control, based on a sample of every 500 units produced. A total of 10 samples close the batch, so 5,000 units per batch (one load). These samples undergo resistance tests and water absorption, according to the procedures of the Brazilian Association of Technical Standards (ABNT), assessing the quality of that particular batch in accordance with the technical standards NBR 8491 and NBR 8492 [23, 24].

To measure the energy consumption the time of each activity and its power consumption (kWh) were measured, but it was found that the engines are not active full-time, switching operation by the operators several and indefinite times with each batch. Thus, the amount considered was 800 kWh/month for the calculation of emissions, projected on the average energy consumption history of the twelve-month payment invoices preceding the study. In this way, it included all the company

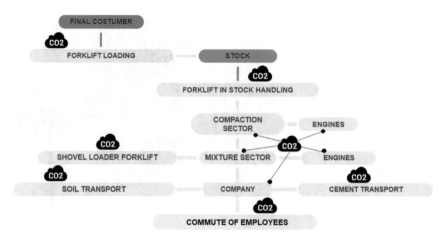

Fig. 8 CO_2 emission points (by the authors)

energy costs, such as lighting, computers, among other sporadic tools and electronic devices.

Carbon Footprint Calculation

Figure 8 illustrates the points found by the authors with CO_2 emissions. Each point was approached individually generating in Table 1 the main reference data for use in calculating the carbon footprint.

To process the data in Table 1, these were entered into the spreadsheet GHG Protocol, an open methodology tool for analysis of GHG emissions. The specifications of the GHG Protocol were adapted to the particularities and conditions of Brazilian organizations by developers. This tool was created by the WBCSD and WRI in 2004 and allowed to ascertain the results shown in Table 2.

5   Discussion of Results

The location of the factory, for being distant from the main suppliers of raw materials, it has a considerable impact on the level of emissions. Especially for soil transportation which contributes 35% of the total CO_2 issued monthly. That said, it is pertinent to study the possibility of a new geographical position of the factory, for the reduction in emissions and, consequently, production costs.

The soil–cement bricks can be made from Construction and Demolition Waste (CDW) [25]. Given the possibility of producing soil–cement bricks with CDW, besides eliminating the cost of raw material (soil) the industry activity would reduce 0.75 ton/month of CO_2 in the atmosphere. There is the possibility for an analysis of the implementation of a collection and processing unit for CDW in the host city.

Table 1 Footprint metric of points (by the authors)

Soil transport

No. Loads	Distance	Displacements along the route/Load	Total km/Trip	Total km/Month
4	132	2	264	1056

Cement transport

No. Loads	Distance	Displacements along the route/Load	Total km/Trip	Total km/Month
2	401	1	401	802

Commute of employees

No. Employees	Distance	Displacements along the route/Day	Total km/Day	Total km/Month
2	4.7	4	37.6	752

Forklift shovel loader

Hours/Day	Liters/Hour	Days/Month	Liters/Month
1	10	20	200

Forklift in stock handling

Hours/Day	Liters/Hour	Days/Month	Liters/Month
1	10	20	200

Forklift loading

Hours/Load	Liters/Hour	Loads/Month	Liters/Month
0.5	10	12	60

Electricity

KWh/Month
800

Table 2 Result GHG protocol (by the authors)

GHG emission factor	Source emission	Unit of measurement	Consumption/Month	Total CO$_2$ (Ton/Month)
Soil transport	Diesel	liters	310.59	0.75
Cement transport	Diesel	liters	235.88	0.58
Commute of employees	Gasoline	liters	20.22	0.03
Forklift shovel loader	Gasoline	liters	200	0.32
Forklift in stock handling	Gasoline	liters	200	0.32
Forklift loading	Gasoline	liters	60	0.10
Electricity	Electricity	Kwh/Month	800	0.05
Total CO$_2$ (Ton/Month)				2.16

Despite electricity has a little impact it is still important to study the deployment of photovoltaic panels, with the possibility of self-sufficiency in energy by eliminating other GHG emission factors.

The authors consider the production of soil–cement bricks to significantly reduce environmental impact, including in the list as a replacement product for the old burnt method. The soil–cement wall, it does not require the burning of the production process and generates carbon credits compared to the burnt ceramic bricks. Carbon Credit is an electronic certificate issued when there is a proven reduction of greenhouse gases released into the atmosphere by industries [26].

The study has a local contribution and at the same time a global contribution, emphasizing the importance of measuring an organization's aggressive impacts. It thus makes it possible to encourage any company to create documents for new strategic visions, to reduce costs, to improve processes and, above all, to contribute to the sustainability of our planet.

6   Proposals for Future Studies

- Location feasibility study to reposition of the company in the city of União da Vitória-PR.
- Economic feasibility analysis of the implementation of a CDW processing plant in the city of Palmas-PR.
- Analysis of the economic feasibility of installing solar panels for clean energy generation in the studied industry.
- Study of proof of CO_2 reduction for certification of carbon credits generation in the soil–cement brick industry.

References

1. Baines, T., Brown, S., Benedettini, O., Ball, P.: Examining green production and its role within the competitive strategy of manufacturers. Journal of Industrial Engineering and Management (2012).
2. Wiedmann, T., Minx, J.: A Definition of Carbon Footprint. ISA-UK Research & Consulting, (2007).
3. MMA - Ministério do Meio Ambiente: Pretendida contribuição nacionalmente determinada para consecução do objetivo da convenção-quadro das nações unidas sobre mudança do clima. República Federativa do Brasil, Convenção INDC (2015).
4. CMMAD, Comissão Mundial sobre Meio Ambiente e Desenvolvimento: Nosso Futuro Comum. 2a ed. Rio de Janeiro, Editora FGV (1991).
5. Reddy, B. V. V., Jagadish, K. S.: Embodied energy of common and alternative building materials and technologies. Energy and Buildings. Elsevier (2001).
6. Baccelli, J. G.: Avaliação do processo industrial da cerâmica vermelha na região do Seridó - RN. Tese (Doutorado em Engenharia Mecânica) - Departamento de Engenharia Mecânica, Universidade Federal do Rio Grande do Norte, Natal (2010).

7. ANITECO Homepage, http://www.aniteco.org.br. Accessed 18 Apr 2019.
8. Castro, M. A. M., Costa F. G., Boba, S. C., Neto, E. F., Rabelo, A. A.: Avaliação das propriedades físicas e mecânicas de blocos de solo-cimento formulados com coprodutos siderúrgicos. Revista Matéria (2016).
9. Kluska, R. A., Duarte, R., Deschamps, F., Pinheiro de Lima, E., Van Aken, E. M.: A new methodology for Industrial Engineering research : Fast Systematic Literature Review. ICPR - 9th International Conference on Production Research – Americas (2018).
10. Bauer, M. W., Gaskell, G.: Pesquisa qualitativa com texto, imagem e som: um manual prático. 3ª ed. Petrópolis, Rio de Janeiro, Vozes (2002).
11. Bardin, L.: Análise de Conteúdo. 70ª ed. Lisboa, Portugal, LDA (2016).
12. Penela, A. C., Doménech, J. L.: Managing the carbon footprint of products: The contribution of the method composed of financial statements (MC3). The International Journal of Life Cycle Assessment (2010).
13. Downie, J., Stubbs, W.: Evaluation of Australian companies' scope 3 greenhouse gas emissions assessments. Journal of Cleaner Production (2011).
14. Oltra, V., Jean, M. S.: The dynamics of environmental innovations: three stylized trajectories of clean technology. Economics of Innovation and New Technology, 14(3), 189–212 (2005).
15. Bontempi, E.: A new approach for evaluating the sustainability of raw materials substitution based on embodied energy and the CO2 footprint. Journal of Cleaner Production 162 (2017).
16. Cucek, L., Klemes, J., Kravanja, Z.: A Review of Footprint analysis tools for monitoring impacts on sustainability. Journal of Cleaner Production 34, 9-20 (2012).
17. Northey, S., Haque, N., Muddb, G.: Using sustainability reporting to assess the environmental footprint of copper mining. Journal of Cleaner Production 40, 118-128 (2013).
18. CTE - Centro de Tecnologia de Edificações Homepage, https://www.eventos/cursos/eficiencia-energetica-da-academia-kingspan-isoeste-bh. Accessed 07 Jun 2019.
19. ALROMA Homepage, https://www.alroma.com.br. Accessed 18 Apr 2019.
20. Slack, N., Chambers, S., Johnston, R.: Administração da Produção. 3ª Ed. São Paulo, Atlas (2009).
21. Pereira, T. A. C.: Concreto auto-adensável, de alta resistência, com baixo consumo de cimento Portland e com adições de fibras de lã de rocha ou poliamida. SET EESC, São Carlos (2010).
22. IAT – Instituto Água e Terra Homepage, http://www.iat.pr.gov.br/Pagina/Licenciamento-de-atividades-especificas. Accessed 07 Jul 2019.
23. ABNT (a) Brazilian Association of Technical Standards: Soil-cement Brick - Dimensional analysis, determination of the compressive strength and water absorption. NBR 8492, ABNT, Rio de Janeiro (2012).
24. ABNT (b) Brazilian Association of Technical Standards: Soil-cement Brick - Requirements. NBR 8491, ABNT, Rio de Janeiro (2012)..
25. Restelli, R. E., Lima, J. D., Batistus, D. R., Restelli, A.: Análise de Viabilidade Econômico da Implantação de uma Fábrica de Tijolos Solo-Cimento. XXXVIII ENEGEP - Encontro Nacional de Engenharia de Produção (2018).
26. Borges, M. S., Borges, K. C. A. S., Souza, S. C. A. S.: Considerações sobre as linhas de Crédito de Carbono no Brasil. Revista Direito Ambiental e Sociedade, v. 6, n. 2 (2016).

Analysis of the Perceptions on Target Costing Using a Pedagogical Case Study

Paulo Afonso⊙, **Valdirene Gasparetto**⊙, **and Cezar Bornia**⊙

Abstract Target costing is an established approach in Japanese companies and outside Japan its implementation is still low but gradually increasing over the years. Cultural aspects as well as normative, organizational and technical issues, among others, may influence the effectiveness of target costing in western companies. Furthermore, there are a set of principles, techniques and tools which should be well understood and applied for the success of target costing; which is mainly focused on the design and development stages of new products but also includes inter-organizational collaboration with suppliers and cost reduction efforts during the production phase. Thus, a pedagogical case study was designed to include most of these aspects. Its objective is twofold: using it to present and discuss target costing to an audience of students, managers or shop floor workers, in an interactive and dynamic manner; and also, to find and understand eventual difficulties, biased perceptions, misunderstandings and different opinions on the principles, tools and process related to target costing. It was applied to students from industrial engineering and accounting degrees. The F-test of equality of variances, the independent two-sample t test and the two-proportion z-test were used for the analysis of the results. In general terms, the responses showed no significant differences in the opinions between the two groups of participants and reflected a reduced alignment with the principles of target costing. More specifically, the students showed less alignment with the assumptions and concepts of target costing than with its tools and approaches.

Keywords Cost management · Target costing · Pedagogical case study

P. Afonso (✉)
Universidade Do Minho, Centro Algoritmi, 4710-057 Braga, Portugal
e-mail: psafonso@dps.uminho.pt

V. Gasparetto · C. Bornia
Universidade Federal de Santa Catarina, Santa Catarina 88040-900, Brazil
e-mail: valdirene.gasparetto@ufsc.br

C. Bornia
e-mail: cezar.bornia@ufsc.br

A. M. Tavares Thomé et al. (eds.), *Industrial Engineering and Operations Management*,
Springer Proceedings in Mathematics & Statistics 367,
https://doi.org/10.1007/978-3-030-78570-3_28

1 Introduction

Cost management approaches based on target costs were developed by Toyota in the early 1960s. Japanese cost management practices have been presented in the literature since the early 1990s through surveys, simple examples or descriptive articles and experiments (e.g. [1, 2]). Cooper [3] presents six Japanese cost management techniques based on twenty-three descriptive cases. Woods et al. [4] present a case in which EVA (economic value added) was incorporated in the target costing (TC) approach adopted in a European company. Complementary, Castellano and Young [5] and Everaert and Bruggeman [6] among others have designed and implemented experiments where these concepts and techniques are explained and tested. There are a few examples of surveys on these issues, for example Dekker and Smidt [7] and Tani et al. [8].

The Japanese approach stresses cost control in three distinct ways [9]. First, upper level management through the efforts of a multidisciplinary team strictly controls the mix of products that are manufactured and sold. Second, the costs of new products are reduced through the techniques of target costing and value engineering. Finally, the costs of existing products are reduced through the kaizen costing. Suppliers are asked to participate in this process as soon as possible supporting effective inter-organizational cost management activities.

The effectiveness of TC may depend on several aspects. Indeed, cultural aspects as well as normative, organizational and technical issues, among others, may influence the effectiveness of target costing, particularly, in western companies. In this research project, a pedagogical case study was designed to include the main assumptions, concepts and approaches related to TC. The case study is complemented with two questionnaires, one on the general assumptions and issues related to TC and the other one about TC in the context of the case study (respectively, 12 and 15 questions, with 4 alternatives each).

It was applied to two groups of students, one from an industrial engineering degree, and the other one from an accounting degree. The participants were asked, firstly, to answer the questionnaire on TC to get valuable information about their sensitiveness to TC. After that questionnaire and an introductory lesson on TC, the participants were asked to answer the questionnaire about the case study.

The remaining of this paper contextualizes the research in the literature on target costing, explains how the research work was designed and implemented and discusses the main results. Several conclusions and interesting opportunities for further research are presented at the end.

2 Target Costing

Target cost management practices play a key role in new product development processes where experience has shown that the majority of the total costs are already fixed in the design stage [10]. Indeed, target costing is a technique for managing product costs during the design stage [11]. Nevertheless, and according to Kato [11, p. 36], "…'target costing' is not actually a form of costing, rather, it is a comprehensive programme to reduce costs […] target costing is a management technique…". Tani et al. [8] also argued that target costing can be part of a wider concept of product cost management.

TC is applied to the development phase of a new product and it is focused on not exceed the maximum allowable cost which is computed considering the product's target price that the market accepts and the margin that the company intends to achieve for that product which should be aligned with the long-term strategic planning of the company [9].

Target costing or *genka kikaku* is a three-stage process [9]. Firstly, the target price is identified, secondly, a target margin is assumed, and thirdly, the target cost is calculated by subtracting the target margin from the target price. Value engineering (VE) and functional cost analysis (FCA) are used to eliminate the excess of the current manufacturing cost over the allowable cost. Value engineering is used to determine allowable costs for each component in every major function of the product and to produce a cost reduction objective for each component. Functional cost analysis is well explained in Yoshikawa et al. [2].

Target costing requires cross functional cooperation. Indeed, it requires intense co-operation in the product development process. Such participative work involves people from production, engineering, design, marketing, accounting and sales [12]. Furthermore, suppliers should be involved in the target costing process and if possible from very early [12].

Target costing is extended through kaizen costing (KC) which involves several procedures that allow costs to be reduced through continuous improvements during the production phase of the product life cycle by involving everyone. Imai introduced the term Kaizen in 1986, defining it as "ongoing improvement involving everyone— top management, managers and workers" [13]. Kaizen costing involves several procedures that allow costs to be reduced through continuous improvements during the production phase of the product life cycle. Internal and external collaborators, i.e. employees and suppliers are involved in this process. KC extends the cost management philosophy of TC to the manufacturing process. However, kaizen costing offers less opportunities to reduce costs because products characteristics cannot be changed so profoundly as in the design and development stages [14].

Kaizen costing follows target costing in timing for it is an approach that goes beyond the design and development stages as it is implemented during the manufacturing phase of the product's life [15]. Kaizen costing activities focus on continual small incremental product cost improvements in the manufacturing phase, as opposed to improvements in the design and development phase. Kaizen costing requires that continuing efforts be made to secure further cost savings. In kaizen costing, management will set the cost reduction targets for the product [16]. Kaizen costing can be applied at three levels: to a specific period, to a specific item (usually a component) or to indirect costs [17].

Finally, inter-organizational cost management (IOCM) is the extension of the cost reduction activities and respective tools to the supply chain [13]. The IOCM is described as a structured approach to coordinate cost management activities which can be generated or lead by buyers or suppliers or even jointly [13]. To put IOCM fully in practice, all companies in the network have to adopt lean buyer–supplier relationships dedicated to produce low-cost products with a high level of functionality and good quality that can meet the needs of the clients or the market. The use of IOCM to coordinate cost reduction projects in supply chains may be useful in three different ways. Firstly, it can contribute to reduce production costs. Secondly, it can help to find new and different ways to develop products at lower costs. And finally, it may be useful to identify ways to increase the efficiency of the customer–supplier interface.

According to Cooper and Slagmulder [18], IOCM is characterised by three main concepts: the quality-functionality-price (QFP) paradigm, inter-organizational cost investigations and concurrent cost management. These three concepts or techniques have, as a common objective, the reduction of costs through changes in the initial specifications of the product or components. The main difference between them is the level of buyer–supplier interaction. In practice, QFP initiatives can be developed with a low level of interaction whilst inter-organizational cost investigations and concurrent cost management demand a higher level of collaboration between the parties [18, pp. 6–9]. QFP is applied to achieve cost reductions in separate components. Inter-organisational cost investigations are useful to reduce costs of groups of components and concurrent cost management in major production functions [18, p. 22].

3 Research Design

3.1 Research Protocol and Data Analysis

The participants in the research were two groups of students, one enrolled in an industrial engineering degree and the other one in an accounting degree. Initially, students answered a questionnaire with 12 questions, each with 4 alternatives to be prioritized, 1 for the one they consider the best alternative or the statement to which they agree more, and 4 for the worst or less significant alternative. Then, they watched a video with a lesson on target costing.

The 17-min lesson covered: (i) the origin of TC as a counterpoint to cost approaches used in Western companies; (ii) the life cycle of products, committed cost curve and costs incurred and TC as a cost planning tool; (iii) market-driven prices versus markup approach; (iv) market-driven costing—market price, estimated margin and target cost equation; (v) product-level TC – allowable and actual cost, cost reduction target, value engineering and cost reductions to achieve TC; (vi) component-level TC—functions/components versus customer characteristics/requirements and the prioritization of cost reduction efforts; (vii) example of VE steps for a similar case, with value indices; and (viii) KC and IOCM concepts. The video was recorded and made available to students in a virtual learning environment, as one of the activities of their courses on cost accounting and management of 3rd year Industrial Engineering class and a 3rd year Accounting class.

Finally, participants read the case study and answered a questionnaire on the case study containing 15 questions: 5 questions with 4 alternatives, where only one of which was correct, and 10 questions with 4 alternatives to be prioritized as in the first questionnaire. For data analysis purposes, the answers with questions to be prioritized were transformed: the first option received a score of 4 and so on, until the last option, which received a score of 1. For right/wrong questions, the correct answers received grade 1 and wrong answers, grade zero. The average score of each alternative was used for analyzing the results. The highest average alternative indicates the respondents' priority and the lowest average alternative indicates the least priority.

The right/wrong questions were analyzed using the proportion of correct answers. The comparison between groups was made using the independent two-sample t test. Before that, the F-test of equality of variances was applied to define which model (similar or unequal variances) would be used. To compare the performance of the groups regarding the questions of right/wrong answers, the two-proportion z-test was used, and a "target costing alignment index" was created considering the answer that would be more aligned with the concepts of target costing for the questions with alternatives to be prioritized.

3.2 Case Study

The case study, adapted from Cooper and Slagmulder [19], refers to the fictitious company ACME Pencil Company, which operates in the writing articles sector. The case reflects the process inherent to target costing, its stages, some of its tools and approaches and allows to discuss some questions and eventual alternative strategies or tensions that may arise during its execution. The product development team has professionals from various areas including engineering and processes, accounting and finance who develop collaborative work to reach the TC. The participant should position himself as a cost specialist, acting as a facilitator for product analysis and cost reduction.

ACME is presented as an important player in the market with a reputation for innovation and competitive pricing and has to deal with major competitors recognized in a similar way. The company intends to conceive and launch in the market a new product (an electric pointer).

After a brief overview of the industry, the market, the company and the product, the case presents the process of target costing that is applied in ACME following the principles found in the literature. In the first phase, the maximum price allowed for the new product is defined, considering the generic characteristics of the product, market conditions and the strategy of the company. In the second stage, the TC is analyzed at the product level (Product-level TC) and in the third stage at the component level (Component-level TC).

In the computation of the Target Price it is studied which price should be practiced with the final consumer. At this point, the reader of the case is required to make or interpret some simple calculations to start thinking about the target cost equation that requires a target price and the estimation of the margin. Estimating the Margin requires a strategic planning exercise for a given time horizon for which the level of sales, costs and expected results are calculated. A simple exercise on the estimation of the margin is made to illustrate it and put the participant aware of the relevance of the corporate strategy in the TC process.

In Product-level Target Costing, the TC is calculated for the new product taking into account the assumptions made in the previous phases, namely the sales price and the margin. In this case, cost savings of 25% must be achieved for the new product to turn it viable. Next, the case study simulates value engineering activities. The information about the functions performed by the product components and the attributes valued by consumers combined with the product's cost structure, allow the calculation of the TC index as presented in Cooper and Slagmulder [19]. On the other hand, value engineering allows the identification of a set of cost reduction opportunities to be obtained internally and from suppliers.

Without exploring inter-organizational cost management and kaizen costing, the case ends with a reference to both because they allow completing the logic of target cost management. But these concepts should be addressed in further research. For this purpose, it is mentioned that after the product be launched, and during the entire production phase, ACME follows a consistent annual cost reduction strategy. It is concluded highlighting that the objective of cost reduction is also imposed on suppliers.

4   Analysis and Discussion of Results

The average obtained in the alternatives indicates the preference of the respondents for the different options and allows us to understand the perception that most participants have on target costing. The averages of the two groups studied—industrial engineering and accounting—were compared using the independent two-sample t test. Initially, the F-test of equality of variances was applied to use the two-sample t test for equal or different variances between groups. In the first questionnaire, it was detected significant differences only in 6 alternatives of 4 questions which are the questions 2, 4, 5 and 6. Table 1 shows the averages of the different alternatives for these questions, globally and by group (engineering and accounting).

Table 1 Answers' average of questions with significant differences

	question 2				question 4				question 5				question 6			
	a	b	c	d	a	b	c	d	a	b	c	d	a	b	c	d
Global	2,91	2,83	2,74	1,62	3,31	2,57	2,88	1,39	3,18	2,74	2,74	1,44	2,47	3,37	2,94	1,22
Engineering	3,12	2,82	2,80	1,43	3,16	2,61	3,08	1,45	3,22	2,92	2,61	1,37	2,27	3,41	3,16	1,14
Accounting	2,60	2,83	2,66	1,89	3,54	2,51	2,60	1,31	3,14	2,47	2,92	1,54	2,75	3,31	2,64	1,33
Δ (Eng-Acc)	0,52*	−0,01	0,15	−0,46	−0,39*	0,09	0,48*	0,15	0,08	0,45*	−0,31	−0,17	−0,48*	0,11	0,52*	−0,20

*0,05 significance

Question 2 asked to indicate which element—costs, price or profit—should be determined according to the others. The alternative (a) (Cost = Price—Margin) was the most indicated by the participants from the engineering group, with average = 3.12, while the accounting group prioritized the alternative (b) (Price = Cost—Margin), with average = 2.83. The reference of the engineering students is compatible with the principles of target costing, while the accounting students gave more importance to the traditional approach of calculating the sales price based on the cost and, secondly, to calculating the profit based on the cost and price (alternative c), which is more distant from the logic of target costing. The accounting students had a course on cost accounting, in which one of the contents is precisely to compute the price based on costs, using a markup, which may have influenced their opinion on the elements of the cost equation. There is, therefore, a normative aspect related to the previous education and training obtained in a given professional context that seems to have an impact on how target costing can be perceived.

Question 4 addressed the trade-off between quality, functionality and price/cost, asking participants to point out the preferred trade-off in the design and development phase of a new product, where these characteristics can be changed. The alternative (a) (changing the cost and/or functionality of the product in order to maintain the quality level initially defined) was considered a priority, with average = 3.31. Although there was a significant difference between the answers in the two groups in this question, there was no difference in the order of priorities defined by the groups. However, the accounting group indicated this alternative more intensely, with average = 3.54, than the engineering group, with average of 3.16. Again, the results indicate that the accounting group may be further away from the principles of target costing than the engineering group. In target costing, the focus is on cost, promoting changes as long as they do not sacrifice globally the perception that the client wants to have of the product and the value it attributes to it.

Question 5 asked about the stage of the product life cycle in which a cost reduction is more significant and sustained. Both groups pointed out the alternative (a) (initial phases) as a priority, with average = 3.18, in line with the assumptions of target costing. The significant difference occurred in the second alternative, where engineering students opted for alternative (b) (during the production phase), with average = 2.92, and accounting students preferred alternative (c) (exploring opportunities with suppliers), with average = 2.92. This difference of opinions maybe explained by the difference in training, since in accounting there is less focus on the production process and so perhaps the suppliers emerged as a stronger possibility than the production phase.

Question 6 asked about the most effective approach to setting cost reduction objectives in the design and development phase of a product. The alternative (b) (applying higher cost reduction rates to the most expensive components and/or with the highest cost reduction opportunities) was the priority for both groups, with average = 3.37. A significant difference was observed in the second priority alternative, where engineering students pointed out the alternative (c) (higher cost reduction rates for components contributing to the lowest customers' perceived value), with average = 3.16, and accounting students pointed out the use of similar cost reduction rates, with average $= 2.75$. Thus, on this issue, engineering students were more consistent with the principles of target costing. The results obtained with accounting students prove the greater preference of this type of professionals to apply general cost reduction measures through indistinct and flat cost reduction rates.

After the video lesson on target costing and having analyzed the case study, the students answered a second questionnaire, which contained 10 opinion questions and 5 evaluation questions. The same statistical tests indicated that, in general, there was no significant differences between groups, in the opinion questions, except for questions 8, 9 and 13, which are presented in Table 2.

Questions 8 and 9 refer to the perception of the target margin and question 13 to cost reduction aspects. It is observed that, in the three questions, the significant differences refer to alternatives that were not those indicated as priorities by the students. Question 8 refers to alternatives to increase the global margin of the business, in the context of target costing and, according to Table 2, the two groups prioritized alternative (c) (increase sales), which among the possibilities presented is the most aligned with the concepts of target costing.

In question 9, the value of the estimated margin to be applied to the new product was questioned, in view of the target costing approach. Accounting students prioritized alternative (b) (it depends on market conditions: price and estimated quantity), with average $= 3.26$, while engineering students prioritized alternative (a) (it depends on the strategic planning of the company), with average $= 3.02$. According to the concepts of target costing, the margin to be considered for the new product depends essentially on the strategic planning of the company. Therefore, the engineering students were again more aligned with the concepts of TC.

Question 13 fits in the context of value engineering, asking about the participants' perception of how to reduce costs per component. Both groups prioritized alternative (b) (opportunities identified by value engineering in each of the components should be prioritized), with average $= 3.32$. These concepts were discussed in the video lesson and presented in the case study, both with details of the stages of value engineering and with an example of the identification of the value indices.

In the second questionnaire there are 5 evaluation questions (right/wrong type). The proportions of correct answers of the groups were analyzed using the two-proportion z-test. Table 3 presents the results.

According to Table 3, on target price, accounting students performed better than the engineering students, however, engineering students performed better on the other questions, on target margin, target cost and cost reduction, what could be more or less expected in this study, given the different normative nature underlying each of the degrees and professional areas.

Respondents' scores were computed (number of right answers) and the respondents were divided into two groups: those with more than 2 right answers (higher scores group) and those with less than 3 right answers (lower scores group). The cited statistical F and t tests detected few significant differences between higher and lowest scores groups. Significant differences were found only in two questions, as showed in Table 4.

Table 2 Answers' average by alternative—second questionnaire

	question 8				question 9				question 13			
	a	b	c	d	a	b	c	d	a	b	c	d
Global	3,04	2,36	3,12	1,41	3,01	2,96	2,58	1,58	2,60	3,32	2,70	1,46
Engineering	3,04	2,13	3,13	1,57	3,02	2,74	2,67	1,65	2,43	3,41	2,74	1,41
Accounting	3,03	2,66	3,11	1,20	3,00	3,26	2,46	1,49	2,83	3,20	2,66	1,51
△ (Eng-Acc)	0,01	−0.53*	0,02	0.37*	0,02	−0.52*	0,22	0,17	−0.39*	0,21	0,08	−0,10

* 0,05 significance

Table 3 Proportion of correct answers—second questionnaire

Proportion correct (general)	0,72	0,72	0,51	0,57	0,84
Proportion correct (Engineering)	0,70	0,78	0,61	0,67	0,93
Proportion correct (Accounting)	0,74	0,63	0,37	0,43	0,71
Difference (Eng-Acc)	−0,05	0,15	0,24	0,25	0,22
Z statistic	−0,47	1,52	2,12	2,21	2,68
p-value	0,64	0,13	0,03	0,03	0,01

Table 4 Averages by alternative—score grouping—second questionnaire

	question 4				question 13			
	a	b	c	d	a	b	c	d
Global	3,05	2,35	3,28	1,25	2,60	3,32	2,70	1,46
Higher scores	3,11	2,25	3,44	1,18	2,43	3,41	2,74	1,41
Lowest scores	2,92	2,54	2,96	1,38	2,83	3,20	2,66	1,51
A (High-Low)	0,19	−0,28	0.47*	−0,20	−0.39*	0,21	0,08	−0,10

*0,05 significance

According to Table 4, in question 4, about what should be adopted if there was a possible price range for the new product, everyone prioritized alternative (c) (a price between the minimum and maximum possibilities), however, students with higher scores indicated this option more strongly than students with lower scores.

In question 13, already discussed, which investigated how cost reduction per component must be done, both groups prioritized alternative (b) (should consider mainly the opportunities identified by value engineering in each of the components), with average = 3.32. However, a significant difference was observed in the indication of alternative (a) (should, whenever possible, be equivalent to the global rate of cost reduction necessary to achieve the TC), as the second priority alternative for the lower group scores and third alternative for the higher group scores.

Finally, in each of the questions, the alternative most aligned with the concepts and assumptions of TC was identified, and by comparing this solution with the answers given by the participants an "alignment with target costing" index was calculated. A general index and an index for each group (engineering and accounting students) and for each questionnaire were created and compared using the two-proportion z-test. There was no significant difference except for question 2 of the first questionnaire. As previously discussed and showed in Table 1, engineering students were more aligned with target costing than accounting students regarding this question. The results of the two questionnaires are presented in Fig. 1.

The general index for the first and second questionnaires were 0.29 and 0.40, respectively. Although this difference of 0.11, the two-proportion z test detected no significance. That is, no evidence was found that alignment with target costing increased among respondents after the video lesson and the case study. However,

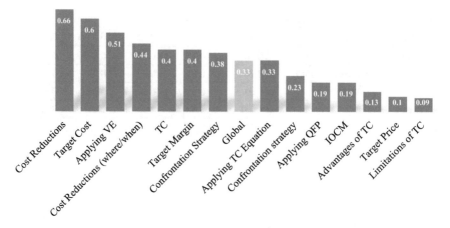

Fig. 1 Target costing alignment index by question (both questionnaires)

there was a much greater alignment with the issues related to the target cost and cost reduction than with the principles related to the target price and margin, and the confrontation strategy.

5 Conclusions

The results of this study showed that industrial engineering and accounting students have different profiles which may explain differences on the perception they have on TC. In fact, in accounting, and particularly financial accounting, there is a focus on recording and analyzing past information and industrial engineering is more focused on the efficiency of production processes. Considering that industrial engineering is more aligned with the logic inherent to effective cost reduction, it was expected that the answers from that group would be more aligned with the principles of target costing. However, this did not occur, in general, when the results of the questionnaires were analyzed, because the answers did not show significant differences in the opinions of the two groups of students. However, we can say that industrial engineering students showed a better understanding of the particularities of the case study, as they scored better in the evaluation questions.

The answers of both groups presented a relatively low alignment with the principles of target costing, which can be verified by the low values of the target costing alignment index in both questionnaires. However, in the second questionnaire, there was a reasonable alignment for the questions related to target cost and cost reduction, although there was low alignment for the ones related to confrontation strategy, target price and target margin. This may indicate that, although not aligned with the assumptions and concepts of target costing, the students are (more) aligned with its tools and approaches.

The pedagogical case study proved to be an interesting tool for the presentation of TC to an audience of people with different profiles and also to promote the understanding and discussion of its main assumptions and tools. However, there are still opportunities to improve some questions in order to clarify the concepts and to capture more accurately the level of alignment of participants with target costing. Also, the protocol and stages of the entire process can be improved namely with a subsequent session for discussing the answers made and results achieved.

The results should also be analyzed with caution because students from both groups are approximately halfway through their academic program. Therefore, their answers may not represent the vision of students at the end of the course and even less of trained and experienced professionals.

Thus, for further studies, it is intended to adjust the questionnaires and the case study, as well as introduce additional activities to be developed with the participants, in order to improve the case study, questionnaires and related materials as pedagogical and research tools. In addition, it can be applied to trained professionals and in other cultural environments. A wider and effective use of TC asks for a better understanding of these issues.

Acknowledgements This work has been supported by FCT—Fundação para a Ciência e Tecnologia, within the R&D Units Project Scope: UIDB/00319/2020 and by the Coordenação de Aperfeiçoamento de Pessoal de Nível Superior – Brasil (CAPES) – Finance Code 001.

References

1. Yoshikawa, T., Innes, J.: A Japanese case study of functional cost analysis. Management Accounting Research 6(4), 415-432 (1995). https://doi.org/10.1006/mare.1995.1029
2. Yoshikawa, T., Innes, J., Mitchell, F.: Applying functional cost analysis in a manufacturing environment. International Journal of Production Economics 36(1), 53-64 (1994). https://doi.org/10.1016/0925-5273(94)90148-1
3. Cooper, R.: Costing techniques to support corporate strategy: evidence from Japan. Management Accounting Research 7(2), 219-246 (1996). https://doi.org/10.1006/mare.1996.0013
4. Woods, M., Taylor, L., Fang, G. C. G.: Electronics: a case study of economic value added in target costing. Management Accounting Research 23(4), 261-277 (2012). https://doi.org/10.1016/j.mar.2012.09.002
5. Castellano, J., Young, S.: Speed Splasher: and interactive, team-based target costing exercise. Journal of Accounting Education 21, 149-155 (2003). https://doi.org/10.1016/S0748-5751(03)00004-6
6. Everaert, P., Bruggeman, W.: Cost targets and time pressure during new product development. International Journal Operations & Production Management 22(12), 1339-1353 (2002). https://doi.org/https://doi.org/10.1108/01443570210452039
7. Dekker, H., Smidt, P.: A survey of the adoption and use of target costing in Dutch firms. International Journal of Production Economics 84, 293-305 (2003). https://doi.org/https://doi.org/10.1016/S0925-5273(02)00450-4
8. Tani, T., Okano, H., Shimizu, N., Iwabuchi, Y., Fukuda, J., Cooray, S.: Target cost management in Japanese companies: current state of the art. Management Accounting Research 5(1), 67–81 (1994). https://doi.org/10.1006/mare.1994.1005

9. Cooper, R., Yoshikawa, T.: Interorganizational cost management systems - the case of Tokyo-Yokohama-Kamakura supplier chain. International Journal of Production Economics 37(1), 51-62 (1994). https://doi.org/10.1016/0925-5273(94)90007-8

10. Wubbenhorst, K. L.: Life-cycle costing for construction projects. Long Range Planning 19(4), 87-97 (1986). https://doi.org/10.1016/0024-6301(86)90275-X

11. Kato, Y.: Target costing support systems: lessons from leading Japanese companies. Management Accounting Research 4, 33-47 (1993). https://doi.org/10.1006/mare.1993.1002

12. Everaert, P., Loosveld, S., Van Acker, T., Schollier, M., Sarens, G.: Characteristics of target costing: theoretical and field study perspectives. Qualitative Research in Accounting & Management 3(3), 236-263 (2006). https://doi.org/10.1108/11766090610705425

13. Imai, M.: Kaizen: the key to Japan's competitive success. Random House Business Division, New York, USA (1986)

14. Ahn, H., Clermont, M., Schwetschke, S.: Research on target costing: pass, present and future. Management Review Quarterly 68, 321-354 (2018). https://doi.org/10.1007/s10997-013-9275-4

15. Monden, Y., Hamada, K.: Target costing and kaizen costing in Japanese automobile companies. Journal of Management Accounting Research 3, 16-34 (1991). https://doi.org/10.1142/978184 8160385_0005

16. Modarress, B., Ansari, A., Lockwood, D.: Kaizen costing for lean manufacturing: a case study. International Journal of Production Research 43(9), 1751-1760 (2005). https://doi.org/10.1080/00207540500034174

17. Cooper, R., Slagmulder, R.: Develop profitable new products with Target costing. Sloan Management Review 40(4), 23-33 (1999)

18. Cooper, R., Slagmulder, R.: Interorganizational cost management and relational context. Accounting, Organizations and Society 29 (1), 1–26 (2004). https://doi.org/10.1016/S0361-3682(03)00020-5

19. Cooper, R., Slagmulder, R.: Supply chain development for the lean enterprise. Productivity Press, Portland, USA (1999)

Bank Mergers and Acquisitions: A Study of Events Regarding the Stock Price in the Hypothesis of Efficient Markets

Vinícius Dalla Vecchia and Ana Paula Etges

Abstract Mergers and Acquisitions (M&A) business strategies are frequently used to increase competitiveness. It is justified by one of the main objectives of these operations, which is to increase scalability and business diversity. Through an analysis of the Study of Events, this article evaluated the impact of the M&A announcement on the performance of the acquiring companies. At acquisition studied between bank and investment broker, it was possible to identify the manner of semi-strong efficiency of the market, while in two mergers involving only acquired and bank acquirers, the hypothesis was rejected, with observation of abnormal returns significant in the period under study. It is suggested for future analyses the study of behavior competing banks' shares during the M&A process, both in the period from the date of announcement and for the period of approval of the operation by the Central Bank, both with longer intervals for the estimation, event and comparison Windows.

Keywords Mergers and acquisitions · Study of events · Efficiency of the market

1 Introduction

The spread of news of events in Brazil and around the world has the power to influence the way people behave, and even to change their plans for the future. In the financial market, the methodology used to understand the repercussions of such events is through Event Testing, which is aimed at checking if the net value of a company has changed due to a certain fact [1]. In other words, the analysis checks whether the normal return expected for a company changes either positively or negatively during the period of study, measuring the effects of the event based on the analysis of the price of its shares. The applicability of this methodology is related to the Efficient Market Hypothesis (EMH), which, in the Semi-Strong form approached in the following paragraphs, would not allow earning abnormal returns (excessive) with the use of technical or fundamentalist analysis, for instance [2].

V. D. Vecchia (✉) · A. P. Etges
Polytechnic School, Pontifical Catholic University of Rio Grande Do Sul - PUCRS, Porto Alegre, Brazil

© The Author(s), under exclusive license to Springer Nature Switzerland AG 2021
A. M. Tavares Thomé et al. (eds.), *Industrial Engineering and Operations Management*,
Springer Proceedings in Mathematics & Statistics 367,
https://doi.org/10.1007/978-3-030-78570-3_29

The processes of Mergers and Acquisitions (M&A) in the banking industry were the events studied in this paper, due to their vast use as strategies for business growth and competitiveness increase, especially after the Real Plan, in 1994, provided greater economic stability to Brazil [3]. M&A have their motivation in the pursuit of business scalability, in the opportunity to enter new markets, in the lack of internal opportunities of growth, in the diversification of the business line, amongst other factors [4].

The Efficient Market Hypothesis, frequently ignored by novice investors, has proven, for years, to be capable of providing a better understanding of the financial market to investors when it comes to seeking a profitable stock portfolio, especially aiming at the long term. Developed by the American economist Eugene Fama, winner of the Nobel Prize in Economics [5], the theory that a market is efficient is based on the fact that the present price of shares reflects all information known until the moment regarding their value [6], something that has been tested by managers who seek to earn consistent returns above the market index.

According to this theory, an efficient market is the one that provides investors with perfect conditions to invest their money, through company assessment based on factors such as how they produce, how they are managed, the competitors' performance and macroeconomic situation [7]. Such market would reflect each and every available information about the data, considering the investors' rationality, leading to the conclusion that there are neither cheap nor expensive shares. This rationality would result in the impossibility of getting high earnings in the long term, adjusted to the risk the company is linked to [2].

The idea of putting money only in passive investments with low costs is more present for investors who believe markets are efficient and follow the profitability of the benchmark as the Ibovespa index, for example. According to this hypothesis, there would not be the market α, nor the possibility of earning returns above the market through analysis, as all investors would get the same average profitability throughout time. Investors who do not believe in market efficiency frequently mention Warren Buffet, deemed as the greatest investor of all time. Therefore, it would be better to invest in active funds with good managers, as they have the capability to overcome the market in the long run.

The econometric model of Study of Events used in this paper is based on the automatic adjustment of prices as new information is disclosed, considering the investor's rationality. It enables quantifying the difference between the expected returns in the price of securities with the actual performance observed, and where the presence of abnormal returns is perceived, the market efficiency hypothesis is not valid, given that the prices were not automatically adjusted and enabled the investor to get significant earnings above the benchmark. This methodology may be applied to several events that require assessment in face of the context they are inserted into, the most frequent being the announcement of profits by a company, issue of debt, mergers and acquisitions, privatization or nationalization and announcement of dividends, for example [8].

2 Objectives

The main objective of the study was assessing the performance of two companies in the Brazilian stock market close to the dates of three M&A operations. The specific objectives were: (i) carry out linear regressions in order to identify standards; and (ii) check how the market reacted to the announcement of M&A and if there was informational efficiency in the Semi-Strong form.

3 Methods

The procedures of a study of events may be described in this sequence: definition of the event, sample selection, measurement of the abnormal return (AR), estimate procedure, test procedure, empirical results, interpretations and conclusions [8].

The choice of the assessed companies was due to the magnitude and impact potential in the economic system of the country. Collection of the rate history of the companies and the respective market index was carried out based on the information available in the website of the Brazilian stock exchange [9]. Three events were analyzed: (i) the acquisition of the broker XP Investimentos by Banco Itaú (2017); (ii) the merger of Banco HSBC and Banco Bradesco (2015) and the merger of Banco Unibanco and Banco Itaú (2008).

First, in order to observe if the behavior of the price of a stock has abnormal characteristics, the control return is defined, which is known as the normal return or expected return, obtained from the logarithmic difference of the stock prices, which represent a continuous capitalization of the shares and is the one that better represents the price expectation of the price assets [10]. The formula of the expected return is described by Eq. 1:

$$R_{i,t} = Ln\left(\frac{P_{i,t}}{P_{i,t-1}}\right) \tag{1}$$

where,

$R_{i,t}$ is the expected return rate,

$P_{i,t}$ is the price of the security at time t and

$P_{i,t-1}$ is the price of the security at time $t - 1$.

The expected return was applied to the shares of the three companies under analysis, as well as to the IFNC financial index, which represents with greater assertiveness the behavior of companies in the banking system, being used as reference portfolio. This index is the result of a theoretical portfolio, used as an average performance index for the prices of the assets of financial, pension and insurance industries [9].

In order to assess the impact of the event, the abnormal return (ARi,t) adjusted to the risk was calculated, being comprised of the performance of the company in the observation period, defined as normal return (Ri,t), and deduced from the difference between α and the product of β and (Rmt). The systematic risk β represents the slope of the straight line in the measurement of the abnormal returns of the market portfolio, while α represents the interception of the straight line in the axis of the abnormal return of the asset. The abnormal return is defined as the difference between the return of the index and the return of the asset in the specified date. The formula of the abnormal return is described by Eq. 2:

$$AR_{i,t} = R_{i,t} - (\alpha - \beta R_{mt}) \tag{2}$$

where,

$ARit$ is the abnormal return rate;

Rit is the expected return rate,

α is the constant of the returns of the asset in relation to the benchmark;

β is the sensitivity of the returns of the asset in relation to the benchmark;

Rm,t is the expected return rate of the market portfolio at the time t.

The calculation of the Accumulated Abnormal Return (CAR), useful to understand the risk assumed by investors in face of the possibilities of loss is a sum of the total of abnormal returns identified during the period of the event under analysis [11]. The formula of the accumulated abnormal return is described by Eq. 3:

$$CAR_i(t_1, t_2) = \sum_{t=t_1}^{t_2} AR_{it} \tag{3}$$

where:

$CAR_i(t_1, t_2)$ is the accumulated abnormal return rate and,

$\sum_{t=t_1}^{t_2} AR_{it}$ is the sum of the abnormal returns.

Regarding the periods in the study, the estimate window represents the standardization of the actual return of the assets and of the market index, independently of the happening of the event. The event window comprises the calculation of the expected returns, including the date when the studied event occurred, based on the market α and β, making it possible to identify if the announcement of an event was anticipated or leaked (period -10 to 0) and how long it took for the information of the event to be absorbed by the market (period 0 to 10). The comparison window, given after the date of the event, enables an analysis on the impact of the happening through time (Fig. 1).

Using the statistic model of linear regression analysis [12], the value of the abnormal return was adjusted to the standard error of the regression, allowing to check if the investors of the mentioned companies could obtain abnormal returns in the M&A period with abnormal returns that indicate or not that the event studied impacted the share price, considering the class these companies are inserted into. The hypotheses tested are presented below, aiming at checking if the informational efficiency, in its Semi- Strong form, is present in the events:

- – H0: the event did not present significant abnormal returns
- – H1: the event presented significant abnormal returns.

Thus, in face of any relevant information publicly disclosed, the price of the assets would be adjusted instantly. Upon the occurrence of significant abnormal returns in relation to the market return, the market inefficiency would be identified.

4 Results

Based on the returns found in the estimate window, linear regression was applied for identification of the factors α, β, standard error and R^2, which explain the inter-relation between the prices analyzed. One of the events analyzed was the merger of Banco Unibanco and Banco Itaú in the year 2008, which formed the largest financial conglomerate in the South Hemisphere, with a market value of R\$ 51.7 billion, standing among the 20 largest in the world. Such association resulted in a Brazilian bank with economic capacity to compete with the largest in the world, turning the company into a powerful booster of national development [13].

Despite an expressive outlier identified 15 days before the zero date in the estimate window, with price variation of 19.00% for the index and 20.89% for the share, the regression of the returns of the historical returns of Itaú in face of IFNC in the merger with Unibanco has proven to be appropriate, as shown in Fig. 2.

An initial abnormal return of −0.03% was observed in the parameter α coming from the regression, a risk below the market represented by the value of 97.38% (lower than one) of parameter β, standard error of 1.38% and, at last, 89.48% of behavior explanation of the shares being explained by the index through the parameter

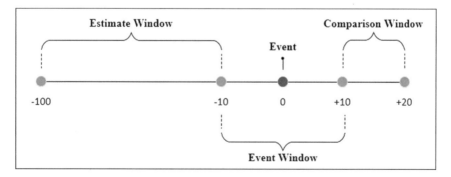

Fig. 1 Periods observed. *Source* Author

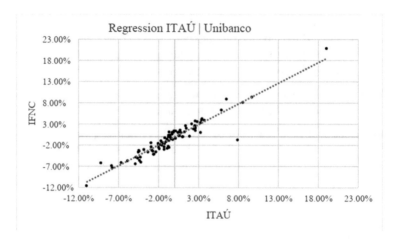

Fig. 2 Regression of the returns in relation to IFNC (Itaú | Unibanco). *Source* Author

R^2. The accumulated abnormal returns were statistically significant in four dates. Three occurrences took place in the pre-event period on the dates −6, −5 and −3. The fourth occurrence took place in the zero date.

Figure 3 contains the event day (11.03.2008) and the analyzed interval, IFNC index, Itaú bank quotation, the variations of both and the abnormal return (AR) calculated with the regression parameters. The t-student statistic was used to understand if the AR are significant and finally the accumulated abnormal returns (CAR).

Date	IFNC	ITUB4	IFNC_VAR	ITUB4_VAR	AR	T-stat	Significant?	CAR
11.17.2008	1.812,97	6,69	-0,19%	-0,15%	0,07%	0,05227228	No	9,12%
11.14.2008	1.816,45	6,7	-1,49%	-1,33%	0,15%	0,111453008	No	9,05%
11.13.2008	1.843,76	6,79	8,16%	8,77%	0,86%	0,622639483	No	8,89%
11.12.2008	1.699,31	6,22	-7,56%	-7,13%	0,26%	0,192192125	No	8,03%
11.11.2008	1.832,81	6,68	-0,12%	-1,04%	-0,89%	-0,647195508	No	7,77%
11.10.2008	1.835,00	6,75	0,93%	3,16%	2,29%	1,658765211	No	8,66%
11.07.2008	1.817,93	6,54	2,88%	3,58%	0,81%	0,588880264	No	6,38%
11.06.2008	1.766,34	6,31	-5,52%	-6,30%	-0,88%	-0,641090689	No	5,56%
11.05.2008	1.866,62	6,72	-8,76%	-9,23%	-0,67%	-0,487409768	No	6,45%
11.04.2008	2.037,43	7,37	6,47%	4,72%	-1,54%	-1,120939128	No	7,12%
11.03.2008	1.909,76	7,03	7,25%	15,18%	8,15%	5,91853056	Yes	8,66%
10.31.2008	1.776,21	6,04	-0,88%	-2,29%	-1,40%	-1,017349483	No	0,51%
10.30.2008	1.791,88	6,18	10,44%	10,03%	-0,10%	-0,069273344	No	1,91%
10.29.2008	1.614,29	5,59	8,33%	12,36%	4,29%	3,112342096	Yes	2,01%
10.28.2008	1.485,32	4,94	11,39%	8,44%	-2,61%	-1,894121975	No	-2,28%
10.27.2008	1.325,48	4,54	-3,97%	0,00%	3,90%	2,832711725	Yes	0,33%
10.24.2008	1.379,18	4,54	-8,22%	-11,24%	-3,20%	-2,323500704	Yes	-3,57%
10.23.2008	1.497,31	5,08	-7,76%	-8,67%	-1,08%	-0,781997584	No	-0,37%
10.22.2008	1.618,12	5,54	-12,85%	-12,86%	-0,31%	-0,224033906	No	0,70%
10.21.2008	1.839,97	6,3	-1,82%	-1,73%	0,08%	0,055049289	No	1,01%
10.20.2008	1.873,76	6,41	6,20%	6,94%	0,94%	0,680682396	No	0,94%

Fig. 3 Merger of Unibanco with Banco Itaú. *Source* Author

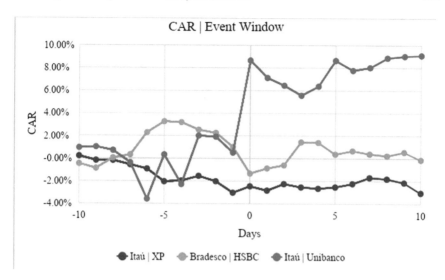

Fig. 4 CAR of the event window. *Source* Author

These calculations were also applied to the comparison and estimate windows to support the method, in the 3 events analyzed.

Figure 4 shows the behavior of the shares in the period observed, with a sharp rise in the accumulated returns from the date −6 until the zero date. From the center of the event, there was a stabilization in the positive sense of the performance of the company.

Using the same methodology applied above, for the case of the acquisition of XP by Itaú, there were three significant abnormal returns in the event window, with an adjustment in the so far trend of low price of the stock, reversed on the zero date, after absorption of the information. For the case of the merger of HSBC with Bradesco, there were also cases of significant abnormal returns, all happening before the zero date of the event, evincing a strong rising trend from the date −5. Such behavior may be explained by the dimension of the events, which generated great expectations in the investors since the announcement of the merger, being rejected the H0 hypothesis, given the presence of significant abnormal returns. Figure 4 shows the three CAR in the same interval of days for comparison purposes in the event window, being possible to note a more abrupt variation in the abnormal returns in events two and three.

It is possible to see, in the comparison window, by Fig. 5, that the events one and two did not present significant changes in the accumulated abnormal returns. Meanwhile, the event three presented a significant abnormal return on the date 14, which may be seen by the table of appendix two at the end of the article.

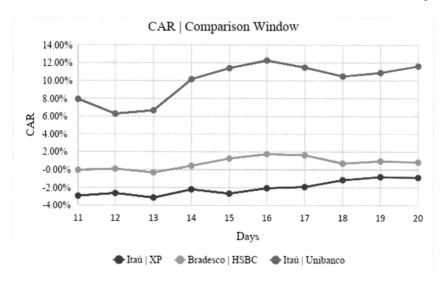

Fig. 5 CAR of the comparison window. *Source* Author

5 Conclusion

The present study has assessed the effects of the announcement of M&A in the price of the shares of two of the largest Brazilian banks, using the procedure of Study of Events to identify abnormal returns obtained in the period. By means of statistic regressions for identification of standards, it was possible to understand how the market reacted to the announcements, with confirmation of the hypothesis of market efficiency only for the case of acquisition of XP Investimentos by Itaú, denying the possibility of obtaining abnormal returns (excessive) with the use of public information on the event. For both mergers, represented by events two and three, the hypothesis of an efficient market in the Semi-Strong form was rejected, given they presented significant abnormal returns. The event two presented a market β above one, which indicates a higher risk than the investment to the benchmark, a fact in contrast with the event three, which presented a market β below one and also a more significant abnormal return, with a variation of $+10.67\%$ in the interval of the days -10 to 20, indicating it is the best investment option. It was also noted, for the three events, that on the last comparison date, the CAR had a positive improvement, evincing that the purchase of shares after the date of the announcement of the M&A can be a valid strategy for those who believe in the inefficiency of the market and look for returns above the market average.

A limitation of this study is the need of further assessment comprising longer periods for the estimate, event and comparison windows, aiming at capturing more precisely the behavior of these companies. This paper suggests, for future analysis, the study of the behavior of shares of competitor banks during the M&A process, both in the period of the date of announcement, as well as in the period of approval of the operation by the Central Bank.

So far, there is no consensus in literature regarding the position of the market during such events, therefore research on the subject is important to help investors to design a strategy that fits better their profile and understand how such operations affect the behavior of companies in the industry.

References

1. 1. McWilliams A., & Siegerl, D. Event studies in management research: Theoretical and empirical issues. Academy of management journal, 40(3), 626–657 (1997).
2. 2. Shleifer, A. Inefficient Markets: An Introduction to Behavioural Finance. Oxford University Press, UK, (2000).
3. Barbedo, F. F. Fusões e Aquisições no Mercado de Capitais Brasileiro: Analisando o Retorno para o Acionista. (2006).
4. Brito, Giovani A.S., Batistella, Flávio D.; Famá, Rubens. Fusões e aquisições no setor bancário: avaliação empírica do efeito sobre o valor das ações. Revista de Administração, São Paulo, v.40, n.4, p.353-360, Oct./Nov./Dec. (2005).
5. Fama, Eugene. Two Pillars of Asset Pricing. Nobel Prize Lecture. (2013).
6. 6. Fama, E. F.; Random Walks in Stock Market Prices. Financial Analysts Journal. Vol. 21, no. 5, pp. 55 -59. (1965).
7. Lima, A. O.; Auge e declínio da Hipótese dos Mercados Eficientes. Revista de Economia Política, 2003. Available at: http://www.rep.org.br/pdf/92-2.pdf. Accessed 9 Sept (2019).
8. Mackinlay, A. C. Event Studies in Economics and Finance. Journal of Economics Literature, 35(1), 13–39, March, (1997).
9. BM&FBovespa, Índice Financeiro (IFNC). Available at: http://www.bmfbovespa.com.br/pt_br/produtos/indices/indices-setoriais/indice-bm-fbovespa-financeiro-ifnc.htm Accessed 11 Nov (2019).
10. Soares, Rodrigo Oliveira; Rostagno, Luciano Martin. Estudo de Evento: o Método e as Formas de Cálculo do Retorno Anormal. Anais eletrônicos ANPAD (2002).
11. Gitman, Lawrence Jeffrey. Princípios de administração financeira. 10ª Edição. São Paulo: Addison Wesley, (2004).
12. Santos, C. M. A.; Estatística Descritiva – Manual de auto-aprendizagem; Edições Sílabo; (2007).
13. Itaú, Fato Relevante, 2017. Available at: https://www.itau.com.br/relacoes-com-investidores/Download.aspx?Arquivo=Sx3ALZZO6zRCJjEOvWXI9g Acessed 11 Nov (2019).

Cleaner Production with Economic and Environmental Assessment: A Case Study in an Aeronautical Workshop

Henrricco Nieves Pujol Tucci⊙**, Geraldo Cardoso de Oliveira Neto**⊙**, Flavio Luiz Rodrigues, Marlene Amorim**⊙**, and João Reis**⊙

Abstract The companies are looking for strategies to improve their relationships with customers and increase their market share with the main objective of obtaining economic benefits. Cleaner Production (CP) practices allow companies to save money through preventive actions in relation to environmental problems. Brazilian companies are under pressure to reduce electricity consumption, otherwise they will pay much more expensive bills or even be without supply in times of greatest need. This research aims to analyze the economic and environmental benefits through the application of CP practices, a case study was developed in an aeronautical engine overhaul workshop and intends to reduce the electricity consumption by replacing lamps and better use of natural light. The results showed that it was possible reduce the annual electricity consumption by 3,744 kWh and save 2,233 USD per year, therefore, the calculations found the return on investment (ROI) would be in 2.63 years. However, through the Mass Intensity Factor tool, it was possible to calculate the environmental gains in relation to the avoided environmental impacts due to not using the amount of electricity saved, these gains were approximately 258,000 kWh per year. Although the calculation of the ROI supported the decision making of the company executives, it drew the attention of everyone involved in the project that the environmental gains were 68 times greater than the economic gains.

Keywords Cleaner production practices · Economic gains · Environmental gains

H. N. P. Tucci (✉) · G. C. de Oliveira Neto · F. L. Rodrigues
University Nove de Julho, Uninove, São Paulo, Brazil

M. Amorim · J. Reis
Department of Economics, Management, Industrial Engineering and Tourism, GOVCOPP, University of Aveiro, 3810-193 Aveiro, Portugal

J. Reis
Industrial Engineering and Management, Faculty of Engineering, EIGeS, Lusofona University, Campo Grande, 1749-024 Lisbon, Portugal

© The Author(s), under exclusive license to Springer Nature Switzerland AG 2021   397
A. M. Tavares Thomé et al. (eds.), *Industrial Engineering and Operations Management*, Springer Proceedings in Mathematics & Statistics 367, https://doi.org/10.1007/978-3-030-78570-3_30

1 Introduction

A common concern across companies operating in Brazil is the pursuit of strategies in order to improve their relationships with customers and increase their market share. Over the years, the adoption of sustainability practices has emerged as one of the most important paths chosen, notably for the expected benefits that are often achieved in the the corporate image, that stand out as competitive advantage towards their direct competitors. On the other hand, companies are increasingly under pressure to take responsibilities for managing the entire life cycle of their products notably for all the environmental impacts that these trigger [1].

The beginning of the journey of sustainability is often traced back to the implementation of the Environmental Management System (EMS), contained in ISO14001. At the same time, companies get acquainted with Cleaner Production (CP) practices and start their application to meet the EMS [2].

CP involves the adoption of preventive actions, aligned with production processes, in order to ensure the continuous application of an environmental strategy for success. CP can be applied to production processes, development of products or services even as it considers the entire life cycle as a function of mitigating environmental impacts [3].

The benefits of a successful implementation of CP include improved efficiency, profitability and competitiveness of enterprises, while also allowing for gains in the safety and preservation of the environment, the consumer and the worker. Among the prevalent CP practices adopted by companies its worth mentions those related with efficiency in energy utilization, the reuse of materials, the reduction of waste as well as the reduction of water consumption levels [4, 5].

The Brazilian Government, through the Law No. 10,295 and the Decree 4059, established a series of measures to ensure that the consumption of electricity was carried out in a rational manner, pointing out the need to create and provide indicators and clear information about the consumption and efficiency [6]. Since, large disparities in energy consumption increases the levels of pollution and encourage inefficient use of energy, as a result the price of energy is high by encouraging further disparity and the generation of pollutants [7].

It is worth noting that Brazil has a high dependence on hydropower for energy production, but this is only possible model to be installed by changes in the environment and use of natural resources. With energy consumption increasing over the years in the country, it appears that the environmental impacts of this power generation model are also increasing [8, 9].

Still, most companies choose not to invest to implement the CP practices. Some of the reasons that are advanced to explain this include uncertainties in the economic returns, and the long periods of return that are associated with this type of investment [10] and the complexity of calculating the economic benefits in order to stimulate businesses adopt practices of CP [11]. In addition, companies are under pressure by various stakeholders to adopt CP practices because of the need to obtain international certifications. De Oliveira et al. [12] Finally, a recent scenario of economic instability and change policies comprise an additional challenge for companies operating in Brazil, forcing them to find solutions that result in economic gains, such as the adoption of CP practices [13].

Few researches were identified in the literature that have addressed the CP practices and gave quantified results related to the consumption of electricity, The works of Wasserman et al. [14] developed in a large pulp and paper industry located in Brazil and the work of work Burritt et al. [15] developed in Southeast Asia in four different small businesses are examples of application of CP practices to reduce energy consumption, although they don't present quantitative evidence about the economic and environmental benefits. Building on this research gap, this paper aims to present and analyze the economic and environmental benefits through the application of CP practices in an engine overhaul workshop in front of a renovation project of the buildings.

This is the overall contextualization of the study the triggering research gap that frame the main objectives of the study. After this introduction, a review of the literature is presented, supporting the research, followed by the description of the methods used. A case study developed in a workshop to review aero engines located in Brazil is presented and discussed in the study as well as the conclusion and contributions of this work, together with future research suggestions.

2 Literature Review

The Sustainable Development and Sustainability were first mentioned in 1987 by the World Commission on Environment and Development (WCED). Sustainable development was introduced as the conditions under which a level of development is achieved without compromising the ability of future generations to reach their needs. Thus, the resultant becomes an equilibrium between economic growth impacts on the environment and human development, as shown in the following Fig. 1. In turn, the sustainability was introduced considering from the regional level to all plans, such as the fact that human activities do not interfere in the natural cycles that underpin

Fig. 1 Sustainable development

all that resilience allows planet and, moreover, not reduce natural capital, ensuring that it will be passed on to future generations [16].

However, traditional means of addressing pollution, also known as end of pipe techniques, do not comply with all the concepts of sustainable development. The tube techniques are devised to treat and control pollutants only after they have been generated, for example in the case of treatment of effluents after water contamination. The most serious problem in applications of tube end techniques is the concept behind the application, since it assumes that natural resources extracted to generate the raw material and energy used in the process are resources with unlimited capacity i.e. inexhaustible [4, 17].

One of the paths chosen by companies that allow for the advancement towards towards sustainable development is the ISO14001 certification and the development of a robust and ongoing action plan, also known as EMS. This management system can be divided according to the engagement of the company and therefore with the practices adopted. The environmental practical applications restricted to a process are operating and can be classified as initial practices. Examples of these practices include proper disposal of waste; pollution control and recycling [2].

Applications of environmental practices that involve the whole company are considered intermediate and examples of these practices are Pollution Prevention (P2) and CP. P2 is an environmental management approach that reduces the source of pollution through efficient use of resources such as energy, materials and water. P2 is most often used for cost reduction seeking to improve the company's financial performance [18].

As for CP, this is an environmental impact prevention strategy that can be applied to products, processes and services. An important difference resides in the fact that CP considers the entire product life cycle. Such holistic view of the entire chain allows CP to improve efficiency, profitability and competitiveness of enterprises while protecting the environment, consumer and the worker [3].

However, the ISO14001 standard does not define the form and the degree to which companies should have to reach the EMS thus allows each company to develop its own solutions for compliance with the standard. Something that allows for customization of the application, but that also opens a gap so that companies do not take actions that address environmental problems indeed. The use of natural resources and exacerbated productions with unreasonable waste has attracted the attention of environmentalists and Non-Governmental Organizations (NGOs), which must press companies for change, as well as the government for controlling shares [12, 19].

The usual business model of companies is through the flow of raw materials, processing and economic gains. Faced with the need to reduce environmental impacts, companies become interested in changing their processes, invest in cleaner technologies and therefore achieving environmental gains. Looking for a tool that is attractive to companies, the application of CP practices is often a good option, from the moment that quantifies the costs and economic gains and highlights the environmental gains through the avoided impacts [11].

The production of any product and the provision of any service currently requires the use of energy. When considering thermoelectric power plants using coal, diesel or natural gas, it is clear that in this model results due to the environmental impact of carbon dioxide emissions into the atmosphere, and using a fossil fuel, therefore non-renewable. However, most of the energy generated in Brazil comes from large hydro, a model that floods large portions of land which does not respect the local vegetation, waterways and much less the species that inhabited the region. In addition, the lack of public control results in poor quality studies that do not even assess the full environmental impacts, much less its proper mitigation [7, 9].

The increasing risk of blackouts due to the problems of excessive energy consumption caused the federal government in 2001 to develop the energy saving program. The program foresaw the gradual exhaustion of the supply of incandescent bulbs in major marketing points and in return, the increase in the supply of fluorescent lamps and currently lamps with Light Emitting Diode (LED) technology, both with superior energy efficiency to 85% and life up 10 times [6].

The application of CP practices encourages the rational use of natural resources in line with efficiency, so that it results in economic and environmental benefits for the company, allowing it to position itself more competitively [12, 18]. One of the natural resources most cited in the literature on the application of CP practices is the power [14, 15].

3 Methods

The methodological procedures adopted in this study fit in the exploratory category with both quantitative approaches, considering numerical data and benchmarks, and qualitative for its major subjective aspects to set the case [20].

It adopted the case study method, understood as an adequate approach to create the appropriate conditions to understand, verify and confirm the theory used in an exploratory study. The case study method is a common research strategy that have the objective of presenting dynamic processes [21].

Most of the case studies highlight the research using three stages: literature review; interviews with people who have experience practices related to the topic or even directly participate in the process study and analysis of examples that contribute to the understanding [20].

In this study, data collection was conducted by a team engaging participant from different sector, this team aimed to develop improvement projects to focus with pieces of recovery workshops in reducing electricity costs. The best solution presented was the renovation of the building for the opening of large windows that allow the use of natural lighting, as well as review of wiring and replacement of traditional lamps for models equipped with LED.

This work has selected the model presented by De Oliveira Neto et al. [22] in order to quantitatively evaluate the economic and environmental benefits that the application of CP practices allowed to achieve. Thus, the following steps were carried out: data collection; economic evaluation; environmental assessment and comparison of earnings.

Table 1 Steps to the calculations of the economic and environmental gains

Elements	Equation	Ref.
Mass Intensity Factor (MIF)	MIF = IF * M	Equation 1
Mass Intensity per Compartment (MIC)	MIC = IF (w compartment residue A) + IF (w compartment residue B) + ...	Equation 2
Mass Intensity Total (MIT)	MIT = MICw + MICv + MICz + MICn	Equation 3
Economic Gain Index (EcGI)	EcGI = MTS / EG	Equation 4
Environmental Gain Index (EnGI)	EnGI = MIT / EG	Equation 5

Data was also collected by means of survey, with the purpose of quantifying residues and emissions resulting from the process for construction of the mass balance. In this case study, the focus was to reduce the power consumption. In addition to collecting data and mass balance of the construction itself, this step also consists of the details of the materials or components and the calculation of the total material saved, in this case, electricity.

The second stage of the study involves the quantification of gains and return on investment through financial quantification of the elements identified in the survey data. The economic evaluation found Economic Gains (EG) for the company and evaluated the Return on Investment (ROI).

In turn, the environmental assessment aimed to measure the environmental gains due to the environmental impacts avoided, to this end, were used the abiotic compartments, biological, water and air, as this makes up the Mass Intensity Factor tool (MIF), considering Mass (M) and Intensity Factor (IF), as shown in Table 1.

The next step was to evaluate the Mass Intensity per Compartment (MIC) in order to measure the reduction of environmental impact by abiotic compartment (w), biotic (x), water (y), air (z) and other (n) as Table 1.

With these steps completed, the next step was to evaluate the Mass Intensity Total (MIT) which allowed account to reduce the total impact, through the sum of the MICs, as Table 1.

The fourth and final step was to compare the economic evaluations with environmental assessments based on EG. Therefore, it was necessary to calculate the Economic Gain Index (EcGI) as well as the Environmental Gain Index (EnGI) as Table 1.

Thus, it was possible to analyze the calculated total related to economic and environmental gains, as these results were presented in the same unit of measurement, it was possible to compare these results and evaluating the representation of each variable. The value of Intensity Factor (IF) for electrical energy corresponds to 1.55 to abiotic material, 66.73 for water and 0.54 for air [23].

4 Case Study

The company in this case study is a global benchmark in the segment of aero engines. Installed in Brazil for over 50 years, the company is considered medium-sized, employing approximately 300 employees in its land across 17 buildings between administrative and operational facilities. The parts recovery workshops alone, occupy 5 buildings. This case study will address the reform of these workshops that was conducted and supported by the concepts of Cleaner Production.

The recovery of aircraft parts is performed according to criteria established by the manufacturer and in accordance with the regulations of the competent bodies. As most of the customers of this engine overhaul workshop are in the United States, it is certified by the Federal Aviation Administration (FAA).

So, saving electricity through Cleaner Production practices in one of the aircraft parts recovery workshops was only possible after a thorough understanding of the regulation on Repair Station Operators (Part 145) establishing, for example, the amount of light through lumens for recovery benches and inspection of aircraft components [24].

The reform was carried out to meet the increased demand for a particular aircraft engine. This type of engine before represented 65% of the production mix of workshops and review, with the increase in demand, now represents 80%. In order to calculate the economic and environmental gains, the team conducted the data collection stage and obtained the values shown in Table 2. All monetary values were presented in US dollars (USD) and energy consumption in kilowatts per hour (kWh).

The reforming embodiment of the workshop (Fig. 2) included the replacement of traditional bulbs T8 models with new LED technology have also replaced the bench lamps with filament lamps old for new LED lamps and new and more potent magnifiers. However, the biggest change was the installation of four large windows

Table 2 Data collect

Elements	Quantities	Units
4 large windows (with glass)	3720.93	USD
LED lamps 50	29.07	USD
6 bench lamps with magnifying glass	186.05	USD
Other electrical materials	81.40	USD
Labor	1860.47	USD
Total invested	5877.91	USD
Before applying electricity consumption practices CP	432 082	kWh/year
After applying electricity consumption practices CP	428 338	kWh/year
Total electricity saved	3,744	kWh/year
Traditional T8 lamps disposed enquiry.c	50	lamps
Old filament lamps disposed enquiry.c	12	lamps
Construction rubble materials sent to recyclers	5,000	kg

Fig. 2 Reformed recovery
workshop

for the best use of natural lighting. To enable energy savings in accordance with the
regulations of the aero engine servicing workshops, switches were installed for each
group of lamps, an improvement on the single switch previously available in this
building.

These activities enabled a reduction in electricity consumption of 3,744 kWh
per year, even considering the increase in production. Additionally, all the work
waste, including lamps and debris of concrete blocks were properly allocated to
recycling companies. However, for purposes of economic viability calculations, it
was only considered the reduction of electricity consumption to analyze the return
on investment (ROI), as shown in Table 3.

From the data collection it was possible to see the total amount invested, and
identify economic gain achieved by reducing the power consumption. Thus, the return
on investment was calculated (ROI) in 2.63 years, or a little less than 32 months.
Therefore, estimates are that from this time the company will make profits because
this reform, this point was essential to justify this reform.

Despite the calculation which estimated the deadline for payback of the investment, it was noted by some members of the team that the scope of this went against
the project with the concepts of Cleaner Production and, from this finding, the
calculations were developed presented in the Table 4 and already presented in the
Methodology section.

Table 3 Economic evaluation

Elements	Quantities	Units
Total invested	5877.91	USD
Electricity cost practices before applying CP	257,651.80	USD/year
Electricity cost after applying practices CP	255,419.24	USD/year
Economic Gain (EG)	2232.56	USD/year
Return of investment	2.63	years

Table 4 Environmental evaluation

Elements	Abiotic materials	Biotic materials	Water	Air
Intensity Factor (IF)	1.55	–	66.73	0.535
Mass (M)	3,744	–	3,744	3,744
Mass Intensity Factor (MIF)	5803.20	–	249,837.12	2003.04
Mass Intensity per Component (MIC) = Total Mass Intensity (MIT)				257,643.36

Table 5 Comparison of gains

Elements	Quantities	Units
Total electricity saved	3,744	kWh/year
Economic Gain (EG)	2232.56	USD/year
Economic Gain Index (EcGI)	1.68	kWh/USD
Avoided environmental impact	257,643.36	kWh/year
Economic Gain (EG)	2232.56	USD/year
Environmental Gain Index (EnGI)	115.40	kWh/USD

The environmental assessment has identified that the reform of recovery workshop focusing on reducing the energy consumption avoided environmental impacts equivalent to annual generation of approximately 258,000 kWh. These calculations considered the environmental problems of responsibility of generating electricity companies, in Brazil's case, the power is generated mainly by hydroelectric plants.

Despite the economic and environmental assessments already carried out, you cannot compare these values because they are not in the same unit of measurement, so the Table 5 presents the calculations that allowed such a comparison.

A comparison by the economic gains and indices Environmental retirement of aircraft components repair shop allowed to establish that the environmental gains are greater than the economic gains. Note that the EnGI was more than 68 times higher than the EcGI, demonstrating that the avoided environmental impacts, and now quantified, allowed much money was saved on the entire system, from the generation of energy.

5 Conclusion

This work concludes that it was possible to apply CP practices and achieve economic gains and calculate the return on investment of a renovation project of recovery workshops aircraft components focusing on reducing the energy consumption. Moreover, it was possible to quantify the environmental benefits and environmental impacts avoided due to the reduction of electricity consumption. Therefore, this paper adds the theory from the time that fills the identified research gap.

The results presented in annual economic gains of at least 2,233 US dollars and return on investment estimated at about 32 months, said that the project to reduce the electricity consumption by replacing lamps and better use of natural light were deployed successfully. Thus, this work adds practice to present to executives of companies that you can get economic gains and calculate the return on investment when applying cleaner production practices.

In addition to the economic results, this work showed a reduction of the annual electricity consumption of 3,744 kWh, however, more important is that this work has quantified that the application of production practices Cleaner avoided the equivalent environmental impact the annual generation of 258,000 kWh. Therefore, this paper adds the company to present a quantitative way to press companies to measure their environmental impacts and establish and publicize reduction targets.

This study has limitations due to be a case study, its results cannot be extrapolated, however, suggests that if future research using these same methods to achieve similar results. It was observed that the proper disposal of the work spoils, especially cement debris and concrete blocks could also have been calculated as the avoided environmental impacts disposal if it had been done improperly, however, these data were not obtained during the data collection stage preventing such continuity, so it is suggested to also consider these elements in future research within this scope.

References

1. Jabbour, C. J. C. Non-linear pathways of corporate Environmental Management: A Survey of ISO 14001 certified companies in Brazil. Journal of Cleaner Production, Vol. 18, pp. 1222–1225 (2010)
2. Hamner, W. B. Relationship among the cleaner production, pollution prevention, waste minimization, and ISO 14000. Proceedings of the First Asian Conference 1996 in CP Chemical Industry, National Center for Cleaner Production (1996).
3. Baas, L. W., & Baas, L. Cleaner production and industrial ecology: Dynamic aspects of the introduction and dissemination of new concepts in industrial practice. Eburon Uitgeverij BV, (2005).
4. Van Berkel, A., Willems, E. & Lafleur, M. The relationship between cleaner production and industrial ecology. Journal of Industrial Ecology, 1 (1), 51–66, (1997).
5. Glavic, P., & Lukman, A. Review of Their definitions and terms sustainability. Journal of cleaner production, 15 (18), 1875–1885 (2007).
6. Nogueira, L. A. H., Cardoso, R. B., Cavalcanti, C. Z. B., & Leonelli, P. A. Evaluation of the energy impacts of the Energy Efficiency Law in Brazil. Energy for sustainable development, 24, 58–69, (2015).
7. Templet, P. H. Energy price disparity and public welfare. Ecological Economics, 36 (3), 443–460, (2001).
8. Chen, S., Chen, B., & Fath, B. D. Assessing the cumulative environmental impact of hydropower construction on river systems based on energy network model. Renewable and Sustainable Energy Reviews, 42, 78–92, (2015).
9. Bermann, C. Energy in Brazil: what for. For those, v. 2, (2002).
10. Gombault, M., Versteege, S. Cleaner production in partnership with SMEs through the (local) Authorities: Successes from the Netherlands. Journal of Cleaner Production, Vol. 7, pp. 249–261 (1999).

11. Oliveira Neto, G. C., Leite, R. R., Shibao, F. Y., & Lucato, W. C. Framework to Overcome barriers in implementation of the cleaner production in small and medium-sized enterprises: multiple case studies in Brazil. Journal of Cleaner Production, 142, 50–62, (2017).
12. De Oliveira J. A., Smith, C. H. A., Ganga, G. M. D., Godinho Filho, M. Ferreira, A. A., Esposto, K. F., & Ometto, A. R. Cleaner Production practices, and motivators performance in the Brazilian industrial companies. Journal of Cleaner Production, 231, 359–369, (2019).
13. De Guimarães, J. C. F., Severos, E. A., and Vieira, P. S. Cleaner production, project management and strategic drivers: An empirical study. Journal of Cleaner Production, 141, 881–890, (2017).
14. Wasserman, J. C., Quelhas, O. L. G., and Lima, G. B. A. Analysis of Cleaner Production Practices in the Printing Company in Brazil. Environmental Quality Management, 26 (2), 45–63, (2016).
15. Burritt, RL, Herzig, C. Schaltegger, S., & Viere, T. Diffusion of Environmental Management Accounting for cleaner production: evidence from some case studies. Journal of Cleaner Production, (2019).
16. Elkington, J. Partnerships from cannibals with forks: The triple bottom line of 21st-century business. Environmental Quality Management, 8 (1), 37–51, (1998).
17. Baas, LW Cleaner production: beyond projects. Journal of Cleaner Production 3 (1–2), 55–59, (1995).
18. And Cagno, Trucco E, Tardini L. Cleaner production and profitability: Analysis of 134 Industrial Pollution Prevention (P2) project reports. Journal of Cleaner Production, (2005).
19. Oliveira JA, Oliveira, OJ, Ometto, AR Ferraudo AS, & Salgado, MH Environmental Management System ISO 14001 factors for Promoting the adoption of Cleaner Production practices. Journal of Cleaner Production, 133, 1384–1394 (2016).
20. Yin, RK Case study research: design and methods. 5th ed. California: SAGE, (2015).
21. Eisenhardt, KM Building theories from case study research, Academy of Management Review, 14 (4), pp 532–550, (1989).
22. de Oliveira Neto, GC Tucci, HNP Pinto, LFR, Costa, I., & Milk, RR Economic and Environmental Recycling Advantages of rubber. In IFIP International Conference on Advances in Production Management Systems (pp. 818–824). Springer, Cham, (2016).
23. Wuppertal Institute. intensity of materials material, fuels, transport services, food, (2014).
24. Part CFR 145 Repair Stations. US Department of Transportation, FAA, Washington, DC.

The Synergetic Effect of Lean Six Sigma and TRIZ on the Improvement of an Electronic Component

Sónia Araújo, João Lopes, Anabela C. Alves, and Helena Navas

Abstract Continuous improvement is a constant concern of companies operating in strongly competitive markets. This means to be constantly unsatisfied with the company status-quo, searching for opportunities to improve. The goal is to increase business efficiency and effectiveness, providing value to the client. In a fast-paced compass industry with complex products, the combination of strategies that bring such value is preferred. To that, the authors proposed to combine synergistically Lean Six Sigma and TRIZ to improve an electronic component of an instrument system in a first-tier automotive industry. This component that had an insignificant cost (1MU maximum) had a high rejection rate provoking complaints and losses that reached a loss of 156 189 MU. This motivated the company to define a project and allocate resources to it. Thus, the project was allocated and developed in the context of an Industrial Engineering and Management master dissertation by the first author of this paper under an action-research strategy. The main results were a decrease of 83% of defects per million opportunities and savings higher than 50 000 MU.

Keywords Lean Production · Lean Six Sigma · TRIZ

S. Araújo · J. Lopes
Department of Production and Systems, Engineering School, University of Minho, 4800-058 Guimarães, Portugal

A. C. Alves
Department of Production and Systems, School of Engineering, ALGORITMI R&D Center, University of Minho, Campus of Azurém, 4800-058 Guimarães, Portugal

H. Navas (✉)
Department of Mechanical and Industrial Engineering, NOVA School of Science and Technology, UNIDEMI R&D Center, Universidade NOVA de Lisboa, 2829-516 Caparica, Portugal
e-mail: hvgn@fct.unl.pt

© The Author(s), under exclusive license to Springer Nature Switzerland AG 2021
A. M. Tavares Thomé et al. (eds.), *Industrial Engineering and Operations Management*,
Springer Proceedings in Mathematics & Statistics 367,
https://doi.org/10.1007/978-3-030-78570-3_31

409

1   Introduction

Lean Production has proved itself as a worthwhile production strategy in many distinct industries by achieving higher levels of performance [1]. Six Sigma, a statistical metric created by Motorola, has been adding cases of success to its curriculum. One of its successes happened in Motorola Corporation, increasing its results from $2.3 billion in 1978 to $8.3 billion in 1988 [2]. TRIZ is known in English as the Theory of Inventive Problem Solving, as known as one of the most effective tools for conceiving engineering designs and solving difficult problems [3]. The importance of TRIZ has been acknowledged in various fields where its applications inspired great results, over the last years [4].

These three different strategies suit a handsome purpose: to get improvements. Over the years, many cases of success of each singular strategy have been shown. The current competitive market does not allow companies to take too long wondering which strategy fits better. Furthermore, the problems and defects are no longer the same as 90's. To overcome this fast-paced and overwhelming phase, companies are invited to seek what master sought but using new strategies that masters did not have [5].

Attending to this, the company of this study, that have been implementing a Lean culture, promoted a project and allocated resources from the quality department to find solutions for a defect in an electronic component. A team was formed and the first author of this paper was integrated in it. Having in mind what needed to be done, and have some knowledge on the mentioned methodologies, she proposed to use it to improve the electronic component, going deeper in finding the root causes and redesign it, if needed. This paper presents partially what was done, reflecting on the synergy of strategies, applying the Define, Measure, Analyse, Improve and Control (DMAIC) phases and the contradiction matrix of TRIZ, to pursuit the Lean Thinking philosophy.

This paper is divided in six sections. The first one introduces the paper itself; secondly, it is presented a brief literature review of the methodologies used; then, the research strategy is explained. It is followed by the fourth section that demonstrates how the case under study was structured, which tools were used and the inputs and outputs of each step; the fifth section represents the results analysis and discussion. Lastly, the conclusions are part of the sixth section.

2   Literature Review

This section presents a brief context of the methodologies used. Firstly, it is mentioned how Lean met Six Sigma in history and, afterwards, how Lean Six Sigma met TRIZ. What they have in common and the successes reached are also part of the description.

2.1 Lean Meets Six Sigma

Lean Six Sigma (LSS) combines elements of Lean Production and Six Sigma into one hybrid strategy. Organizations have been adopting Lean and Six Sigma to improve their performance and competitiveness. Six Sigma rely on a methodology called DMAIC in which each step of the series has various tools. Through these steps, Six Sigma seeks to eliminate defects of critical importance to customers [6, 7]. The term "Six Sigma", or 6 σ, refers to a statistically derived performance target of operating with only 3.4 defects per million opportunities (DPMO) [8, 9].

Lean is an improvement strategy focused to do more with less while providing customers exactly what they want, when they want (just-in-time) [10]. The term "Lean" was originated to describe Toyota Production System in 1988 by Krafcik [11] and later on, in a study performed by Womack in 1990 [12]. Lean Thinking, the philosophy behind Lean Production is nowadays recognized as an organizational culture [13]. Five principles should be followed to transform a company in a lean company: (1) Define value; (2) Map the value stream; (3) Create flow; (4) Establish pull production; (5) Seek perfection.

On 2000s, the first synergy occurred. Lean and Six Sigma merged into what was called Lean Six Sigma. Its aim was to maximize shareholder value, by achieving the fastest rate of improvement in customer satisfaction, cost, quality, process speed and invested capital through the use of the tools and principles of both approaches [14]. Six Sigma improves productivity through variation reduction. Lean improves productivity through process design and the elimination of wasted activities. Both focus on process, but with different perspectives [15]. The implementation of Lean Production in many companies [1] along with the use of Six Sigma methods [16] were highlighted because its countless savings and advantages. For instance, on 1999, Ford began to implement Six Sigma and the results were notable. Since its very beginning, it was possible to recognize how the two strategies could complement each other perfectly. On 2002, the synergy of Lean and Six Sigma enabled Ford to save $675 million [17].

2.2 Lean Six Sigma Meets TRIZ

Most of the applications of LSS used in the improve phase rely on psychological brainstorming tools that make the improvement inadequately. In some specific cases, it might happen that Lean Six Sigma tools are no longer enough, especially in the improve step of DMAIC. It is in these cases that TRIZ could be a strong partner.

TRIZ is a systematic methodology that allows creative problems in any field of knowledge to be revealed and solved, while developing creative (inventive) thinking skills and a creative personality. Often at the root of a problem's solution lies what seems at first glance to be a wild idea. TRIZ gives one the ability not only to be prepared for such ideas, but to create them [18].

TRIZ, the Russian acronym for "Theory of Inventive Problem Solving", is a toolbox of techniques for problem solving. The basis for TRIZ was developed by Genrich Altshuller, a Soviet naval patent clerk, in the 1950s [19].

TRIZ was defined as a methodology for solving problems based on logic, data and research and a toolkit of methods to support systematic creativity. It was created by analysing more than 3 million patents to discover the patterns, which provide innovative solutions to problems. Therefore, it was based on past knowledge and ingenuity of many thousands of engineers, it brought the ability to accelerate the project team to resolve issues creatively based on the structured and algorithmic approach [20].

These 40 inventive principles are one of the tools proposed by TRIZ methodology. Among other techniques and analytical tools of TRIZ the following stand out: Thinking in Time and Space, Eight Trends of Technical Evolution, Contradictions, Forty Principles, Seventy-six Standard Solutions, Resources, Ideality, Functional Analysis, Smart Little People, and Size-Time-Cost [21].

Lean Six Sigma and TRIZ share a purpose: to design and deliver products that customers want. To achieve it, both use principles and tools that aim to reduce waste/contradictions or minimize the use of new resources, and both follow a continuous improvement mind-set [22]. TRIZ focus on individual elements to improve, while Lean focus on the whole system to find potential efficiencies. In this context, TRIZ can be useful to find solutions that use available resources and decrease waste [20].

Modifying a process means knowing the process. Also, means to know how it works, which imply getting feedback on its behavior. Principle 23, Feedback, has this concern. Statistical Process Control (SPC) can help on the measurement of this process. In this context, Six Sigma, or more recently, Lean Six Sigma is also combined with TRIZ methodology to improve quality and service process performance [21]. This synergy came up with a plenty of positive results [22, 23].

3   Research Strategy

The research strategy of this paper was based on Action Research methodology. Action Research (AR) originates primarily in the work of Kurt Lewin and his colleagues and associates. In the mid-1940s, Lewin and his associates conducted AR projects in different social settings [24]. Action research aims to contribute both to the practical concerns of people in an immediate problematic situation and to further the goals of social science simultaneously. Thus, there is a dual commitment in action research to study a system and concurrently to collaborate with members of the system in changing it in what is together regarded as a desirable direction. Accomplishing this twin goal requires the active collaboration of researcher and client, and thus it stresses the importance of co-learning as a primary aspect of the research process [25]. This methodology has five main steps: (1) Diagnosing; (2)

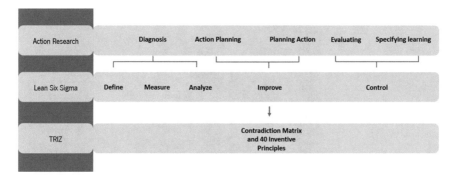

Fig. 1 Integrated use of action research, Lean Six Sigma and TRIZ

Action Planning; (3) Taking Action; (4) Evaluating; (5) Specifying learning [26]. The following case study used the AR, LSS and TRIZ structured as shown in Fig. 1.

4 Description, Critical Analysis and Proposals

This section, briefly, presents the case study, component characterization, critical analysis, and the define, measure and analyze part of DMAIC. It describes the path of the three defect types until the root cause was found for each of them.

4.1 Case Study and Component Characterization

This section refers to a real case study performed in an automotive multinational company. This company is a supplier of automotive industry, has around 3800 collaborators in Braga location. It is focused on mobility solutions and it is well known by creating technology for life. This factory is mostly focused on producing instrument systems and drive infotainment.

This study was raised due to the high rejection rate related to the assembly of an electronic component—foil (see Fig. 2). This component ensures the transmission of date between PCB board and display and allow customers to see an image in the screen.

The research of the defects on foil component followed the structure of Fig. 3.

Fig. 2 Electronic component: foil

Fig. 3 DMAIC applied to the foil incorrectly assembled

4.2 Critical Analysis

The research started around the "foil damaged" topic. Then, it was split into three different branches, according to defect type: Lifted Pads, Broken Pads and Foil Incorrectly Assembled. Once found the root causes of each defect, the research was merged to get improvements, with TRIZ support. The control phase would be the project's closure and would help monitoring the results.

4.3 DMAIC Application

This section presents the define, measure and analyze of each of the three main foils' defect types. It demonstrates the tools and strategies used to achieve the root causes.

Define phase

The define phase was crucial to understand what was the problem—identify problem statement; set the target—process sigma calculator; elaborate priorities—use of Pareto chart; and get to know the process—go to the Gemba. Based on the Fig. 4, Foil Incorrectly Assembled is the defect with more occurrence. Anyhow, the three main defect types were covered in the investigation.

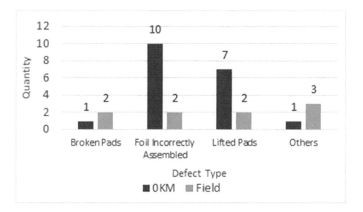

Fig. 4 External claims by defect type

To understand and enable others to do so, a project charter and a turtle diagram were elaborated. For each defect, a timeline with claims were performed and an exploratory data analysis allowed the team to understand three relevant facts:

1. During 2019, the loss generated by foils defects summed up a total of 156 189 monetary unit (MU).
2. The current state is 23,4 defects per million opportunities related to foils; the goal would be to decrease it by 50%.
3. Foils assembly issues affect 40% of the overall production.

In each specific case, some certain tools were needed. In Lifted Pads case, a trend analysis was performed to identify were the defect was occurring. The Broken Pads case led the team to create an IS/IS NOT, in order to get to know why it happened in those specific products but never in others quite similar. Lastly, the Foil Incorrectly Assembled made use of the simplest and, quite often, the most effective tool— discussing directly with collaborators to understand their doubts and knowledge area.

Measure phase

The Measure phase was pointed out as a fundamental step to perform line walks to every production line with foils, with the help of the multidisciplinary team. Afterwards, a plenty of process maps were elaborated and countless factors and variables entered the study through three Ishikawa Diagrams, that were generated by the following general one (see Fig. 5).

Analyze phase

The analyse step was wholly distinct for each defect. Regarding Lifted Pads, an incoming inspection, a process risk analysis and a lab simulation were carried out. From these activities, the team apprehended that the root cause was the use of a wrong assembly angle, scratching on the lower connector pins.

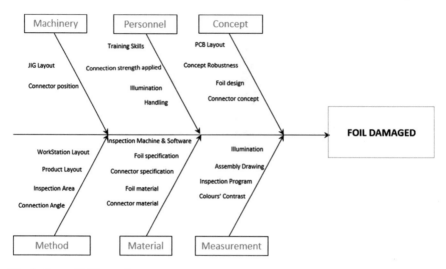

Fig. 5 General Ishikawa diagram

Regarding Broken Pads, a bunch of reliability tests were fulfilled, mostly temperature and vibration tests. With supplier's support, the team realized its root cause: bending during assembly.

Last, but not least, the Foil Incorrectly Assembled carried out a long-detailed research. In this case, around 100 internal rejected foils were analysed carefully-specifications, scratches, any trend among them. Added to that, an attribute agreement analysis was performed with more than 100 collaborators. This activity consists in analysing 30 units (15 with defects and 15 good), in which the one performing the activity knows exactly which defects are there and in which units. The collaborators (appraisers) were asked to classify each unit as good or bad three times, in three different days to mitigate bias. With this, three relevant conclusions were taken: if the collaborator was evaluating units in the correct way; if the collaborator had the same opinion over the same units during the three classifications; and if appraisers from the same group/line detected the same defects. This tool helped understand the root cause: lack of standardization during foils assembly related to lack of detailed assembly instructions.

4.4 Proposals Presentation

This section presents the Improve and Control phases of DMAIC. It includes all the improvements under study, its cost/benefit analysis and how to monitor them in a long-term plan.

Table 1 Improvement description and motivation

Improvement description	Motivation
1. Foils assembly course	Lack of foils assembly standardization; There is no detailed instructions of how to assemble;
2. Replacement of one type of foils	High rejection rate due to fragile material;
3. Connector with 2-point contact	Contaminations claims sum a total of 135 000 UM loss during 2018 and 2019;
4. Inspection with robot	Insufficient current inspection;
5. Automatic assembly	Human margin of error as one of the biggest causes;

Improve phase plus TRIZ

Since this problem related to foils were an old one in this company, various teams had tried to solve it in the past, without success. Knowing this, the team faced the Improve phase with the feeling that it would happen the same as before, hopeless. The company knew that doing the same all over again would not bring new conclusions and eventually the team would get stuck. TRIZ played a crucial role here. When repeating the same process of getting improvements over and over again, creativity is no longer present. TRIZ allowed the team to overcome this.

Through contradiction matrix and the 40 inventive principles, the multidisciplinary team, quickly, reached more than 10 completely new ideas. Through brainstorming sessions, some ideas jumped out, while some others were transformed into more feasible ones. The improvements under study ended up being (see Table 1).

1. Foils Assembly Course: a course oriented to detail with three main phases— diagnoses evaluation (attribute agreement analysis), theoretical and practical training session, final evaluation (attribute agreement analysis). The course was performed with more than 100 collaborators. During the diagnosis's evaluation, 46% of collaborators failed, and after the training session, only 6% were enabled to perform foils assembly station.
2. Replacement of one type of foils: the flexible flat cables (foils FFC) were way less robust than other foil types and most of the internal rejection happened with FFCs. Based on this, the team performed a cost/benefit analysis, in which concluded that this replacement would mean an increase of costs in the proof of concept used. The savings on claims would not pay off the investment. Some intangible variables should be, also, considered such as the company's image and the long-term relation with the client. Also, this cost/benefit analysis must be performed to all products, since each case is different from the other.
3. Introduction of a connector with 2-point contact: During 2019, the contaminations on foils summed up a loss of 135 000 MU. During the Define, Measure and Analyse, this problem was not tackled, since another team was already investigating its root cause. This two-point contact technology would avoid almost 100% of contamination external claims. The cost/benefit analysis indicated that

the replacement of the current connector by this one would generate a saving of around 49 000 UM and a decreasing of 0,2% claims.

4. Inspection with robot: Currently, after the manual assembly, there was always an inspection station. These stations can vary from 18 000 MU to 66 000 MU. Anyhow, collaborators cannot trust this station because of illumination contrasts and reflections leading to wrong conclusions. To avoid it, the team thought of a new inspection station with a collaborative robot, a laser camera and an extracting tool. The suggested station would cost around 54 000 MU. Comparing the station's costs, the replacement itself would already mean a saving in some cases. The cost/benefit result was quite positive. Anyway, to ensure this inspection savings, a measurement system analysis should be performed.

5. Automatic Assembly: The human margin of error was a critical topic that everyone struggled to mitigate. The last improvement suggestion would do it perfectly. The automatic assembly would be carried out with a collaborative robot and a camera/sensor. This would eliminate the higher waste: inspection station; would be robust, adaptable to any product, would decrease dramatically claims, would increase repeatability and reproductivity, and it would be the first plant from this multinational company using such technology. The introduction of the automatic assembly would mean a reduction of 83% of defects per million opportunities, which would be translated into a reduction of 44 000 MU; a reduction of 15,9 s in cycle time; and a reduction of the manual time used by operator of 24 000 MU. It all sums up a saving of 69 000 MU during the lifecycle of a product. After this proof of concept, the plan would be to Yokoten this to all new products in every plant.

Control Phase

In the control phase, specific team members took care of each improvement, assuring its monitoring through control charts and periodical line walks. The plan was to extend improvements if they were getting positive results, and on the opposite side, reverted them in case they are no longer creating the expected impact.

5   Results Analysis and Discussions

The above improvements suggested achieved very positive results. Foils assembly course enabled more 40% of people to perform foils assembly station. Inspection with robot had also a great impact. Anyhow, due to this improvement low maturity, it was not possible to come up with the exact savings portion. The highest impact was caused by automatic assembly, as shown in Table 2.

Table 2 Results analysis

Improvement	DPMO reduction (%)	Savings (MU)
Automatic assembly	83	69 000
Connector with 2-point contact	0,2	49 000
Replacement of one type of foils	50	

6 Conclusions

Companies have been evolving in a fast-paced compass, and along with it, the products are becoming more complex with tiny components and advanced human skills. Taken that, the use of the same methods used in 90s may not be enough. Engineers are expected to be not only problem solvers, but also problem finders, so that the customer would not even have time to carry out the issue. To reach that, the combination of past strategies must be a faster and more effective way to get results. The current trend aims at the joint use of several methodologies and different approaches in order to fill the gaps and enhance the strengths of each one. In the real case study, Lean Six Sigma and TRIZ were used together and the result achieved was a reduction of 83% DPMO, while the goal was to decrease it by 50%.

Due to constraints on the size of the article, the authors focused only in the strategies used, its tools and the synergy among them. This way, a plenty of work remained unrevealed. Even, for the new cycles of the action-research some works must be done in the future such as the implementation of automatic assembly, the elaboration of the measurement system analysis of robot with inspection, and the continuation of foils assembly course, currently, included in a new company position called "training on a job".

Acknowledgements The authors acknowledge the company where this project was developed.
    This work has been supported by FCT—Fundação para a Ciência e Tecnologia within the R&D Units Project Scope: UIDB/00319/2020.

References

1. Amaro, P., Alves, A. C., & Sousa, R. M. (2019). Lean thinking: A transversal and global management philosophy to achieve sustainability benefits. In Lean Engineering for Global Development. Springer Nature Switzerland AG 2019. https://doi.org/10.1007/978-3-030-135 15-7_1.
2. Taghizadegan, S. (2006). Essentials of Lean Six Sigma. Butterworth-Heinemann 2006. https://www.elsevier.com/books/essentials-of-lean-six-sigma/taghizadegan/978-0-12-370502-0.
3. Navas, H. (2013). TRIZ. Guia de Empresas Certificadas, January 2013 (in Portuguese).
4. Sheu, D. D., Chiu, M.-C., & Cayard, D. (2020). The 7 pillars of TRIZ philosophies. Computers & Industrial Engineering, 146, 106572. https://doi.org/10.1016/j.cie.2020.106572.

5. Robinson, C. (2009). Using ASQ's Knowledge Center to answer common questions: Kano on Customers. Journal for Quality and Participation, 32(2).
6. Jiju, A., & Ricardo, B. (2002). Key ingredients for the effective implementation of Six Sigma program. Measuring Business Excellence, 6(4), 20–27. https://doi.org/10.1108/136830402104 51679.
7. Yahia, Z. M. (2011). Six-Sigma: methodology, tools and its future. Assembly Automation, 31(1), 79–88. https://doi.org/10.1108/01445151111104209.
8. Linderman, K., Schroeder, R. G., Zaheer, S., & Choo, A. S. (2003). Six Sigma: a goal-theoretic perspective. Journal of Operations Management, 21(2), 193–203. https://doi.org/10.1016/S0272-6963(02)00087-6.
9. Pande, Neuman, & Cavanagh. (2007). The Six Sigma Way. Gabler. https://doi.org/10.1007/978-3-8349-9320-5_24.
10. Womack, J. P., Jones, D. T., & Roos, D. (1990). The machine that changed the world.
11. Krafcik. (1988). Triumph of the Lean Production System.
12. Daniel, J., & Micaela, M. (2009). The performance effect of HRM and TQM: a study in Spanish organizations. International Journal of Operations & Production Management, 29(12), 1266–1289. https://doi.org/10.1108/01443570911005992.
13. Amaro, P., Sousa, R., & Alves, A. (2020). Lean Thinking: From the Shop Floor to an Organizational Culture. IFIP WG 5.7 International Conference, APMS 2020, Novi Sad, Serbia, August 30 – September 3, 2020, Proceedings, Part II, 592(Agosto), 406–414. https://doi.org/10.1007/978-3-030-57997-5.
14. Bonome, L., Costa, M., Godinho, M., Fredendall, L. D., & Miller, G. (2021). Lean six sigma in the food industry: Construct development and measurement validation. International Journal of Production Economics, 231(November 2019). https://doi.org/10.1016/j.ijpe.2020.107843.
15. Gleeson, F., Coughlan, P., Goodman, L., Newell, A., & Hargaden, V. (2019). Improving manufacturing productivity by combining cognitive engineering and lean-six sigma methods. Procedia CIRP, 81, 641–646. https://doi.org/10.1016/j.procir.2019.03.169.
16. Hilton, R. J., & Sohal, A. (2012). A conceptual model for the successful deployment of Lean Six Sigma. International Journal of Quality and Reliability Management, 29(1), 54–70. https://doi.org/10.1108/02656711211190873
17. Harvin, H. (2019). Six Sigma Implementation & its Benefits in Ford Motor Company.
18. Wang, F. K., & Chen, K. S. (2010). Applying Lean Six Sigma and TRIZ methodology in banking services. Total Quality Management and Business Excellence, 21(3), 301–315. https://doi.org/10.1080/14783360903553248.
19. Bligh, A. (2006). The Overlap Between TRIZ and Lean. Manufacturing Systems, 1–10.
20. Navas, H. & Cruz Machado, V. (2013). Systematic innovation in a lean management environment. IIE Annual Conference and Expo 2013, May, 2138–2147. https://doi.org/10.13140/RG.2.1.2046.4083.
21. Maia, L. C., Alves, A. C., & Leão, C. P. (2015). How could the TRIZ tool help continuous improvement efforts of the companies? Procedia Engineering, 131, 343–351. https://doi.org/10.1016/j.proeng.2015.12.412.
22. Alves, A. C., Leão, C. P., Maia, L. C., & Navas, H. V. G. (2016). Understanding if and how TRIZ is used in the Portuguese reality. International Conference on Systematic Innovation, 1990.
23. Boangmanalu, E., Abigail, T., Sembiring, A., & Tampubolon, J. (2020). Minimizing damage of product using six sigma and triz methods. IOP Conference Series: Materials Science and Engineering, 801(1). https://doi.org/10.1088/1757-899X/801/1/012101.
24. Coughlan, P., & Coghlan, D. (2002). Action research for operations management. International Journal of Operations and Production Management, 22(2), 220–240. https://doi.org/10.1108/01443570210417515
25. O'Brien, R. (1998). An overview of the methodological approach of action Research. University of Toronto, 1–15. http://www.web.ca/~robrien/papers/arfinal.html.
26. Susman, G. (1983). Action Research: A Sociotechnical System Perspective.

The Coordination of Packaging Recycling Chain in Brazil

Ticiana Braga De Vincenzi⬭, Roberta Souza Piao⬭,
Diego Vazquez-Brust⬭, and Marly M. de Carvalho⬭

Abstract This paper analyses the coordination of the packaging recycling chain established by Company A. Company A developed a technical solution for coordinating the packaging recycling chain and provides a certificate of recycling for its clients (industry). The paper also highlights the initiatives raised by the private sector for addressing the problem of waste management and social inclusion. To understand and allow a broad view of the phenomenon, data was collected from Company A and members of the packaging recycling chain, as waste picker cooperatives, through a mix of desk research with qualitative methods, including interviews, observation research, and participatory workshop. The content of the transcripts and desk research was analysed based on themes identified in the literature and real terms used by the interviewees and by the lecturers. Theoretically, it contributes to the literature of waste management shedding light on the coordination of activities by a private company. The participation and sharing of responsibilities with local institutions are essential to increase the efficiency of waste management. As practical implications, the outcomes indicate the importance of links with waste picker cooperatives for increasing the formalization in the recycling packaging chain and promoting social inclusion.

Keywords Recycling chain · Circular economy · Social inclusion

1 Introduction

Urban waste management is a big challenge for society, considering the increasing urbanization and industrialization [1]. According to the Living Planet Report [2], the world is using 50% more resources than the Earth can provide. The worlds' cities

T. B. De Vincenzi (✉) · R. S. Piao · M. M. de Carvalho
Universidade de São Paulo, Av. Pr. Luciano Gualberto, 1380, SP 05508-010, Brazil
e-mail: ticivin@usp.br

D. Vazquez-Brust
University of Portsmouth, Portsmouth PO1 3DE, UK

© The Author(s), under exclusive license to Springer Nature Switzerland AG 2021
A. M. Tavares Thomé et al. (eds.), *Industrial Engineering and Operations Management*,
Springer Proceedings in Mathematics & Statistics 367,
https://doi.org/10.1007/978-3-030-78570-3_32

generated 2.01 billion tons of solid waste in 2016 and it is expected a 70% increase by 2050. Developing countries are more affected, especially considering the urban poor population. This population is deeply affected by unsustainably managed waste, which contributes to flooding, air pollution, and public health impacts [3]. In 2018, Brazil generated 79 million tons of waste, and only 3% of dry waste was recycled [4].

In Brazil, the management of urban solid waste is regulated by the Federal Law n. 12,305/2010, which instituted the National Policy on Solid Waste (PNRS). The Law determines an integrated and shared management of solid waste among different actors as industry, importers, distributors, retailers, and society. The main objective of PNRS is to reduce the amount of residue destined to landfills and dumps. One important point addressed by PNRS is the selective collection regarding packaging. In 2015, representatives of the private sector and the Ministry of the Environment signed the Sectorial Packaging Agreement that sets the goal to reduce at least 22% of the packaging of solid urban waste disposed of in landfills [5]. In 2018, the Board of CETESB (Environmental Company of the State of São Paulo) authorized recycling certificates pegged to invoices to demonstrate compliance with the goals of the Sectorial Packaging Agreement [6].

Taking this, we propose to analyse the coordination of the packaging recycling chain established by Company A (Comp A). Comp A developed a technical solution for coordinating packaging recycling and provides a certificate of recycling for its clients (industry).

As far as we are concerned, most studies are dedicated to analysing public actions regarding waste management [7–9]. In this paper, we highlight the initiatives raised by the private sector for addressing the problem of waste management. In addition to this, there is a lack of studies concerning social effects [10–12].

After this introduction, a brief background is provided about waste management, followed by a concise description of the regulation in Brazil. Then we describe the methodological approach employed in this research. Follow the description of results and analyses. Finally, some conclusions are provided.

2   Theoretical Background and Regulation in Brazil

Previous studies indicate that the participation of local authorities in the implementation of waste management is essential for reaching better results [13]. In the case of packaging waste management in the Netherlands, the main changes are conducted by industries and are related to resource efficiency (reduction of resources such as eliminating the excess of packaging), energetic efficiency, and increasing product shelf life [14]. According to [10], the main challenges in solid waste management are (a) knowledge creation, awareness, and participation of society "to segregate waste at source, door-to-door collection and disposal in appropriate collecting bin" [10, p.766], (b) infrastructure and urbanization; (c) legal policies and institutions; (d) financial sustainability; (e) citizen engagement; (f) social inclusion; (g) health

Table 1 Main environmental and social issues regarding waste management

Environmental	Sources
Environmental labelling	PNRS
Efficiency of resources	[14]
Energetic efficiency	[14]
Health and safety	[10]
Reduction of GHG emissions	[3]
Social	Sources
Participation and coordination from local authorities	PNRS, [13]
Integrated management responsibility	PNRS, [8]
Citizen engagement	[10]
Infrastructure and urbanization	[10]
Financial sustainability	[10]
Social inclusion	[10–12]

and safety. The author also pointed out the participation of the private sector in waste management, contributing to the regularization of waste pickers' initiatives, introducing technical innovations, and financing capital investments.

Regarding social issues, previous studies indicate that disadvantaged groups (as informal waste pickers) play an essential role in the recycling chain in developing countries [11, 12]. From an environmental perspective, studies show a strong link between urban solid waste generation and GHG emissions [3]. Table 1 summarizes the studies.

In 2010, the Brazilian government instituted the National Policy on Solid Waste (PNRS). PNRS has two fundamental aspects: (i) the establishment of shared responsibility for product life cycle, encompassing manufacturers, importers, distributors, traders, consumers, and solid urban waste cleaning and management public services; and (ii) the conditioning of incentives, financing and fund transfers from the Federal Government to states and municipalities provided that they elaborate solid waste plans. The shared responsibility takes place via a sectorial agreement signed between public authorities and the private sector. Besides, the mandate prohibits in natura solid waste from being released in the open, into the sea or other water bodies, or burned in an unlicensed manner. Among the objectives of PNRS, the main highlights are the integration of waste collectors, integrated management and sustainability, and environmental labelling [8].

In November 2015, representatives of the private sector and the Ministry of the Environment signed the Sectorial Packaging Agreement that sets the goal to reduce by 2018, through reverse logistics, at least 22% of the packaging of solid urban waste disposed of in landfills [5]. The measurement of the goal considers all packaging-involving products, including transport packaging (primary, secondary, and tertiary). In April 2018, the Board of CETESB stated that to obtain the issuance or renewal of the environmental licensing, the private sector must demonstrate compliance with the goals of the Sectorial Packaging Agreement, and this proof can be done through invoices or equivalent document [6].

In May 2018, the Term of Commitment for Reverse Packaging Logistics (TCRL) was signed between FIESP (Federation of Industries of the State of São Paulo), CIESP (Centre of Industries of the State of São Paulo), CETESB, Secretariat of the Environment of the State of São Paulo, unions and associations representing the private sector, including Comp A among the signatories [15]. The Term of Commitment authorized recycling certificates pegged to invoices and defined that a Management Council will carry out the system governance. This Council is formed by representatives of companies, adhering operators, and signatory entities [15], and must establish the responsibilities of each participant.

The first resolution of the Management Council deals with the qualification and the minimum requirements of the certifier and the packaging recycling certificates. The certifier must have the technology for automatic capture and validation of invoices with external systems; anti-fraud system, with minimum sampling; security plan to guarantee the confidentiality and uninterrupted system operation; results and processes endorsed by an independent auditor, among others. The first resolution also defines the rules and origins of invoices that can generate PRC (Packaging Recycling Certificates), as well as the types of packaging materials allowed for issuing PRC.

3 The Methods of Analysis

This is qualitative research whose purpose is to explore and analyse [16] the coordination of the packaging recycling chain established by a private company. To understand and allow a broad view of the phenomenon, data was collected through a mix of desk research with qualitative methods, including interviews, observation research, and participatory workshop with a private company (Comp A) representatives and members of the packaging recycling chain, as waste picker cooperatives from the States of São Paulo (SP), Rio de Janeiro (RJ) and Paraná (PR). More than one researcher was involved in the process of collecting and analysing data as suggested by [17] to reduce subjectivity that could interfere with the results.

Desk research comprised of sector and Comp A annual financial reports, webinars promoted by Comp A, news, internal bulletins, websites, and other studies about the research topic [17]. The content of the transcripts and desk research was analysed based on themes identified in the literature and real terms used by the interviewees and by the lecturers, allowing a comparison between the data and shedding light on the possible patterns that could emerge from the findings [18]. Table 2 presents the data collection sources.

Table 2 Data collection

Primary data—interviews			
Company A	Codes	Position	Date of the interview
Interviewed A	Inter_A	Head of Projects & Business Development	Jun/2019
Interviewed B	Inter_B	Operations Manager	Jun/2019 e Jul/2020
Recycling Cooperatives	Codes	Position	Date of the interview
Cooperative A (RJ)	Coop_A	President	Aug/2020
Cooperative B (PR)	Coop_B	Finance Director	Jul/2020
Cooperative C (SP)	Coop_C	Board Director	Jun/2020
Cooperative D (SP)	Coop_D	President	Jul/2020
Primary Data—participatory workshop			
Workshop	Inter_B	Operations Manager at Comp A	Jul/2020
	Coop_B	Finance Director at Cooperative B	
	Coop_C	Board Director at Cooperative C	
Observation research			
Event		Promoted by	Date of the event
Recycling certificate auction 1		FIESP	Feb/2020
Recycling certificate auction 2		FIESP	Jun/2020
Secondary data			
Event	Codes	Position	Date of the event
Results presentation	Partic_A	New Business Director at Comp A	Dec/2019
	Partic_B	CEO at Comp A	
Webinar 1	Partic_A	New Business Director at Comp A	May/2020
	Partic_C	General Packaging Reverse Logistics System Coordinator	
	Partic_D	Environmental Engineer at the State General Accounting Office	
Webinar 2	Partic_A	New Business Director at Comp A	May/2020
	Partic_E	Sustainability Manager at Partner Company E	
Webinar 3	Partic_F	Customer Experience Leader at Comp A	Jun/2020

(continued)

Table 2 (continued)

Observation research			
Event	Codes	Position	Date of the event
	Partic_G	CEO at Partner Company G	
	Partic_H	CEO at Partner Company H	
International forum	Partic_A	New Business Director at Comp A	Jul/2020
	Partic_B	CEO at Comp A	
	Partic_I	Sales & General Director at Company I—Germany	
	Partic_J	European Director at Company J—France	
	Partic_K	Innovation Manager at Company J—France	
	Partic_L	General Director at Comp A—Chile	

4   Description of Results

4.1   Company A

Comp A was founded in 2014. The firm initial capital came from a US$ 70 thousand award from the Kellogg School of Management, in Chicago (USA). In 2016, Comp A developed a technical solution that allows the tracking of solid waste. It also started offering a "recycling seal" so that adherent companies can communicate compliance with the current legislation on packaging recycling to their consumers. With the development of this technical solution, Comp A started to work as a certifier, providing recycling certificates to the market (Inter_A). Comp A business is based on the environmental compensation model (by material equivalence). The first customers were small companies that were concerned with sustainability and therefore hired the Comp A solution to prove the reverse logistics of packages discarded by consumers (Inter_B).

According to Inter_B, for Comp A's business, it was important the moment that CETESB linked the proof of reverse logistics to the renewal and issuance of environmental licenses in São Paulo and recognized the recycling certificate as proof. When this type of regulation is established, associations (such as FIESP and CIESP) and unions analyse and direct how companies and the sector should act, what the possibilities are and what type of solution they should use (Inter_A).

FIESP opened a public notice seeking a solution to packaging recycling. Comp A applied, and it was the beginning of Comp A's partnership with FIESP. During one year, FIESP and Comp A developed a project defining system rules and governance (Inter_B). Comp A developed a specific software technical solution for this (Inter_A).

According to Inter_A, the partnership with FIESP and CIESP is important for Comp A, as a solution provider, and for FIESP, since through the technical solution developed by Comp A is possible to integrate firms and associations. FIESP does not receive any value. It only works as a facilitator, approving a reliable solution for companies that need to prove reverse logistics.

According to the firm´s report, Comp A has become a certifier, and today the solution is available to members of more than 50 associations and unions that have signed the TCRL. Inter_B mentioned that all associations have voting power and there is a rotation so that the participating institutions can build the system together. All minutes and resolutions are published, including the list of cooperatives, consenting parties, and signatories.

From the field research, law and enforcement are essential for the model's existence. The sectoral agreement is national, but the inspection takes place depending on the action of the Public Ministry or the State environmental agency (Inter_A). Comp A managed to develop a quantifiable solution to the supervisory agency, in which companies can show that they have reached the established target of 22%. Before that, some large companies implemented reverse logistics processes involving cooperatives, but, as invoices did not back them, they could not prove it. Comp A's technical solution enabled this traceability (Inter_B). Currently, Comp A has 40 employees and operates throughout Brazil, with around 1500 adherent companies (Partic_B).

The pillar of the recycling certificate is environmental compensation, which encouraged the development of the recycling chain (Partic_B). The chain steps and relationships details are: (i) The filler/brand owner receives the packaging from the packaging manufacturer and fills the product; (ii) The customer consumes the product and discards the packaging after consumption; (iii) The discarded packaging is collected and taken to a sorting operator; (iv) The packaging goes through a sorting operator (cooperative or private), which separates it into different types of materials (paper, plastic, metal, and glass); (v) The different materials go through recyclers, which make physical or chemical transformations in the material, transforming it again into raw material; (vi) The transformed raw material is sold to packaging manufacturers, who produce new packaging from recycled raw material and sell it to the fillers/brand owners; (vii) Comp A links the parties, tracking the invoices and remunerating operators for the service provided.

Some materials do not close the cycle because other chains use them. However, the goal of the solution is to close the loop. This chain was already established before the PNRS, with actors (iii), (iv), and (v) providing the correct disposal service for some materials just for their value. With PNRS, actors (i), (ii), and (vi) became legally responsible for the correct destination of packaging. The law thus created a demand. On the other hand, actors (iii), (iv), and (v) used to perform reverse logistics without being paid for it. Therefore, the supply of reverse logistics services could meet the demand created by the legislation (Partic_B).

Comp A (actor vii) connected these two parts through a technical solution, which allows the chain to be traced via invoices and the remuneration of operators (cooperatives and sorting centres) for the reverse logistics service. With this solution, instead of the filler/brand owner deploying its own reverse logistics structure, it may outsource this service to a chain that already exists. Thus, the filler/brand owner subsidizes the recycling chain, so that it can increase its capacity, allowing the filler/brand owner to achieve the goals determined by law (Partic_A).

Recycling credit can be considered as a financial instrument by which the filler/brand owner internalizes the environmental cost (externality) that it causes in the chain. There are credit buyers ("polluters") and credit sellers, who provide the "cleaning" service and correct destination for buyers. This type of market has existed in Europe for many years, and this was the source of inspiration for the solution developed by Comp A (Partic_B).

Comp A acts as a link between producing companies, players in the recycling chain, and government institutions. The environmental compensation applied to reverse packaging logistics consists of disposing of, in an environmentally correct way, a mass of waste equivalent to the mass of packaging that a company places on the market (Partic_A).

Comp A's technical solution follows the entire chain, tracking the path of invoices at each stage, and certifying the circular economy. There are currently four ways to collect and sort packaging: (i) at voluntary delivery points; (ii) through its own reverse logistics, when the company takes its packaging from the market (e.g. returnable glass); (iii) through cooperatives using waste pickers or using city collection; and (iv) by private companies that do selective collection or the normal collection and have adequate machines that separate organic and recycled material (Inter_A).

Technical knowledge is another fundamental issue of the Comp A model, since a good part of the chain is informal, and the challenge is to include this part in the system. The technology allows information crossing to prove that the sorted material did not end up in a dump. The system uses blockchain to add security and transparency to the process and follows the invoices until they arrive at the receiver, allowing the backing tracking (Inter_A).

The system stores and processes the files, which undergo integrity checks and statistical analysis with machine learning to detect fraud. After validation, the flows of each type of material are quantified by year and region in blockchains and become marketable certificates (Partic_B).

Also, Comp A points out that the technical solution performs an anti-fraud check with the Internal Revenue Service and analyses price statistics to validate the notes. Comp A only approves a note, and only issues a certificate, if the operator proves that he sold that volume for final recycling or the processing industry, ensuring the correct destination and the reinsertion of the material in the production chain (Inter_B). The system controls that the note is only used once (Inter_A). The technology has enabled operations and certificate auctions to continue even during the Covid-19 pandemic (Inter_B).

To link operators and customers, Comp A carries out an audit process at the recycling operators' facilities. Through the size of the webs, the degree of mechanization, the number of members, and other several checks, it is possible to calculate the cooperative productivity in tons per month and monitor the activities (Inter_A). If there is any quantity variation, Comp A seeks to know what happened (Inter_B).

Comp A checks other aspects, as following: the post-consumption origin of the recyclable material; documentation, operational and labour conditions; personal protective equipment use; and compliance with the prohibition of slave and child labour. Comp A also checks the destination of recyclable materials through the invoices that operators make available in the system, which must be mandatory for processing industries or final recyclers. An external audit validates the Comp A processes.

As soon as the recycling operators sell the packages, they report the invoices, which are available in the system. As Comp A sells these notes (which are the basis of the certificates) to companies that need to reach the goal determined by legislation, operators are paid. There is an exchange between demand and supply for certificates (Inter_A).

As Comp A customers' demands arise regarding the declared mass of their packaging, upon confirmation of payment, the system selects the most appropriate certificates to carry out the compensation. Each customer receives a certificate at the end of the process listing the operators, the invoices, and the mass involved in the reverse packaging logistics chain (Partic_A).

The industry pays an amount for the notes to Comp A; on the other hand, Comp A keeps a percentage (remuneration for the service) and passes on the rest of the amount to cooperatives and other organizations that sort and sell recycled material. The invoice generates the recycling certificate that the company uses for proof with the regulatory agency.

This model generates incentives and extra income for cooperatives, which becomes more important the less the material is worth. This is because invoices are sold by type of material. Aluminium, for example, is a highly valued material. Despite generating extra income, the sale price of the aluminium invoice is low compared to the sale price of the material itself. On the other hand, in the case of other less valued materials, such as glass, the sale of the invoice practically doubles the revenue from the sale of this material. This means that the cooperatives are encouraged to recycle other materials, such as glass and paper, and not just aluminium and PET bottles, which are traditionally the most valued (Inter_A).

From July 2018 to June 2019, Comp A certified a total mass of 93 thousand tons of recyclable material, of which 5.5 thousand tons of metal, 6.6 thousand tons of glass, 11.6 thousand tons of plastic, 14.1 thousand tons of paper, and 55.1 thousand tons of returnable glass. Of these 93 thousand tons, 37.9 thousand tons were compensated using material equivalence. Therefore, plastic packaging is compensated by recycling plastic packaging, and so on. In this period, just over 68 thousand invoices were received, benefiting 49 sorting operators (of which 27 are waste picker cooperatives) (Partic_B).

The interviewees observe that the model established by Comp A generates social inclusion, as the cooperative members decide what to do with the extra income (Inter_A). Besides, the model encourages formalization. Many cooperatives started to issue invoices when realized the possibility of extra earnings (Inter_B).

Cooperatives have a pioneering and essential role in the Brazilian recycling market. So far, they are responsible for the social reintegration of many people who live in a peripheral setting and who informally collect waste, often without remuneration and in living conditions with little or no quality. In Brazil, 17 thousand employees work for operators approved by Comp A (Partic_B).

To measure the financial impact of the Recycling Certificates on the operators' businesses, Comp A compared the revenues from the certificates and the sale of the wastes. The certificates represent at least a 14.79% increase in revenue from these operations, as following: plastic: 16.23%; paper: 7.90%; metal: 10.36%; disposable glass: 30.74% (Partic_B). A survey to identify the destination of the remuneration received by partner operators identified that investments were essentially in machinery and trucks to improve the recyclables collection and sorting capacity; remuneration distribution to employees and members to increase income; labour charges payment; and investment in protective equipment (Partic_A).

During the interviews, some recycling cooperatives mentioned that Comp A is a necessary intermediary in the value chain because many companies do not want to deal with cooperatives (Coop_A; Coop_D). On the other hand, the cooperatives consider that they provide an important environmental service and criticize the low amounts paid by the invoices (Coop_A; Coop_B). Another issue stressed by cooperatives is the difficulty of formally organizing themselves to issue invoices and participate in auctions (Coop_A; Coop_C).

4.2 Analysis of the Results

The analysis of results was based on the main environmental and social issues raised by the literature. The results indicated that Comp A is linked with local institutions for implementing the coordination process; institutions as FIESP and CIESP—which elaborated with Comp A system rules and governance for the development of a specific technical solution. From the institutions' perspective, the solution allows integration between companies and associations to provide a tool for businesses to prove packaging recycling. From Comp A's perspective, it was possible to develop its solution considering all the important information provided by the main local institutions' stakeholders. Therefore, the technical solution provided by Comp A integrates distinct social actors, from companies to institutions and cooperatives of waste pickers (integrated management responsibility) with the same objective—increase the efficiency of waste management, which is aligned with PNRS objectives and [8] results.

Comp A also provides a recycling certificate to industries creating environmental labelling. The certification allows companies to prove recycling rates demanded by regulation, and support the formalization of waste picker cooperatives. This could lead to better health and safety conditions of disadvantaged groups, as waste pickers as highlighted by [10] as one of the main challenges of waste management. So, environmental labelling is related closely to integrated management responsibility in this case.

Regarding financial sustainability, we highlight that recycling certificates provide extra income to cooperatives, called recycling credit. Therefore, cooperative usually

Table 3 Main environmental and social issues regarding waste management and actions identified through research

Environmental	Actions identified
Environmental labelling	√
Efficiency of resources	
Energetic efficiency	
Health and safety	√
Reduction of GHG emissions	
Social	Actions identified
Participation and coordination from local authorities	√
Integrated management responsibility	√
Citizen engagement	√
Infrastructure and urbanization	
Financial sustainability	√
Social inclusion	√

invests in machinery and trucks to improve the process. The extra income is also an incentive for cooperatives that are not on the system to become formalized and promote social inclusion. This is a huge issue in Brazil. According to interviews, Comp A provided support to cooperatives paperwork regarding formalization. Financial sustainability is also highlighted in the literature [10].

Regarding end consumers, Comp A has a strong action through social media aiming to increase the awareness of end consumers about recycling. Comp A elaborates easy recommendations to end-users about the best recycling practices.

In the Netherlands, [14] identified actions from industries for increasing the efficiency of resources and energy efficiency regarding packaging. In this study, these types of actions were not identified. Table 3 presents the main issues raised by the literature and actions identified in the field research.

5 Final Considerations

The objective of this paper was to analyse the coordination of the packaging recycling chain established by Comp A. Comp A developed a technical solution for coordinating the packaging recycling and provides a certificate of recycling for its clients (industries). To address the objective, it was conducted desk and observation research, participatory workshop, and interviews with Comp A representatives and members of the packaging recycling chain, as waste pickers cooperatives.

The outcomes have practical and theoretical implications. Theoretically, it contributes to the literature of waste management shedding light on the coordination of activities by a private company. In these matters, the participation and sharing of responsibilities with local institutions are essential to increase the efficiency of waste management. As practical implications, the outcomes indicate the importance of links with waste picker cooperatives for increasing the formalization in the recycling packaging chain and promoting social inclusion. It was also pointed out the importance of the industry to work on the design of packaging for increasing resource and energy efficiency. As future studies, it is suggested to identify the main actions of Comp A clients regarding packaging and how they are addressing the challenges of waste management.

Acknowledgements This research was supported by São Paulo Research Foundation (FAPESP), process number 2018/03191-5, and process number 2019/21292-6 and by the University of Portsmouth (UK) QR-GCFR Social Inclusion and Circular economy Project.

References

1. Viva, L., Ciulli, F., Kolk, A., Rothenberg, G.: Designing Circular Waste Management Strategies: The Case of Organic Waste in Amsterdam. Advanced Sustainable Systems 4(9), 2000023 (2020). https://doi.org/10.1002/adsu.202000023.
2. WWF: Living Planet Report - 2018: Aiming Higher. Grooten, M. and Almond, R. E. A. (Eds). WWF, Gland, Switzerland (2018).
3. Hoornweg, D. A., Bhada-Tata, P.: What a waste? A global review of solid waste management. The World Bank 68135, 1–116 (2012). Available at: https://openknowledge.worldbank.org/handle/10986/17388.
4. ABRELPE - Brazilian Association of Public Cleansing and Waste Management Companies. Panorama dos resíduos sólidos no Brasil (2019), https://abrelpe.org.br/download-panorama-2018-2019/, last accessed 2020/07/08.
5. SINIR - National Solid Waste Management Information System. Sectoral agreement to implement the reverse packaging logistics system (2015), https://sinir.gov.br/images/sinir/Embala gens%20em%20Geral/Acordo_embalagens.pdf, last accessed 2020/05/26.
6. CETESB - Companhia Ambiental do Estado de São Paulo. Decisão de diretoria n° 076/2018/C (2018), https://cetesb.sp.gov.br/wp-content/uploads/2018/04/DD-076-2018-C.pdf, last accessed 2020/04/03.
7. Rubio, S., Ramos, T. R. P., Leitão, M. M. R., Barbosa-Povoa, A. P.: Effectiveness of extended producer responsibility policies implementation: The case of Portuguese and Spanish packaging waste systems. Journal of Cleaner Production 210, 217–230 (2019). https://doi.org/10.1016/j.jclepro.2018.10.299.
8. Maiello, A., de Paiva Britto, A. L. N., Valle, T. F.: Implementation of the Brazilian national policy for waste management. Revista de Administração Pública 52(1), 24–51 (2018). http://doi.org/https://doi.org/10.1590/0034-7612155117.
9. Silva Filho, J. C. L., Küchler, J., Nascimento, L. F., de Abreu, M. C.: Gestão ambiental regional: usando o IAD Framework de Elinor Ostrom na "análise política" da gestão ambiental da Região Metropolitana de Porto Alegre. Organizações & Sociedade 16(51), 609–627 (2009).
10. Rajesh, P.: Solid waste management-sustainability towards a better future, role of CSR – a review. Social Responsibility Journal 15(6), 762–771 (2019). https://doi.org/10.1108/SRJ-11-2018-0286.
11. Sehnem, S., Vazquez-Brust, D., Pereira, S. C. F., Campos, L. M.: Circular economy: Benefits, impacts and overlapping. Supply Chain Management 24(6), 784–804 (2019). https://doi.org/10.1108/SCM-06-2018-0213.
12. Siman, R. R., Yamane, L. H., de Lima Baldam, R., Tackla, J. P., de Assis Lessa, S. F., de Britto, P. M.: Governance tools: Improving the circular economy through the promotion of

the economic sustainability of waste picker organizations. Waste Management 105, 148–169 (2020). https://doi.org/10.1016/j.wasman.2020.01.040.

13. Demajorovic, J., Massote, B.: Sectoral agreement on packaging: Assessment based on extended producer responsibility. Revista de Administração de Empresas 57(5), 470–482 (2017). http://doi.org/https://doi.org/10.1590/s0034-759020170505.

14. Van Sluisveld, M., Worrell, E.: The paradox of packaging optimization: A characterization of packaging source reduction in the Netherlands. Resources, Conservation and Recycling 73, 133–142 (2013). https://doi.org/10.1016/j.resconrec.2013.01.01.

15. CETESB - Companhia Ambiental do Estado de São Paulo. Termo de compromisso para a logística reversa de embalagens em geral (2018), https://cetesb.sp.gov.br/logisticareversa/wp-content/uploads/sites/27/2018/06/Termo-de-Compromisso-Embalagens-em-Geral.pdf, last accessed 2020/04/03.

16. Richardson, R. J., de Souza Peres, J. A., Wanderley, J. C. V., Correia, L. M., de Melo Peres, M. H. Social research: methods and techniques. Atlas, São Paulo (1999).

17. Denzin, N. K.: The research act: A theoretical introduction to sociological methods. McGraw-Hill, New York (1978).

18. Bauer, M. W., Gaskell, G. (Eds.): Qualitative researching with text, image and sound: A practical handbook for social research. Sage, London (2000).

Enabling the Use of Shop Floor Information for Multi-criteria Decision Making in Maintenance Prediction

Rolando J. Kurscheidt Netto, Eduardo de F. R. Loures, and Eduardo A. P. dos Santos

Abstract In the era of industry 4.0, the focus of data analysis is on predicting future possibilities and events using information obtained from current system and/or asset conditions to apply early actions to solve problems before they occur. With the development of the two main bases of I4.0, Cyber-Physical Systems (CPS) and Internet of Things (IoT), more assets in the industry are equipped with sensors and information systems. In the context of asset maintenance scheduling, the decision of the optimal time to stop and perform the repair actions results in increased availability and reduced costs. To support the maintenance scheduling prediction, we propose a framework for extracting information collected by FIS systems using Process Mining algorithms and aggregating them in a Multicriteria Decision Model to estimate the optimal period of maintenance execution. The aim is to schedule the maintenance action on a production line within a time window, estimated based on information extracted from factory floor event records, subject to conflicting criteria.

Keywords Delay time modelling · Condition-based monitoring · Process mining · MCDM · Maintenance forecasting

1 Introduction

In the era of industry 4.0, the focus of data analysis is on predicting future possibilities and events, using the information obtained to understand the current conditions of systems and/or assets to apply early actions to solve problems before they occur [1, 2].

In the context of industrial maintenance, it perceives the value of the data collected to reveal important information that can improve production processes [3]. With the development of the two main bases of I4.0, Cyber-Physical Systems (CPS) and Internet of Things (IoT), more production machines in the industry are equipped with various sensors and information systems. Thus, Maintenance Management seeks to

R. J. Kurscheidt Netto (✉) · E. de F. R. Loures · E. A. P. dos Santos
Pontifical Catholic University of Parana, Curitiba, Paraná, Brazil
e-mail: rolando.k@pucpr.edu.br

© The Author(s), under exclusive license to Springer Nature Switzerland AG 2021
A. M. Tavares Thomé et al. (eds.), *Industrial Engineering and Operations Management*,
Springer Proceedings in Mathematics & Statistics 367,
https://doi.org/10.1007/978-3-030-78570-3_33

identify the most efficient and effective strategies in order to continuously improve operational capacity, reduce maintenance costs and increase its competitiveness [4]. According [5], the cost of maintenance activities in production varies between 15 and 40%, being the second highest cost of the operating budget, behind only the cost of energy. In this way, the correct planning and execution of maintenance actions implies significant gains. Two general objectives of maintenance management are to estimate the evolution of equipment conditions over time [6] and/or to predict or diagnose faults and their relationship with future failure modes [7].

According [8], Condition Based Maintenance (CBM) is the main maintenance policy researched, corresponding to 40% of published articles, with a prominent use of predictive and probabilistic models. In contrast, most companies still focus on Time Based Maintenance (TBM) [5, 9], making equipment condition monitoring a demand to optimize maintenance and increase availability, facilitated by the increasing availability of data provided by the adoption of Industry 4.0 bases [3]. Based on the CBM strategy, we can adopt preventive maintenance if the level of deterioration exceeds an M limit, which is a decision point, or we can adopt an age-based maintenance, from the perspective of TBM, where the action of maintenance is performed at a time T relative to the age of the asset [9].

In this context, decisions involve probabilistic estimates of the state of a system, based on uncertain information, available from measurements or other observations of the components of that system [10]. In addition, considering a temporal dimension, component degradation, evolution of symptoms corresponding to deterioration mechanisms, the impact of preventive maintenance, actions on degradation and the influence of conditions and the effects of operating conditions on the evolution of components state should be considered [7]. Thus, maintenance optimization decisions are complex problems, needing to meet multiple and conflicting criteria [11].

To support managers in the context of maintenance planning, being a strategic decision linked to the highest level of the hierarchical organizational structure, where the consequences are characterized by multiple and less tangible objectives, the Multicriteria Decision Making (MCDM) methods are applied [12–14]. MCDM methods have been used in the context of planning and scheduling preventive maintenance actions, evaluating the performance of two or more conflicting criteria in relation to a set of time windows to determine the optimum moment for the asset repair action [11, 15]. For these evaluations, each criterion is mathematically modeled, and its parameters estimated based on information from experts, equipment manuals, maintenance histories, among others.

To deal with the interaction between plant performance and inspection interventions, the concept of Delay Time Modeling (DTM) has been emphasizing the optimization of inspection intervals due to its explicit modeling between plant failures and inspection intervals and its relationship with condition monitoring [5, 16]. The focus for using the DTM in methods of optimizing maintenance intervals for the case of production lines is to obtain the parameters $\lambda(u)$ (Failure arrival rate) and the Delay Time h. Based on the current literature reviews surveyed [5, 16], the theme of obtaining the necessary parameters for the use of DTM modeling is highlighted, in order to extract the characteristics of the system under analysis for assertiveness of the optimization of maintenance scheduling. In this domain space, and considering the advances obtained by the diffusion of the pillars of Industry 4.0, we highlight the Factory Information Systems (FIS).

FIS systems are responsible for the acquisition, processing, storage, and availability of data related to production processes [17], recording events of scheduled, unscheduled stops and production times in the form of event records for later consultation and calculation of performance indices [18, 17]. Using information extracted from equipment event logs to predict failures can help to detect potential problems in advance as a low-cost and not yet fully explored alternative [19].

We propose a framework for extracting information collected by FIS systems for use in modeling maintenance scheduling based on the DTM method. By using Process Mining algorithms, the proposed framework operationalizes the extraction of information from event records generated by the process, that is then used as a reference for identifying the parameters used with MCDM methods to estimate the optimal maintenance window conditioned to conflicting criteria, and to identify possible deviations and deterioration trends, to finding the starting point of failure u. The parameterization of the proposed Framework occurs in an Off-Line Step and the update of the identification of the initial point of failure in an On-Line Step.

This paper is organized as follow. Relate works are presented in Sect. 2 resuming the actual research related with the presented problem. In Sect. 3 a background is presented, to facilitate the understanding of the basic concepts used in the proposed framework. The Proposed Framework formulation is presented in Sect. 4, with a theorical example of application, using a real manufacturing DB. A discussion about the gains obtained on use of our proposal are presented in Sect. 5. Finally, a conclusion about the found issues and discussions about the results, and future works are presented in Sect. 6.

2 Related Works

Maintenance is of vital importance in production processes, in which unexpected equipment failures result in process inefficiency and safety problems, making monitoring equipment conditions of great value, contributing to the reduction of unplanned interruptions and maintenance prevention actions do not require [3]. In today's industry, the focus is on how to use the data collected to identify and understand current conditions and detect failures [2].

Specifically, in the area of Industrial Maintenance, [8] elaborated a systematic review focusing on tools and methods of decision making, explaining that Condition Based Maintenance (CBM—Condition Based Maintenance) is the main maintenance policy researched, corresponding to 40% of published articles, with prominent use of models predictive and probabilistic. In addition, there is considerable demand for the development and improvement of real-time monitoring technologies.

Basri et al. [20] conducted a review in Preventive Maintenance (PM) planning and methods used in industry, where could be highlighted that most of the analyses and methods suggested in the published literature were based on mathematical computation rather than on solutions derived from real problems experienced by industries. In practice, PM activities tend to be planned based on cost, time, or failure.

An extensive systematic review of the literature on the state of the art in Process Mining (PM) was prepared by [21], where an exponential trend was identified in articles published in this area, with a strong focus on model discovery and compliance checking. The use of PM techniques applied in industrial activities corresponds to 13% of published articles. However, there are few contributions in the literature that deal with the application of Process Mining techniques in the area of Industrial Maintenance [8, 19, 22].

Syan and Ramsoobag [11] developed a review of MCDM/Multi-Attribute Decision Making (MADM) models with many criteria applied to solve Multi-Criteria Optimization (MCO) problems in the maintenance area. More than 100 criteria, 21 alternatives and 24 restrictions were identified when defining multicriteria optimization problems for maintenance. Problems defining the frequency of maintenance actions correspond to half of the researched articles. Three criteria are widely adopted: Cost, availability, and reliability. Additionally, it highlights the strengths and weaknesses of the MCDM/MCO tools and techniques applied in maintenance.

DTM modeling was reviewed by [16] and [5], covering the presentation of the main fundamentals of the method, maintenance modeling and use cases, covering the period from 1984 to 2018. The current development of the method and modeling for multicomponent systems stands out. Also are presented the mathematical models for the criteria commonly used in MCDM approaches and directions for future research.

Regarding articles relating to the DTM and MCDM method applied to the industrial maintenance area, in a search in research bases with the terms "Delay Time Modeling" AND "MCDM", there is a few works in this line, and those who apply the method for scheduling maintenance, define the parameters of the DTM model in subjective information from specialists and technical information from maintenance manuals [15, 23, 24].

3   Background of Base Concepts

To assist the understanding of this work, it is necessary to introduce some concepts and techniques that support the approach and facilitate the understanding of the results obtained and their analysis.

Factory Information Systems (FIS) are responsible for the management of data related to a manufacturing system, where the information is typically received in the form production events [17]. The data acquired by FIS are stored in the form of sequential events into defined structures of data, entries from operators on the factory floor or from sensors installed on monitored machines. Thus, we consider that FIS may provide a lot of information that can be used to support decision making in the context of maintenance management.

In our work we use a discovery process mining algorithm called Heuristic Miner (HM) [25]. One of the main characteristics of HM is that it considers sequences and frequencies of events when constructing a process model. Thus, infrequent paths may not be incorporated into the model. HM builds a process model named Causal Nets. A causal net is a graph where nodes represent events (activities) and arcs represent causal dependencies. Each event (activity) has a set of possible input bindings and a set of output bindings. Using both information the number of times one activity is directly followed by another activity and the dependencies measure between activities in an event log, it is possible to build the dependency graph, where the arcs are only considered if they are over a threshold.

Many types of equipment failures present some kind of change in their condition that is detectable [16]. When the equipment is in a good condition, it has performance or quality factors with known mean μ and variance δ^2 values [26]. However, in a wear condition or in the eminence of a failure, the mean, the variance or both values might change. We used in this research the reasoning presented by [27], who performed a correlation assessment between cycle time and time between failures of a CNC machine, where it was concluded that an increase in cycle time caused the maintenance stop.

DTM is a modeling tool for problems with planned maintenance interventions and defines that the failure of an asset occurs in two stages [28]. It defines two intervals where maintenance decisions can be made: (1) from normal operation to the point where an eventual defect can be identified, defined as the starting point u of the failure and (2) the time between the identification of the failure until the functional failure of the asset, defined as failure delay time h [5]. The time between the identification of the defect and the occurrence of the failure $(h, u + h)$, being characteristic of the asset and the type of defect, it is a window of opportunity for maintenance action. Thus, it is possible to identify the failure arrival rate $\lambda(u)$ and the associated delay time h, the DTM method can model the relationship between the frequency of inspections and the number of plant failures [28]. By knowing the distribution function and its parameters for the initial time $g(u)$ and for the delay time $f(h)$, time maintenance scheduling estimates can be obtained by applying the distribution functions in the mathematical models of evaluation criteria to optimize their values together [16].

From the perspective of the Time-Based Maintenance (TBM) technique, preventive maintenance actions are decided based on the estimated lifetime of the asset, depending on failure or user data [29]. Thus, for the establishment of a preventive maintenance policy, it is necessary to determine the optimal time and frequency of inspection [30], conditioned to conflicting criteria that need to be optimized simultaneously.

MCO are mathematical optimization problems where two or more conflicting criteria need simultaneous optimization, to minimize a given objective function [11], as presented in Eq. 1:

$$\text{Minimize}\{f_1(x), f_2(x), \ldots, f_k(x)\}, \text{ subject to } x \in S \text{ onde } k \geq 2 \qquad (1)$$

To deal with these problems, the Multicriteria Decision Making Methods (MCDM) has been applied, with great emphasis in the area of maintenance scheduling [11].

4   Process Aware Maintenance Decision Making Framework

The main objective is, using the arrival failure probability $g(u)$ and the delay time $f(h)$ extracted from the event log history, estimate the time window time, based on MCDM methods, restricted by conflicting criteria. Frequent inspections or corrective maintenance actions when equipment is in good operating do not add information about its condition, unnecessarily increasing costs of operation. On the other hand, if the machine already has a relevant time of use or its condition is deteriorated, to defer the inspection can increase the occurrence of unexpected failure and economic losses [31].

To carry out the present study, data collected by an FIS system, from a main equipment or bottleneck were used. The data comes from sensors connected to the active (activation status, temperature, vibration, among others) and/or with manual entry performed by the operator, such as indication of reason for stop, start of shift, among others. This data set is grouped into a data collector ① and sent in the form of a record structure ② each event produced (including timestamp) by the monitored equipment. This structure of events is based on the recommendations suggested in [32]. This structure is sent to a cloud server ③ and stored in a Structured Database ④ [18] in order to compose a history of event records ⑥. Online data ⑦ is extracted through a server running SQL scripts ⑤ which extract, for example, cycle time from the last event record received, production activity and/or event occurred, additional measurement data, among others. With the data extracted, the proposed framework is applied, to check the current condition of the asset and indicate to schedule maintenance on the appropriate stop window.

4.1   Proposed Framework and Explanation of Use

The proposed framework, to obtain the optimal maintenance window based on the history of events of an asset, is based on three methods: Process mining, Time Delay Modeling and Multicriteria Decision Making. Figure 1 presents the framework structure and main steps.

The Framework Stage **FIS Knowledge Discovery** refers to obtaining the parameters and models that will be used to estimate the average value and standard deviation (SD) of the cycle time under normal conditions, the starting point of failures and delay

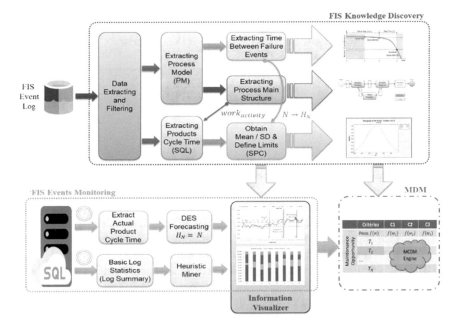

Fig. 1 Structure and steps of the proposed MDM framework

time, estimates of time to repair and definition of the horizon of analysis and estimation of initial maintenance windows. This Step is based on historical event record data (FIS Event Log). Stage **FIS Events Monitoring** aims to monitor the cycle time of the asset, conditioned to the parameters obtained in the previous step, to identify the initial time of the failure trend. This step uses N values obtained from events generated in a sampling time period, for example, 1 day of collection. In the event of a failure trend, the maintenance schedule is signaled for the window obtained in the step **Maintenance Decision-Making** (MDM). The following sequences presents the actions that are performed by each stage of the proposed framework, for detailing and description of the tools used. A theoretical example, based on data collected by a FIS system from a machining machine (CNC lathe), is used in each stage to understand the proposed framework, exploring the obtaining of parameters for DTM modeling.

FIS Knowledge Discovery (Off-Line): This stage occurs in the implementation of the proposed framework, performing the necessary parameterization for estimating the maintenance window and thresholds for detecting the time of failure start.

(I) The event records stored in BD FIS are extracted and the table with the events ordered sequentially according to their start timestamp are converted to CSV format, which then are imported into process mining software.

(II) Using the Heuristic Miner Mining tool, the graph of the sequences of activities and events occurred, and their respective frequency values are obtained. In this step, the Main Activities structure of the process is defined, which

Table 1 Events for main structure

Event	Occurrency	Rel. Freq (%)	Cumulative (%)	Class
Machine working	25,142	29.3	29.3	Activity
Finished goods	20,601	24.0	53.3	Quality event
Product removal	20,437	23.8	77.1	Activity
Short stoppage	18,662	21.7	98.9	Activity

are the activities with the highest occurrence of absolute frequency in the record. In this step, the activity related to the working action of the asset is identified, which will be used to obtain the cycle time t_{cycle}. Follows is the result after application of steps I and II in the FIS DB. The main structure was extracted and then converted to the MXML format, using the DISCO [33] application. After using the HM mining tool, the events corresponding to the main structure were obtained, shown in Table 1.

(III)   The time between maintenance events (Maintenance Events) is obtained and the mean and standard deviation values are calculated. This information will be used to estimate the delay time $f(h) = T_h$; Extracting from FIS DB, it obtained $f(h) \rightarrow \mu_{Th} = 131.6/\sigma = 28.32$ (In this case in days).

(IV) N samples of the cycle time of the activity (Machine Working) are obtained after each moment identified for the maintenance stop event (Step III) and the average and standard deviation of the cycle time of the machine in its normal condition is calculated. It is suggested $N > 30$, if possible. For this example, the values are $\mu_N = 50.45s$, $\sigma = 4, 2$.

(V)   With the values of Mean and SD of the cycle times of step IV, the limit $SupLim$ for monitoring are defined, based on the SPC method (Statistical Process Control) [34]: Considering the value of (N) 10 and k equal to 3, one can obtain the limit $SupLim = 54,5s$. The Functional Failure Limit (FF_{Lim}) is determinate by use of specialist´s information.

(VI)   To obtain the start of fault u, a set of N cycle time values is taken after each maintenance event (Time T_k), then the DES (Double Exponential Smoothing) [35] trend analysis method is applied and it is verified if the estimated value for a Horizon H_N keeps within the limits defined in step V. If deviation is identified, it obtains the value of the time between the a priori maintenance event and the time relative to the measurement of the cycle value $(T_{U_1} = T_{Tcycle_k} - T_k)$; otherwise, it is repeated again for a new data set to $k = k + 1$;

(VII)   The probability distribution of the failure occurring $\lambda(u)$ (Expected number of failures) is estimated based on the values obtained for $T_{U_{1 \rightarrow k}}$, for the period T_h. This will be the parameter to be used in the criteria estimation equations in relation to time [28]. Using de information extract from FIS DB, it is found $\lambda(u) \rightarrow \mu_{\lambda u} = 95.6/\sigma = 29.41$ (In this case in days). After extracting information from the onset of failure and delay time estimates,

Table 2 Estimated values for $u, h, \lambda(u)$ e $f(h)$

ProdLot 1	ProdLot 2	ProdLot 3	ProdLot 4	ProdLot 5	DTM	Mean	SD
96	115	156	127	164	$f(h)$	131.6	28.32
87	67	135	117	72	$\lambda(u)$	95.6	29.41

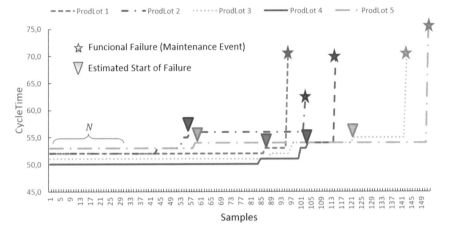

Fig. 2 Data used to obtain the information with steps III to VII

the values shown in Table 2 are obtained, obtaining the estimates for $f(h)$ e $\lambda(u)$.

Figure 2 presents the data used to obtain the information following the steps III to VII, for five production batches of product 118, in this example.

(VIII) Finally, the time windows for evaluation are estimated based on the difference in the average delay time T_h, obtained in step III, and the mean time to failure $T_{U_{mean}}$, obtained in Step VI. In this period, the time resolution for the windows to be evaluated in the criteria estimates is defined (*SampleTime* T_S), for example, in days, weeks or months.

FIS Events Monitoring (On-Line) In this Step, scripts are used to extract information from the event records stored during the sample period, defined in step VIII of the previous step. Thus, the analysis is performed after the occurrence of a set of events, provided that $T_s \ll T_h$.

(I)   For each sample period, the main structure is extracted dynamically, using the HM algorithm; obtaining values of the relative frequencies, for use in the analysis of any deviations not correlated with events of unscheduled maintenance stops, and which should not occur with the application of the framework;

(II) The cycle times of each record of the sample period are extracted;

(III) The DES method is applied to this data set to obtain the variation trend;

(IV) It is evaluated whether the information obtained remains within the limits defined in V of the Off-Line Stage. In case of observation of failure trend

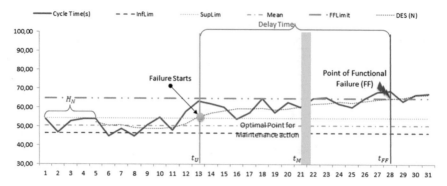

Fig. 3 Parameterized on-line stage illustration

($Trend_{slope} > SupLim$), signals for scheduling maintenance in the estimated window, in MDM Stage. Then, it returns to step I. With the extracted
values, it is then possible to parameterize the framework for the On-Line
Step, in order to be able to identify variations in the monitored variable, in
the example, the cycle time of the CNC machine. Figure 3 illustrates the
Online Stage, where the window of opportunity to perform maintenance can
be seen.

Maintenance Decision-Making (MDM) In this step, the maintenance window
is estimated, based on the probability of failure $\lambda(u)$ obtained in step VII of the Off-
Line Step applied to mathematical models of the desired criteria for analysis with
MCMD methods [11], evaluated for a set of time windows defined in step VIII of
the Offline Line. In this step, the decision maker can add his expertise, defining the
weights for the criteria and cost values, among other parameters directly related to
the type of process and/or product.

5 Discussion

In this paper a key problem in production systems related with planning the stoppage is highlighted and we propose a Process Aware Maintenance Decision Making
Framework as a solution to aid the decision on defining the optimal window to do
the maintenance actions.

Although a Predictive Maintenance policy has attracted attention with the
advancement of Industry 4.0, presenting an exponential growth in research [36],
Preventive Maintenance has been widely used in the industry, benefiting from the
advances and gains caused by the use of AI, ML, MCDM and other techniques [20].
In this context, many studies have been proposed in the literature to optimize maintenance scheduling ([5, 8, 11, 20, 37]). However, most of the proposed solutions are
based on mathematical formulations and/or estimates for model parameters using

data from manuals and/or consultations with specialists. Additionally, although the use of AI techniques and mathematical models show significant, but theoretical, results, it is difficult to be applied in the reality of the industries [20].

Our proposal for a framework for scheduling maintenance seeks to fill this gap, integrating process information with expert knowledge in decision making when scheduling production downtime. In this respect, the use of process mining techniques in industrial maintenance is still poorly explored in the literature [8, 20, 22] and its use associated with DTM models has rarely been developed [15, 16]. The main gains expected from our approach are obtaining information with low computational cost, easy implementation in the industry and support for maintenance decision making models with real process information.

6 Conclusion and Future Works

FIS Systems are a data source being generated and stored in the form of event records rich in information that can be used for Maintenance Management, helping to monitor the condition of the process and/or assets and for use in scheduling models of the maintenance action. In this article, a proposal for a framework for extracting information was presented to parameterize models in DTM for use in the optimization of conflicting criteria, additionally providing an expectation of the time window for preventive maintenance. It was possible to verify by means of a theoretical example based on real data from a CNC machine the application of the proposed steps and steps, obtaining values for the estimates of probability of failure occurrence $g(u)$ and estimated delay time $f(h)$.The developed framework proves to be a feasible alternative for obtaining, in an objective way and based on real information, the parameters necessary for use in the time estimates for the optimum moment of the maintenance action. As works yet to be developed in this line of research, it stands out: (1) Validation of the proposal with a real database; (2) Application of the DTM model parameters in the criteria estimation equations; (3) Application in an MCDM method for estimating the maintenance time window and analyzing compliance with the real case and (4) Evaluation/definition of criteria weights based on DTM models;

This study was financed in part by the Coordenação de Aperfeiçoamento de Pessoal de Nível Superior—Brasil (CAPES) Finance Code 001.

References

1. Bousdekis A., Lepenioti K., Ntalaperas D., Vergeti D., Apostolou D., Boursinos V. A RAMI 4.0 View of Predictive Maintenance: Software Architecture, Platform and Case Study in Steel Industry. In: Proper H., Stirna J. (eds) Advanced Information Systems Engineering Workshops. CAiSE 2019. Lecture Notes in Business Information Processing, vol 349. Springer, Cham. (2019).

2. Sami Sivri M., Oztaysi B. Data Analytics in Manufacturing. In: Industry 4.0: Managing The Digital Transformation. Springer Series in Advanced Manufacturing. Springer, Cham. (2018).
3. Wu W., Zheng Y., Chen K., Wang X., Cao N. A Visual Analytics Approach for Equipment Condition Monitoring in Smart Factories of Process Industry. IEEE Pacific Visualization Symposium (PacificVis), 140–149. Kobe. (2018).
4. Kumar, U., Galar, D., Parida, A., Stenström, C., Berges, L. Maintenance performance metrics: a state-of-the-art review. Journal of Quality in Maintenance Engineering 19 (3). 233–277. (2013).
5. Wang, W. An Overview Of The Recent Advances In Delay-Time-Based Maintenance Modeling. Reliability Engineering & System Safety (106), 165–178. (2012).
6. Sand, K., Aupied, J., Spruyt, F. IEEE 11th International Conference on Probabilistic Methods Applied to Power Systems. IEEE. Singapore. (2010).
7. Weber, P., Medina-Oliva, G., Simon, C., Iung, B. Overview on Bayesian networks applications for dependability, risk analysis and maintenance areas. Engineering Applications of Artificial Intelligence, 25(4), 671–682. (2012).
8. Ruschel, E., Santos, E. A. P., Loures, E. de F. R. Industrial maintenance decision-making: A systematic literature review. Journal of Manufacturing Systems 45, 180–194. (2017).
9. De Jonge, B., Teunter, R,. Tinga, T. The influence of practical factors on the benefits of condition-based maintenance over time-based maintenance. Reliability engineering & system safety, 158. 21–30. (2017).
10. Bensi, M., Der Kiureghian, A., Straub, D. Efficient Bayesian network modeling of systems. Reliability Engineering and System Safety 112, 200–213. (2013).
11. Syan, C. S., Ramsoobag, G. Maintenance applications of multi-criteria optimization: A review. Reliability Engineering & System Safety 190, 106520. (2019).
12. Ngai, E. W. T., Peng, S., Alexander, P., Moon, K. K. L. Decision Support and Intelligent Systems In The Textile And Apparel Supply Chain: An Academic Review Of Research Articles. Expert Systems with Applications 41, 81–91. (2014).
13. Abdelhakim A. Maintenance scheduling based on PROMETHEE method in conjunction with group technology philosophy. International Journal of Quality & Reliability Management 35 (7), 1423–1444 (2018).
14. Almeida, A.T de., Cavalcante, C. A. V., Alencar, M. H., Ferreira, R. J. P., Almeida-Filho, A. T de., Garcez, T. V. Multicriteria and multiobjective models for risk, reliability and maintenance decision analysis. International Series in Operations Research and Management Science. Springer. New York. (2015).
15. Da Silva, J. G., Lopes, R. S. An integrated framework for mode failure analysis, delay time model and multi-criteria decision-making for determination of inspection intervals in complex systems. Journal of Loss Prevention in the Process Industries 5, 17—28. (2018).
16. Werbinska-Wojciechowska, S. Delay-Time-Based Maintenance Modeling for Technical Systems—Theory and Practice. Advances in System Reliability Engineering, 1–42. (2019); Wang W. Delay Time Modelling. In: Complex System Maintenance Handbook. Springer Series in Reliability Engineering. Springer, London. (2008).
17. Santos E. A. P., De Freitas R. L., Deschamps F., De Paula M. A. B. Proposal of an Industrial Information System Model for Automatic Performance Evaluation. Emerging Technologies and Factory Automation, 2008. ETFA 2008. IEEE International Conference on, pp.436–439. (2008).
18. Pierezan, R., Santos, E. A. P., Loures, E. F. R., Busetti de Paula, M. A., Ferreira, L. R. Machine simulation for operational decision support using colored Petri nets. In: 21st International Conference on Production Research, 21st International Conference on Production Research. Stuttgart. (2011).
19. Sipos, R., Fradkin, D., Moerchen, F., Wang, Z. Log-based predictive maintenance. In: Proceedings of the 20th ACM SIGKDD international conference on knowledge discovery and data mining. 1867–1876. (2014).
20. Basri, E. I., Razak, I. H. A., Ab-Samat, H., Kamaruddin, S. Preventive maintenance (PM) planning: a review. Journal of Quality in Maintenance Engineering, Vol. 23, Iss 2, (2017).

21. Dos Santos Garcia, C., Meincheim, A., Junior, E. R. F., Dallagassa, M. R., Sato, D. M. V., Carvalho, D. R., Santos, E. A. P., Scalabrin, E. E. Process mining techniques and applications-A systematic mapping study. Expert Systems with Applications. Elsevier. (2019).
22. Horn, Richard and Zschech, Patrick. Application of Process Mining Techniques to Support Maintenance-Related Objectives. In: 14th International Conference on Wirtschaftsinformatik. Siegen, Germany. (2019).
23. Emovon, I., Norman, R.A., Murphy, A.J., An integration of multi-criteria decision-making techniques with a delay time model for determination of inspection intervals for marine machinery systems. Appl. Ocean Res. 59, 65–82. (2016).
24. Ferreira, R.J.P., de Almeida, A.T., Cavalcante, C.A.V. A multi-criteria decision model to determine inspection intervals of condition monitoring based on delay time analysis. Reliab. Eng. Syst. Saf. 94 (5), 905–912. (2009).
25. Weijters A. J., Van der Aalst W. M. P., Medeiros A. K. Process mining with the heuristics miner algorithm. Technische Universiteit Eindhoven, Tech. Rep. WP. Vol. 166. pp.1–34. (2006).
26. Venkatasubramanian, V.; Rengaswamy, R.; Kavuri, S. N.; Yin, K. A review of process fault detection and diagnosis: Part III: Process history based methods. Computers and chemical engineering 27 (3), p.327–346. (2003).
27. Kurscheidt Netto R. J., Santos E. A. P., de Freitas Rocha Loures E., Pierezan R. Using Overall Equipment Effectiveness (Oee) To Predict Shutdown Maintenance. In: Amorim M., Ferreira C., Vieira Junior M., Prado C. (eds) Engineering Systems and Networks. Lecture Notes in Management and Industrial Engineering. Springer, Cham (2017).
28. Wang W. Delay Time Modelling. In: Complex System Maintenance Handbook. Springer Series in Reliability Engineering. Springer, London. (2008).
29. Ahmad, R. Kamaruddin, S. An overview of time-based and condition-based maintenance in industrial application. Computers & Industrial Engineering 63, pp. 35–149. (2012).
30. Cavalcante, V., Alexandre, C., Pires Ferreira, R., De Almeida, A. T. A preventive maintenance decision model based on multicriteria method PROMETHEE II integrated with Bayesian approach. IMA Journal of Management Mathematics 21 (4). 333—348. (2010).
31. Abeygunawardane, S. K.; Jirutitijaroen, P.; Xu, H. Adaptive Maintenance Policies for Aging Devices Using a Markov Decision Process. Power Systems, IEEE Transactions on, Vol. 28, N°3, p.3194–3203. (2013).
32. Kurscheidt Netto, R., Santos, E., Loures, E. R. Restrictions on use of The Factory Floor Information in Maintenance Management. In: Brazilian Symposium on Information Systems (SBSI), 11. Goiania. Proceedings of the 11th Brazilian Symposium on Information Systems. Porto Alegre: Sociedade Brasileira de Computação, p. 447–454. (2015).
33. DISCO, https://fluxicon.com/book/read/reference/, last accessed 2020/02/25.
34. Kurscheidt Netto, R. J.; Santos, E. A. P.; Loures, E. R.; Pierezan, R. Condition-Based Maintenance Using OEE: An Approach to Failure Probability Estimation. In: 7th International Conference on Production Research - Americas 2014, Lima. Proceedings of 7th International Conference on Production Research - Americas (2014).
35. Ahmad, R., Kamaruddin, S. Maintenance Decision-making Process for a Multi-Component Production Unit using Output-based Maintenance Technique: A Case Study for Non-repairable Two Serial Components. Unit International Journal of Performability Engineering, Vol. 9, No.3, May, pp.305–319. (2013).
36. Carvalho, T. P., Soares, F. A. A. M. N., Vita, R., Francisco, R. da P., Basto, J. P., Alcalá, S. G. S. A systematic literature review of machine learning methods applied to predictive maintenance. Computers & Industrial Engineering. Vol. 137, pp. 106024. (2019).
37. Ruschel, E., Santos, E. A. P., Loures, E. de F. R Establishment of maintenance inspection intervals: an application of process mining techniques in manufacturing. Journal of Intelligent Manufacturing. Vol. 31, N°1, pp. 53–72. (2020).

Exploitative Model for Dynamic Project Management Based on the Project Management Body of Knowledge

Luciano Sales and Sanderson Barbalho

Abstract System Dynamics (SD) is a useful tool for improving traditional project management performance, however existing barriers prevent further integration between these approaches. As such a barrier is the communication one, hence system dynamics project models do not talk to the main background of project management professionals. The main goal of this article is to present a holistic dynamic project management model, integrating the mental models and vocabulary from traditional project management into the SD approach, and investigating if this could lead to a better acceptance of this model by project management professionals. The project management knowledge areas are modeled for this research. As a result, from validation interviews, the newly developed dynamic model improved the understanding of SD by project managers. In addition, a strong interest in using SD as a project management simulation tool has been demonstrated by these public.

Keywords System dynamics · Project management · Stakeholder engagement

1 Introduction

System Dynamics (SD) is a method confirmed to be effective to model and to analyze the interaction of complex, dynamic and non-linear variables [1]. SD has been applied to numerous knowledge fields, with the intents of understanding how the main components of a specific system interact among themselves over time, allowing for the analysis of non-intuitive or non-evident structures of this system as well as its decision rules [2].

Project management (PM) is one of the fields that present the most success when making use of SD, with proven results in terms of development of new concepts and theories, number of professionals applying the built models, as well as in terms of generated value to organizations and customers [3, 4]. As discussed by Love et al. [5] and Lee et al. [6],the traditional project planning tools use a static approach that can

L. Sales (✉) · S. Barbalho
Brasilia University, DF, Brazil

© The Author(s), under exclusive license to Springer Nature Switzerland AG 2021
A. M. Tavares Thomé et al. (eds.), *Industrial Engineering and Operations Management*,
Springer Proceedings in Mathematics & Statistics 367,
https://doi.org/10.1007/978-3-030-78570-3_34

lead to poor realistic estimates to its users, as they ignore various feedback processes and the non-linear relationships of a project, rendering them to be inadequate to the challenges of today's projects.

Despite the growing importance given to project management by organizations, considering that almost 30% of the world economy originates from projects [7], and even with the development of several best practice guides to the PM area; paradoxically, the poor performance of projects and the dissatisfaction of their stakeholders seems to have become the rule rather than the exception nowadays [8].

In spite of its results and of the recognition of the importance of the SD for the improvement of the PM discipline, this method is still used in very few projects. Hence, there is a need to develop strategies to make it more common in organizations [4]. According to these researchers, three strategies can improve the diffusion of the SD practices: the publication of more success stories in the PM area; the effort to make the SD models easier and less costly to develop; and improvement of the integration of dynamic models into the traditional project management tools.

Jalili and Ford [9] stated that the professionals in the SD community have failed to communicate adequately all the potential that this area has to solve the traditional project management problems. That is, there seems to be a barrier that separates the community that uses the SD to improve the performance of some projects and the project management community. This community doesn't know (or doesn't trust) SD, and moreover, it recognizes the existance of problems in the outcomes of their projects.

Rumeser and Emsley [10] when analyzing the key challenges to SD implementation in the project management area, showed how the lack of understanding and confidence in the dynamic models are among the root causes for the low dissemination of the dynamic project management, suggesting that the use of the project management traditional vocabulary in the dynamic models could be one of the strategies for revert this scenario.

Therefore, it seems to be appropriate to incorporate in the dynamic structures of project management, the mental models and vocabulary of the traditional approaches, which has already established a de-facto recognized vocabulary in the project management area, thus contributing for the development of an holistic and more accessible structure to the project management community.

The aim of this study thus is to develop a new dynamic model for project management, integrating the mental models from traditional approaches to the dynamic structures already enshrined in the scientific literature, while investigating if there would be an increase in the acceptance of use of this type of tool by the project management community. The PM knowledge areas as stated on Project Management Body of Knowledge [11] are approached in this model, as they are the main elements for project delivery, according this standard.

To achieve this goal, we present our theoretical background in Sect. 2; in Sect. 3, the methodology in use; in Sect. 4, the results and discussions; and finally, in Sect. 5, the conclusions of this work.

2 Theoretical Background

2.1 Problems in the Traditional Project Management Frameworks

Lyneis et al. [3], allege that the projects are complex systems, but the traditional tools to project management continue to use mental models inadequate to cope with this complexity, and as a result, project managers continue to make many mistakes, and the projects continue to face serious performance issues.

Project management is one of the fields that present the most success when making use of SD, with proven results in terms of development of new concepts and theories, number of professionals applying the built models, as well as in terms of generated value to organizations and customers [3, 4]. As discussed by Love et al. [5] and Lee et al. [6], the traditional project planning tools use a static approach, that can lead to not so realistic estimates to its users, as they ignore various feedback processes and the non-linear relationships of a project, rendering them inadequate to the challenges of today's projects. In a recent research, the Project Management Institute (PMI) reported that despite the efforts of the organizations, less than 50% of projects finished on the time and costs that were initially agreed [12].

SD, due to persistent problems with delays and cost overruns in the projects, and even with the evolution of project management techniques, has shown to be an effective tool to minimize these issues, acting in a complementary manner to the traditional project management approaches [13]. Both SD and project management scientific literature reported some problems in the traditional approaches, which are described below. One of the issues cited by several researchers to justify the need to include the dynamic paradigm to Project management is the "90% syndrome". Abdel-Hamid and Madnick [14] discuss the aforementioned phenomenon, demonstrating that it is a common problem in traditional project management, leading to projects with a much longer duration than initially planned.

A second issue of the traditional approaches, explained by Mingers and White [15], is the formation of system archetypes in the PM decision sctructure. To these researchers, the structures of traditional project management can hide system archetypes, leading projects to collapsse, even if apparently, all plans are being followed thoroughly. According to Braun [16], the system archetypes describe common behavioral patterns in the organizations and should be viewed as diagnostic tools that provide information and alert managers to unexpected future consequences brought by their decisions.

Sales and Barbalho [17] and Kontogiannis [18], showed that the limitation in the knowledge of a complex system can lead experienced professionals to act inappropriately, as their mind models can contain distortions, which lead to wrong decisions due to system archetypes. As observed by Wolstenholme [19], the identification of archetypes improves decision making, because organizational boundaries cause action and reaction to be found in separate sources within organizations.

A third issue of the traditional approaches, explained by Rodrigues and Bowers [20], is the redutionist view of management through the decomposition of elements of a project. This problem stems from the mental model itself used for the development of PM practices, based, precisely, in the decomposition paradigm.

Senge [21], claims that, from early, we learned to divide and fragment the world, to make tasks and complex subjects more manageable, however, by doing this, we pay a high price, because we can no longer realize the consequences of our actions, since we lose the notion of connection with the whole. To Williams [22], even though the project management tools are based on the decomposition of their fundamental parts, during the project execution systemic effects will occur, and that should be taken into account during planning.

2.2 Dynamic Models to the Project Management

Rodrigues and Williams [23], state that the first dynamic model built for project management was proposed by (Roberts [24]), introducing a workflow based on the applied resources as well as on the productivity of such resources taking into account the human factor in the actions and decisions of a project.

Lyneis and Ford [4], when evaluating the main dynamic models used in projects, realized that all of them had four fundamental structures that seem to solve the main problems presented by traditional approaches: the **project features**, that include the development processes, the mind models for decision making and the project components; **the rework cycle**, the most important characteristic belonging to the project dynamics, because it includes in such dynamic models the recursive nature in which the work may generate more work; the **project controls,** for the comprehension of the performance of the variable system; and the **ripple and knock-on effects**, in other words, the unintentional consequences of the decision-taking process that can generate system archetypes.

In addition, these researchers, in a paper that discusses the SD development applied to project management, presented the various contributions that have enabled to leverage Roberts' initial model, as shown in Fig. 1, where it is possible to perceive the systemic effects of the overtime, of the pressure in the project schedule, and of the hiring of professionals during project execution.

Rodrigues and Bowers [25] and Ford and Sterman [26] show how common processes that occur in projects, such as the overlap of tasks and delays in the discovery of rework, can create non-planned iterations, generating delays, higher costs and inferior quality, in other words, these iterations, through time, make the project work to generate more rework, which constitutes the cause of systemic problems in traditional project management.

As shown in Fig. 1, the errors in the project flow, generate more errors (within the rework structure), generate more work (increase the scope) and put pressure on the project manager, which leads to the development of the work in a different way from the planned sequence, further aggravating the errors, which quickly leads to

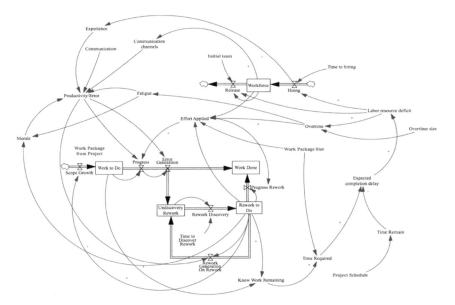

Fig. 1 Policy resistance via "knock-on" effects to controlling feedback to improve schedule performance. Fonte: Adapted by the authors from Lyneis e Ford [4]

degradation of project performance. This systemic understanding is important to the project team, because it allows minimizing the errors, such as the occurrence of system archetypes.

2.3   The Mental Model of Traditional Project Management

There are several project management standards, which seek to increase the successful rate of projects. In spite of this, the many frameworks available, the PMBOK Guide impact more than 900,000 certified professionals in the world [11], working as a guide able to provide a set of standards and a globally accepted vocabulary that can be applied to any project, regardless of their type, size, application area and organization [8]. The sixth edition, released in 2017, is the latest interation of the PMBoK Guide, which also incorporated into its set of practices, the so called "agile practices". For one to better understand the structure of PMBOK, it is necessary to track the flow of workpackages through the several different processes brought by PMBoK. Only then it would be possible to draw a comparison between this workpackage flow and the workflow already mapped by the dynamic models themselves.

To the Project Management Institute [11], the listing of work packages can be found in the scope baseline, one of the documents that constitute a project management plan. In Fig. 2, below, the mapped flow of work packages is presented.

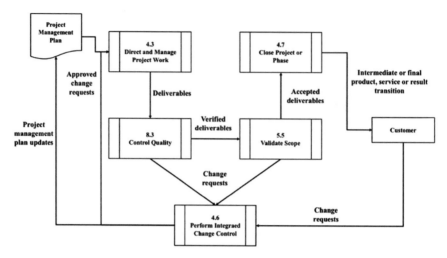

Fig. 2 PMBOK 6 work package flow

The work packages are executed progressively through process "4.3 Direct and Manage Project Work". While the work package execution is occurring, the project deliverables are being produced, and this flows to the process "8.3 Control Quality", before being released to customer acceptance tests, which is part of the process "5.5 Validate Scope". If the deliverables are accepted, they are cleared to transition to the operational environment through the process "4.7 Close Project or Phase". In the case of errors or of any deliverable being rejected, the "4.6 Perform Integrated Change Control" process is triggered to evaluate needed changes and their impacts, and as soon as changes are authorized, the work packages in the form of rework, return to the process "4.3 Direct and Manage Project Work".

In addition, before the project closes, there is a possibility that the customer will issue a change request with the intents to modify some intermediate deliverable (PMI, 2017), thus generating a new rework cycle, since this change request will go to the process "4.6 Perform Integrated Change Control". In summary, three distinct rework cycles were identified through the analysis of the PMBOK 6 work packages flow.

3 Methodology

In this section we present the steps proposed by this research and that should be taken in order to improve the PM dynamic models, through the employment of PMBOK 6 mindset and vocabulary. For this purpose, we describe, in a detailed way, the steps of methodology, which were inspired in the research of Luna-Reyes and Andersen [27].

3.1 Conceptualization

This is the first step of the modeling focused on the system design to be developed. For this, we analyzed the work packages flow in the traditional approach (Fig. 2) and the work packages flow in the proposed dynamic approach (Fig. 1).

The Fig. 2 structure was chosen, because, as presented in the theoretical framework, the PMBOK 6 is amongst the most used frameworks and is a guide able to provide a set of globally accepted of standards and recognized vocabulary to the project management profession. The option for the Fig. 1 model, occurred for two reasons: it was identified that this model was the most cited on dynamic project management publications found on Scopus and Web of Science databases. This model also represents well the progressive nature of SD development to the PM discipline.

3.2 Models Integration

The second step aimed to quantify the dynamic model formulation, that is, the flows and stocks diagrams, from the integration of the traditional to dynamic models, preserving the four fundamental structures identified by Lyneis and Ford [4]. This form of modeling, oriented towards integration, was developed through the employment of the "VENSIMPLE" software.

The integration occurred in two phases: in the first phase, we conducted a focus group with 4 SD experts. The objectives were to define the **project features** and **the rework cycles**. In second phase were defined the **project controls** and the **ripple and knock-on effects**. Thus, in this phase, interviews were conducted in unstructured way, until the theoretical saturation were achieved. This means that, when the interviewees answers converge to the extent that no new concept or category may be derived in the analysis, the sampling of interviewees can be interrupted.

Each interviewee was presented to the reference model (Fig. 1), and the narrative of the model relations, after which the respondent had the opportunity to comment on the presented relations, suggesting improvements to variables and flows present in the new structure. If a new variable, outside of the reference model, was cited by one interviewee, this would only be incorporated into the new model if that variable was cited again by another interviewee. Ten experts in project management from 8 different different organizations belonging to seven different domains (such as civil construction, oil, consulting, IT, government) were interviewed on this phase.

3.3 Testing

With the flows and stocks diagram developed in the previous step, this step aimed to test the developed dynamic model (behaviors and outcomes) through computer simulations and semi structured interviews with ten project management experts (same experts from previous step). These interviews, according Luna-Reyes and Andersen [27], aim to check if these experts agree with the model developed and to identify reasons for agreeing or disagreeing.

3.4 Implementation

The fourth step aimed at strengthening the theoretical validation of our developed model through the use of a case study that provided data about application policies, the stories and experiences of the involved project team, thus allowing us to analyze patterns and outcomes that were later subject of discussion and analysis.

This case study was provided by a telecommunications project, namely "Infovias Project", whose goal is to allow the construction of fiber optics networks, laid down in river beds in the interior of Brazilian Amazon cities. Data from the third phase (2016–2017) of this project was used. Interviews were conducted in a non-structured way with the project initial team (four professionals), where, after the simulations presentation, were asked to freely present their perceptions about the understanding of the new dynamic model.

The input parameters used in the simulation were obtained in the project third stage plan (Table 1). In addition, the behavior used in variables of type "lookup" were obtained from the team who worked on the project, as well as from related literature.

Table 1 Case study inputs used

Input variables	Data from project	Comments
Work package from project plan	187 work packages	Work package from work breakdown structure
Project planned duration	8 months	
Initial team	5 professionals	
Work package size	40 h	
Team availability	160 h/month	
Time to hiring	1 month	Delay when hiring a new professional
Max overtime size	+30% for month	
Stakeholder engagement	0.99	Stakeholder engagement level maintained during the project

In spite of having been planned with PM best practices, this project, scheduled to be over in 8 months, took almost 16 months. The project team, during the execution, had to be nearly doubled in terms of the total number of involved people.

4 Results and Discussion

4.1 Characteristics of the Proposed Dynamic Model

The Figs. 1 and 2, after being integrated by SD experts, spawn the Fig. 3, with the project features and rework cycles.

The narrative for the flow of workpackages was the following: every workpackage, whether they come from the project plan or from rework, is grouped into the stock "Product Backlog". These workpackages are progressively elaborated and arrive at the stock "Control Quality". The issues identified will be the target for a change request to be analyzed by the stock "Perform Integrated Change Control".

When approved in Quality Control, the workpackages flow into the stock "Validate Scope", being object of inspection by the stakeholders. If the deliverable is accepted (group of work packages), this flows to the stock "Close Project or Phase", where it will standby for transition into the customer's operational environment so that the work is considered finalized (stock "Intermediate or final product or service"). There are two additional rework cycles, trigerred when: (a) the delivery is not acepted from the stock "Validate Scope"; (b) or if there is a change request after final delivery in the phase. In both cases, the work packages are forwarded to the stock "Perform Integrated Change Control".

Figure 4 presents the model developed in this research, according discussion with specialists. The model illustrated was the result of the cycle of discussing to specialists. For example, the variables "Quality assurance" and "Time to discover rework" variables. The interviewed professionals emphasized that the rework coordination rate also depends on the proactive work of quality assurance, which may mean a different time for the discovery of rework, depending on the project size and on the

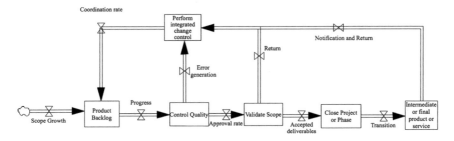

Fig. 3 Structure and rework cycles of the developed model

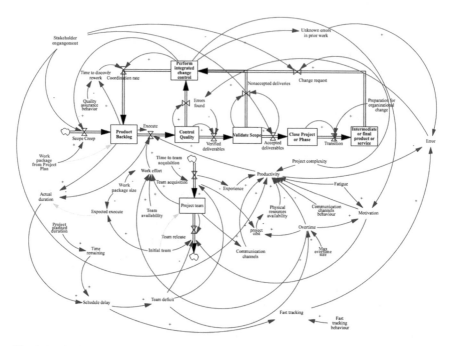

Fig. 4 Model for dynamic project management

"Product backlog" size. In other words, if there is still too much work in the backlog, the team tends to pay less attention to the discovery and coordination of rework.

4.2   Model Testing and Final Evaluation by Specialists

Ten PM specialists were interviewed and compared the model from Fig. 4 with the model from Fig. 1, in order to analyze the understanding and the alignment of the developed model with their PM practices.

Only four interviewed specialists used some simulation tool in their projects. Only two specialists affirmed to vaguely know dynamic project management, through reading, without any practical application in the projects they conducted. One of them mentioned the "Lean Project Management" as a model where SD seems "to be in alignment with one of its principles: to optimize the whole".

Beside this, only one professional stated that the model of Fig. 1 was more coherent to the project management structured he was used to employ, also informing that he uses the project management tool in a very simple way. The other nine professionals pointed Fig. 4 as the most coherent. Next, in the Table 2, we show parts of the answers from those specialists when justifying the choice for Fig. 4.

Table 2 Parts of answers by interviewees

Specialist	Commentary
Specialist 1	"[…] has a greater number of mapped and correlated variables, that will allow for a better precision within the situations found in a project"
Specialist 2	"[…] the elements are much better related, allowing for a better comprehension of a Project Manager's day-to-day"
Specialist 3	"[…] looks alike to the reality of Project management"
Specialist 4	"[…] looks more complete regarding variables and relationships"
Specialist 5	"[…] it has more returns to previous phases, rework, replanning, which seems to represent better what happens with projects in present day"
Specialist 6	"[…] This image uses a language that is more adequate, initiating with work packages and following through the stages of Project management, allowing for a more precise evaluation"
Specialist 7	"I found it is easier to follow the work package flow and the language used is closer to my reality"
Specialist 8	"Absolutely, this image makes more sense to me, as it reflects the logic of Project management, despite being more complex"
Specialist 9	"The work flow is better aligned with what happens in project's day-to-day activities, and also seems more clear to me. Besides this, it could help in the development of lessons learned"

Of the nine specialists that agreed that the model in Fig. 5 is more coherent with the structure of project management, seven showed great interest in incorporating the dynamic models as a project management practice in their projects, for the improvement of decision making.

4.3 Evaluation of the Project Used as Case Study, By Its Own Project Team

The developed and refined dynamic model was presented to the "Infovias" project team. Of the five members of the initial project team, four participated on the interviews and were also presented with the proposed models as well as with the simulation results.

The results of the simulation showed that the project was completed between months 16 and 17, in other words, very close to what happened in reality. The project team members realized that the productivity and the issues led to the identifiable delays.

When investigating the reduction in project productivity, besides other factors, one can realize that the number of communication channels was growing while the project experience was decreasing. This is a "compatible" reaction with the increase in the project team size during the execution in other to compensate delays.

4.4 Discussions

Through interviews, one can realize that within the project management community, the knowledge about SD is quite limited, which corroborates the very few cases about the use of this method. In order to make the model easier to be accepted and better integrated with the traditional tools, we had to prioritize the communication aspects, by employing in the dynamic model the same vocabulary already in use in the best practices adopted by project management teams around the world.

As stated by Rumeser and Emsley [28], the lack of comprehension and trust in the model are the main challenges for the use of SD in a more ample manner. The majority of the interviewed professionals stated that since they understood the dynamic model developed by this research, this would lead to an increase in the possibility of the use of SD as a tool for simulation in their projects. It seems to be the first step for the publication of more stories of successful application of SD in the project management field, since before anything else, it is necessary to convince project managers and their teams about the benefits of the use of this tool.

5 Conclusion

The model developed by this reasearch integrated the different approaches (the dynamic and traditional), that were built from specific mental models, in an holistic and integrative approach, which improved the comprehension of SD by representatives of the traditional project management community. Besides this, as a result of the improvement in comprehension, the interviewed professionals demonstrated great interest in using SD as a simulation tool in the future projects they will work.

There are limitations in this research, such as the small number of interviewees and the application of a single case study, which does not allow for the generalization of its results. As a suggestion for future work, it should be verified if there is feasibility in the integration of dynamic models with program management and the agile management strucutre. More project typologies must also be studied.

References

1. X. Zhang, Y. Wu, L. Shen and M. Skitmore, "A prototype system dynamic model for assessing the sustainability of construction projects," *International Journal of Project Management, 32 (1)*, pp. 66–76, 2014.
2. H. Yuan, A. R. Chini, Y. Lu and L. Shen, "A dynamic model for assessing the effects of management strategies on the reduction of construction and demolition waste," *Waste Management, 32 (3)*, pp. 521–531, 2012.
3. J. Lyneis, K. Cooper and S. Els, "Strategic management of complex projects: a case study using system dynamics," *System Dynamic Review, 17 (3)*, pp. 237–260, 2001.

4. J. M. Lyneis and D. N. Ford, "System Dynamics Applied to Project Management: a survey, assessment, and directions for future research," *System Dynamics Review, 23(2),* pp. 157–189, 2007.

5. P. Love, G. Holt, L. Shen, H. Li and Z. Irani, "Using systems dynamics to better understand change and rework in construction project management systems," *International Journal of Project Management, 20(6),* pp. 425–436, 2002.

6. S. H. Lee, F. Peña-Mora and M. Park, "Dynamic planning and control methodology for strategic and operational construction project management," *Automation in Construction,* pp. 84–97, 2006.

7. D. Parker, J. Charlton and A. Ribeiro, "Integration of project-based management and change management: Intervention methodology," *International Journal of Productivity and Performance Management, 62(5),* pp. 534–544, 2013.

8. J. Varajão, R. Colomo-Palacios and H. Silva, "ISO 21500:2012 and PMBoK 5 processes in information systems project management," *Computer Standards & Interfaces, 50,* pp. 216–222, 2017.

9. Y. Jalili and D. Ford, "Quantifying the impacts of rework, schedule pressure, and ripple effect loops on project schedule performance," *System Dynamic Review, 32,* pp. 82–96, 2016.

10. D. Rumeser and M. Emsley, "Key challenges of system dynamics implemantation in project management," *Procedia - Social and Behavioral Sciences, 230,* pp. 22–30, 2016b.

11. Project Management Institute, PMBOK Guide - Sixth Edition, Newton Square, PA: Project Management Institute, 2017.

12. Project Management Institute, "PMI's Pulse of the Profession," Project Management Institute (PMI), Newton Square, PA, 2018.

13. F. Peña-Mora, S. Han, S. Lee and M. Park, "Strategic-operational construction management: Hybrid system dynamics and discrete event approach," *Journal of Construction Engineering and Management, 134 (9),* pp. 701–710, 2008.

14. T. Abdel-Hamid and S. Madnick, "Lessons learned from modeling the dynamics of software development," *Communications of the ACM, 32 (12),* pp. 1416–1438, 1989.

15. J. Mingers and L. White, "A review of the recent contribution of system thinking to operational research and management science.," *European Journal of Operational Research, 207,* pp. 1147–1161, 2010.

16. W. Braun, *The System Archetypes,* 2002.

17. L. Sales and S. Barbalho, "Identifying System Archetypes in Order to Comprehend and Improve the Program Management Practices in Organizations," *IEEE Transactions on Engineering Management, 67 (1),* pp. 163–173, 2020.

18. T. Kontogiannis, "Modeling patterns of breakdown (or archetypes) of human and organizational processes in accidents using system dynamics," *Safety and Science, 50 (4),* pp. 931–944, 2012.

19. E. Wolstenholme, "Towards the definition and use of a core set of archetypal structures in system dynamics," *Syst. Dyn. Rev., 19,* pp. 7–26, 2003.

20. A. Rodrigues and J. Bowers, "The role of system dynamics in project management," *International Journal of Project Management, 4 (14),* pp. 213–220, 1996a.

21. P. Senge, The Fifth Discipline: The Art & Pratice of the Learning Organization, New York, NY: Currency Doubleday, 1990.

22. T. Williams, "Assessing and moving on from the dominant project management discourse in the light of project overruns," *IEEE Transactions on Engineering Management, 52 (4),* pp. 497–508, 2005.

23. A. Rodrigues and T. Williams, "System dynamics in project management: assessing the impacts of client behaviour on project performance," *Journal of the operational research society, 49 (1),* pp. 2–15, 1998.

24. E. Roberts, The Dynamics of Research and Development, New York: Harper&Row, 1964.

25. A. Rodrigues and J. Bowers, "System dynamics in project management: A comparative analysis with traditional methods," *System Dynamics Review, 12 (2),* pp. 121–139, 1996b.

26. D. Ford and J. Sterman, "Overcoming the 90% Syndrome: Iteration Management in Concurrent Development Projects," *Concurrent Engineering: Research and Applications, 11 (3),* pp. 177–186, 2003.

27. L. Luna-Reyes e D. Andersen, "Collecting and analyzing qualitative data for system dynamics: methods and models," *System Dynamics Review, 19 (4)*, pp. 271–296, 2003.
28. D. Rumeser and M. Emsley, "Key Success Factors in Implementing System Dynamics in Project Management: Coping with lack of understanding and trust," in *Proceedings of the 34th International Conference of the System Dynamic Society*, 2016a.

Exploratory Analysis of the Impacts of Digital Transformation on Supply Chain Management Processes

Lia Denize Piovesan⊙, Daniel Pacheco Lacerda⊙, and Antônio Márcio Tavares Thomé⊙

Abstract Digital transformation presents itself as a new paradigm in supply chain processes. The development of technologies with a high capacity for data capture, storage, processing and application coupled with cost reduction is allowing this transformation. Digital technologies were organized into three classes, enabling, integrating and applying. Based on multiple case study research incorporated in two large companies, empirical evidence of the occurrence of digital transformation. As an additional contribution to this research, a case study protocol was developed, with guidelines for selecting cases, collecting and analyzing data. It became evident that the digital transformation is taking place, however, at different levels, depending on the case and the process analyzed. The cases converge in the use of some technologies and the main competitive requirements aimed at the implementation of technology that is the cost. The difference is most evident in the digital transformation strategy and in the processes in which digital technologies are applied.

Keywords Digital transformation · Supply chain · Competitive criteria

1 Introduction

Competition between organizations is increasing as globalization and customer demands grow. Thus, organizations no longer operate in isolation, needing to cooperate with the Supply Chain (SC) to boost the businesses in which they are involved. Supply Chains can pursue this cooperation through digital transformation [1].

L. D. Piovesan (✉) · D. P. Lacerda
Engenharia de Produção, Universidade Do Vale Do Rio Dos Sinos, São Lepoldo, Brazil
e-mail: ldpiovesan@hotmail.com

D. P. Lacerda
e-mail: dlacerda@unisinos.br

A. M. T. Thomé
Departamento de Engenharia Industrial, Pontifícia Universidade Católica Do Rio de Janeiro, Rio de Janeiro, Brazil
e-mail: mt@puc-rio.br

Digital transformation is promoted by the exponential development of digital technologies and their combined use, allowing for new applications [2]. Reducing the cost of acquisition allows access to higher storage, processing and connection capacity. In this context, companies start the digital transformation process, looking for an interconnected, agile and intelligent SC, with information flow and integrated products throughout SC. Integration occurs in the internal processes of organizations, in the Supply Chain and network of organizations and throughout the product's life cycle, from development to final disposal [2, 3].

Organizations that do not consider the changes brought about by digital transformation and "do not adopt digital technologies will have great difficulty in remaining competitive and, consequently, in the market" [4, p. 2], since their businesses, the market in which they operate, and even the SC they are inserted in will be transformed by digital technologies. As a consequence, traditional organizations will have to rethink their business models to remain competitive [2], understanding what benefits they can obtain if they choose to implement digital technologies.

However, the lack of technical skills makes companies unaware of the potential of the technologies. In a study carried out in Germany, Hammerman and Stettes, 2016 apud Gurria [2, p. 100] suggest that the "ability to plan, organize and act autonomously", combined with the experience of the company and workers, are crucial to the success of companies' digital transformation.

Thus, for digital transformation to occur, technologies need to be known in companies. However, there is no convergence between researchers and professionals of what Industry 4.0 (I4.0) digital technologies are. The technologies attributed to I4.0 are not always the same, creating a conceptual inaccuracy [5].

Frank et al. [6] state that the lack of empirical evidence and evaluation of the effect of technologies on industrial performance are limitations of the research done and suggest that future work overcome these limitations. The authors declare that the real benefit of I4.0 technologies is still a concern for professionals, and an empirical study that evaluates the actual performance of the technologies can be useful for theory and practice [6].

Considering the importance of digital transformation for the integration of supply chains and realizing the lack of understanding by researchers about the definitions of I4.0 technologies, this work will provide empirical evidence to evaluate digital transformation and the technologies adopted. The objective is to identify the level of adoption of digital technologies in companies and how they impact SC processes and capabilities.

2   Digital Transformation

The digital transformation of SC processes will bring efficiency and connection from product development, purchasing, manufacturing, logistics, suppliers to the delivery of products or services to customers [7]. In the digital SC, there is potential for interactions by all members of the network. Thus, communications are multidirectional, with digitalization being the core of this network, generating interconnections between processes traditionally disconnected from SC [8].

The digital transformation of Supply Chains will contribute to the management of the complexity of SC, accelerate the responsiveness to the market, with better performance of the flow of products and information constituting a key factor for SC to obtain a competitive advantage [9]. The internal integration of data and external with suppliers, customers and partners through digital technologies will be of significant importance in SC [10].

In terms of digital technologies, three classes were established: (i) Enabling Technologies, which serve as the basis for the others, represented by Big Data Analytics, Cloud Computing and Internet of Things; (ii) Integrating Technologies, Artificial Intelligence/machine learning, simulation, cyber-physical systems and blockchain; and (iii) Application Technologies, additive manufacturing, self-guided drones and vehicles and advanced and collaborative robots [2].

3   Methods

In methodological terms, a multiple case study was carried out. The selection of multiple cases reduces the need for further investigation compared to a single case. In multiple case studies, external validity is more significant than in single cases [11].

An additional contribution of this research consists of the development of a case study protocol, with the definition of the research vision and purpose, procedures for data collection, case study questions and procedures for analysis. The case study questions were organized into five interview scripts. These being the complexity of SC, stage of adoption and level of investment in technologies, impact on SC processes, impact on capabilities and evaluation of competitive criteria in relation to competition and importance given by the client in the opinion of the interviewees.

The protocol was submitted to face validation with specialists from GMRG (Global Manufacturing Research Group) in Brazil. A pilot test was applied, aiming at improving the case study protocol. After the pilot test, two large companies were contacted and demonstrated to have initiatives for digital transformation.

Thus, looking for literal replication of the results, the case studies were in these companies in which it was considered that the phenomenon of digital transformation is occurring. Case A supplies parts to the automaker while Case B produces and delivers mobility equipment, as well as performing maintenance services.

4   Case Study

This section provided the descriptions of Cases A and B. From that description, the analyzes that expose the digital transformation in both companies will be accomplished. Finally, the cases were comparatively analyzed in order to contrast and explain convergences, divergences, complementarities, and infer possible explanations.

4.1   Case A

Case A supplies parts to vehicle manufacturers. Parts are supplied on request, and the company has a demand forecast of approximately four months.

The corporation has a global roadmap with five phases for digital transformation. Roadmap I4.0 proposes deadlines up to 2024 for the implementation of digital technologies and highlights the corporation's vision for digital transformation. Despite having the Roadmap I4.0 and being provided, during the interviews, a synthesis of the concepts of digital technologies, the analysis of the responses on the stage of adoption of digital technologies identified a low level of agreement among the interviewees, demonstrating the lack of clarity about the concepts of digital technologies.

Some technologies, mentioned during the interviews, do not match the concepts studied, for example, the MES (Manufacturing Execution System) information system, was mentioned as being the technology of Internet of Things (IoT), Cyber Physical System, Big Data Analytics, Simulation and Artificial Intelligence. However, it was necessary to reclassify as MES.

Furthermore, in the analysis of the results, technologies were observed that did not belong to the previously defined classes. In this way, the Digital Infrastructure class was created to contemplate the technologies mentioned by those in the interviews. Digital Infrastructure class was subdivided into three subclasses: data acquisition technologies, information systems and micro automation. The technologies observed, augmented reality and image recognition, were added to the class of Application technologies, since they were not provided for in the protocol but present a physical concept and which modify the processes in which they were employed.

Thus, Case A presented, mainly, digital infrastructure technologies in the information system subclass. The transformation (production) process is the one with the most considerable number of related technologies.

The highest expectation of benefits comes from the use of MES. Current use provides improvements in production capacity, improving the flexibility to increase and decrease production volume. The company hopes to connect it with sensors and information systems aiming at greater internal integration of the processes and thus increase flexibility, minimize operating costs and improve delivery reliability.

4.2 Case B

Case B produces and performs the maintenance service of mobility equipment. Thus, it delivers the product and service directly to the end-user. The customer does not provide a demand forecast for the product. The service is provided when the equipment obstructs the operation.

The company does not have a formalized strategy for digital transformation, but the application of digital technologies in SC processes has become evident, mainly related to contact with the customer.

The case study showed that the company has a product with the IoT technology implemented, with real benefits in cost and quality. Currently, the technology operates as corrective maintenance, and the equipment itself communicates to the company when a failure interrupts its operation. The future expectation is that it can operate with predictive analysis of the operation, promoting quality in the provision of the service.

However, most of the technologies mentioned refer to micro automation. This micro automation is related to the performance of applications and the Internet to carry out activities that used to be manual.

4.3 Comparative Case Analysis

The cases differ in terms of the strategy for digital transformation. Case A has a corporate strategy defined globally and evidenced through Roadmap I4.0. Case B did not present any defined strategy, and the technologies pare implemented according to the needs of the SC processes.

In both cases, it was observed that most of the technologies adopted belong to the Digital Infrastructure class. Of the other classes, Application technologies are the ones with the highest level of agreement among respondents, in addition to some adoption stage.

Despite these adoptions, in order to achieve more significant benefits, companies need to overcome barriers to implementation. Thus, the cases converge on the difficulty of implementing digital technologies due to the obsolescence of facilities and equipment. The high cost to acquire and maintain the technologies operating is also an obstacle for both, so the technologies of collaborative robots and augmented reality, currently implemented, will not have expanded use. The legislation in Brazil makes it difficult to use the technologies as designed or prevent the use of those that are not widely disseminated. Thus, it was necessary to isolate the operation performed with a collaborative robot and, tests with drones for short-distance delivery in inhabited places could not be performed. In this limitation or prohibition, the objective is to preserve the physical integrity of individuals. In Case A, investments in technologies must be justified with a return in two years. Also, corporate policies require the company to implement standard technologies defined in Roadmap I4.0, which can become costly in the economic context of Brazil. For Case B, the lack of ability to analyze data makes it impossible to extract value from Big Data.

Despite the barriers, the cases have digital technologies implemented, under development, pilot stage and others are still being designed. In Case A, technologies are concentrated in the transformation (production) process, while in Case B, technologies are dispersed in the processes. However, there is a tendency to adopt technologies that improve the service provided and, consequently, contact with the client.

Cross-analysis of results showed that the cases converge in the adoption and diverge in the use of some technologies, for example, the additive manufacturing and the drones and self-guided vehicles. Additive manufacturing is used in Case A for production aid parts, as clamping devices, and in Case B, for replacement parts that will be used in providing the service. However, in Case A, there is a forecast of use for spare parts, proving that this is a proper use of additive manufacturing. In this application, additive manufacturing provides reduced inventory costs, quick delivery and flexibility of a variety of items.

For the use of drones and self-guided vehicles, Case A provides for the use of AGVs (Automated Guided Vehicles), and Case B, drones for measuring the location of equipment installation and short distance deliveries. Case A does not have the expected benefits. Case B expects, with the measurement of installation locations, to improve quality, delivering products as specified, equipment delivery speed, in addition to reducing labour costs with rework. For the other technologies observed, the competitive cost criterion was the most targeted, being in case A related to the product and in Case B, to the service. In addition, this criterion is in line with the criterion of most significant importance to the client.

The synthesis of the cross-analysis of digital transformation in cases A and B are presented in Table 1.

Thus, it can be concluded that the two cases analyzed have digital technologies aimed at digital transformation. However, differences related to the management of the corporation to which they belong, the final product offered to customers, and the SC processes in which they employ digital technologies to deliver value to customers were observed.

Table 1 Synthesis of digital transformation

Criteria	Case A	Case B
Strategy for digital transformation	Yes	No
Concentration of digital technologies in an SC process	Yes	No
Process with more related digital technologies	Transformation (production)	Transformation (service)
Competitive criteria focused on technologies	Cost	Cost
Barrier to the implementation of digital technology	General	Specific

5 Discussion

As in the literature, in both cases, the interviewees demonstrated different under-standings about what characterizes each digital technology presented in the interview script. These are recent digital technologies, and those that make up the Enabling and Integrating class do not have a physical concept that characterizes them, making understanding more complex. Sorkun [12, p. 177] established that "the lack of a clear understanding of these buzzwords and their interrelations is a barrier to determining an accurate roadmap for the digitization process in companies." However, even with the roadmap defined in Case A, the interviewees presented divergences of concepts with the literature.

Based on what was identified in the field as digital technologies, it was necessary to create a fourth category, advancing in relation to research on Gurría [2]. The company considers that the digital transformation origins with technologies from system information, data collection, and micro automation and denote those as the most prominent technologies observed in the cases studies.

Case A presented some technologies implemented as a pilot, corroborating with the literature that foresees that these implementations serve to evaluate the initial results, verify the potential result and determine the necessary infrastructure to capture benefits on a larger scale [13]. In Case B, as most technologies have simpli-fied implementation, for example, smartphone application, the company does not need a pilot project for the adoption of technology.

It remains evident that the digital transformation will not be sudden. The integra-tion of new technologies with those implemented in organizations should be consid-ered, enabling the old systems to operate in real-time [3]. However, the information systems implemented in the analyzed cases are rigid, making it difficult to integrate with new information systems and technologies that perform data acquisition.

The two cases analyzed presented ERP (Enterprise Resource Planning) and MES information systems implemented, but no integrations were observed between them. However, for more significant benefits to be observed, MES data should be integrated into the ERP. In this way, the ERP will receive production information in real-time, making it possible to respond adequately to the planning and minimize the failure conditions [14]. The lack of integration means that information regarding the order's progress—and mostly the delivery date—is not available until the entire order has been completed and entered into the system [14].

Case A adopts the strategy of integrating and monitoring processes. This strategy is mainly followed by companies with a physical product and, which serve other businesses [15], as well as the company in Case A, which supplies its products to vehicle manufacturers. Soon, the transformation in internal processes began, seeking improvements in operations.

Case B seeks to improve services while making changes to SC processes. This path is considered a hybrid between the focus on the service or the process and, it can be the most time-consuming and difficult since at the same time that it seeks to innovate in the product, it makes changes in the processes [15].

As for the technologies adopted in the cases, additive manufacturing is the only Application technology implemented in both cases. The uses implemented, prototypes and parts for visual analysis, fixation devices, and spare parts, corroborate with studies that show that by 2022, the percentage of spare parts produced by additive manufacturing reaches 85% of production [16]. For the manufacture of prototypes and parts for visual analysis, the technology currently represents 45% of the manufacturing processes, and it is estimated that by 2022 this process will reach 100% of production [16].

The expected benefits with the application of additive manufacturing in SC are increased manufacturing flexibility, allowing for a greater variety of items, reduced delivery times, product customization, and reduced inventory [17]. Thus, the use implemented for spare parts, with repercussions on cost through reduction of stock and obsolete parts, is supported in the literature.

The interviewees did not mention the sustainability predicted by the use of additive manufacturing. However, when using the exact amount of material, waste is eliminated [18], reducing material consumption and impacting sustainability. It can be inferred that as in both cases, only the polymeric material is used, this benefit was not significant since this is not a scarce resource.

The literature postulates that advanced robots are machines with intelligence, automated and incorporated resources [19] and that collaboration in an environment without fence will improve the productivity and efficiency of resources, combining the flexibility of human beings and the precision of the machines [20].

Unlike the literature, the capacity for self-awareness and self-maintenance in robots presented was not evidenced. Thus, the technology was redefined as cobot. Cobot has part of the characteristics of an advanced and collaborative robot since the need for fencing is provided by the legislation in Brazil, but it does not have the characteristics of self-awareness and self-maintenance.

However, the most significant divergence from the literature is related to the expected benefit from the use of robots. Robots will be used to reduce the cost of the operation and supply the demand for operators, in addition to enabling faster customer service [21]. However, in the analyzed use, the robots do not increase the speed of production, since the same operation can be carried out by two operators and are not significant in reducing costs since the implementation and maintenance parts make the use costly and the labour cost is not high in the country.

Likewise, both cases project the use of drones and self-guided vehicles. In Case A, the studies are limited to internal logistics, with the use of AGV, while in Case B, the expected use of drones will be for measuring locations for the installation of equipment. This use projected in Case B is not foreseen in the literature, since it provides for the use of drones for logistical activities, such as for short-distance deliveries internal to organizations or external, being called "last mile" logistics [22]. The use of self-guided vehicles for deliveries, pointing out the decrease in fuel consumption, reduction in the cost with operators and accidents is proposed in the literature [23]. Tests for the use of drones in "last mile" logistics were mentioned in Case B; however, due to legislation in Brazil, these tests and subsequent use were not allowed.

The IoT installed in Case B needs to evolve so that more significant benefit is acquired. Four layers make up the IoT [24]. It can be concluded in this case that the first three were reached, firstly the layer that integrated with the existing hardware, secondly the network layer that supports the transfer of information, and finally the layer that creates and manages services and applications that satisfy users. It must also develop a layer that provides methods of interaction with the user and other applications.

However, the inclusion of IoT in the product/service cannot be used by the company as a way to increase the value of the service provided since the service customer considers the cost to be a competitive performance of the utmost importance. Thus, digital transformation should not be considered as a source of income as customers expect digital offers to be free, so the creation of digital resources should be seen as an investment [25].

In Case B, despite having equipment operation data, they are not used for product development. However, IoT data is expected to be used to improve development or design new products, since operating data, operating conditions of parts and proactive wear recognition are available [26].

Augmented reality applications can be found in the context of quality control, maintenance, assembly, material separation in warehouses [27]. In Case B, augmented reality is used for maintenance, as provided in the literature. The use of augmented reality glasses for maintenance allows the superimposition of the virtual image with a real image, contributing to the provision of services as it facilitates the identification of the fault. The projected use, in Case A and by the corporation in Case B, for material separation in the warehouse, facilitates the operation by freeing the operator to consult printed lists or displays, is also proposed in the literature. However, in the company in which the augmented reality is implemented, there are disadvantages in its use, since the augmented reality glasses add material for employees to carry when carrying out maintenance.

The use of RPA (Robotic Process Automation) is not considered artificial intelligence. However, the respondents classify it as such. The RPA, as it is applied in companies, does not learn from the data and is not flexible to adapt to unscheduled situations, characteristics of artificial intelligence and Machine Learning [28]. The digital technology of RPA was not previously in the protocol. However, it was identified in the two case studies, demonstrating that this is an essential technology for the digital transformation. Full-time availability, quality and accuracy of activities are benefits of RPA [29] that were not cited by respondents.

Despite being discussed in the literature to guarantee the origin of products and veracity in SC information, the blockchain technology [30–32], was not pointed out by the interviewees, nor did they present studies or projects for use. Since blockchain is a recent technology and "few companies have adopted its full deployment in SC" [33, p. 2026], it can be concluded that companies are unaware of the technology and its possible applications.

The most applied or studied technologies in the analyzed cases are in the Application class. However, the theory studied defined that the implementation should start with the Enabling technologies, followed by the Integrators and finally the Application ones [2]. However, it became evident that, in the case of advanced and collaborative robots, what has been implemented are cobots with collaborative capacity and not with intelligence to learn and adapt themselves to situations that were not programmed. In the same way for AGV, the applied technology is not characterized as driverless vehicles in the studied concept, but towing car guided by magnetic strips and tracks. For additive manufacturing, the applied and studied technology is utilized as in the definition.

In Case B, no data security issues were identified. However, in Case A, it is essential to prove data security to implement technologies. It is worth mentioning that data security issues, data theft, attacks, dependence on information technology, correct functioning of working algorithms, among others, should be considered as a priority by organizations when choosing a technology [26].

Finally, it is concluded that for digital transformation to happen, companies need to expand the use of digital infrastructure technologies, as these will serve as a basis for other technologies. Consequently, it is expected that efforts will be maintained and expanded, both in the implementation of digital technologies and in the integration of SC processes.

6   Conclusion

The digital transformation is taking place in companies, but digital technologies are more related to information systems, use of automation and smartphones. Thus, there is evidence that most technologies have a positive impact on processes, allowing competitive criteria for the Supply Chain. However, other technologies, despite having effects in the processes, due to barriers, will have their use discontinued, and the adoption is not to be expanded, at least in the short term.

From the theoretical point of view, it is understood that the research presents applied contributions on the existing theory about digital transformation, it is visualized that it presents the empirical results of the occurrence of these two companies. The case studies make it possible to deepen the assessment of the adoption of digital technologies in companies and to understand the interviewees' understanding of the concepts of recent technologies, also identified as buzzwords. The technologies that present the best developments are those that have a physical concept, belonging to the class of Application technologies.

For future work, it is suggested to carry out empirical studies evaluating a complete supply chain, making it possible to assess the perception of the members of the supply chain about digital transformation, and not only the focal company.

Finally, it established that digital transformation is discussed in companies and is changing how processes are performed. Companies need to be prepared and know the benefits of adopting digital technologies. In the same way, failure in the adoption of digital technologies must be disclosed. Thus, the investments made in technologies that are known to have had no significant return for the SC will be informed to other companies. This knowledge will allow companies to plan their digital transformation.

References

1. S. Vachon, A. Halley, and M. Beaulieu, "Aligning competitive priorities in the supply chain: The role of interactions with suppliers," *Int. J. Oper. Prod. Manag.*, vol. 29, no. 4, pp. 322–340, 2009.
2. A. Gurría, "The Next Production Revolution," OECD Publishing Press, 2017.
3. H. Kagermann, W. Wahlster, and H. Johannes, "Recommendations for implementing the strategic initiative INDUSTRIE 4.0," *Final Rep. Ind. 4.0 WG*, no. April, p. 82, 2013.
4. CNI, "Sondagem especial," 2016.
5. C. O. Klingenberg, M. A. V. Borges, and J. A. V. Antunes, "Industry 4.0 as a data-driven paradigm: a systematic literature review on technologies," *J. Manuf. Technol. Manag.*, 2019.
6. A. G. Frank, L. S. Dalenogare, and N. F. Ayala, "Industry 4.0 technologies: implementation patterns in manufacturing companies," *Int. J. Prod. Econ.*, vol. in press, 2019.
7. M. Brettel, N. Friederichsen, M. Keller, and M. Rosenberg, "How Virtualization, Decentralization and Network Building Change the Manufacturing Landscape: An Industry 4.0 Perspective," *Int. J. Mech. Aerospace, Ind. Mechatron. Manuf. Eng.*, vol. 8, no. 1, pp. 37–44, 2014.
8. A. Mussomeli, D. Gish, and S. Laaper, "The Rise of the Digital Supply Chain," *Deloitte*, vol. 45, no. 3, pp. 20–21, 2015.
9. D. Schulman, M. Hajibashi, R. Narsalay, and S. Sreedharan, "Is your supply chain in sleep mode?," Accenture, 2018.
10. R. Schmidt, M. Möhring, R.-C. Härting, C. Reichstein, P. Neumaier, and P. Jozinović, "Industry 4.0 -Potentials for Creating Smart Products: Empirical Research Results," *Int. Conf. Bus. Inf. Syst.*, 2015.
11. C. Voss, N. Tsikriktsis, and M. Frohlich, "Case research in operations management," *J. Oper. Manag.*, vol. 22, no. 2, pp. 195–219, 2002.
12. M. F. Sorkun, "Digitalization in Logistics Operations and Industry 4.0: Understanding the Linkages with Buzzword," in *Digital Business Strategies in Blockchain Ecosystems: Transformational Design and Future of Global Business.*, Switzerland: Springer International Publishing AG, 2020.
13. C. Schmitz, A. Tschiesner, C. Jansen, S. Hallerstede, and F. Garms, "Industry 4.0: Capturing value at scale in discrete manufacturing," *McKinsey & Company.* 2019.
14. J. Kletti, *Manufacturing execution systems - MES*. New York: Springer Berlin Heidelberg, 2007.
15. S. Tripathi and M. Gupta, "Transforming towards a smarter supply chain," *Int. J. Logist. Syst. Manag.*, vol. X, pp. 1–24.
16. R. Geissbauer, J. Wunderlin, and J. Lehr, "A look at the challenges and opportunities of 3D printing," 2017.
17. D. Ivanov, A. Dolgui, and B. Sokolov, "The impact of digital technology and Industry 4.0 on the ripple effect and supply chain risk analytics," *Int. J. Prod. Res.*, pp. 1–18, 2018.
18. A. Angeleanu, "New Technology Trends and Their Transformative Impact on Logistics and Supply Chain Processes.," *Int. J. Econ. Pract. Theor.*, vol. 5, no. 5, pp. 413–419, 2015.
19. B. Bayram and G. İnce, "Advances in Robotics in the Era of Industry 4.0," in *Industry 4.0: Managing The Digital Transformation*, Switzerland: Springer, 2018.
20. L. Wang, M. Törngren, and M. Onori, "Current status and advancement of cyber-physical systems in manufacturing," *J. Manuf. Syst.*, vol. 37, pp. 517–527, 2015.
21. M. Merlino and I. Sproge, "The Augmented Supply Chain," *Procedia Eng.*, vol. 178, pp. 308–318, 2017.

22. M. Heutger, M. Kuckelhaus, D. Niezgoda, and S. Endriß, "Unmanned Aerial Vehicles in Logistics." 2014.
23. T. Stock and G. Seliger, "Opportunities of Sustainable Manufacturing in Industry 4.0," *Procedia CIRP*, vol. 40, no. Icc, pp. 536–541, 2016.
24. L. Da Xu, W. He, and S. Li, "Internet of things in industries: A survey," *IEEE Trans. Ind. Informatics*, vol. 10, no. 4, pp. 2233–2243, 2014.
25. G. Oswald and M. Kleinemeier, *Shaping the digital enterprise: Trends and use cases in digital innovation and transformation.* 2017.
26. D. Ivanov, A. Tsipoulanidis, and J. Schönberger, "Digital Supply Chain, Smart Operations and Industry 4.0," in *Global Supply Chain and Operations Management: A decision-oriented introduction into the creation of value*, no. 2, Springer Nature, p. 2019, 2019.
27. H. Correa, "Industry 4.0 and its implications for global supply chains," in *Administração de Cadeias de Suprimentos*, Atlas, 2019.
28. J. L. Hartley and W. J. Sawaya, "Tortoise, not the hare: Digital transformation of supply chain business processes," *Bus. Horiz.*, no. Article in press, 2019.
29. A.-W. Scheer, *Enterprise 4.0 - From disruptive business model to the automation of business processes*, 1st ed. Saarbrucken: AWSi Publish, 2019.
30. H. M. Kim and M. Laskowski, "Toward an ontology-driven blockchain design for supply-chain provenance," *Intell Sys Acc Fin Mgmt*, vol. 25, pp. 18–27, 2018.
31. N. Kshetri, "Blockchain's roles in meeting key supply chain management objectives," *Int. J. Inf. Manage.*, vol. 39, pp. 80–89, 2018.
32. F. Tian, "An Agri-food Supply Chain Traceability System for China Based on RFID & Blockchain Technology," *2017 Int. Conf. Serv. Syst. Serv. Manag.*, pp. 1–6, 2017.
33. S. Kamble, A. Gunasekaran, and H. Arha, "Understanding the Blockchain technology adoption in supply chains-Indian context," *Int. J. Prod. Res.*, vol. 57, no. 7, pp. 2009–2033, 2019.

Industry 4.0 Technologies for Improving Functional Machinery Safety Management from the Interoperability Point of View in the Automotive Industry

Marcio Lazai Junior, Álvaro dos Santos Justus,
Eduardo de Freitas Rocha Loures, Eduardo Alves Portela Santos,
and Anderson Luis Szejka

Abstract The functional machinery safety is currently one of the most relevant perspectives in industrial machinery design and processes in Brazil. Poor management strategy of it can lead to problems such as project delays, overuse of resources and even work-related accidents. A point of attention in processes management is interoperability, which can be defined as an organizational function necessary for proper communication and understanding of information between computational and human agents. In order to identify elements that influence the good performance of machinery safety management (GSFM) from the perspective of interoperability, a study was conducted in an automotive company of European origin located in Brazil. With the aid of enterprise interoperability frameworks (EI) and AHP (Analytic Hierarchy Process) multicriteria analysis method, a diagnostic approach is proposed in order to identify the main barriers on GSFM performance by pointing out weak, stable and strong criteria from organizational perspectives, processes, services and information. The results allow better targeting of improvement actions through industry 4.0 technologies linked to digital transformation initiatives. Thus, a review of the literature on the industry's 4.0 enablers is presented to suggest effective ways to improve the diagnosed weakened criteria in the GSFM while minimizing incident interoperability barriers. Through the PROMETHEE II decision support method, a prioritization of the main technologies is presented, consuming the diagnosis obtained by the AHP method in the automotive company under analysis.

Keywords Functional machinery safety · Enterprise interoperability · Industry 4.0 · Management · AHP · PROMETHEE II

M. L. Junior (✉) · Á. dos Santos Justus · E. de Freitas Rocha Loures · E. A. P. Santos · A. L. Szejka
Pontifícia Universidade Católica Do Paraná (PUC-PR), Rua Imaculada Conceição, Curitiba, PR, Prado Velho 1155, Brazil

© The Author(s), under exclusive license to Springer Nature Switzerland AG 2021
A. M. Tavares Thomé et al. (eds.), *Industrial Engineering and Operations Management*,
Springer Proceedings in Mathematics & Statistics 367,
https://doi.org/10.1007/978-3-030-78570-3_36

1 Introduction

Functional machinery safety is defined by elements that work safely, reliably and productively, as well as solutions that are technically and commercially viable [1]. The problem addressed refers to the poor management of the safety of functional machines (GSFM), which can result in the excessive use of technical and human resources and, in the worst case, work-related accidents. Both cases result in additional, direct and indirect, costs for the company.

The inability of effective communication is one of the main points that weaken the GSFM and can be assessed from the point of view of interoperability, which represents the ability of two or more systems to exchange information and consume this information [2]. In this context, a study was conducted in a Brazilian subsidiary of a European automotive company in order to evaluate the enterprise interoperability in the functional machinery safety management and this article is the continuation of this study [3].

Another point that can contribute to the GSFM are technologies, mainly the most recent ones addressed in industry 4.0. Industry 4.0 (I4.0) has been considered a new industrial stage in which several emerging technologies are converging to provide digital solutions [4] and it is one of the most trending topics in professional and academic fields [5] providing a broad branch of study to serve as a basis for the purpose of this study.

Therefore, this article aims to select I4.0 technologies that assist in improving or eliminating barriers of enterprise interoperability (EI) for the functional machinery safety criteria evaluated in the previous study. To do that, resources will be used as a survey with specialists and multicriteria decision making analysis (MCDA). As a result, a hierarchical classification of I4.0 technologies that will most contribute to the improvement of functional machinery safety criteria will be proposed.

2 Methodological Procedure

The article will be developed in three main parts, such as literature review, evaluative space, and results. This review will be the basis for understanding the GSFM criteria with great barriers of enterprise interoperability which will be the focus of improvement addressed in this study.

Industry 4.0 is understood as a new industrial stage in which there is an integration between manufacturing operations systems and information and communication Technologies [6]. In the GSFM scenario, this represents an aid in the collecting, transmission, processing and understanding of data related to functional machinery safety. For these reasons, the technologies of industry 4.0 were chosen to enable improvements to the GSFM criteria.

To evaluate I4.0 technologies to serve as a basis for improving the GSFM criteria from the perspective of enterprise interoperability, a multicriteria decision analysis approach is used. The method PROMETHEE II was chosen because it presents an adequate structure for the crossing and weighting between the GSFM criteria and the technologies.

As enablers, I4.0 technologies will be assessed individually supported by the RAMI 4.0 framework, to guide the area of implementation of the technology for better performance when applied to the GSFM context. A similar approach is presented by Ramos et al. [7] where an evaluation methodology is introduced between legacy and local systems using MCDA in order to verify if they meet the specifications of the organization's digital transformation projects.

Finally, the proposed MCDA method will be applied based on the GSFM criteria from the perspective of interoperability, relating them to I4.0 technological enablers applicable to the context. As a result, it is expected that this weighting will more adequately list the technologies that most contribute to the improvement of the process in the organization.

3   Conceptual Basis and Literature Review

3.1   GSFM from the Perspective of Interoperability

Previous results show that when it comes to functional machinery safety, depending on the installation location, different legal regulations that make it necessary to apply different standards may come into question [8]. Internationally, ISO (International Organization for Standardization) standard is available for anyone to consult, that entity is an independent, non-governmental international organization with a membership of 164 national standards bodies [9]. However, some countries develop their own regulatory standards, and Brazil is one of those countries. Having 36 regulatory standards (NR), Brazil has made it mandatory to comply with them, including the NR12 standard that specifically deals with safety at work in machinery and equipment [10]. Therefore, this standard dictates the minimum safety requirements that machinery and equipment must follow in Brazil [11], which is why it was chosen as the basis for obtaining safety criteria.

3.2   Technologies and Framework of I4.0

To start evaluating I4.0 technologies, it is necessary to list and understand these technologies. According to Franka et al. [4] I4.0 can be divided into 5 dimensions, each of which covers a range of technologies aimed at a common goal. Within each dimension there are subdivisions into categories, it is worth mentioning that some types of technologies can be used in one or more dimensions such as the types of technologies aimed at automation and flexibility [12].

The dimensions addressed by the authors [4] are base technologies, smart working, smart supply chain, smart products e smart manufacturing. For the management of functional machinery safety, all of these dimensions are relevant throughout the production process where interoperability requirements are fundamental for the performance of the processes involved. In highlight, we have the dimension of smart manufacturing as it is the centralizer of the processes involving machines and equipment.

It is in this dimension that categories such as energy management (i), flexibility (ii), traceability (iii), automation (iv), virtualization (v) e vertical integration (vi) are addressed and help to understand the pillars that make up the dimension [4]. Each category has its own characteristics that put together different I4.0 technologies.

In energy management there are technologies aimed at monitoring [13] and improving systems [14] of energy management. This dimension is at the top of organizations and is contained in a business sphere. In the flexibility category, technologies that promote agility in line setups, rapid prototyping and agile physical real-locations are addressed. Some technologies to be listed are additive manufacturing [15] e flexible and autonomous lines [16]. In traceability, the focus is on identification and traceability of raw materials and final products [17] avoiding losses related to stocks and storage.

The automation category has the greatest focus on the industrial scenario involved because it is dependent on machines and equipment to existing, thus demanding a high commitment to functional safety. In it, you can find technologies such as Machine-to-machine communication (M2M), robots and Automatic non-conformities identification in production [13].

The virtualization category addresses the idea of bringing the factory into a virtual environment, being used in several stages of design and series life, as in the simulation of processes for optimizations [14], artificial intelligence for predictive maintenance [18] and artificial intelligence for the planning of production [13].

Factory's vertical integration comprises advanced ICT systems that integrate all hierarchical levels of the company—from shop floor to middle and top-management levels—helping decision-making actions to be less dependent of human intervention [19]. The technologies that are addressed for this category in the literature are sensors, actuators and programmable logic controllers (PLC), supervisory control and data acquisition (SCADA), enterprise resource planning (ERP) [14], manufacturing execution system (MES) [20], machine-to-machine communication (M2M) [13] and virtual commissioning [21].

Considering the divisions proposed by Franka [4], in the smart manufacturing dimension and its categories, the most relevant technologies are:

- Energy efficiency improving system;
- Energy efficiency monitoring system;
- Flexible and autonomous lines;
- Additive manufacturing;
- Identification and traceability of final products;
- Identification and traceability of raw materials;

- Automatic nonconformities identification in production;
- Robots;
- Machine-to-machine communication (M2M);
- Artificial Intelligence for planning of production;
- Artificial Intelligence for predictive maintenance;
- Simulation of processes;
- Virtual commissioning;
- Enterprise Resource Planning (ERP);
- Manufacturing Execution System (MES);
- Supervisory Control and Data Acquisition (SCADA);
- Sensors, actuators and Programmable Logic Controllers (PLC);

These technologies addressed in the smart manufacturing dimension are just a branch of many others (smart product, supply chain, smart work and base technologies), that is why several frameworks focused on I4.0 are being developed in order to define, manage, evaluate and among other objectives. RAMI 4.0 (Reference Architectural Model for Industrie 4.0) is one of the most cited frameworks in the literature [22].

With the main objective of defining relevant prerequisites to start the implementation process in the industry, RAMI 4.0 aims to define a communication structure and a common language with its own vocabulary, syntax, grammar, semantics and culture in the company [23].

The Reference Architectural Model Industrie 4.0, abbreviated RAMI 4.0, consists of a three-dimensional coordinate system that describes all crucial aspects of Industrie 4.0. In this way, complex interrelations can be broken down into smaller and simpler clusters. The model can be seen in the Fig. 1.

Briefly, observing the model, the left horizontal axis represents the life cycle of facilities and products, based on IEC 62890 for life-cycle management. On the right horizontal axis are hierarchy levels from IEC 62264, these hierarchy levels represent the different functionalities within factories or facilities. The six layers on the vertical axis serve to describe the decomposition of a machine into its properties structured layer by layer [23].

Fig. 1 RAMI 4.0 framework [23]

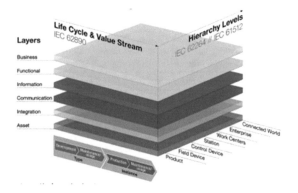

In this article, the main purpose is to correlate GSFM with enterprise interoperability and I4.0 technologies, or RAMI 4.0 to assist in structuring technologies hierarchically for those that are directed to the correct way of each sector responsible for the organization where the study is conducted. Therefore, the hierarchy levels of the layer will be a compass that will guide or direct the technologies.

4 Evaluative Space

4.1 GSFM Criteria with Major Interoperability Barriers

The attributes of functional machinery safety come from previous work [3] based on NR12. These attributes are filtered by Keeney's approach [24] which suggests that the criteria should be essential (i), controllable (ii), complete (iii), measurable (iv), operational (v), isolable (vi), non-redundant (vii), concise (viii) and understandable (ix), the attribute remains only if it meets all of these requirements. The attributes selected after the filtered, also called criteria, can be seen in Table 1.

After these definitions, the study of the criteria started from the perspective of enterprise interoperability. For this, the FEI framework (Framework for Enterprise Interoperability) was chosen [25]. This framework works on business interoperability in three dimensions, being concerns, barriers and approaches. However, the

Table 1 GSFM attributes

ID	Attributes
A1	Safety on machinery uses
A2	Use of international standards in the absence or omission of NR 12
A3	Application of standards on equipment purchased before the law that imposes NR12
A4	Collective protection measures
A5	Clean, level, unobstructed, signposted roads that allow safe traffic for employees
A6	Minimum space between machines and protective barriers
A7	Suitable location for tools
A8	Grounding in accordance with current regulations
A9	Start and stop pushbuttons, bimanual and emergency stops must be in sufficient quantity and in accordance with current regulations
A10	Mobile protection when frequent access
A11	Technical documentation developed by the owner when did not come from the manufacturer
A12	Pressurized components according to current standards
A13	Machines that respect ergonomics (NR 17)
A14	Surfaces without sharp corners

'concerns' and 'barries' dimensions are more significant for the objective of classifying the criteria with a focus on projects and serial life rather than products. The distribution of the GSFM criteria was made by allocating each one of them in the quadrants ('concern' vs 'barrier') of the FEI. Thus, each criterion had twelve evaluation points of view, which can be attributed to one or more of the quadrants of the matrix from the FEI.

The modeling of the evaluation space was done using the MCDA AHP method. As the name of the method already suggests, AHP does the hierarchical assessment for weighted quantification. This hierarchy is top-down, with the objective layers (i), perspectives (ii), barriers (iii), criteria (iv), levels of interoperability (v).

With the proper hierarchical assignments and their respective weights, pairwise comparisons are made with specialists from the organization where the study is conducted and with know-how in industrial automation and functional machinery safety. As a result, after the calculation inherent to the AHP model, the diagnosis on the organization level was obtained.

From this view on the company's position in the GSFM, it can be concluded that the result places the company at 53% at level 3, 39% at level 2 and 8% at level 1, which is the lowest [3]. In addition to this evaluation, in a more granular way, it was possible to identify the criteria with the greatest barriers to enterprise interoperability.

This means that there is a great potential for evolution, based on this diagnostic approach provided by the AHP, in the identification of improvements in the gaps identified through a plan for the adoption of technological enablers. The present work aims to suggest I4.0 technologies, guided by the RAMI4.0 framework, for these criteria impacted with barriers of business interoperability in order to contribute to a better performance of GSFM.

4.2 I4.0 Technologies Applicable to GSFM

In the methodological approach related to the literature review, the crossing of keywords related to functional machinery safety and I4.0 returns us few articles that in fact cover safety aimed at machinery and equipment and I4.0 technologies. The most discussed field at this intersection is that of cybersecurity, one of the main pillars of I4.0 focused on the security of the data that the organization generates. However, this type of safety is not the only representative of functional machinery safety in its entire scope of industrial activity.

In view of such a gap in relation to the connection of the two themes, 'functional machinery safety' and 'industry 4.0', a survey was developed in order to question machine design engineers with expertise in functional machinery safety, as proven by TUV [26], about which I4.0 technologies previously raised in Sect. 2.2 could contribute to the theme.

As a result of the survey, eight technologies stood out among the 17, as enabling functional machinery safety by more than 50% of the consulted specialists. These technologies are:

Hierarchy Levels (RAMI 4.0)	Smart manufacturing Technologies
Conected World	
Enterprise	Simulation of processes
Work Centers	Virtual commissioning Machine-to-machine communication (M2M)
	Flexible and autonomous lines
Station	Supervisory Control and Data Acquisition (SCADA) Automatic non conformities identification in production
Control Devices	
	Robots
Field devices	Sensors, actuators and Programmable Logic Controllers (PLC)
Product	

Fig. 2 Technologies distributed by the axis of the RAMI 4.0 framework

- Flexible and autonomous lines;
- Automatic non conformities identification in production;
- Robots;
- Machine-to-machine communication (M2M);
- Simulation of processes;
- Virtual commissioning;
- Supervisory Control and Data Acquisition (SCADA);
- Sensors, actuators and Programmable Logic Controllers (PLC);

Returning to the 'hierarchy levels' axis of the RAMI 4.0 framework, which presents a structure attributed to industry 3.0 defined by ANSI/ISA-95 [27], the first layer 'field devices' refers to technologies used in the field, objects that define what information should be exchanged. Above, the 'control devices' layer defines centralizing information technologies that exchange data between systems, the 'station' layer describes and compares different production areas in a standardized way, 'work centers' defines the models in which the information from the previous layer must be exchanged and the last layer 'enterprise' connects and organizes the operations and production of the previous levels. This whole structure connects the 'product' to the 'connected world'.

Thinking about the smart manufacturing technologies listed, it is possible to attribute them to this axis of RAMI 4.0, this relationship is illustrated in the Fig. 2.

4.3 MCDA Assessment of I4.0 Technologies as Enablers for GSFM Criteria

Having the attributes allocated to the interoperability quadrants, a total of 65 qualifying attributes are obtained, each composed of the GSFM attribute, a 'concern' and a 'barrier'. It is possible, therefore, with the aid of the PROMETHEE II multicriteria method and experts from the group who have knowledge of I4.0 technologies and

Scenario1	A1 BC	A1 PC	A1 PT	A1 PO	A1 SC	A1 SO	A1 DC	A1 DO
Unit	impact	impact	impact	impact	impact	impact	impact	impact
Cluster/Group	◆	◆	◆	◆	◇	◆	◇	◆
Preferences								
Min/Max	max	max	max	max	max	max	max	max
Weight	0,17	0,09	0,25	0,12	0,17	0,10	0,17	0,20
Preference Fn.	Usual	Usual	Usual	Usual	Usual	Usual	Usual	Usual
Thresholds	absolute	absolute	absolute	absolute	absolute	absolute	absolute	absolute
- Q: Indifference	n/a	n/a	n/a	n/a	n/a	n/a	n/a	n/a
- P: Preference	n/a	n/a	n/a	n/a	n/a	n/a	n/a	n/a
- S: Gaussian	n/a	n/a	n/a	n/a	n/a	n/a	n/a	n/a
Statistics								
Minimum	3,0000000000	2,0000000000	1,0000000000	3,0000000000	2,0000000000	2,0000000000	3,0000000000	4,0000000000
Maximum	5,0000000000	5,0000000000	5,0000000000	5,0000000000	5,0000000000	5,0000000000	5,0000000000	5,0000000000
Average	4,2500000000	3,8333333333	3,5000000000	4,1428571429	3,5000000000	3,8571428571	4,0000000000	4,7500000000
Standard Dev.	0,6614378278	0,8975274679	1,1180339887	0,8329931278	1,2583057392	0,9897433186	0,7071067812	0,4330127019
Evaluations								
☑ Flexible and auto... ☐	high	high	high	very high	n/a	high	n/a	n/a
☑ Automatic non c... ☐	moderate	n/a	very high	n/a	n/a	n/a	high	very high
☑ Robots ☐	very high	high	high	high	low	high	n/a	high
☑ Machine-to-mach... ☐	high	n/a	high	moderate	moderate	very high	moderate	n/a
☑ Simulation of pro... ☐	high	high	moderate	very high	very high	moderate	n/a	n/a
☑ Virtual commissio... ☐	very high	very high	high	very high	very high	low	n/a	n/a
☑ Supervisory Con... ☐	high	high	very low	high	high	very high	very high	very high
☑ Sensors, actuato... ☐	very high	low	moderate	moderate	low	high	high	very high

Fig. 3 Representative model of PROMETHEE II

understanding the dimensions of interoperability, to create a structure where qualitative weights can be attributed individually regarding the relevance of each I4.0 technology for each attribute.

Each resulting attribute that makes up the method is represented by a column, returning a high dimensionality and extension of the model. For this reason, the attributes derived from A1 (Safety on machinery uses) are shown in Fig. 3 as a means of exemplifying the model as a whole, as the other attributes followed the same behavior.

Figure 3 shows the decision matrix of the Promethee II method implemented on the Visual PROMETHEE II platform. The first line, present in the section 'A' of the figure, represents the 65 qualifying attributes, positioned in the evaluation quadrants of interoperability. In other words, there is the representation of the intersection of the functional machinery safety attribute with the concern and barrier of the enterprise interoperability. Therefore, 'A1 BC' means the attribute A1 (Safety on machinery uses) positioned at the intersection between the concern 'Business' and the barrier 'Conceptual'. Still in 'A' is where Clusters and groups are defined so that the influence on each criterion can be calculated.

In the 'Preferences' session, represented by the 'B' session, the positive trend of the criterion is defined, how relevant it is to the study, signaled by 'max', for greater relevance if the value is higher and signaled by 'min' if the relevance is higher if the value is lower.

The weights must also be assigned to 'B' session, in this study these weights had their origin in the result of the first MCDA method, the AHP. As the main purpose of the study is to prioritize attributes with the greatest barriers to enterprise interoperability, the weights resulting from the AHP were inverted and normalized in order to prioritize the technologies over the most fragile criteria diagnosed in the AHP multicriteria method. Therefore, it is possible to say that the weights assigned in PROMETHEE II are the result of the formula 1 and 2.

$$MF = Wc * Wb \tag{1}$$

$$Wp = \frac{1 - (MF * Wahp)}{\sum ACBi} \tag{2}$$

- MF = Multiplication Factor
- Wc = Weight Concern
- Wb = Weight Barrier
- Wp = Normalized Weight Promethee II
- Wahp = Resulting weight from AHP analysis
- ACBi = Attribute weight on enterprise interoperability

The 'Preference fn.' And 'Thresholds' fields were defined with 'usual' and 'absolute' because these are future assignments on a qualitative scale in an impact assessment. The 'Evaluations' section represented by 'D' in Fig. 3 is where each of the 'actions' is assigned a value, in this case, the 8 I4.0 technologies listed for the criteria to be addressed. The 'C' session is automatically completed by the mathematical method. After all these assignments are made for all 65 attributes, it is possible to analyze the results.

5 Results

In order to verify the technology with the greatest contribution to the organization as a whole, addressing all the attributes of all the interoperability quadrants, the method is requested considering all 65 criteria and the result can be seen in the Fig. 4, which illustrates the order of relevance of each technology against each criterion allocated in the quadrants of enterprise interoperability.

From the resulting table that represents the values of 'Phi', it is possible to conclude that the technology that involves the base of the hierarchical pyramid of automation,

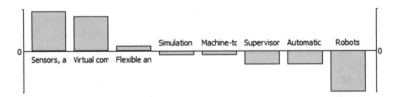

Fig. 4 Relevance of each 4.0 technology to GSFM

in this case, 'Sensors, actuators and Programmable Logic Controllers (PLC)' are the ones that present the greater contribution potential to eliminate interoperability barriers in the organization where the study has been conducted. However, with a very close result, 'Virtual commissioning' is an outstanding technology, pointing to a higher organizational approach because it is a technology applied in projects and of a strategic nature.

Some factors contributed to these two technologies in particular to have such prominence. The first factor is the maturity with which the organization already conducts these technologies in its routine, this factor has great importance because it implies low costs in relation to the potential for gain. Technologies that are already present in the know-how of the technical staff tend to have a faster deployment speed, with fewer errors and less need for investment in tools, whether hardware or software.

The other technologies were closer to or below zero, this implies a low contribution compared to the two technologies highlighted above. This does not imply a negative contribution, but a little one, which requires greater organizational, financial and knowledge development efforts to generate expressive results for this case study.

These results corroborated the understanding that I4.0 technologies directly impact the functional machinery safety management, however, some stand out and sometimes for different reasons such as deployment speed, cost of implementation, maturity of the technical team with the technology and among others. In addition, this approach of assessing the functional machinery safety from the perspective of interoperability and having a purposeful space supported by I4.0 technologies is new, and it fills a literary gap.

6 Conclusion

Based on an already developed study that addressed the functional safety management of machines through an assessment of business interoperability, this study had the propositional direction supported by I4.0 technologies in order to list their relevance in improving interoperability barriers in order to improve GSFM in the organization where the study was conducted.

From the literature review, several technologies aimed at smart manufacturing were selected and filtered through a survey with specialists. The resulting technologies were related to the GSFM attributes and their respective interoperability concerns and barriers in order to structure another multicriteria method, PROMETHEE II, resulting in a hierarchical organization of the technologies according to their potential for improving the criteria of the GSFM from the perspective of enterprise interoperability.

In the end, two technologies stood out for having a high maturity in the organization where the study was conducted and presenting a potential superior to the others in relation to the evolution of GSFM, these technologies were 'Sensors, actuators and Programmable Logic Controllers (PLC)' and 'Virtual commissioning'.

This study represents the filling of a scientific gap that involves a purposeful space for improving the management of functional safety of machines, in addition to a contribution in the industrial context, listing the technologies with the greatest potential for improving management in the topic addressed.

In the next steps of this study, the attributes will be separated according to the clusters and groups so that there is a purposeful analysis of each technology highlighted for each hierarchical level of the organization since it has these well-defined and individualized levels among themselves. Still, a more in-depth evaluation using other tools provided by PROMETHEE II is considered, to evaluate rankings, networks and dimensions.

References

1. TUV Rheinland, "Functional Safety Technician Training Program," TUV Rheinland, 2012. [Online]. Available: http://www.tuv.com. [Accessed 15 feb 2020].
2. Institute of Electrical and Electronics Engineers, "IEEE standard computer dictionary: A compilation of IEEE standard computer glossaries," 1990.
3. M. L. Junior, E. d. F. R. Loures, E. A. P. Santos e A. L. Szejka, "Interoperability analysis in the functional machinery safety management," vol. 6, n° 1, pp. 3009–3023, january 2020.
4. A. G. Franka, L. S. Dalenogare e N. F. Ayala, "Industry 4.0 technologies: Implementation patterns in manufacturing," *ELSEVIER,* pp. 15–26, 2019.
5. F. Chiarello, L. Trivelli, A. Bonaccorsi e G. Fantoni, "Extracting and mapping industry 4.0 technologies using wikipedia," *ScienceDirect,* pp. 244–257, 2018.
6. L. S. Dalenogare, G. B. Benitez, N. F. Ayala e A. G. Frank, "The expected contribution of Industry 4.0 technologies for industrial performance," *International Journal of Production Economics,* vol. 204, pp. 383–394, 2018.
7. L. Ramos, E. Loures, F. Deschamps e A. Venâncio, "Systems evaluation methodology to attend the digital projects requirements for industry 4.0," *International Journal of Computer Integrated Manufacturing,* 2019.
8. Pilz GmbH & Co. KG, "Resumo de normas básicas (normas A+B)," 08 december 2019. [Online]. Available: https://www.pilz.com/pt-BR/knowhow/law-standards-norms/iso-standards.
9. International Organization for Standardization, "iso," 25 november 2019. [Online]. Available: https://www.iso.org/about-us.html.
10. Coordination of Technical Editions of the Federal Senate of Brazil, "Consolidation of Labor Laws - CLT," december 2017. [Online]. Available: https://www2.senado.leg.br/bdsf/bitstream/handle/id/535468/clt_e_normas_correlatas_1ed.pdf.
11. Portaria SIT n.° 197, "NR-12 – Segurança no Trabalho em Máquinas e Equipamentos," 2010.
12. Y. Liao, F. Deschamps, E. d. F. R. Loures e L. F. P. Ramos, "Past, present and future of Industry 4.0 - a systematic literature review and research agenda proposal," *International Journal of Production Research,* 2017.
13. A. Gilchrist, Industry 4.0: The Industrial Internet of things, Bangken, Nonthaburi Thailand: Apress, 2016.
14. S. Jeschke, C. Brecher, T. Meisen, D. Özdemir e T. Eschert, Industrial Internet of Things and Cyber Manufacturing Systems, Switzerland: Springer International Publishing, 2016.
15. C. Weller, R. Kleer e F. T.Piller, "revisited, Economic implications of 3D printing: Market structure models in light of additive manufacturing," *ScienceDirect,* pp. 43–56, 2015.
16. S. Wang, J. Wana, D. Zhang, D. Li e C. Zhanga, "Towards smart factory for industry 4.0: a self-organized multi-agent system with big data based feedback and coordination," *ScienceDirect,* pp. 158–168, 2016.

17. R. Angeles, "Anticipated IT infrastructure and supply chain integration capabilities for RFID and their associated deployment outcomes," *Elsevier,* pp. 219–231, 2009.
18. F. Tao, Q. Qi, A. Liu e A. Kusiak, "Data-driven smart manufacturing," *Elsevier,* pp. 157–159, 2018.
19. G. Schuh, R. Anderl, J. Gausemeier, M. t. Hompel e W. Wahlster, "Managing the Digital Transformation of Companies," *acatech STUDY,* 2017.
20. A. Telukdarie, E. Buhulaiga, S. Bag, S. Gupta e Z. Luo, "Industry 4.0 implementation for multinationals," *ELSEVIER,* pp. 316–329, 2018.
21. S. T. Mortensen e O. Madsen, "A Virtual Commissioning Learning Platform," *ELSEVIER,* pp. 93–98, 2018.
22. K. Schweichhart, "Reference Architectural Model Industrie 4.0 (RAMI 4.0)," Platform Industrie 4.0, 2015. [Online]. Available: https://ec.europa.eu/futurium/en/system/files/ged/a2-schweichhart-reference_architectural_model_industrie_4.0_rami_4.0.pdf. [Accessed 17 feb 2020].
23. ZVEI, "The Reference Architectural Model Industrie 4.0 (RAMI 4.0)," April 2015. [Online]. Available: https://www.zvei.org/fileadmin/user_upload/Themen/Industrie_4.0/Das_Referenzarchitekturmodell_RAMI_4.0_und_die_Industrie_4.0-Komponente/pdf/ZVEI-Industrie-40-RAMI-40-English.pdf.
24. R. L. Keeney, Value-Focused Thinking: A Path to Creative Decisionmaking, London: Harvard University Press, 1992.
25. D. CHEN, "Framework for Enterprise Interoperability," Université Bordeaux, Bordeaux, 2006.
26. TUV Rheinland, "Functional Safety," 13 january 2020. [Online]. Available: https://www.tuvasi.com/.
27. ISA95, "Enterprise-Control System Integration.," [Online]. [Accessed 10 feb 2020].
28. M. Papa, D. Kaselautzke, T. Radinger e K. Stuja, "Development Of A Safety Industry 4.0 Production Environment," *Annals of DAAAM & Proceedings,* 1 January 2018.
29. B. Dieber, A. Schlotzhauer e M. Brandstötter, "Safety & Security – Erfolgsfaktoren von sensitiven Robotertechnologien," *Springer Vienna,* pp. 1613–7620, 2017.

A Model to Integrate the BPM Life Cycle and the Design Thinking Process

Carolina Sierra Vargas⬡, João Francisco da Fontoura Vieira⬡,
Marlon Soliman⬡, Érico Marcon⬡, and Arthur Marcon⬡

Abstract The current scenario has been pressuring companies to innovate in order to stand out in the market. Innovation is not just about products, but also about improving processes. In this sense, this research aimed to propose an integrated life cycle model for process improvement, inspired both on the Business Process Management and on Design Thinking, in order to lead to more effective and creative process improvements. To propose the life cycle model, a literature review on the subject was conducted, and grounded on this result an integrated model is proposed. The initial proposition was then reviewed and improved based on opinions from specialists in the field. Finally, a revised life cycle model consisting of seven stages is proposed, integrating the main stages of the Business Process Management life cycle and the Design Thinking process.

Keywords Business process management · Design thinking · Life cycle model

C. S. Vargas (✉)
University of the Sinos Valley, Av. Dr. Nilo Peçanha, 1600, Boa Vista, Porto Alegre, Rio Grande do Sul 91330-002, Brazil

J. F. da Fontoura Vieira · É. Marcon · A. Marcon
Department of Industrial and Transport Engineering, Federal University of Rio Grande Do Sul, Av. Osvaldo Aranha, 99, Bom Fim, Porto Alegre, Rio Grande do Sul 90035-190, Brazil
e-mail: joao.francisco@ufrgs.br

A. Marcon
e-mail: arthur.marcon@ufrgs.br

M. Soliman
Department of Industrial Engineering and Systems, Federal University of Santa Maria, Av. Roraima, 1000, Camobi, Santa Maria, Rio Grande do Sul 97105-900, Brazil
e-mail: marlon.soliman@ufsm.br

© The Author(s), under exclusive license to Springer Nature Switzerland AG 2021
A. M. Tavares Thomé et al. (eds.), *Industrial Engineering and Operations Management*,
Springer Proceedings in Mathematics & Statistics 367,
https://doi.org/10.1007/978-3-030-78570-3_37

1 Introduction

Due to the constant changes in the economic scenario, companies are currently being pressured to innovate to survive. In this context, companies must seek new technologies, markets, and forms of management through innovation, to develop competitive advantages.

The traditional organizational structure is no longer sufficient to provide companies with the necessary agility to respond to market demands. Due to this phenomenon, business processes have been receiving more attention than the functional sectors that make up an organization, since these processes are the link between the company and customers. Business Process Management (BPM) is aligned to this context, as it aims to supervise processes in an organization to ensure consistent results [1]. This concept allows creating mechanisms to structure, evaluate, measure, and control business processes focused on results following the company's strategy [1, 2]. However, the methods for process improvement based on the BPM are still conservative in their approach, hardly producing innovative ideas and not always fulfilling the client's desires and needs. The BPM process improvement techniques are focused on incremental improvements, where it is difficult to obtain disruptive ideas.

On the other hand, design thinking is another approach that may be adequate for current market needs. It is an approach focused on the human-being, grounded on designers' mindset for solving business problems [3]. It is an approach that focuses on the client and seeks to deeply understand them, through an exploratory and empathic effort [3]. In this sense, Design Thinking promotes a collaborative and experimental agreement between the actors involved in the process improvement initiative to design real value-added solutions. Design Thinking translates observations into insights, and insights into innovation, through an exploratory, iterative and non-linear process, leading to unexpected discoveries [3–5]. Thus, Design Thinking and BPM have complementary attributes, which jointly can potentially create the competitive advantage through innovation that companies desperately need. However, little is known about how to apply the principles of Design Thinking to improve business processes. Consequently, literature has not described how an integrated approach between BPM and Design Thinking should be, in order to obtain the benefits of both approaches. In this sense, the following research question emerged: how can business processes be improved through an integrated BPM—Design Thinking approach? To answer this question, we proposed an integrated life cycle model that relates BPM and Design Thinking. The proposal was further evaluated by specialists and refined. We describe the final version that includes specialist's considerations.

2   Background

2.1   Business Process Management (BPM)

The BPM was created due to the increase in competitiveness and the need to improve the performance of organizations [6]. Similarly to the evolution of other approaches in process improvement, such as Lean Manufacturing, Six Sigma and Process Reengineering, BPM has evolved into an important management discipline [7, 8]. BPM is seen by many academics as a convergence of the process-focused disciplines aforementioned in a single knowledge field [9].

Baldam et al. [10] mention some decisive factors for a successful BPM adoption, such as: top management support; alignment with organization's strategy; people with the necessary experience and skills; clear and objective BPM-oriented structure, among others. The implementation of a BPM solution is complex, as it intersects departments and organization's boundaries, including customers and suppliers in the processes [11]. BPM adoption allows modeling existing processes, testing different configurations, and managing improvements and innovations [12]. Such activities aim to address a business objective [13].

BPM life cycle models. The BPM life cycle is the representation of the process stages. It aims to achieve service quality, in a structured, monitored, and optimized way. This leads to lower operating costs and greater customer satisfaction, which, provides better results for the organization. However, the phases need to be well defined in order to maintain an alignment of the process improvement project with the strategic objectives [2].

Life cycles begin with a strategy alignment. According to Weske [13], the alignment has the purpose of guaranteeing the successful implementation of the business process in the organization, before starting to define business process improvements.

Literature proposes several BPM life cycle models. Although the number of stages and their labels are different, the models have similar characteristics [14], as shown in Table 1. Dumas et al. [1] proposes a model focused on the process's organizational context through its positioning in the process architecture. Van der Aalst [15] and Netjes et al. [16] models are similar, however, the former has a greater focus on process automation, while the latter emphasizes monitoring improvements. Weske [13] offers a simplified life cycle model, covering BPM concepts and technologies, shedding light on the importance of using a Business Process Management System (BPMS). The ABPMP model [2] has a broader scope, which covers the stages of the models mentioned above.

Table 1 BPM life cycle models

ABPMP [2]	Dumas et al. [1]	Weske [13]	Netjes et al. [16]	Van der Aalst [15]
Planning	Process identification			
	Process discovery	Design	Design	Process design
Analysis	Process analysis			
Design	Process redesign	Configuration	Configuration	System configuration
Implementation	Process implementation	Enactment	Execution	Process enactment
Monitoring and control	Process monitoring and control	Evaluation	Control	Diagnosis
			Diagnosis	
Refining				

2.2 Design Thinking

Design Thinking is a systematic approach centered on the human-being and focused on problem-solving [3, 4, 17]. The purpose of Design Thinking is to obtain an in-depth customer understanding, aimed at meeting their emotional demands, with a reliable and creative analytical process to design the solution making the best use of the resources available [3, 18]. Design Thinking encourages innovation in different workplaces, so that people feel involved in the problem-solving process [19]. For Brown [3], Design Thinking is an abstraction from designers' mindset to bring new ideas to life, and anyone can absorb its concept and apply it in a diverse business or social scenarios.

As it is fundamentally an exploratory process [20], Design Thinking leads to unforeseen discoveries [3–5]. The information must be collected early in the process to generate insights and, subsequently, opportunities for innovation. The insights arise from deepening in the opportunity identification stage, which should be motivational and inspiring for people during the idea generation process [3, 21, 22].

Design Thinking process models. Similarly to the BPM, Design Thinking models vary according to the author, but they follow a similar logic. In Design Thinking, although the models show sequential steps, there is an emphasis on the process nonlinearity, presenting several comings and goings between steps [3].

Vianna et al. [20] propose the stages of immersion/analysis and synthesis, ideation, and prototyping, highlighting the need for a team with different profiles to generate innovative solutions to a problem. IDEO [21] presents similar steps, focusing, however, on human-beings' needs to maximize the impact of new solutions on individuals. Liedtka and Ogilvie [17] present steps with a different approach, through questions, aiming at stimulating creativity.

Table 2 shows a comparison between the models. We adopted the model from Brown et al. [3] as the basis for this study because, in addition to being an established

Table 2 Comparison between design thinking process models

Brown et al. [3]	Vianna et al. [20]	IDEO [21]	Liedtka and Ogilvie [17]
Inspiration	Immersion-analysis-synthesis	Hear	What is?
Ideation	Ideation	Create	What if? What wows?
Implementation	Prototyping	Deliver	What works?

model in literature and companies, it comprises all the other models in a simplified way, as depicted in Table 2.

3 Methodology

To conceive the proposed model, we adopted a Design Science Research (DSR) approach in this study. The DSR aims to create a solution to real problems with practical and theoretical contributions [23–25]. DSR encompasses the proposition and evaluation of a solution [26].

In our article, the proposition stage consisted of a literature research in the BPM and Design Thinking life cycle models (presented, respectively, in Sects. 2.1 and 2.2), where the ABPMP [2] and Brown et al. [3] models were selected as base article to build the proposal. Then, an initial version of an integrated model was developed based on a comparison between the objectives of the stages from the ABPMP [2] and Brown et al. [3] models, in order to understand the complementarity between them. We also organized a focused group composed of specialists in both themes to evaluate the integrated model proposal. Finally, grounded on the specialist's review, a second version of the model was proposed.

4 Results

4.1 Proposition of the Integrated BPM—Design Thinking Life Cycle Model

Initially, we identified the objectives of each BPM (Table 3) and Design Thinking (Table 4) stage. Subsequently, a crossed analysis was performed between the stages of the two approaches (Fig. 1). The analysis was carried out by questioning how the stages of Design Thinking could contribute to achieving the objectives of each BPM stage.

The first version of the proposed integrated model is shown in Fig. 2 and further described in Table 5. The model starts with the "Immersion and empathy" step, and then follows a clockwise direction. Nonetheless, it is noteworthy that the "Immersion

Table 3 Steps and objectives of the ABPMP [2] life cycle model

ABPMP [2]	Objective
Planning	To plan the process improvement project
Analysis	To understand the process and its problems
Design	To identify improvement opportunities for the process and design its future state
Implementation	To implement the designed process improvements
Monitoring and controlling	To compare the actual process to the designed process
Refining	To make incremental improvements in the post-implementation phase

Table 4 Steps and objectives of the Brown et al. [3] Design Thinking life cycle model

Brown et al. [3]	Objective
Inspiration	To understand clients' needs and aspirations, the problems faced by them, and the possible causes.
Ideation	To generate solutions to the problem
Implementation	To implement the solution through agile thinking

and empathy" stage can be resumed at any time. So, whenever necessary, it supports the decisions of the other stages, maintaining the principle of human-centered design in the model. Besides, the "Idea generation" stage is expected to follow the principle of creativity, and the "Practice," "Inspection," and "Enhancement" stages follow the principle of agile development.

5   Discussion

The initial version of the integrated model was presented to three specialists in the subject of this study. Table 6 presents the profile of each specialist. We conducted a focus group with these specialists to discuss the proposed model's applicability and identify opportunities for improvement.

All specialists agreed that the model is interesting and that it would contribute to their professional practices. Besides, they emphasized the the "Immersion and empathy" stage is the greatest differential from other process improvement approaches, which usually oversimplify studies on the problem and its causes. They also stressed that the relationship between the people involved in the process and the problems is rarely studied. However, specialist 2 pointed out that how the "Immersion and empathy" step relates to the other steps was not clear. Therefore, operational guidance on this step is required when applying the model.

The experts raised a concern on how specific the model was for business processes. Specialist 1 pointed out that the stages' labels were not explicitly referring to

		Steps and objectives – Design Thinking [3]		
		Inspiration	**Ideation**	**Implementation**
Steps and objectives - BPM [2]	Planning	The in-depth study on the client in the "Inspiration" stage helps to accurately identify the objectives in the "Planning" stage of the process improvement project	-	-
	Analysis	The in-depth understanding of clients' and actors' needs in the "Inspiration" stage can contribute to identifying process problems in the "Analysis" stage of the BPM life cycle	-	-
	Design	-	The creative and collaborative way of generating ideas in the "Ideation" stage can be useful for identifying improvements in the "Design" stage	-
	Implementation	-	-	The agile implementation (i.e., in a cycle of implementation, monitoring and adjustment) can bring a cyclical dynamic to the BPM "Implementation" and "Monitoring and control" steps
	Monitoring and control	-	-	
	Refinement	As the BPM "Refinement" stage consists of short cycles of analysis, improvement, and implementation, aiming at incremental improvements, the three stages of Design Thinking contribute to it.		

Fig. 1 Relationship between the BPM and design thinking approaches

processes, and thus it was not adequately contextualizing the model to this purpose. Specialist 3 agreed and pointed out that the model could be applied to any type of problem in its first version. Finally, specialist 1 pointed out it would be interesting to include in the explanation of the "Practice" stage the possibilities of applying both in back-end processes and in front-end processes.

Another discussion was whether the model could be applied to the development of new business processes or it would be focused only on improvements to existing processes. Specialist 2 signaled that the "State of the things" stage is not adequate for developing new processes, given that there is no current situation to be represented.

Fig. 2 First version for the integrated BPM—design thinking model

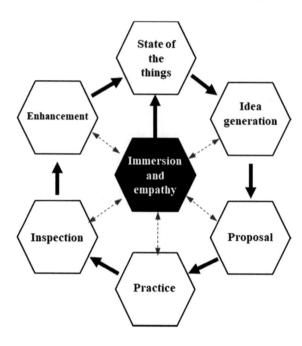

Table 5 Steps and objectives of integrated BPM—design thinking model

Integrated model	Objective
Immersion and empathy	To gather, in rich detail, the stories of actors and clients in the process to understand their needs and aspirations
State of the things	To analyze the data gathered to understand in an empathetic and profound manner the causes of problems in the current process. To draw the AS IS model of the process to be improved
Idea generation	To generate ideas to solve the problems in the process, considering its impacts on the context and people involved
Proposal	To select the best ideas and design a TO BE model with these ideas in mind
Practice	To implement the improvements in the process, aiming to maximize positive impact on the actors and clients
Inspection	To monitor the performance of the implemented process
Enhancement	To make incremental adjustments in the process based on the monitoring in the inspection stage

As a solution to this, expert 3 pointed out that, to develop a new process, one could go from the "Immersion and empathy" stage directly to "Idea generation." At this point, the authors made it clear to the specialists that the "State of the things" stage aims to analyze the data gathered in the "Immersion and empathy" stage, which is a relevant matter for both new and existing processes. However, when a new process is

Table 6 Specialists' profile

Specialist	Job title	Area	Experience (years)
Specialist 1	Industrial engineer	BPM	10
Specialist 2	Industrial engineer	BPM	2
Specialist 3	Industrial engineer	BPM+DT	5

being developed, modeling the AS IS process during the "State of the things" stage would not be appropriate.

Based on the specialists' perceptions in the focused group, a discussion was held about the opportunities for improvement to the proposed model. Thus, a second version for the model was built, which we present in Fig. 3. In this version, each stage's outcomes and the relationship between the "Immersion and empathy" stage and the others are explained.

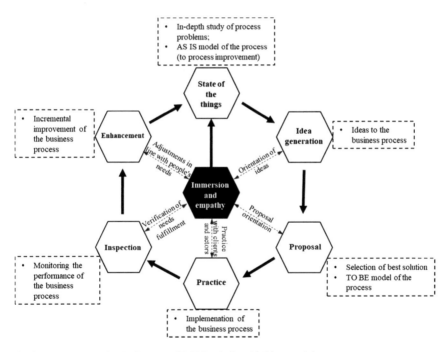

Fig. 3 Final version for the integrated BPM—design thinking model

6 Conclusion

This paper aimed to propose a new artifact to integrate the BPM life cycle model and the Design Thinking model. To this end, the existing BPM life cycle and design thinking models were studied, exploring their relationships and convergences. After reviewing the cycles and processes, an integrated model was proposed and presented to specialists. The specialists contributed to the analysis of applicability and to refine the model.

The initial version of the model was composed of seven stages, being "Immersion and empathy," "State of the things," "Idea generation," "Proposal," "Practice," "Inspection," and "Enhancement." After being presented to three specialists in the fields, the relations between the "Immersion and empathy" stage and the other stages were made explicit. Specialists evaluated the model as viable and applicable in front-end and support (back-end) processes.

As limitations of this work, the artifact was conceived from a bibliographic review, without a systematic method. Thus, not all existing models in literature may have been covered. In addition, there was no application of the proposed model, and therefore is not possible to evaluate the model's performance in a practical context.

References

1. Dumas M, La Rosa M, Mendling J, Reijers H a. (2013) Fundamentals of Business Process Management. Springer Berlin Heidelberg, Berlin, Heidelberg
2. ABPMP (2013) BPM Cbok. ABPMP
3. Brown T (2009) Change by Design: How Design Thinking Transforms Organizations and Inspires Innovation. HarperCollins e-books
4. Leverenz CS (2014) Design Thinking and the Wicked Problem of Teaching Writing. Comput Compos 33:1–12. https://doi.org/10.1016/j.compcom.2014.07.001
5. Jackson S (2015) Design Thinking in Argumentation Theory and Practice. Argumentation 29:243–263. https://doi.org/10.1007/s10503-015-9353-7
6. Damian IPM, Borges LS, Pádua SID de (2015) A Importância das Tarefas e os Fatores Críticos de Sucesso para o Gerenciamento de Processos de Negócios. Rev Adm da UNIMEP 162–185. https://doi.org/10.15600/1679-5350/rau.v13n2p162-185
7. Van Den Bergh J, Thijs S, Viaene S (2014) Transforming Through Processes: Leading Voices on BM, People and Technology. Springer, London
8. Hammer M (2015) What is Business Process Management? In: vom Brocke J, Rosemann M (eds) The Handbook of Business Process Management. Springer, pp 3–16
9. vom Brocke J, Rosemann M (2015) Handbook on business process management 1: Introduction, methods, and information systems. Handb Bus Process Manag 1 Introd Methods, Inf Syst 1–727. https://doi.org/10.1007/978-3-642-45100-3
10. Baldam VSR, Valle R, Pereira H, et al (2009) Gerenciamento de processos de negócios: BPM - Business Process Management, 2nd ed. Érica, São Paulo
11. Jeston J, Nelis J (2008) Business Process Management: Practical Guidelines to Successful Implementations. Elsevier
12. Smith H, Fingar P (2007) Business Process Management: The Third Wave. Meghan-Kiffer Press
13. Weske M (2007) Business Process Management: Concepts, Languages, Architectures. Springer

14. Houy C, Fettke P, Loos P (2010) Empirical research in business process management – analysis of an emerging field of research. Bus Process Manag J 16:619–661. https://doi.org/10.1108/14637151011065946
15. Aalst WMP Van Der (2009) Process-Aware Information Systems: Design, Enactment, and Analysis. In: Wah B (ed) Wiley Encyclopedia of Computer Science and Engineering. Wiley Online Library, pp 1–31
16. Netjes M, Reijers HA, Aalst WMP Van Der (2006) Supporting the BPM life-cycle with FileNet Configuration Design Diagnosis Control Execution
17. Liedtka J, Ogilvie T (2012) Helping Business Managers Design Thinking Discover Their Appetite for Design Thinking. Des Manag Inst 6–13
18. Jiao J, Zhang R (2015) Design thinking : a Fruitful Concept for Strategic Enterprise Management. In: International Conference on Education, Management and Computing Technology (ICEMCT 2015). Tianjin, pp 1591–1594
19. Liedtka J (2015) Perspective: Linking Design Thinking with Innovation Outcomes through Cognitive Bias Reduction. J Prod Innov Manag 32:925–938. https://doi.org/10.1111/jpim.12163
20. Vianna M, Vianna Y, Adler IK, et al (2014) Design Thinking Design Thinking Inovação em negócios. MJV press
21. IDEO (2012) The Human-Centered Design Toolkit. 1–200
22. Sato S, Lucente S, Meyer D, Mrazek D (2010) Design Thinking to Make Organization Change and Development More Responsive. Des Manag Rev 21:44–52. https://doi.org/10.1111/j.1948-7169.2010.00064.x
23. March ST, Smith GF (1995) Design and natural science research on information technology. Decis Support Syst 15:251–266. https://doi.org/10.1016/0167-9236(94)00041-2
24. Lukka K (2003) The Constructive Research Approach. In: Ojala L, Himola O-P (eds) Case study research in logistics. pp 83–101
25. Holmström J, Ketokivi M, Hameri A-P (2009) Bridging Practice and Theory: A Design Science Approach. Decis Sci 40:65–87. https://doi.org/10.1111/j.1540-5915.2008.00221.x
26. Dresch A, Lacerda DP, Jos, Antunes Jr. JAV (2015) Design Science Research: A Method for Science and Technology Advancement. Springer, London

An Internal Efficiency Benchmarking Investigation for CEF's Sustainable Banking and Response to the COVID-19 Crisis

Thyago Celso Cavalcante Nepomuceno⬥, Susane de Farias Gomes⬥, and Túlio Fidel Orrego Rodriguez⬥

Abstract The Brazilian Federal Savings Bank (*Caixa Econômica Federal*—CEF) is the biggest government-owned bank in Latin America and the fourth globally in terms of assets. This work reports an empirical assessment of best practices conducted in one of the bank's units situated in Pernambuco. This empirical investigation is developed with a combination of Time-Series Data Envelopment Analysis for efficient (time) peers, human interactions, and the PROMETHEE methodology for outranking the best practices and strategies in accordance with the support of employees, sustainability impact and response to minimizing COVID-19 propagation during open hours.

Keywords Data envelopment analysis · Time-series data · Multicriteria decision aid · PROMETHEE · Sustainability · Banking · COVID-19

1 Introduction

According to Zari [1], Benchmarking's modern use has become synonymous with the means of identifying a standard or point of reference for industry best practices that can lead to superior performance at strategic and operational levels. One of the most recurrent tools used for this identification is the so-called Data Envelopment Analysis (DEA). This non-parametric methodology introduced by Charnes et al. [2] and Banker et al. [3] provides an efficient frontier which represents the maximum industry production capacity for any Decision-Making Unit (DMU) operating under similar technology. Efficient units on this frontier conduct the most valuable best

T. C. C. Nepomuceno (✉) · S. de F. Gomes · T. F. O. Rodriguez
Universidade Federal de Pernambuco, Av. Prof. Moraes Rego, Recife, Pernambuco 1235, Brazil
e-mail: thyago.nepomuceno@ufpe.br

T. C. C. Nepomuceno
Università Degli Studi Di Roma La Sapienza, Piazzale Aldo Moro, 5, 00185 Roma, Italy

S. de F. Gomes
Universidade Federal de Alagoas, Av. Beira Rio, s/n - Centro, Penedo, Alagoas, Brazil

© The Author(s), under exclusive license to Springer Nature Switzerland AG 2021
A. M. Tavares Thomé et al. (eds.), *Industrial Engineering and Operations Management*,
Springer Proceedings in Mathematics & Statistics 367,
https://doi.org/10.1007/978-3-030-78570-3_38

501

practices and strategies, serving as references for the remaining inefficient units gauging efficient prospects. Many model extensions, computations, and theoretical contributions have been proposed since the first DEA formulations with thousands of empirical applications in all sectors of economic activity [4–6]. Some engaging assessments in Banking, Healthcare, and Policing can be found in [7–9].

In regards to benchmarking, one of the biggest challenges is how to have access to strategic information about competitors' performances. This issue many times limits some DEA applications to measuring technical and scale efficiencies and related discussions on potentials for improvements, spillovers, and associated factors, i.e., defining how much needs to be improved in reducing used resources or elevating generated products, without clearly defining what to do to reach this quantitative objective. To assist managers in this regard, Nepomuceno et al. [10, 11], proposed a DEA formulation using time-series data instead of cross-sectional comparisons to aid resource allocation and knowledge benchmarking in service units. In this linear formulation, the service unit operating in a given month is compared to a benchmark (reference) month located in the efficient parametric frontier. Because the reference units are efficient months when best practices were conducted, information about relevant strategies and significant prospects can be easily retrieved.

How effective are those practices and information can be another matter of concern during the benchmarking process. The concept of effectiveness in this context is related to specific goals or objectives [9] of relevance for society's organization and interest. Today in Brazil, two main topics are drawing the attention of individuals, families, companies, and public authorities: Sustainability (primarily because of Amazon's forest fires) and the COVID-19. Day by day, more and more people and organizations are engaging with small but essential sustainable practices to mitigate risks on the environment. *Caixa Econômica Federal* (CEF), the Brazilian Federal Savings Bank, is one of those organizations. The biggest 100% government-owned bank in Latin America and fourth in the world in the number of assets, is also one of Brazil's public companies with exciting sustainability programs and funding engagement. Multiple Criteria Decision Aid (MCDA) approaches can help choose the best strategies to improve financial performance based on the potential impact in the business objectives, sustainability, and minimizing risks associated with the pandemic outbreak in the service units during daily operations [12–14].

Reporting sustainable banking strategies brings, in addition to the institution's social responsibility, a valuable benefit for the local and macro current development without jeopardizing the possibilities of future generations [15]. The current COVID-19 pandemic has made this position even more evident. Struggling with providing the best possible financial service in a safe and sustainable environment for customers is the challenge of these new times, which can use the support of decision methods and heuristics from academia. This work aims at providing a combined DEA/MCDA methodology for this purpose. The reference periods resulted from the Time-Series DEA application are investigated and prioritized using the PROMETHEE outranking method [16], offering a set of the best-adopted strategies by the service unit in terms of the managerial objectives, sustainable practices, and potential response to mitigate the effects of the pandemic in the bank operations. Details of this methodology are presented in the next section. The third section is dedicated to applying this benchmarking approach to one of Pernambuco's CEF's branches. The conclusion summarizes and comments the main remarks.

2 Methodology

The Internal Benchmarking procedure begins with applying the reformulated directional output-oriented variable returns DEA to time-series data in order to identify the Benchmarking Domain of the service unit, i.e., the set of efficient peers (periods) to investigate the most critical adopted strategies. The traditional approach provide non-radial measures for the technical or scale efficiencies considering a feasible contraction g_{x_i} $\forall$ $i = 1, 2, \ldots, n$ for the input or a feasible expansion g_{y_r} $\forall$ $r = 1, 2, \ldots, s$ for the output of a given decision unit j from m decision units operating under a known production technology.

Nepomuceno et al. [11] reformulation provides an absolute non-radial measure for the technical inefficiency of a given decision unit j operating $i = 1, 2, \ldots, n$ inputs to produce $r = 1, 2, \ldots, s$ outputs under a known production technology set $\Psi_{(x,y)} \in R^{n+s}$ in a specific period $t = 1, 2, \ldots, p$ in the most favorable direction for the input reduction g_{x_i} or output improvement g_{y_r} defined by the data. A short discussion on the objective and subjective options for the choice of potential directions is provided by Nepomuceno et al. [14]. An output-oriented directional efficiency can be defined as follows:

$$D_t(x, y, g_{x_i}, g_{y_r}) = max\beta$$

$$s.t. \sum_{t=1}^{p} z_t y_{tr} \geq y_{t'r} + \beta g_y, \forall r = 1, 2, \ldots, s; \tag{1}$$

$$\sum_{t=1}^{p} z_t x_{ti} \leq x_{t'i}, \quad i = 1, 2, \ldots, n;$$

$$\sum_{t=1}^{p} z_t = 1, \quad z_t \geq 0, \quad t = 1, 2, \ldots, p$$

where z_t are the time optimum weights for the linear combination and estimation of the efficiency frontier for comparing the same unit with itself in different moments. As a result, we have a relative measure for the technical efficiency β and an absolute measure for the technical inefficiency $x_{t'i} - \beta g_{x_i}$, which defines the set of time-sorted benchmarks for retrieving the strategies. The periods presenting $\beta = 1$ (*and* $(x_{t'i} - \beta g_{x_i}) = 0$) are the efficient periods to be internally investigated. For all the remaining inefficient periods, the programming provides linear combinations toward the efficient frontier's closest points to benchmarking the best practices.

The analyst uses the Benchmarking Domain (set with efficient periods to be investigated) to discover valuable information from the employees, supervisors, and managers of the branch unit about what they believe to be the reasons for reaching efficient prospects on those specific periods. Interviews in three dimensions characterize this process: strategic (general manager or director of the institution), tactical (sector managers), and operational (remaining personnel). The interviews are applied in two manners: spontaneous, i.e., seeking clear answers with no exogenous influences, and directed, allowing the access of the other employees' responses. The analyst and stakeholders can take part in this process.

As a result of this process, we have a set of strategies that require prioritization. There is a sort of multicriteria methods in the literature that uses numeric techniques to assist the decision-makers in prioritizing a set of available alternatives, according to preference structures [17] based on the consequences and perceived utility. The PROMETHEE (Preference Ranking Organization Method for Enrichment Evaluations) family is classified as outranking methods [18], based on paired comparisons of alternatives in a criterion to analyze how strong the preference is for the choice of an alternative "a" over another alternative "b" through the determination of partial binary relations [19]. The accessible building, understanding, and interpretation are the main advantages of the method, enabling the decision maker evaluates all the alternatives of the issue.

The method uses functions $Fi(a,b)$ into a preference degree ranging from 0 to 1 that translates the difference among the performances of the alternatives "a" and "b" on a particular criterion i. It assumes 1 when "a" is better in performance than "b" or 0 on the opposite case. For this method, the weight represents the degree of importance of the criteria. Each one of the functions is multiplied by the weights for calculating the outranking degree, represented by the following expression:

$$\pi(a, b) = \sum_{i=1}^{n} W_i F_i(a, b) \tag{2}$$

It is important to note that $Fi(a,b)$ can take intermediate values. There are six basic kinds of preference functions: usual function, U-shape function, V-shape function, level function, linear function, and Gaussian function [16]. The outranking flows are calculated based on the preference function. It indicates the preference intensity of one alternative over another (in pairs) by analyzing all criteria. In a set of alternatives A, each alternative "a" has a π(a,b) that represents a preference index of "a" over another alternative "b" observing all criteria. The value function is ϕ(a). A high value of ϕ(a) represents a high attractiveness of the alternative "a". The positive ϕ + and negative ϕ – outranking flows are presented in the following equations [16, 19]:

$$\phi + (a) = \frac{1}{n-1} \sum_{b \in A} \pi(a, b) \tag{3}$$

$$\phi - (a) = \frac{1}{n-1} \sum_{b \in A} \pi (b, a) \tag{4}$$

Then, a liquid outranking flow is obtained as presented in (5) for the prioritization:

$$\phi(a) = \phi^+(a) - \phi^-(a) \tag{5}$$

Figure 1 illustrates this entire procedure of Internal Benchmarking. In the end, this methodology provides a complete ranking from all retrieved strategies from the best to the worst best practices using the liquid flow to rank the alternatives, offering the most effective actions to attain efficiency.

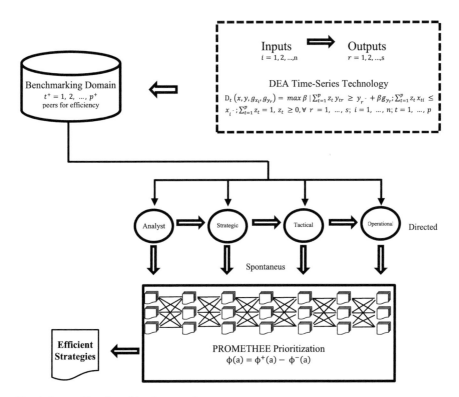

Fig. 1 Internal benchmarking framework

3 Application

3-years data on the number of employees per month (input) and the number of business transactions, social programs, and teller services (output) for a specific branch of CEF located in Pernambuco was used to develop the first DEA results. The number of employees diverges from one month to the other due to vacations, maternity leave, and other medical issues. From the assessment, eight months resulted in efficient peers for benchmarking of best practices (July/16, August/2016, December/2016, March/2017, January/2017, October/2018, April/2019 and May/2019). The most relevant strategies adopted during these months were investigated through interviews with the bank personnel, which resulted in the set of strategies/actions reported in Table 2.

We have defined four decision criteria for the prioritization of those strategies. Sustainability is one of the qualitative criteria because of the bank's institutional mission and sustainable profile. The potential impact of each of those strategies on mitigating the propagation of COVID-19 in the unit during open hours is the other qualitative criteria given the importance of customers' safety. The other two criteria are quantitative, representing the frequency more than one employee mentions the strategy, and the interaction this strategy may have with other sectors of the service unit, i.e., a potential positive impact in producing more services in more than one sector. This last criterion is defined by the Total Link Strength (TLS) [20], a bibliometric network methodology serving as a proxy for the volume of interaction among the referred statements of best practices. Table 1 summarizes the decision criteria.

Because Sustainability and COVID-19 are the subjective criteria, the decision-maker evaluates them according to a 5-points Likert scale: Very Low (VL), Low (L), Average (A), High (H) and Very High (VH), considering all criteria the same importance (weight). Effective management of bank resources consists of good planning,

Table 1 Decision criteria for the problematic

Criterion	Definition	Unit
Occurrences	The frequency of a strategy reported by the employees	Absolute value
Total Link Strength	Proxy for the interactions of a strategy with other strategies	Absolute value
Sustainability	The degree in which the strategy is positively related to mitigating environmental risks associated with resource usage	1- Very low 2- Low 3- Average 4- High 5- Very High
COVID-19	The degree in which the strategy is positively related to mitigating the virus's spread (inside or outside the service unit) during operation	1- Very low 2- Low 3- Average 4- High 5- Very High

development, management, and optimization, focusing on some aspects such as availability, demand, internal conflicts, mitigation approaches, combating the degradation of the environment. Table 2 reports the matrix of consequences, in which all alternatives and consequences are elicited. In Table 3, the PROMETHEE II provides an outranking of alternatives from the best to the worst according to the decision-maker preferences. All the criteria are usual, which means there is no defined parameter.

Table 2 The matrix of consequences for the problematic

Action (strategy)	Occurrences	Total link strength	Sustainability	COVID-19
Account document digitization	2	4	A	H
Account issues	4	8	L	A
Accounts qualifying strategy	6	22	VL	VL
Automatic FGTS credit	2	4	VH	VH
Business division	2	4	L	L
Calls	6	20	L	VL
"Cartão Cidadão"	2	4	VH	VH
Client base expansion	2	6	VL	VL
Credit card analysis	4	10	H	H
Day offs	2	10	VL	VL
Deposit envelope strategy	2	4	VH	L
Employee re-allocation	2	4	L	L
"Expressinho"	4	8	VH	VH
Faster services	4	14	VH	VH
"FGTS"	2	10	VH	VH
Insur. premium (installments)	4	16	L	L
Life certification	2	4	VH	VH
Meetings	4	20	VL	VL
"Mundo Caixa em dobro"	2	10	VL	VL
Offering	4	16	VL	VL
Opportunity	2	4	L	VL
Phone services	6	18	A	VH
Quita Fácil	4	14	VL	VL
Real estate agencies	2	10	L	VL

(continued)

Table 2 (continued)

Action (strategy)	Occurrences	Total link strength	Sustainability	COVID-19
Reducing overtime	2	4	VH	VL
Registration	2	10	H	L
Scheduling	2	10	H	A
System update	4	14	VH	H
Technology	2	10	VH	H
Whatsapp groups	2	4	A	VL

Fifteen strategies have a positive liquid flow reported in Table 3 by the PROMETHEE II application based on the criteria Occurrences, TLS, Sustainability, and COVID-19 impact. The most relevant adopted strategy during the efficient months according to the support of employees, and the one that can favorably impact the prospects of sustainability and COVID-19 is the Phone Services strategy. This practice is about recurrent recommendations by the bank employees for the clients to use the phone banking for a range of financial transactions which involve ordering credit cards, citizens card (*cartão cidadão*, a particular card for the government social programs), real estate loan amortizations using FGTS (Employee Guarantee Fund), applying for loans, among others. This strategy has prevented the need to visit a bank branch or ATM for many clients, impacting complimentary resources.

Many of those strategies with positive liquid flow reported by terms in Table 3 are strongly correlated with each other such as 'Faster Services', 'System Update', 'Expressinho', 'FGST', 'Technology', 'Automatic FGTS credit', 'Cartão Cidadão', and 'Life Certification', which basically refers to changes of computer technologies (not necessarily of the unit's responsibility), changes of processes, social program integration with online banking, and reformulation of service positions all to reduce the waiting time inside the bank unit, reducing queues and positively impacting the mitigation of COVID-19 and the use of office resources. Other interesting strategies are related to the bank business transactions such as increased phone calls and new approaches to selling life and auto insurances.

Lastly, Table 4 reports a Sensitivity Analysis of results in three classes by changing each decision criteria's importance level for different managerial perspectives. In the first class, the four criteria have the same set up at 25% (weight) from Table 3; this means that each criterion has an equal decision-maker preference. The following two classes are simulations made on 50% of preference for one criterion, keeping the others with 16.66% equally and 75%, keeping the other three at 8.33%. This sensitivity analysis has the purpose of assessing from a standard equally distributed prioritization to a subjective preference of one compared to the other decision-criteria. This information can be crucial to identify which actions can become more convenient to reach the company goals on each perspective.

When the manager is more interested in the strategies with the higher support from the strategic, tactical, and operational staff, or when he/she is more interested in those strategies with higher interaction among the bank sectors, Phone Services,

Table 3 The rank of alternatives (strategies)

Rank	Action	φ	φ +	φ -
1	Phone services	0.6379	0.7759	0.1379
2	Faster services	0.6293	0.7069	0.0776
3	System update	0.5345	0.6724	0.1379
4	"Expressinho"	0.4483	0.6207	0.1724
5	Credit card analysis	0.3276	0.5776	0.2500
6	"FGTS"	0.3017	0.4828	0.1810
7	Calls	0.2241	0.5259	0.3017
8	Insurance premium	0.2155	0.5259	0.3103
9	Technology	0.2069	0.4483	0.2414
10	Accounting qualify	0.1293	0.4828	0.3534
11	Automatic FGTS credit	0.0948	0.3707	0.2759
11	"Cartão Cidadão"	0.0948	0.3707	0.2759
11	Life Certification	0.0948	0.3707	0.2759
14	Accounting issues	0.0517	0.4569	0.4052
15	Scheduling	0.0431	0.4052	0.3621
16	Meetings	0.000	0.3879	0.3879
17	Registration	−0.0172	0.3621	0.3793
18	Offering	−0.0517	0.3621	0.4138
19	"Quita Fácil"	−0.0948	0.3362	0.4310
20	Deposit envelope	−0.1121	0.2759	0.3879
21	Account document	−0.1638	0.2845	0.4483
22	Reducing overtime	−0.2586	0.1724	0.4310
23	Real state agencies	−0.3017	0.1724	0.4741
24	Business division	−0.3621	0.1638	0.5259
24	Employee re-allocation	−0.3621	0.1638	0.5259
26	Days off	−0.4224	0.1121	0.5345
26	"Mundo Caixa em dobro"	−0.4224	0.1121	0.5345
26	Whatsapp groups	−0.4224	0.1207	0.5431
29	Opportunity	−0.5086	0.0603	0.5690
30	Client base expansion	−0.5345	0.0862	0.6207

Faster Services, System Updates, and Calls should be prioritized. When the manager is more interested in providing sustainble financial services or mitigating the potential COVID-19 propagation in the unit, he/she should invest in reducing queues. This can be done by adopting the *Expressinho* strategy (a new sector destined to timely demands that do not need a specialized knowledge/service) and high technology-oriented operations for automatic FGTS credits, faster services, among others.

Table 4 Sensitivity analysis

	Rank	Occurrence	Total link strength	Sustainability	COVID-19
Criterion (25%)	1°	Phone services			
	2°	Faster services			
	3°	System update			
	4°	Expressinho			
	5°	Credit card analysis			
Criterion (50%)	1°	Phone services	Phone services	Faster services	Phone services
	2°	Faster services	Faster services	System update	Faster services
	3°	System update	System update	Expressinho	Expressinho
	4°	Expressinho	Calls	Phone services	System update
	5°	Calls	Accounting qualify	FGTS	FGTS
Criterion (75%)	1°	Phone services	Phone services	Faster services	Phone services
	2°	Calls	Accounting qualify	System update	Faster services
	3°	Accounting qualify	Calls	Expressinho	Expressinho
	4°	Faster services	Meetings	FGTS	FGTS
	5°	System update	Faster services	Technology	Automatic FGTS credit

4 Conclusion

This work proposed a methodology for internal benchmarking of best practices using a combination of Data Envelopment Analysis adapted for Time-Series data, providing the reference periods for investigation of efficient strategies, and Multiple Criteria Decision Aid through the PROMETHEE outranking method for the prioritization of the set of strategies. The PROMETHEE II applied in this case study, especially considering the current pandemic situation, offers managers a ranking of possible sustainable action alternatives. This framework provides rapid response to changes in the business structure according to the decision-maker preferences, offer robust flexible for improvments and support allocation strategies such as implemented by [10, 21, 22]. The reimplementation of those solutions, however, can find budget limitaions.

Thirty alternatives were obtained from this internal benchmarking with employees in three administrative dimensions: strategic, tactical, and operational. Phone services were the strategy set up at the first level in most simulations, except the Sustainability criterion, which reinforces the importance of customer reeducation concerning other service channels. From a sustainable point of view, the Brazilian Federal Savings Banks always seek to reinvent their process and activities toward sustainable perspectives, offering particular loan clauses and financial service conditions for sustainable clients, funding innovative initiatives for environmental protection, and seeking directions to reduce wastes. In this alignment, rearranging processes and structures such as the *Expressinho* and investing in technology and strategies to provide faster services and reduce resources is the more relevant reported in this assessment and exciting potential for future analyses.

References

1. Zairi M., Leonard P: Origins of benchmarking and its meaning. In: Practical Benchmarking: The Complete Guide. Springer, Dordrecht (1996). https://doi.org/10.1007/978-94-011-1284-0_3.
2. Charnes, Abraham, William W. Cooper, and Edwardo Rhodes: Measuring the efficiency of decision making units. *European journal of operational research* 2.6 (1978): 429-444. https://doi.org/10.1016/0377-2217(78)90138-8
3. Banker, R. D., Charnes, A., & Cooper, W. W.: Some models for estimating technical and scale inefficiencies in data envelopment analysis. *Management science*, 30(9), 1078-1092, (1984). https://doi.org/10.1287/mnsc.30.9.1078
4. Daraio, C., Kerstens, K. H., Nepomuceno, T. C. C., & Sickles, R: Productivity and efficiency analysis software: an exploratory bibliographical survey of the options, *Journal of Economic Surveys*, Vol. 33(1), pp. 85–100 (2019). https://doi.org/10.1111/joes.12270
5. Daraio, C., Kerstens, K., Nepomuceno, T., & Sickles, R. C: Empirical surveys of frontier applications: a meta-review, *International Transactions in Operational Research*, Vol. 27, No. 2, pp. 709-738, (2020). https://doi.org/10.1111/itor.12649
6. Nepomuceno, T. C. C., Costa, A. P. C. S. & Cinzia Daraio: Theoretical and Empirical Advances in the Assessment of Productive Efficiency since the introduction of DEA: A Bibliometric Analysis. Forthcoming in *Int. Journal of Operational Research*, (2021). https://doi.org/10.1504/IJOR.2020.10035180.
7. Sherman, H. D., & Ladino, G: Managing bank productivity using data envelopment analysis (DEA). *Interfaces*, 25(2), 60–73 (1995).
8. Nepomuceno, T. C., Silva, W., Nepomuceno, K. T., & Barros, I. K: A DEA-Based Complexity of Needs Approach for Hospital Beds Evacuation during the COVID-19 Outbreak. *Journal of Healthcare Engineering*, (2020). doi.org/https://doi.org/10.1155/2020/8857553
9. Nepomuceno, T. C. C., Santiago, K. T. M., Daraio, C. & Costa, A. P. C.: Exogenous Crimes and the Assessment of Public Safety Efficiency and Effectiveness. *Annals of Operations Research*. (2020). doi: https://doi.org/10.1007/s10479-020-03767-6
10. Nepomuceno, T. C. C, & Costa, A. P. C: 'Resource allocation with time series DEA applied to Brazilian Federal Saving banks', *Economics Bulletin*, Vol. 39(2), pp. 1384-1392 (2019).
11. Nepomuceno, T. C. C., de Carvalho, V. D. H., & Costa, A. P. C. S.: Time-Series Directional Efficiency for Knowledge Benchmarking in Service Organizations. In World Conference on Information Systems and Technologies, pp. 333–339 (2020). Springer, Cham. https://doi.org/10.1007/978-3-030-45688-7_34

12. Jan, A., Marimuthu, M., bin Mohd, M. P., & Isa, M.: The nexus of sustainability practices and financial performance: From the perspective of Islamic banking. *Journal of Cleaner Production*, 228, 703–717 (2019).
13. Lin, A. J., & Chang, H. Y.: Business sustainability performance evaluation for taiwanese banks—A hybrid multiple-criteria decision-making approach. *Sustainability*, 11(8), 2236 (2019).
14. Nepomuceno, T. C. C., Daraio, C., & Costa, A. P. C. S: Combining multicriteria and directional distances to decompose non-compensatory measures of sustainable banking efficiency, *Applied Economics Letters*, Vol. 27(4), pp. 329-334 (2020).
15. Bouma, J. J., Jeucken, M., & Klinkers, L. (Eds.): Sustainable banking: The greening of finance. Routledge, New York (2017).
16. Brans, J. P., P. Vincke, and B. Mareschal: How to Select and How to Rank Projects: The PROMETHEE Method. *European Journal of Operational Research* 24 (2): 228–238 (1986).
17. Triantaphyllou, E: Multi-criteria decision making methods: a comparative study, Springer: Lousiana, 2013.
18. Vincke, P: Multicriteria decision-aid. John Wiley & Sons, Bruxelles, 1992.
19. Mergias, I., Moustakas, K., Papadopoulos, A. & M. Loizidou: Multi-criteria decision aid approach for the selection of the best compromise management scheme for ELVs: The case of Cyprus, *Journal of Hazardous Materials* 147, pp. 706–717 (2007).
20. Perianes-Rodriguez, A., Waltman, L., & Van Eck, N. J.. Constructing bibliometric networks: A comparison between full and fractional counting, *Journal of Informetrics*, Vol. 10(4), pp. 1178-1195 (2016).
21. Nepomuceno, T., Silva, W. M. D. N., & Silva, L. G. D. O. (2020). PMU7 Efficiency-Based Protocols for Beds Evacuation during the COVID-19 Pandemic. *Value in Health*, 23, S604. https://doi.org/10.1016/j.jval.2020.08.1219
22. Nepomuceno, K., Nepomuceno, T., & Sadok, D. (2020). Measuring the Internet Technical Efficiency: A Ranking for the World Wide Web Pages. *IEEE Latin America Transactions*, 18(06), 1119-1125. DOI: https://doi.org/10.1109/TLA.2020.9099750

Paradox of Firm Theory and Sustainable Development in the Mining Industry: Approximation Through the Managing Leader

Ana Paula Braga Garcez⓿, **Josilene Aires Moreira**⓿,
Ricardo Moreira da Silva⓿, **Mário Franco**⓿,
and Fernando Bigares Charrua Santos⓿

Abstract The Mining Industry, in general, involves large economic capital and causes a huge socio-environmental impact, aspects both linked to the Firm Theory and to Sustainable Development. In this relationship between these theoretical constructs, the leading manager must daily decide to balance his decisions to meet the expectations of the stakeholders. This article used the theoretical data triangulation methodology and aimed to map the paradox experienced by the leading manager within the Mining Industry sector according to (a) Costs, (b) Value creation around the mining area, (c) Image together to society, (d) Establishment of ethical standards in community relations, (e) Information sharing and reporting, (f) Investments and customer relations, and (g) Digital transformation. It is concluded that the leaders of the companies in this sector are always in a dilemma, in the paradox, between attending to inherent variables of the Firm Theory or the sustainable development agenda, since there is a trade-off in these two theoretical constructs, so, the leader being in charge balance the relationship between economic growth and care for natural ecosystems.

Keywords Firm theory · Sustainable development · Mining industry · Leader

A. P. B. Garcez (✉) · F. B. C. Santos
University of Beira Interior (C-MAST—Centre for Mechanical and Aerospace Science and Technologies), Calçada Fonte do Lameiro, 6200-358 Covilhã, Portugal
e-mail: ana.garcez@ubi.pt

F. B. C. Santos
e-mail: bigares@ubi.pt

J. A. Moreira · R. M. da Silva
Federal University of Paraíba-Cidade Universitária, João Pessoa 58051-900, Brazil
e-mail: josilene@ci.ufpb.br

M. Franco
Department of Management and Economics, CEFAGE-UBI Research Center, University of Beira Interior, Estrada Do Sineiro, 6200-209 Covilhã, Portugal
e-mail: mfranco@ubi.pt

© The Author(s), under exclusive license to Springer Nature Switzerland AG 2021   513
A. M. Tavares Thomé et al. (eds.), *Industrial Engineering and Operations Management*,
Springer Proceedings in Mathematics & Statistics 367,
https://doi.org/10.1007/978-3-030-78570-3_39

1   Introduction

The Mining Industry is very important worldwide because it directly generates a high number of jobs and income [1] and indirectly moves other sectors related to technology, steel and the like. Usually, the focus of mining industry is the extraction of non-renewable resources, ore, where once removed from the soil, it cannot be regenerated. This depletion of natural mineral resources is a major concern in the dimensions of sustainable development [2].

In addition, when mines are closed, their residual value and environmental liabilities end, and often, with traumatic economic and social implications, for residents of the territorial surroundings [3]. For this reason, the structure and business plan of the mining industry has a complex management [4], mainly due to the use of limited natural resources [5], which was not observed in the Fordist paradigm.

This Fordist paradigm served as support for the birth of the firm theory [6]. The companies in this sector are organizations that produce and sell goods and services, contract and use factors of production [7], thus, working in expectation of the market offer, that is, with the minerals products that they will offer to consumers. The firms gather capital and labor to carry out production and are responsible for adding value to the raw materials used in this process, using technology.

Lozano et al. [8] suggest that the firm is a profit-generating entity in constant evolution, being composed of resources and networks. The employees are responsible for representing the firm, managing its resources, and empowering its stakeholders to ensure that the company complies with the laws, maintains its "license to operate", increases its competitive advantage and contributes better to promote evolution of more sustainable societies.

In this mining industry, the primary raw materials are the ore itself, and the resources should be used efficiently, without causing losses or waste [3]. In this aspect, when thinking about Sustainable Development, one should think not only about financial gains, but also should take care of nature and human beings.

In this sense, companies need to develop more sustainable management models, supported by strategies aimed at the environment and society, avoiding the exclusive ambition to maximize profits [9], incorporating these aspects in their operational processes, however, there is still a significant gap between the ideals promoted in the existing literature and real practice [10].

Thus, managers in this sector should consider social and environmental aspects, for sustainable development, in their decision-making process, while also making possible economic gains, crucial for the growth of companies [11].

For example, Laing [12] showed a current social and health aspect: "the Covid-19 global pandemic has not only caused infections and deaths, but it had the potential to destroy individual livelihoods, businesses, industries and entire economies. The mining sector is not immune to these impacts, and the crisis has the potential to have severe consequences in the short, medium and long-term for the industry".

So, it is possible to state that the negative impacts of the mining sector's production process, include geographic, cultural and environmental and many others problems, such as air and land pollution, resulting from toxic substances released during extraction and the contamination of water, among others [1].

Even within this complex framework, Cragg and Greenbaum [3] argue that mining industry is not incompatible with environmental, social and economic sustainability, since, sustainable responsibility is built with the participation of many interconnected entities, that is, suppliers, manufacturers, retailers and others, when a relationship is established by trust between members of the value chain.

The objective of this work is to verify the approximations between Firm Theory and Sustainable Development in the Mining Industry, realizing the influence of the leader like as a catalyst.

2 Methodology

For verification of the possible interconnection of two theoretical constructs mediated by the presence of the leader, a systematic literature review approach was used built from articles on science direct. According to Rodrigues and Mendes [13], literature reviews generally pursue two main aims: (a) to summarize current research by identifying patterns, themes and issues, and (b) to identify the conceptual content of the field, and to contribute to theory development.

Concerning the literature review development process, this study followed the way proposed by Tranfield et al. [14], who highlight three level for conducting a systematic literature review: (i) review planning, (ii) review development articles selection and data synthesis and, (iii) results communication and dissemination. In addition to these steps, we add the construction of Fig. 1 that summarizes the entire framework linking firm theory and sustainable development, mediated by a manager leader, through a data triangulation, show to Fig. 1.

The key words used in the screening were: (i) mining industry (ii) leader (iii) Stakeholder (iv) mining impacts (v) sustainability, which enabled the retention of 86 papers to form the body of analysis. Than, we undertook an in-depth content analysis to capture the main trade-off that Mining Industry leaders have between serving their companies and being ethical and correct with the surrounding community.

For it, we used the VOS Viewer version 1.6.4 software for visualizing bibliometric networks, enabling the extraction of the seven investigated variables: (a) Costs, (b) Value creation around the mining area, (c) Image together to society, (d) Establishment of ethical standards in community relations, (e) Information sharing and reporting, (f) Investments and customer relations, and (g) Digital transformation.

In this proposal of approximation beetwen the Firm Theory and Sustainable Development in the Mining Industry, through the managing leader, each element of Fig. 1 was interconnected following the logic of the commitment to sustainable development in order to provide future best practices. This framework is expected to deal

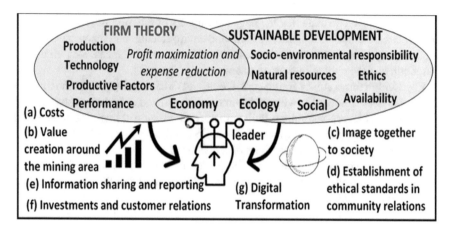

Fig. 1 Approximation between firm theory and sustainable development and the paradox of the leading manager of mining industry companies. *Source* By the authors from data triangulation

with the nature and the social in mining, reinforcing dialogue, instead of denouncing and building consensus, instead of conflict and confrontation.

3 Mining Industry and Its Responsibilities

With a view of the economic domain, inherent Firm Theory, the mining industry are perceived as a positive influence. It is important for economic development, as it contributes not only in their region of operation, but also at the national level [15].

In fact, with this exclusive look of firm theory, Li et al. [16] evidenced the benefits of the activities of exploration and mineral production: (1) provision of revenues to local and national governments in the form of taxes, royalties and license fees; (2) increased employment for the local population, both in mining and related industries; (3) indirect contribution to the local economy through multiplier effects due to its links with other sectors (4) airstrips, railways, roads, water supply, sanitation and electricity systems, these infrastructures and services established for exploration and production mineral benefit communities inside and outside the mining area.

However, in the view of sustainable development, the productive resources of the mining industry are limited, due to their origin, so they must be used efficiently, maintaining the commitment to sustainability, balancing economic, social and environmental issues, to satisfy the needs of the present without harming the future [5].

Rodrigues and Mendes [13] found that some mining industries have broken their promise of economic development, due to the mining process inducing large-scale and lasting impacts, on human habitat and at the natural, which includes changes in topography and landscapes; surface disturbances or groundwater systems; collapse of the soil, as well as above-ground structures; disposal of residual rocks and tailings, acidic drainage of rocks, (emission of heavy metals and other toxic substances; emissions of greenhouse gases such as sulfides, carbon dioxide and other harmful particles.

In this sense, communities in the vicinity of the mining industries can be heavily polluted by mineral processing, smelting, refining and fuel burning activities and it may face other serious environmental problems [16]. That is why stakeholders must perceive the mining industry not only from the perspective of the firm theory with a focus on economic development, but also from the perspective of sustainable development, which considers the balance of the three dimensions: economic, environmental and social [17].

In this sense, the mining industry is subject to external pressure to increase its responsibilities, adding to its decision-making structure, in addition to economic responsibilities, social and environmental responsibilities [17], and in this role the industry's leading manager can influence, decide and make a difference, recognizing the needs of stakeholders; effectively protecting the environment; prudently use natural resources; and maintaining high and stable levels of economic and employment growth.

4 Leader's Mediation in the "Firm Theory" and "Sustainable Development" Paradox

(a) **Costs**: Although cost reduction is a premise of Firm Theory, Cragg and Greenbaum [3] state that the viability of sustainable development requires an analysis of costs and benefits of the mining project, where these costs must be accounted for by the leader, in the direction of being shared fairly, with the surrounding society. This situation makes mining companies identify all stakeholders direct ones: employees, shareholders, government, suppliers and buyers and the indirect ones: local communities, employee families and others.

(b) **Value creation around the mining area**: It is important to identify and understand the needs and interests of stakeholders, so that the leader can see the possibilities of generating value [18]. The mining industry must balance the diverse demands of communities, generate jobs and income, protect the environment, and offer financial returns to shareholders [17]. In this sense, organizations in the mining sector have more responsibilities than just meeting shareholders' expectations [19].

(c) **Image together to society**: There is a trade-off between reducing costs, maximizing profits and interacting positively with the surrounding community, for this reason, the leader is permanently in a context the paradox of the mining industry, where the company's reputation towards society should be improved, demonstrate an ethical position to interested parties and without forgetting the financial needs of investors [20]. For this, it is important to guarantee the integrity of the mining company's image, where business goals and values must be aligned with all parties involved [18].

(d) **Establishment of ethical standards in community relations**: In the leader's effort to create a positive image of the sector, Cragg [21] found that clarity, honesty and transparency are crucial to facilitate communication, confidence and respect building for the entire mining sector. In fact, the growing activism into the community imposes additional costs on mining companies, however the current competitive environment is changing, in order to favor companies that take their community relations seriously [4].

Invest in good relations with the community, which includes workers, customers, communities and the general public, addressing issues such as (i) human rights, (ii) the well-being of workers and (iii) climate change [17] can produce benefits for the company in the long run [4]. This attitude seeks to boost the company's good reputation and increase sales, consequently also increasing results for shareholders [22].

It is a search for a relationship of reciprocity and mutual trust, where the leader does not forget technical rigor, but works in the direction of sustainability [4], that is, this relational assumption, assures the public and other users that the information processed is true and fair [20].

(e) **Information sharing and reporting**: In the pursuit of fair, healthy and reciprocal relationships with the community, the leader of the mining industry, subjected to pressure from internal and external stakeholders, which creates the need for reporting environmental, social and financial performance [23].

The structure of the reports varies according to the levels of assurance required by the stakeholders. When pre-existing levels of trust are high, the guarantee of fidelity of information can be achieved through self-certification and internal review [20]. If the confidence level is low, the verification of the information must be done by independent auditors, to increase the credibility of the information [24].

The leader must present sustainability reports, developed to respond to the expectations, pressures and criticisms of stakeholders, who wish to be better informed about the social and environmental impacts of business activities [24]. In this sense, the dissemination of reliable and standardized information from the mining industry companies is a way to guarantee reliable data [13].

(f) **Investments and customer relations**: With the disclosure of secure, coherent information, integrated with the company's commercial practices, and based on authentic values, managed and provided by the leader, a change in customer behavior is generated in one direction positive [22]. This change in behavior drives the disposition of new investments in companies. Biggemann et al. [18] consider that these new investments in the mining sector should be focused on strengthening the integrity of the value chain, as these increase organizational performance.

(g) **Digital transformation**: The disclosure of reliable information from the leader to stakeholders is not just the company's will. It is necessary to understand that currently, the universe of possible knowledge is so large that, without the incorporation of digital computerized systems, it would be impossible to have information. Thus, in addition to being a new skill required of the leader, working in the digital environment is a necessity.

The advancement of information technology has raised the power of consumers, considering that they can expose, in short periods, any problem that a company may have [18], thus, the dialogue with the market in in line with the company's mission, it is a key factor in the leader's strategy, who works with corporate businesses [11], where mining companies are located.

The digital transformation will allocate time for the development and massive implementation of four well-defined areas of technology: (i) Internet of Things (IoT), (ii) Artificial Intelligence, (iii) Robotics, and (iv) Blockchain [25, 26, 27], which will imply the virtualization of practically everything that can move, turn on, turn off, that is, a virtualized world as recommended in science fiction films of 20 years ago:

This set of technologies will enable the so-called "Internet of Skills", where human capabilities can be transferable, interconnected and work in synergistic ways. There may be a negative impact, reducing jobs; however, this problem is an old one, observed since the 50 s when technology started to advance [28].

The digital transformation enables new consumer groups to emerge who are proactively seeking information about business practices, as well as probing the behavior of the suppliers of these companies by networks, as well as investigating the behavior of its members [18]. For all these reasons, [29] found that the number of companies that publish environmental reports has grown over the past few years.

5 Conclusion

Business practice in the mining industry is not a unidirectional process and the growing importance of environmental and social issues, it creates pressure for mining companies to adopt a more systematic approach to presenting their results [4].

Mining companies constantly place their leaders in a paradox where, according to the firm theory, they must maximize profits, but at the same time, they are forced to live with an entire community in the surroundings where they work. Moreover, there are many workers who live in the surroundings, with their families.

The leader of mining companies is not sufficiently independent for decision making, and they have no real power to criticize the executive management and the shareholders [30]. However, their decisions they make must consider the interests of all stakeholders all those who are affected by the consequences of the decisions and the company's actions. Salehi et al. [30], that is, leaders live in constant conflict.

In addition, leading managers must be alert for the situations of interaction of informal actors and rival players to assess the legitimacy of the claims [31].

In fact, sustainable development should identify and understand the needs and interests of all stakeholders, for create value when interacting with them. However, the evaluation of value is difficult, especially due to the importance given to economic [18].

In another direction, communities expect the leader manager to demonstrate commitment to social and environmental issues through high quality sustainability actions and reports and the public commitment letter [32]. Therefore, it is demanding that the leader manager synchronize social and commercial purposes, admitting that business and society are inextricably linked [11].

In fact, mining is an industry where ethical issues directly overwhelm managerial leaders. The whole mining sector brings it to its core the problem of balancing costs and spending production with the aggregated social benefits costs. The distribution of these costs and benefits among the different groups is difficult to balance [3], and all this impacts on the relationship with the community, but above all it impacts on clients' investments too.

For all these difficulties, reports are needed, which require a great deal of skill from the managing leader. The constant flow of information between stakeholders in the mining industry, requires management skills from leading. They must be able to recognize and respond appropriately to the economic, social and environmental issues of the local community in which they operate [4].

The purpose of these mining reports is to evaluate in a transparent way the performance of sustainability and, consequently, analyzed the main impacts, positive or negative [24], in this perspective, the reports should enhance the communication process between mining companies and the community through the leader [24].

This includes the role of digital transformation currently in progress that came to streamline the flow of information, collect reliable data, and find patterns of behavior, previously unimaginable. However, for Clarkson et al. [33] there are still flaws in the controls of intangible environmental data, such as: inadequate sample selection and inadequate performance measures, which compromises the decision of the managing leader.

Such reports flawed and/or correct impact the issue of the image and the acceptance of the leader. Clarkson et al. [33] supports his analysis in the theory of legitimacy, as environmental activists face political pressures to change the community's perceptions of real performance. Boiral [24] found narcissistic trends in sustainability reports towards an overemphasis on companies' positive achievements, and also, in another hand, the negative impacts are minimized and are not mentioned.

Finally, the leading manager of a mining company must balance the relationship between economic growth and care for natural ecosystems, responsibly and consciously. For this, Humphreys [4] suggests companies some alternatives. They must: (i) implement income generation policies to develop community activities (ii) assume responsibility for supporting the creation of companies, outside the mine that leverage local development, (but that do not depend on existence of the mine), strengthening local community institutions (iii) creating physical and/or social infrastructure for the community.

In these actions, the interference of leaders with personal attributes and technical skills, will be extremely important, for the development of employees and increase the relationships of trust between those involved in the process.

As a possibility for future research, it is suggested to investigate new variables that map production, ethical and social conflicts that leaders suffer from companies in the mining industry. It is hoped that this innovative research can stimulate other studies that allow a better understanding of the conflicts experienced by managers of mining companies on a daily basis.

Acknowledgements This work has been supported by Fundação para a Ciência e Tecnologia (FCT) and C-MAST- Centre for Mechanical and Aerospace Science and Technologies, under project UIDB/00151/2020 and the projectsCentro-01-0145-FEDER-000017—EMaDeS—Energy,Materials and Sustainable Development, project 026653, POCI-01-0247-FEDER-026653.

References

1. Alves, W., Ferreira, P., & Araújo, M. (2018). Sustainability awareness in Brazilian mining corporations: the case of Paraíba state. Environment, Development and Sustainability, 20(s1), 41–63.
2. Jenkins, H. (2004). Corporate social responsibility and the mining industry: Conflicts and constructs. Corporate Social Responsibility and Environmental Management, 11(1), 23–34.
3. Cragg, W., & Greenbaum, A. (2002). Reasoning about responsibilities: Mining company managers on what stakeholders are owed. Journal of Business Ethics, 39(3), 319–335.
4. Humphreys, D. (2000). A business perspective on community relations in mining. Resources Policy, 26(3), 127–131.
5. Rifai-Hasan, P. A. (2009). Development, power, and the mining industry in Papua: A study of Freeport Indonesia. Journal of Business Ethics, 89(2), 129–143.
6. Tigre, P. B., & Brasileira, R. (2005). Paradigmas Tecnológicos e Teorias Econômicas da Firma. Revista Brasileira de Inovação, 4(1), 187–223.
7. Coase, R. (1937). The nature of the firm. In The Economic Nature of the Firm: A Reader, Third Edition (pp. 386–405).
8. Lozano, R., Carpenter, A., & Huisingh, D. (2015). A review of "theories of the firm" and their contributions to Corporate Sustainability. Journal of Cleaner Production, 106, 430–442.
9. Silva, C., Magano, J., Moskalenko, A., Nogueira, T., Dinis, M. A. P., & e Sousa, H. F. P. (2020). Sustainable management systems standards (SMSS): Structures, roles, and practices in corporate sustainability. Sustainability, 12(15), 1–24.
10. Borgert, T., Donovan, J. D., Topple, C., & Masli, E. K. (2020). Impact analysis in the assessment of corporate sustainability by foreign multinationals operating in emerging markets: Evidence from manufacturing in Indonesia. Journal of Cleaner Production, 260, 120714.
11. Lorenc, S., & Sorokina, O. (2015). Sustainable development of mining enterprises as a strategic direction of growth of value for stakeholders mining science, 22(2), 67–78.
12. Laing, Timothy. (2020) The economic impact of the Coronavirus 2019 (Covid-2019): Implications for the mining industry. The Extractive Industries and Society 7, 580–582.
13. Rodrigues, M., & Mendes, L. (2018). Mapping of the literature on social responsibility in the mining industry: A systematic literature review. Journal of cleaner production, 181, 88-101.
14. Tranfield, D., Denyer, D., Smart, P., (2003). Towards a methodology for developing evidence-informed management knowledge by means of systematic review. Br. J. Manag. 14 (3), 207e222.

15. Raufflet, E., Cruz, L. B., & Bres, L. (2014). An assessment of corporate social responsibility practices in the mining and oil and gas industries. Journal of Cleaner Production, 84(1), 256–270.
16. Li, Z., Nieto, A., Zhao, Y., Cao, Z., & Zhao, H. (2012). Assessment tools, prevailing issues and policy implications of mining community sustainability in China. International Journal of Mining, Reclamation and Environment, 26(2), 148–162.
17. Jenkins, H., & Yakovleva, N. (2006). Corporate social responsibility in the mining industry: Exploring trends in social and environmental disclosure. Journal of Cleaner Production.
18. Biggemann, S., Williams, M., & Kro, G. (2014). Building in sustainability, social responsibility and value co-creation. Journal of Business and Industrial Marketing.
19. Erdiaw-Kwasie, M. O., Alam, K., & Shahiduzzaman, M. (2017). Towards Understanding Stakeholder Salience Transition and Relational Approach to 'Better' Corporate Social Responsibility: A Case for a Proposed Model in Practice. Journal of Business Ethics.
20. Dando, N., & Swift, T. (2003). Transparency and Assurance: Minding the Credibility Gap. Em Journal of Business Ethics, Vol. 44, pp. 195–200.
21. Cragg, W. (1998). Sustainable Development and Mining: Opportunity or threat to the industry? Metals and the Environment, 191–205.
22. Hoffmann, J. (2018). Talking into (non)existence: Denying or constituting paradoxes of Corporate Social Responsibility. Human Relations, 71(5), 668–691.
23. Tsang, S., Welford, R., & Brown, M. (2009). Reporting on community investment. Corporate Social Responsibility and Environmental Management, 16(3), 123–136.
24. Boiral, O. (2018). Sustainability reporting and transparency: A counter-account of GRI reports. Academy of Management Proceedings, 2012(1), 12066.
25. Kitano, H. (2017), The future of blockchains lies in linkage to artificial intelligence. Diamond Harvard Business Review, August 2017. (in Japanese).
26. Warburg, B. (2016). How the blockchain will radically transform the economy. TED Talk Homepage https://www.ted.com/talks/bettina_warburg_how_the_blockchain_will_radically_transform_the_economy? last accessed June 2016
27. Ølnes, Svein; Ubacht, Jolien & Janssen, Marijn. (2017). Blockchain in government: Benefits and implications of distributed ledger technology for information sharing. Government Information Quarterly.
28. Rekimoto, J. (2016), From IoT to IoA, a network to extend humankind, Nikkei Electronics, January. (in Japanese).
29. Vintró, C., Sanmiquel, L., & Freijo, M. (2014). Environmental sustainability in the mining sector: Evidence from Catalan companies. Journal of Cleaner Production, 84(1), 155–163.
30. Salehi, M., Tarighi, H., & Rezanezhad, M. (2017). The relationship between board of directors' structure and company ownership with corporate social responsibility disclosure: Iranian angle. Humanomics, 33(4), 398–418.
31. Yakovleva, N., & Vazquez-Brust, D. A. (2018). Multinational mining enterprises and artisanal small-scale miners: From confrontation to cooperation. Journal of World Business, 53(1), 52–62.
32. Sethi, S. P., Martell, T. F., & Demir, M. (2016). Building corporate reputation through corporate social responsibility (CSR) reports: The case of extractive industries. Corporate Reputation Review, 19(3), 219–243.
33. Clarkson, P. M., Li, Y., Richardson, G. D., & Vasvari, F. P. (2008). Revisiting the relation between environmental performance and environmental disclosure: An empirical analysis. Accounting, Organizations and Society, 33(4–5), 303–327.

Implementation of Takt Time in the Development of a New Value Stream Mapping in the Production of Coffee Powder

Érik Leonel Luciano, Marcelo Tsuguio Okano, Rosinei Batista Ribeiro, Thulio Cesar Ferreira Rocha, and Wagner Alexandre Dias Chaves

Abstract This work consists of the analysis of the coffee powder production system using the Value Flow Mapping (MFV) tool in a small factory, located in the interior of Vale do Paraíba, in the state of São Paulo, which is undergoing a moment of transition. The practice of the MFV tool was identified in the Toyota Production System, in which the production of value maps was seen as routine. As a general objective, we seek to survey the Takt Time tool and point out its relevance in production planning. For this, the MFV will be elaborated and proposed to identify possible flaws or waste in the search for improvements in the production line. As a specific objective, it focuses on evaluating and investigating supply chain management and, therefore, proposing a map of the production process using the Takt Time tool. To achieve the presented objective, an exploratory quantitative review of the literature was initially carried out with the applicability of the current and proposed Value Flow Mapping tool. The adopted methodology is classified as exploratory quantitative with applicability of the MFV tool. To carry out the study, the methods adopted were the delimitation of the theme, on-site visits, data collection and the delimitation of the problem. As an expected result, the goal is to be able to contribute significantly to the improvement of the company's production process and future new studies using the tools discussed.

Keyword Lean production · Supply chain management · Applied production management tools · Value stream mapping · Coffee productions

É. L. Luciano (✉) · R. B. Ribeiro · T. C. F. Rocha · W. A. D. Chaves
FATEC, Cruzeiro Unit, Cruzeiro, São Paulo, Brazil

M. T. Okano · R. B. Ribeiro
State Center for Technological Education Paula Souza—CEETEPS—Postgraduate, Extension and Research Unit, Rancharia, São Paulo, Brazil

R. B. Ribeiro
Teresa D'Ávila University Center, UNIFATEA, Lorena, São Paulo, Brazil

Department of Aerospace Science and Technology—DCTA, Technological Institute of Aeronautics—ITA, Institute of Advanced Studies—IEAv, São José Dos Campos, São Paulo, Brazil

© The Author(s), under exclusive license to Springer Nature Switzerland AG 2021
A. M. Tavares Thomé et al. (eds.), *Industrial Engineering and Operations Management*,
Springer Proceedings in Mathematics & Statistics 367,
https://doi.org/10.1007/978-3-030-78570-3_40

523

1 Introduction

Currently, Brazil is the world's largest producer of coffee and also the main consumer of this drink. Coffee did not originate in our country, but in the mountainous lands of the Middle East, however, it adapted to the Brazilian soil, where it was responsible for the growth of the economy and the hiring of labor and the arrival of immigrants of different nationalities, who came to work in the coffee fields centuries ago.

Due to the relevance of Coffee to the Brazilian economy and because it is a drink served in the daily life of Brazilians, this work aims to analyze a family-owned company that produces powdered coffee, through a case study, where the survey will be carried out by Takt Time tool that will point out its relevance in production planning.

The specific objective is to evaluate and investigate the supply chain management and, thus, propose a map of the production process using the Takt Time tool. To achieve the presented objective, an exploratory quantitative review of the literature was initially carried out with the proposed applicability of the Value Flow Mapping tool. To carry out the study, the methods adopted were the delimitation of the theme, the on-site visits, data collection and the delimitation of the problem. As an expected result, this research will use the value flow mapping to diagnose possible flaws or waste and, through it, use the tools that make up the MFV to propose solutions that meet the needs of the company.

It is known that nowadays it is no longer possible to operate competitively in the market just to buy products from suppliers and offer them to consumers. This is because the competitive advantages at the tip of consumption do not depend only on the retailer, but on aggregated sources throughout the supply chain. The correct way to act in a competitive way is to seek continuous improvements with the other elements of the chain, in order to reduce costs, improve the quality of products and the level of service to the final customer, consumers. Therefore, any improvement that may be obtained along the supply chain, there is an element that allows the systematic analysis of the process that is called value chain. With that in mind, in order to seek continuous improvement, the commitment of everyone in the organization is necessary for it to be successful in everything it produces.

The question that guides the research is what is the impact of MFV on the production process and how to find possible problems. How to solve new obstacles, so that it is possible to indicate the appropriate management tool for the solution of the problem in a way that expands the path of quality in the company as a whole, thus reaching the goal of reducing waste and consequently costs and being able to offer a better working environment and everyone's involvement. In response to the development of the current MFV, a proposal arises in the development of a new MFV, using the Takt Time management tool that can contribute to the powder coffee manufacturing process.

The justification of the present study is to identify the main challenges for the improvement of work processes, for that purpose, it is intended to analyze the company, investigate its needs and thus propose new strategies, in a way that contributes to continuous improvement of the productive process.

As an expected result, this research will use the value flow mapping to diagnose possible flaws or waste and, through it, use the tools that make up the MFV to propose solutions that meet the needs of the company.

2 Theoretical Reference

2.1 Lean Manufacturing

The origins date back to the Lean Manufacturing Toyota Production System (also known as Just-in-time production). The Toyota executive Taiichi Ohno began in the 50 s, the creation and implementation of a production system whose main focus was the identification and subsequent elimination of waste, in order to reduce costs and increase the quality and delivery speed product to customers.

To Werkema [1], Lean Manufacturing is an initiative that seeks to eliminate waste, is, exclude what is of no value to the customer and print speed to the company.

To Mauricio and Sousa [2] the concept of Lean Manufacturing System goes further, aims to eliminate or reduce the so-called "seven wastes" most common in companies: overproduction, time/waiting, quality defects, unnecessary stocks, specialized processing, excessive transport of goods, and necessary movement of people.

2.2 Just in Time

To Slack [3], Just in Time aims to meet demand instantly, with perfect quality and without waste.

According to Cheng and Podolsky [4] Just in Time is a Production Management system that determines that nothing should be produced, transported or purchased before the exact time. It can be applied in any organization, to reduce stocks and the resulting costs.

2.3 Supply Chain Management (SCM)

Campos [5], explains that in the logistics management process, we have the so-called Supply Chain Management (SCM), which is nothing more than the correct Supply Chain Management.

2.4 Matrix S.I.P.O.C.—Suppliers, Inputs, Process, Outputs and Customers

According to Rotandaro [6], the name S.I.P.O.C corresponds to the initial combination of each aspect analyzed by the tool (Supplier, Input, Process, Outputs and Customers). Analyzing all these factors, it is possible to better understand the work performed and act on specific points in the process, promoting continuous improvement.

In addition, because it is an easy template to understand and summarizes much information in a simple way, S.I.P.O.C matrix helps to communicate all relevant information of a process for staff and other users of the process. In addition to the five letters that form the acronym S.I.P.O.C, it is important to remember that within the processes must be defined the beginning of the process, its end and the steps that take place during its execution.

2.5 Lead Time

According to Tubino [7], Lead Time or crossing time or flow is defined as "a measure of the time spent by the production system to transform raw materials into finished products. You can either consider this time broadly, calling it the customer's lead time, when you want to measure the time from the customer's request for the product to its actual delivery, as it can be considered in a restricted way, production lead time, taking into account only the internal manufacturing activities".

Since the lead time a measure of time, it is related to the flexibility of the production system to meet a client's request, i.e. the lower the product manufacturing time, the lower the production system costs in meeting the needs of customers.

Tubino [7], reinforces that lead time should not be confused with cycle time. Lead time is the time required to transform raw materials into finished products, while cycle time is the time interval between the output of finished products. You can have short cycle times with long lead times, as long as you produce based on stocks.

2.6 Takt Time

Takt Time is basically the working time that a certain piece of equipment has divided by the customer's demand for a certain item, that is, how much the company planned to manufacture to serve the customer in a certain item. It also shows how many items will be produced that day. It is necessary to produce at the speed at which the customer consumes the item, with the objective of leveling production step by step.

In the view of Tybel [8], Takt Time defines "the rate of market demand, that is, the pace of the market. Thus, the rhythm of the production line must be established so that there is not over or under production, establishing a continuous flow. To control Takt Time, it is possible to manipulate the available production time, through the production overtime or the creation of new shifts".

2.7 Value Stream Mapping (MFV)

The Value Stream Mapping (MFV) is a powerful communication and planning tool, this tool has the ability to detail the entire manufacturing process so that people know the step-by-step of creating and developing a product and thereby being able to identify possible failures or points to be improved. With it, a common language is established among employees, subsequently initiating an improvement process. The MFV will be a driver for improvements in the processes responsible for the transformation of a product. Once the map of the current and future status has been drawn up, it will be possible to realize that many processes can be eliminated from the company. The great advantage of MFV is to find possible problems and/or failures at a certain point in a process or production of the product, and thereby implement new solutions for the production process.

Rother and Shook [9] mentions that (MFV) is a tool that seeks to assist in the visualization of the company's manufacturing processes. A value stream is every action (adding value or not) necessary to be executed on a product seeking its complete manufacture. In this way, the prospect of the value stream means to take into account the widest scenario (that is, including materials and information), not just the individual value-added processes, aiming at improving the whole production flow, from the raw material to the costumer.

3 Methodology

According to Lakatos and Marconi [10], bibliographic research is defined as the survey of all "state of the art" already published in various forms, such as books, magazines, articles and websites. Its objective is to put the researcher in direct contact with everything that has been written about it, in order to allow the parallel reinforcement in the analysis of his research.

To Yin [11], a case study is defined as a research strategy that encompasses a method that uses data observation, collection and analysis.

According to Severino [12], explains that for this, the data must be collected in field research, keeping their characteristics and following all the collection processes and, after analyzing the case, reports must be generated to present the conclusions.

This research was characterized in two phases, the first was carried out through bibliographic research and the second a case study. The survey was conducted by

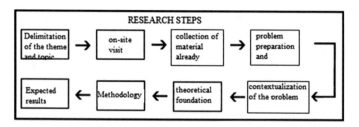

Fig. 1 Steps for conducting the research. *Source* Authors

quantitative exploratory method that involves literature search, interviews and on-site visit in order that it can be done applicability of Value Stream Mapping Tool (MFV) current and proposed.

In the on-site visit, information was collected to perform the current MFV. Through the analysis of the MFV, data was observed that guided the need for improvement in the process. The place for carrying out the implementation of the project was in a family business that manufactures coffee grounds since 1943 and passing through a time of transition in administrative management. Figure 1 shows the steps to perform the search.

The steps shown in the figure above show the methods that were carried out to carry out this research. For the development of the research, the choice of the theme was initiated according to the need of the company under study, with that it was chosen to map the manufacturing process.

3.1 Case Study

The company under study operates in the business of manufacturing and marketing coffee powder, has been in the market for more than 75 years and its main characteristic is being a family company. Currently the company has eight employees, however it has a job plan, with the tasks of each one, however, as it is a small company, they often mix the tasks. The physical space of the company has 400 m2, adding where the manufacturing and store for sale to small customers takes place. The factory is open from Monday to Friday from 07:00 to 17:00. As for its location is in the city of Cruzeiro, in the state of São Paulo.

The production cell consists of a production layout in a sequential manner, that is, in the manufacture of a particular product it passes through various stages, each stage there is a cycle time. The production cell may contain a small number of employees, as there is a possibility that it may be multipurpose, that is, be able to work on other machines.

In the production cell, there is total participation of the team, so there is a better focus on the final quality of the product, thus generating a smaller number of errors.

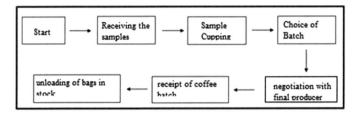

Fig. 2 Negotiation, purchase and supply of coffee beans. *Source* Authors, 2020

A great advantage in the production cell is that there are possibilities to place several productive resources in a well-defined space so that it can add value to the product. Another relevant point is when there is a sequential production cell that the cycle time between one stage and the next is close, as there will be no possibility of bottleneck formation, that is, an accumulation of stock between one cycle and another.

The coffee powder production cell is composed of several machines, which are the roaster, the cooler, the coffee elevator, storage silo, mill, date, conveyor and scale.

As for the manufacture of coffee powder, the company is committed to producing always offering freshly made products to ensure more quality and freshness in the final product on offer to the customer. Among the products manufactured there are three types of presentation, they are gourmet grains, ground almost at the time that are produced through orders and the ground at the time so far the most recommended by them. Initially, the coffee bean is negotiated, purchased and supplied, as shown in Fig. 2.

The procedures mentioned in Fig. 2 show the steps since the receipt of the coffee bean sample, the cupping test that in turn generates the choice of the lot to be negotiated with the supplier, obtains the final delivery of the flow product and unloading the bags in stock. To carry out the coffee powder production step, the following procedures are necessary, as shown in Fig. 3.

The steps mentioned in Fig. 3 starts with the roasting procedure, although it is a simple step, it requires some care such as the roasting point and the ideal roasting temperature, after leaving the roaster the coffee rests in a cooler, the next step is to go through a coffee elevator that assists its cooling and leads it to the storage silo. After this procedure, the grinding is done according to the order demand, as it interferes with the final quality and flavor of the drink, it is interesting to note that the more recent the grinding, the better the product intended for customers and after grinding the coffee, it is weighing and packed in packs of 500 (five hundred) grams or 1 kg and grouped in 5 kg bales as described in the figure.

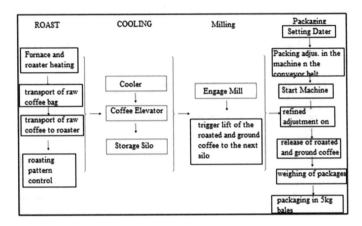

Fig. 3 Production stage of coffee powder. *Source* Authors, 2020

4 Results and Discussion

Through an interview with those responsible for the company, information was collected on the stages of the coffee powder manufacturing process. Initially during the site visit, it was noted that it is very organized, based on the 5S philosophy in the company. As a factor of great value for this research through the collection of information, the tool S.I.P.O.C. to assist in the development of the MFV.

Figure 4 illustrates the S.I.P.O.C tool in the coffee powder manufacturing process.

The steps of the S.I.P.O.C. details all the procedures that are performed in each group. With all the information indicated, we have a detailed summary of the manufacturing process, from the supplier to the final recipient who is the customer. Thus, based on the collection of information obtained, the current Value Flow Mapping (MFV) of the company under study was presented in Fig. 5.

The current MFV starts with the supply of a sample of the coffee batch, when it arrives at the company, the samples are tested (cupping) for the selection and purchase of the batch, then delivery and unloading is carried out in the stock. For

Fig. 4 S.I.P.O.C. of the coffee powder manufacturing process. *Source* Authors, 2020

		S.I.P.O.C.		
Supplies	Entries	Processes	Outputs	Customers
Supplier A	coffee beans	Checking Merchandise	Separate organic and recyc waste	Physical person
Supplier B	laminated packaging	Weighing		Supermarket
Supplier C	date coils	Weekly invetory	Final product	Bakeries
Supplier D	Plastic Packing (input)	Manufacturing steps	Issue invoice	Coffee Shop
Supplier E	Wood	Toaster	Dispatch	Hotels
		Cooler		
		Milling		
		Packing		

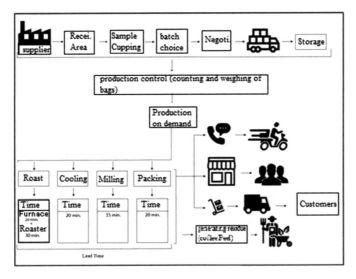

Fig. 5 Value stream mapping (MFV) current. *Source* Authors, 2020

production control, coffee bags are counted and weighed, and the same procedure is carried out for other suppliers.

In the production of coffee powder there is a division into four stages, roasting, cooling, grinding and packaging. It is important to highlight that in the roasting procedure, the generation of waste (coffee bean film) arises, and it is provided to the rural producer for fertilizing the soil in the cultivation of vegetables.

Each of these steps was provided the cycle time, that is, how long it takes to perform each step in either machine time or operating time. As for the Lead Time, we can understand it in several ways, among them is the Manufacturing Lead Time, which is the time that the product takes to transform the raw material into a finished product. Thus, it is an important indicator of time, where it is calculated by adding the times of individual cycles of each stage of the process. Manufacturing is done on demand, so it uses the same JIT procedures as producing only what is necessary. The production starts with the roasting process, then moves on to cooling and storage. After grinding and packaging is carried out as requested by the customer, it can be in pillow-type or vacuum packages of 500 (five hundred) grams or 1 kg. To finish the production stage, a bundle (bales) of 5 kg (five kilos) is made.

The sales procedure divides between the store that is the retail sale and it is also worth noting that due to the current world scenario in terms of the COVID pandemic, delivery has emerged. Wholesale sales vary between bakeries, coffee shops, hotels and supermarkets.

In order to use this tool, it is necessary to collect data regarding the quantity demanded, the production capacity, the available production time, the number of hours worked per day and the number of days worked per month. Through this information it is possible to calculate the Takt Time as shown in Fig. 6.

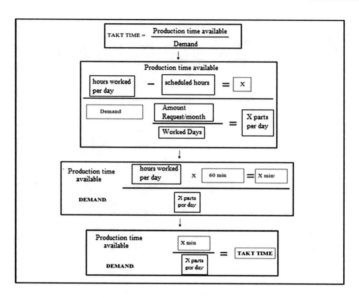

Fig. 6 Calculation to get the takt time. *Source* Authors, 2020

Takt Time is the available production time divided by demand, that is, the quantity of products to be manufactured.

Initially we need to know the available production time to share with the demand. The available production time will be the number of hours worked per day subtracted by the number of hours scheduled (lunch, work gym and even a meeting on the day).

The demand will be the division of the forecast to be produced and the number of days worked in the month, thus obtaining the result of how many parts will be needed to be produced per day.

Subsequently, the hours worked are transformed into minutes and divided by the quantity of products that will be manufactured per day. Finally, divide the number of minutes worked by the number of products to be manufactured per day. This will calculate the Takt Time of what time it will take to produce a certain product.

After the implementation of Takt Time, a better production schedule is obtained, in order to make it faster and more accurate in productivity. As a sequence in the process of the aforementioned tool, there is the production management procedure, which is the Lead Time, this occurs at the time of entry into the raw material in the manufacturing process. For each stage of product development, there are so-called production cycles, that is, how long the machine uses to perform a certain activity. Every activity performed in a cycle time is known as work in process represented by the English acronym WIP (Work in Process).

It is important to highlight that the shorter the process cycle time, the better the Lead Time. There are four stages in the coffee powder production process, each stage being a production cycle. The Lead Time is the sum of all the steps from the entry into the raw material until the end of its manufacture, as shown in Fig. 7.

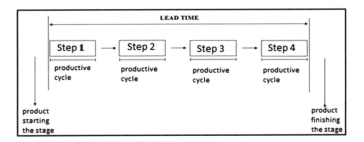

Fig. 7 Lead time. *Source* Authors, 2020

Measuring and knowing the Lead Time is very important to improve the manufacturing stage. Lead time being reduced without interfering with product quality results in customer satisfaction and cost savings.

As a proposal for the present study, the MFV was developed with the implementation of the Takt Time tool, as previously explained throughout the calculation procedure. Its implementation will provide better management in production planning. Accordingly, it achieves a better pace and leveling of production, thus it will constitute an adequate speed in the process according to the speed of demand. Figure 8 clarifies the proposed Value Stream Mapping.

With the defined Takt Time, it is favorable to diagnose which machines or collaborators are causing delays in the production line, also with this implementation will

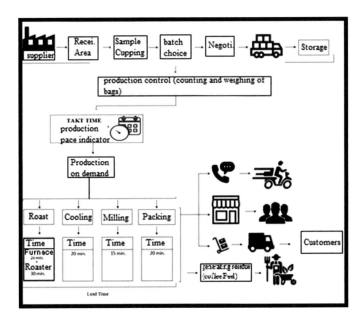

Fig. 8 Proposed value stream mapping. *Source* Authors, 2020

have a productive standardization, thus, it will increase industrial efficiency in a short period of time. Takt Time is advantageous for industries to maintain constant standardization while preserving wasted time and labor.

5　Final Considerations

In view of the current MFV, it was possible to identify the need to implement the Takt Time tool to better manage the production line. The advantages of using Takt Time are the ability to level the production in a considered way and, also to identify if there is a machine operating out of the standard or accumulation in one step of the process.

Takt Time determines the expected pace and thus enables the accurate identification of delays, always guided by the basis of demand, with the consequent demand for quick solutions to correct them. The Takt Time tool is crucial for the optimization of productive capacity and with the use of this tool it is possible to establish a better pace at each stage. In the face of process management, better results are obtained and gains in speed in manufacturing and thus lead to on-time delivery of final products without loss of quality and cost reduction. Anyway, this work can contribute to future studies in the field of production engineering and logistics and serve as a basis for practical studies in the industrial area.

References

1. Werkema, M.: Lean seis sigma: introdução às ferramentas do lean manufacturing. Belo Horizonte: Werkema Editora, 2006.
2. Mauricio, F. and Sousa V.: Implementação do SMED em uma empresa de autopeças: um caso francês. XXXIV Encontro Nacional de Engenharia de Produção. Curitiba, Brasil, 2014.
3. Slack, N.: Vantagem Competitiva em Manufatura. São Paulo: Atlas, 1993.
4. Cheng, T. and Podolsky, S.: Just-in-time manufacturing: an introduction [Em linha]. 2ª ed. London: Chapman & Hall, 1996. Available in: http://books.google.com/books?id=WL95yz pj1TIC ISBN 978–0–412–73540–0. Accessed in: 16 Feb. 2021.
5. Campos, L.: Supply Chain: uma visão gerencial. Editora Ibpex. 2010.
6. Rotandaro, R.: Seis Sigma: Estratégia gerencial para a melhoria de processos, produtos e serviços. São Paulo: Atlas, 2006.
7. Tubino, D.: Sistemas de produção: a produtividade no chão de fábrica. 1ª reimpressão. Porto Alegre, 2004.
8. Tybel, D.: Guia da monografia: o que é estudo de caso. 2017. Available in: < https://guiada monografia.com.br/estudo-de-caso>. Accessed in: 23 Jun. 2020.
9. Rother, M. and Shook, J.: Learning to See - Value Stream Mapping to Add Value and Eliminate Muda. The Lean Enterprise Institute, 1998.
10. Lakatos, E. and Marconi, M.: Metodologia do trabalho científico. 5 ed. rev. ampl. São Paulo: Atlas, 2001.
11. Yin, R.: Estudo de caso: planejamento e métodos. 2ª Ed. Porto Alegre. Editora: Bookmam. 2001.
12. Severino, A.: Metodologia do Trabalho Científico. 23ª. ed. São Paulo: Cortez, 2007.

Planning Intensive Care Resources: A Forecast and Simulation Approach Due COVID-19 Pandemic in Rio de Janeiro City

Daniel Bouzon Nagem Assad⊙, Javier Cara⊙, Miguel Ortega-Mier⊙, Thaís Spiegel⊙, and Luana Carolina Farias Ramos⊙

Abstract The COVID-19 pandemic recently challenges worldwide health system. Many countries have been establishing several rules in order to fight against the expected exponentially growth of affected people. After some weeks hospital bed occupancy started to grow because of infected people and this demand can potentially overlap hospital bed capacity. In this scenario, doctors would have to decide who should be treated and patients without adequate treatment can increase number of deaths. Thus, in order to avoid that patient demand overlap hospital bed capacity, this research propose a forecast and simulation approach in order to provide the number of intensive care beds and health care professionals that could be available in a short-term of 21 days ahead taking into account 3 patient demand scenarios. This research is applied to Rio de Janeiro city (Brazil) and all records were retrieved from Rio de Janeiro public database. Hospital length of stay (LOS) records were obtained with Rio de Janeiro City Health Department (RJHD). We concluded that to open the number of new intensive care unit (UCI) beds proposed by the government is lower than expected admission on the worst scenario.

Keywords Health care operation management · Simulation · Forecast · Resource planning

1 Introduction

Since HIV in the 1980s and until Zika virus in 2015, emerging infectious diseases challenges health systems driving then to public health emergencies [1]. First time reported by China in December 2019, COVID-19 out-break is both similar to the prior severe acute respiratory syndrome (SARS; 2002–2003) and Middle East respiratory syndrome (MERS; 2012-ongoing) outbreaks [2]. Over the last few decades,

D. B. N. Assad (✉) · J. Cara · M. Ortega-Mier
Universidad Politécnica de Madrid, Madrid, Spain

D. B. N. Assad · T. Spiegel · L. C. F. Ramos
Universidade Do Estado Do Rio de Janeiro, Rio de Janeiro, Brazil

© The Author(s), under exclusive license to Springer Nature Switzerland AG 2021
A. M. Tavares Thomé et al. (eds.), *Industrial Engineering and Operations Management*, Springer Proceedings in Mathematics & Statistics 367, https://doi.org/10.1007/978-3-030-78570-3_41

mathematical models of disease transmission have been helpful to gain insights into the transmission dynamics of infectious diseases and the potential role of different intervention strategies.

Over the last few decades, mathematical models applied over infectious diseases growth have been helpful to gain insights into the transmission dynamics [3] allowing case numbers forecasts and control options evaluation [1]. Although still showing numerous limitations and pitfalls to their use [4], often driven by data scarcity and delay, tighter integration of infectious disease models with public health practice and development of resources has the potential to increase the timeliness and quality of responses. For instance, Walker et al. [5] has published a research warming United Kingdom (UK) about the expected results according to each constrains established by the UK government.

However, Roosa and Chowell [6] enhance that mathematical modeling successful application to investigate epidemics depends upon our ability to reliable estimate key parameters which are subject to two uncertainty sources: noise in the data and assumptions built in the model. The same authors state that ignoring this uncertainty can potentially result in public health policy decisions.

In order to avoid assumptions assumed by model selection, recently, Adhikari et al. [7] proposed a neural network using a deep learning approach to help US policymakers on implementation of effective countermeasures to control both seasonal and pandemic influenza outbreaks. However, due COVID-19 pandemic data scarcity, sub-notification and considering that neural network requires a large data set to work well, at this moment, it seems to be an unwise choice.

According to Smirnova et al. [8], approaches to investigating transmission rates of diseases are based on system design with deterministic, stochastic, network or hybrid (combination) models. Their research pointed out two common practices: to assume a constant transmission rate or to assume that transmission rate be- haves as some pre-set periodic, exponential or other function with a finite number of parameters. These parameter values can be estimated by least squares data fitting or optimization and statistical approaches.

In addition, analyzing only influenza forecasting models in human population, Chretien et al. [9] propose a framework to classify research as follows: Population-based forecasting studies (seasonal or pandemic), forecast type (temporal or Spatial-temporal) and forecasting method (mechanistic, Statistical).

Focusing on forecasting method, they classified them as follows: compartmental model, regression tree, generalized linear model, agent-based model, survival analysis, Bayesian network and time series model.

Jombart et al. [10] propose a complementary approach to estimated needed hospital beds combining hospital LOS in China and expected hospital patient admission in UK some days ahead. In this study they state that Weibull distribution fitted better Chinese LOS.

In this context, the current research can be classified as a pandemic Population-based forecasting studies where we provide a temporal forecasting without a spatial component using time series models to forecast COVID-19 intensive care unit patient admission (ICUPA) in Rio de Janeiro city and thereafter provide the number of hospital intensive care unit (ICU) beds and health care professionals that could be available in the short-term.

2 Problem Statement

The ED's efficiency expressed by waiting time and length of stay (LOS) are related to how better is balance between emergency departments (ED) resources (doctors, nurses, beds, medical equipment, etc.) and demand (patient arrivals). Linking them, accurate forecasting of ED patient demand can support long and short-term staffing policies in order to better prepare ED for the coming demand variations [11]. The COVID-19 total and daily ICUPA in Rio de Janeiro city (TICUPA and DICUPA) were retrieved from a public dashboard available in Painel COVID-19 RJ [12] website. Both time series are presented in Fig. 1.

Taking into account the exponential growth expected in pandemic environment we generate a "third" time series where we convert the TICUPA into a logarithmic scale (LS). Thus, we will analyze TICUPA time series over LS and original scale (OS). In all time series we applied two Unit root tests (p-value results on Table 1) with null hypothesis of stationary:

1. Augmented Dickey Fuller Test (ADF) test;
2. Kwiatkowski–Phillips–Schmidt–Shin (KPSS) test for level and trend.

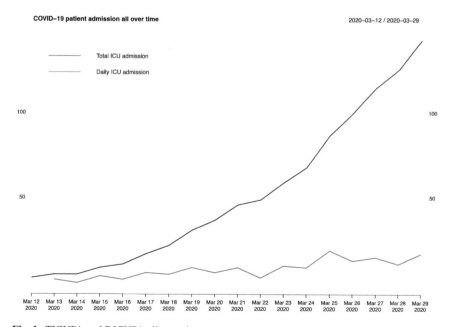

Fig. 1 TICUPA and DICUPA all over time

Table 1 Stationarity test results by time series. Source: the authors

Time series	ADF test	KPSS level	KPSS trend
TICUPA OS	0.03	0.016	0.021
TICUPA LS	0.955	0.015	0.022
DICUPA	0.092	0.1	0.032

Assuming a significance level of 0.05 we can reject the stationary null hypothesis on all TICUPA OS tests and level stationary on all of them while in TICUPA LS and DICUPA we cannot reject stationary condition using Unit root tests.

3 Forecasting

In current section we evaluate how forecasting models fits TICUPA OS and TICUPA LS by noise analysis according to criteria widely used in time series analysis [13].

3.1 Forecasting Model Selection

To forecast TICUPA 21 days ahead (up to 21/04) we used univariate models from the literature (for more details about them see Hyndman & Athanasopoulos [12]) and linear regression presented below:

1. Exponential smoothing state space (ETS);
2. Auto Regressive Integrated Moving Average (ARIMA);
3. Linear regression (LR).

Applying these models over TICUPA LS and TICUPA OS we used the lowest small-sample corrected Akaike Information Criterion (AICc) to select the best model and parameters for each class of ETS and ARIMA models. For LR we found the best coefficients of intercept and trend.

Before comparing accuracy measures between all models applied, to select one or a combination of them we must first take into account if found parameters catches well the time series signal. It can be done by evaluating the noise (residuals) in each model.

Therefore, in Tables 2 we evaluate forecasting models residuals according to four statistical tests criteria:

1. True mean is equal to 0 using Student test;
2. Can be fitted by a normal distribution (ND) using Shapiro–Wilk normality test;
3. Are Independent and identically distributed (i.d.d) using Ljung-Box test at lag 1–10;

Table 2 Residuals evaluation (p-value) per forecast model

Null hypothesis	Mean equal to 0	Fit ND	i.d.d	No ARCH effect
ETS OS	0.171	0.238	0.001	0.394
ARIMA OS	0.037	0.980	0.205	0.303
LR OS	1	0.123	0.001	0.052
ETS LS	0.925	0.233	0.131	0.065
ARIMA LS	0.277	0.650	0.552	0.089
LR LS	0.092	0.104	0.001	0.045

Table 3 Accuracy measures by model

Model	RMSE	MAPE (%)	AICc
ETS OS	3.553	10.103	112.665
ARIMA OS	2.212	5.956	86.575
LR OS	11.592	58.021	145.294
ETS LS	2.348	6.166	−17.553
ARIMA LS	4.368	7.341	−22.174
LR LS	17.697	18.874	4.235

4. Has no Autoregressive Conditional heteroscedastic (ARCH) effects (ef.). In other words, if variance remains constant on residual series (homoscedastic condition) at lag 1–10.

If any model residuals fulfil all these criteria Hyndman and Athanasopoulos [13] state that it is a white noise. So, considering a significance level of 0.05 only ETS LS and ARIMA LS models meets all white noise condition. ETS OS model was rejected at lag 3–10 on Ljung-Box test while LR OS and LR LS were rejected at all lags on the same test. ARIMA OS was rejected on t-student test. Only LR LS was rejected on ARCH test on lag 3.

The AICc (out-of-sample prediction error) and accuracy measures between each model compared with each TICUPA time series are summarized in Table 3 and on last step we conclude by noise evaluation that only ETS LS and ARIMA LS model meets all proposed conditions. Even though accuracy measures analysis pointed out better root mean square error (RMSE) and mean absolute percentage error (MAPE) results to ARIMA SO model.

3.2 Forecasting Results and Discussion

On last section we presented six models and evaluated their residual and calculate their accuracy measures we choose ARIMA SL (1, 2, 0) and ETS SL (A,Ad,N) models. The parameters of each model are presented below:

1. ETS SL: Additive error (A), additive damped trend (Ad) without seasonality (N). Smoothing parameters: alpha = 1e-04, beta = 1e-04 and phi = 0.920. Initial states: level = 0.597 and trend = 0.4858;
2. ARIMA SL: Auto-regressive at lag 1 with coefficient -0.9515 and double differentiation.

To generate the three scenarios of Fig. 2 we follow the four steps presented below. The percentage growth between our last observation (143) and 1, 2 and 3 weeks can be seen on Table 4.

1. Define a percentile to pessimistic (0.9), expected (0.5 or median) and optimistic (0.1) scenarios.
2. Run 10,000 possible scenarios (replications) by adding to each model signal their respective error component at each lag randomly sorted using a ND (as we expected from a white noise and ensured by Shapiro-Wilk normality test).
3. Get at each lag between all replications the pessimistic, expected and optimistic values.

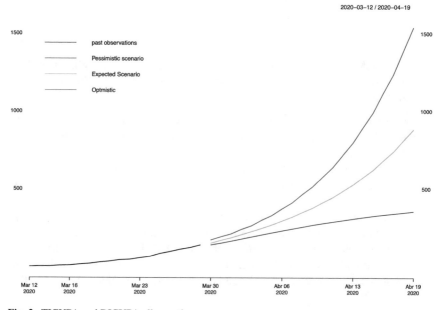

Fig. 2 TICUPA and DICUPA all over time

Table 4 Patient demand forecasting by scenario

Date	PES	EXS	OPS
05/04/2020	132.168	90.210	54.545
12/04/2020	404.895	241.958	110.490
19/04/2020	979.021	519.580	150.350

4. Combine ETS SL and ARIMA SL results by choosing between both forecasts the highest value to pessimistic scenario (PES), the average between expected scenario (EXS) and the lowest value to optimistic scenario (OPS).

4   Planning Intensive Care Resources

In Brazil, according to ordinance [14] the number of health care professionals is given by a proportion (at least one doctor per 10 beds, one nurse per 8 beds and 1 nurse technician per 2 beds) from the number of ICU beds which mainly depends on two variables:

1. Patient daily admission: covered on this research assuming forecasted TICUPA for all scenarios presented in last section. On this step we must convert forecasted TICUPA into forecasted DICUPA by differentiating the first one on lag 1;
2. Patient LOS in ICU: random variable fitted by exponential probability distribution (p-value = 0.083) with rate of 0.1608.

The estimated rate was obtained applying over histogram observed data the Chi-square Goodness of Fit Test (Fig. 3). Only exponential distribution got p-values higher than significance level of 0.05. Assuming a significance level of 0.01 we would not reject a Weibull distribution (0.031) which is the same distribution used in Jombart et al. [10].

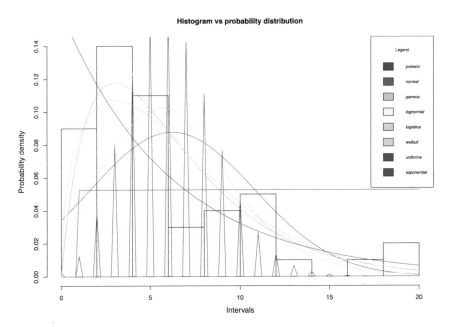

Fig. 3 Chi-square goodness of fit test

In order to schedule each patient (p) on the day expected (d) all over LOS period (l) on ICU bed (b) on Timeline (T) in each scenario (s) all over each replication (r) where the ICU total capacity (B) is the number that we want to know, we follow the five steps bellow and their results are presented in Fig. 4:

1. Convert DICUPA of s into a list of patients where each p must be assigned to any b since their arrival d until discharge on d + 1.
2. Calculate T where T will be the last discharge date.
3. Assign each p to a b in day d and block b schedule ahead until d + 1 comes. If all beds B has a patient than B becomes B + 1.
4. If s patients list is over storage B on tuple s,r.
5. Run the method above until fulfil B for every tuple s,r.

The Fig. 4 results is summarized in Table 5. In addition, the number of health care professionals according to [13] is provided.

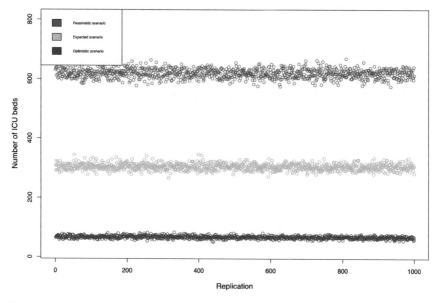

Fig. 4 Number of ICU needed beds 21 days ahead

Table 5 ICU beds and health care professionals needed

Resources	PES	EXS	OPS
ICU beds	669	305	51
Doctors	67	31	6
Nurses	84	39	7
Nurses technician	335	153	26

5 Conclusion

In this research we propose a forecast and simulation approach to deals with both uncertainties: patient admission infected by COVID-19 and their ICU LOS.

Besides other studies pointed out many types of techniques that consider for example contamination rate, transmission dynamics and spatial–temporal models among others, our research presented an approach that do not starts from a set of assumptions that, eventually, cannot be confirmed. Thus, worked with a short period time series to better estimate in short-term how many ICU resources could be necessary to provide adequate treatment to COVID-19 infected patients.

Thus, even though a quite different proposal from studies like [5], using forecast techniques we showed in next weeks is expected a significantly UCI patient admission growth that probably will have a huge impact on public and private health systems of Rio de Janeiro City.

References

1. C. Metcalf, J. Lessler, Opportunities and challenges in modeling emerging infectious diseases, Science 357(6347), 149 (2017).
2. Z. Wu, J.M. McGoogan, Characteristics of and important lessons from the coronavirus disease 2019 (covid-19) outbreak in china: summary of a report of 72 314 cases from the chinese center for disease control and prevention, Jama (2020).
3. G. Chowell, L. Sattenspiel, S. Bansal, C. Viboud, Mathe- matical models to characterize early epidemic growth: A review, Physics of Life Reviews 18, 66 (2016).
4. A. Smirnova, G. Chowell, A primer on stable parameter estimation and forecasting in epidemi- ology by a problem- oriented regularized least squares algorithm, Infectious Disease Modelling 2(2), 268 (2017).
5. P.G. Walker, C. Whittaker, O. Watson, M. Baguelin, K. Ainslie, S. Bhatia, S. Bhatt, A.
6. K. Roosa, G. Chowell, Assessing parameter identifiabil- ity in compartmental dynamic models using a computa- tional approach: application to infectious disease trans- mission models, Theoretical Biology and Medical Mod- elling 16(1), 1 (2019).
7. B. Adhikari, X. Xu, N. Ramakrishnan, B.A. Prakash, in Proceedings of the 25th ACM SIGKDD International Conference on Knowledge Discovery & Data Mining (2019), pp. 577–586.
8. A. Smirnova, L. deCamp, G. Chowell, Forecasting epidemics through nonparametric estimation of time-Dependent transmission rates using the seir model, Bulletin of mathematical biology 81(11), 4343 (2019).
9. J.P. Chretien, D. George, J. Shaman, R. Chitale, F. McKenzie, Influenza forecasting in human populations: A scoping review, PLoS ONE 9(4) (2014).
10. T. Jombart, E.S. Nightingale, M. Jit, O.l.P. de Waroux, G. Knight, S. Flasche, R. Eggo, A.J. Kucharski, C.A. Pearson, S.R. Procter. Forecasting critical care bed requirements for covid-19 patients in england (2020). URL https://cmmid.github.io/topics/covid19/current-patterns-tra nsmission/ICU-projections.html.
11. Boonyasiri, O. Boyd, L. Cattarino, et al., The global impact of covid-19 and strategies for mitigation and suppression, On behalf of the imperial college covid-19 response team, Imperial College of London (2020).

12. Portal do Rio de Janeiro. Painel COVID-19 RJ kernel description (2020). URL:http://pcrj. maps.arcgis.com/sharing/rest/content/items/ae85fc84a9b244108d96c7072be4d3d3/data.
13. R.J. Hyndman, G. Athanasopoulos, Forecasting: principles and practice (2018).
14. Brasil. Resolução rdc No 7, de 24 de fevereiro de 2010. dispõe sobre os requisitos mínimos para funcionamento de unidades de terapia intensiva e dá outras providências (2010). URL: https://bvsms.saude.gov.br/bvs/saudelegis/anvisa/2010/res000724022010.html.

Genetic Algorithm Applied to Packaging Shape Models Using Sustainability Criteria

Tarcisio Costa Brum, Marcos Martins Borges,
Afonso Celso de Castro Lemonge, and Lino Guimarães Marujo

Abstract The design and choice of materials are important steps in the process of developing a product and it's supposed to be designed according to consumers' requirements and sustainability criteria. Managing these issues can be associated with Kansei Engineering (KE) methodology, which consists to translate costumers´ emotions into products' specifications. The paper aims consists to find an optimum product packing design considering customers' requirements and sustainability criteria, applying Kansei Engineering theory quantification type I technique and triangular fuzzy number to measure costumers´ requirements collected by questionnaires and genetic algorithm to perform the evaluation process combining costumers´ requirements and sustainability criteria by a fitness function. Considered in this study material type and specifications recycling rate, decomposition time, CO_2 emission rate and reuse financial return as sustainability criteria. The proposed model based on KE and its result, a package sketch that considers costumers' requirements and, at the same time, sustainability criteria. Implications for theory and practice: the model can be applicable and enhanced as a tool for generating improvements in the product design area.

Keywords Kansei engineering · Costumers' requirements · Sustainability criteria

1 Introduction

Managing the product development process to reach positive results involves an organization identifying costumers' needs, translating that into concrete actions and developing a product/service to meet these needs, always aiming at quality, agility and product cost [1]. Therefore, environmental matters have become an important factor for customers when purchasing a product [2]. Consequently, companies need

T. C. Brum (✉) · L. G. Marujo
Federal University of Rio de Janeiro, Rio de Janeiro-RJ, Brazil

M. M. Borges · A. C. de Castro Lemonge
Federal University of Juiz de Fora, Juiz de Fora-MG, Brazil

© The Author(s), under exclusive license to Springer Nature Switzerland AG 2021
A. M. Tavares Thomé et al. (eds.), *Industrial Engineering and Operations Management*,
Springer Proceedings in Mathematics & Statistics 367,
https://doi.org/10.1007/978-3-030-78570-3_42

to study alternatives to respond to global and market changes by considering the development of new and more sustainable products and services that meet costumers' requirements.

These demands planning and strategies to identify and translate costumers' needs into product requirements. There are tools and techniques which can integrate costumers' input into the product development process, such Kansei Engineering (KE), developed by Mitsuo Nagamachi, seeks to translate customers' emotions into product requirements [3]. Many works using KE apply computational techniques to translate customers' needs into product requirements ([4, 5, 6]). The main computational techniques reported in the literature are neural networks, support vector machines and genetic algorithms (GA) [7]. Among these, GAs are more effective at finding the solution in the whole search space. That is interesting due to possibility of choosing an optimal product design model where there is no strong need to generate a new product or a new design in the short term.

In this study, GA are used to optimize packaging design based on assessment criteria (linguistic variables, costumers' preferences and sustainability criteria). As [8, 9] describe, building a fitness function requires combinations of different methods such as Quantification Theory Type I and Fuzzy Logic. The Kansei Engineering Method was applied to collect and measure costumers' preferences about product shapes and materials. Afterwards, each product was evaluated and selected using the genetic algorithm. The guidelines adopted to execute a general KE methodology were described by [3] and adapted for this research as follow:

- Product domain: In this first step the outcomes are the product and target audience choice by market analysis. The basic characteristics of the product can also be defined is this step as well.
- Semantic field: In this step linguistic variables and adjectives associated with the product are defined in accordance with the product selected in the domain.
- Characteristics field: The characteristics of the product are chosen, such as possible shapes, colors, materials and uses. The designers can develop mock-ups, prototypes or sketches to represent them.
- Synthesis: This is the main step and where the connection between linguistic variables and product characteristics occurs.
- Model building: A model may be created to express the products selected according to costumers' preferences. This step is described as engineering or scientific method [10].

2 Theoretical Fundamentals of the Kansei Engineering

2.1 Quantification Theory Type I

The quantification theory type I it is one of the main techniques applied in the Kansei Engineering studies to identify the relation between linguistic variables and product characteristics [9]. Each shape element is named as an item and each variation of an item is named as a category. Qualitative data are quantified to the dummy variables (0 or 1), indicating the absence or presence of a category in each item. The relationship between each adjective and the categories of items is obtained by using an adapted multiple linear regression analysis, as shown in Eq. (1):

$$\bar{y} = \sum_{i=1}^{m} \sum_{j=1}^{n} a_{ij} x_{ij} + \epsilon \tag{1}$$

where $\bar{y}$ is the predictive value (linguistic variable average), a_{ij} is the category weight or contribution degree, x_{ij} is the dummy variable, m is the total number of items, n is the total number of categories, and ϵ is the model error. The results are the weights of the categories for each linguistic variable, and these weights are used to make up the Fitness function.

2.2 Triangular Fuzzy Numbers

The fuzzy sets theory proposes to measure natural language subjectivity, such as "approximately" and "around", for example in [11] working with non-extreme answers and hence closest to the real world. A fuzzy number can be shown in different ways and a triangular representation is the most frequently used one. A triangular fuzzy number shows for $A = (a_1, a_2, a_3)$ a pertinence function denoted by Eq. (2):

$$\mu_A(x) = \begin{cases} 0 & if \ x \leq a_1 \\ \dfrac{x - a_1}{a_2 - a_1} & if \ a_1 \leq x \leq a_2 \\ \dfrac{a_3 - x}{a_3 - a_2} & if \ a_2 \leq x \leq a_3 \\ 0 & if \ x \geq a_3 \end{cases} \tag{2}$$

The function $\mu_A(x)$ can be interpreted as the possibility of occurrence of a value in ranges defined between a_1 and a_3.

Research on Kansei Engineering require human involvement and, consequently, it deals with subjectivity. In this regard, fuzzy logic is applied to handle subjectivity, mainly semantic variables, as discussed in [12, 13] for example, which applied triangular fuzzy numbers to obtain the linguistic variables weights.

2.3 Genetic Algorithms

Holland in 1975 was the first researcher to make use of Darwin's theory as a computational resource. Genetic Algorithms work as an evolutionary process, where each solution is an individual of the population. In each generation, there is the crossover and mutation process, where a new individual is created combining elements from its parents. Each candidate solution is evaluated by the fitness function and the fittest individual is selected for the next generation. First, it is necessary to encode the solutions into chromosomes and, in this study, the binary alphabet is adopted. A generic genetic algorithm can be described as [14]. The final process consists of decoding the optimum chromosome into a product generated. In terms of the Kansei Engineering methodology, this step can be interpreted as the final building model.

3 Methodology Description

3.1 Target Product Selection

For product selection, the criteria adopted consider whether the product is commonly consumed and has a high negative environmental impact. In Brazil, the amount of domestic waste represents a significant negative impact. According to [15] show that one third of total waste volume comes from packages, and 1/3 of them come from food packaging. Based on this information and considering a common product consumed in Brazil, the sweet milk package was selected as a target product in this study.

3.2 Characteristics of Target Product

The basic items identified for a sweet milk package are body and lid. In most of the packages found, there are four combinations of materials, defined as a category of the item Material. For each item, the categories were defined in Fig. 1 based on a morphological classification in the study described in [16].

The drawings were made using the software SolidWorks, where it is possible to measure the weights and dimensions of any kind of product with different materials. All the packages have a fixed capacity of 500 g (the main sweet milk volume traded). In the next step, a group of products that represents all the categories was determined, as per the recommendations proposed in [3]. In total, 17 packages were drawn (Fig. 2) to represent all the categories and reflect customers' ideas regarding shape and material.

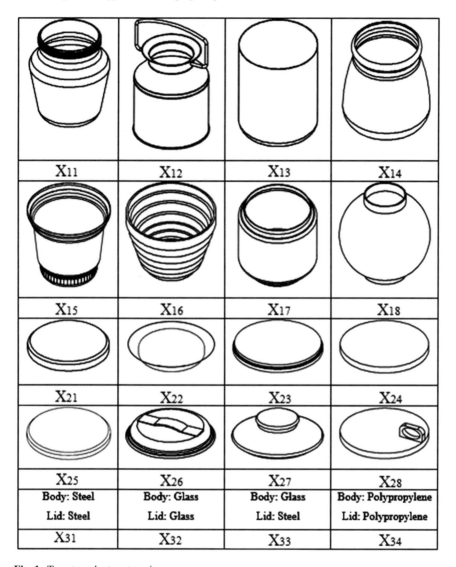

X_{11}	X_{12}	X_{13}	X_{14}
X_{15}	X_{16}	X_{17}	X_{18}
X_{21}	X_{22}	X_{23}	X_{24}
X_{25}	X_{26}	X_{27}	X_{28}
Body: Steel Lid: Steel	Body: Glass Lid: Glass	Body: Glass Lid: Steel	Body: Polypropylene Lid: Polypropylene
X_{31}	X_{32}	X_{33}	X_{34}

Fig. 1 Target products categories

3.3 Linguistic Variables

Words are the main medium of human communication. Linguistic variables are one of way to obtain costumers' impressions and emotions. In this sense, according to the Brazilian Association of Packaging (ABRE), the words/adjectives that represent costumers' impressions about package shapes are: **Beautiful, Practical, Attractive, Sustainable, Safe and Protect of product**. The first three are directly related to

Fig. 2 17 samples representing product domain

package shape, whereas sustainable packages are a trend in the packaging industry. The functions Safety and Product Protection, the last two linguistic variables, are essential functions in food packaging [17].

The option to work with few adjectives is defended in [8], which says that working with a small number of words is better than with a large number of words, because it can imply in human error upon responding questionnaires.

3.4   Material Sustainability Criteria

In Brazil, the main materials used in sweet milk packaging design are polypropylene plastic (PP), steel and glass. Figure 3 shows three examples commonly found in the Brazilian market.

The sustainability criteria adopted were based on the 3R Strategies: Reduction at source, Reuse and Recycling, according to the manufacturing strategies aimed at the end of life (EoL) product life cycle, as described in [18]. According to [19], reduction at source consists in material consumption reduction in the industrial process, mainly changing package shape and/or thickness. In this sense, the criteria CO_2 emission rate in Brazil" and "financial return rate in Brazil" can be measured proportionally with the material mass in each package. Furthermore, the reuse financial return is a criterion that can be associated with the material reuse. The data about the recycling rate in Brazil and material decomposition time account for the third 'R' of Recycling. Another reason to use this criterion is the quantitative data that contribute to the evaluation function used in the model. The values of each criterion are displayed in Table 1.

The data regarding reuse financial return, recycling rate of glass and steel were collected from [20]. The carbon emission rate for glass was collected from [21]. The carbon emission rate for steel was collected from [22]. The carbon emission rate for the polypropylene production considered was obtained from the Braskem industry (the only company that produces polypropylene in Brazil) described in [22].

Fig. 3   Basic models of sweet milk packages in the Brazilian market

Table 1   Sustainability criteria for each material

Material	Recycling rate (%)	Decomposition time (years)	CO_2 emission rate (kg CO_2/kg Steel)	Reuse financial return (R$/kg)
Glass	47	4000	0.17	0.120
Steel	70	10	1.85*	0.127
PP	47	100	1.33	1.163

*Average emissions from CSN

The sustainability evaluation for each product, here denoted by *M factor* and based on the data provided in Table 2, was calculated following the scoring criteria defined as Recycling rate (RR): Number 1 was assigned to steel because, among the selected materials, it has the highest recycling rate. For the other materials, the scores were assigned proportionally, thus adding the score to form the *M factor*. Decomposition time (DT): Number 1 is assigned to glass because, among the selected materials, it has the highest decomposition time in nature. For the other materials, the score is assigned proportionally, thus reducing the score to form the *M factor*. CO_2 Emission rate (ER): This criterion is directly measured by material volume in each package, reducing the score in the *M factor* by multiplying the rate ER and materials masses in each product. Reuse financial return (FR): This criterion is directly measured by the material volume in each package, adding the score in the *M factor* by multiplying the FR rate and material masses in each product.

Subsequently, the *M factor* was normalized by the material volume of each category, to benefit products with less material employed. The relationship between the criteria regarding material is shown in Eq. (3), where one Equation is for Body (M_{pb} as example) and another equals for Lid (M_{pl}).

$$M_{pb} = RR_{pb} - DT_{pb} + m_{pb} * (FR_{pb} - ER_{pb})/ \sum_{i=1}^{t} m_{pb}(FR_i - ER_i) \quad (3)$$

Table 2 Average scores of linguistic variables

Package	Beautiful	Practical	Attractive	Sustai-nable	Safe	Protective of product
1	3.6167	3.5500	3.3500	3.2000	3.033	3.9667
2	2.8667	3.0000	2.9000	3.0667	3.650	3.8167
3	2.6833	2.5500	2.6000	2.9000	3.283	3.6000
4	3.9000	3.5500	3.7167	3.2500	2.783	3.5000
5	2.3333	3.7500	2.3167	2.2833	3.466	3.3000
6	3.5333	3.8833	3.3667	3.2833	3.383	4.0333
7	1.8667	1.8500	1.9667	2.5000	2.900	3.8000
8	3.6667	3.6833	3.4500	2.9157	3.316	3.7333
9	2.8000	2.3667	2.8000	3.2167	2.566	3.6500
10	2.0000	1.7167	1.9333	2.7000	2.933	3.7000
11	3.7833	3.5833	3.8500	3.3167	3.083	3.5000
12	2.6500	3.1333	2.7333	2.9833	3.716	3.7000
13	2.6833	3.6167	2.6500	2.4000	3.666	3.5000
14	2.6333	3.1333	2.7000	3.1667	2.916	3.5833
15	2.8667	3.3333	2.9000	3.1500	3.450	3.4500
16	3.5167	3.2167	3.6167	3.0833	2.983	3.5167
17	2.3167	3.1333	2.2833	3.1500	3.766	3.8333

These Equations for body and lid, normalized by volume, make up the final Equation *M factor*) as shown in Eq. (4):

$$M_p factor = M_{pb} * \left(1 - \frac{V_{pb}}{\sum_{i=1}^{tb} V_i}\right) + M_{pl} * \left(1 - \frac{V_{pl}}{\sum_{i=1}^{tl} V_i}\right) \tag{4}$$

where M = Material Sustainability Score; m = Category Mass; RR = Recycling Rate (Brazil)/100; DT = Decomposition Time in Nature; FR = Reuse Financial Return (Brazil); ER= CO_2 Emission Rate in Production (Brazil); t = Total Number of Categories; l = lid b = body; p = Product Index; V = Category Volume.

The *M factor* is added to the fitness function, aggregating the final product evaluation with costumers' preferences aspects and sustainability aspects.

4 Experiments and Results

4.1 Synthesis: Linguistic Variables and Representative Products Connection

The relationship between linguistic variables and the representative products in the domain was obtained using a two-part questionnaire, each one in a 1–5 scale, where the target audience chooses a level of intensity in each answer to a question. The first part of this survey consists of the only question about the information of how much the interviewee considers it important for each linguistic variable. In the second part, the interviewee analyzes each product by linguistic variable. The scale adopted was based in the examples mentioned in [9].

The average scores for each product are shown in Table 2. The questionnaires were answered between 10/17/2019 and 10/25/2019, with a total of 89 people being interviewed (44% women and 56% men).

Contribution of product design scores: the Table 3 shows the numbers of categories composing each package (sample). The scores of the categories are shown in Table 4 (calculated applying the Quantification Theory Type I, from the average scores in Table 2 and the numbers of categories shown in Table 4). In this study, the code developed for Shigenobu Aoki, Gunma University, was applied in this stage.

Linguistic variables weights: the linguistic variables data were collected from the first question in the questionnaire. Firstly, triangular fuzzy number intervals were determined for each answered questionnaire and, subsequently, the average values for each variable (crisp values) were used to rank the adjectives, as shown in Table 5.

The ranking was obtained by weighing the high value from a_3 as 1, with the other values being weighted proportionally.

Table 3 The 17 samples of solutions in a morphological analysis

Sample	Body	Lid	Material	Sample	Body	Lid	Material
1	1	1	3	10	6	4	1
2	2	2	1	11	1	8	2
3	8	2	1	12	2	6	1
4	7	7	2	13	7	8	4
5	5	3	4	14	5	5	2
6	3	1	3	15	4	7	1
7	3	4	1	16	6	5	2
8	4	3	3	17	3	6	1
9	8	5	2	–	–	–	–

Table 4 The contribution scores of each category

Categories	Beautiful	Practical	Attractive	Sustainable	Safe	Protective of product
x_{11}	0.117	−0.104	0.093	−0.042	−0.223	0.049
x_{12}	0.183	0.096	0.310	0.041	0.085	0.078
x_{13}	0.033	0.229	0.110	0.041	0.127	0.116
x_{14}	−0.143	−0.121	−0.083	−0.023	0.143	−0.066
x_{15}	−0.442	0.212	−0.465	0.074	0.081	−0.061
x_{16}	0.350	0.229	0.327	0.074	0.152	−0.080
x_{17}	0.067	−0.167	−0.108	−0.225	−0.157	−0.051
x_{18}	−0.183	−0.488	−0.240	0.041	−0.273	−0.043
x_{21}	0.126	0.137	0.106	0.060	−0.013	0.030
x_{22}	0.208	0.080	0.159	0.059	0.230	0.042
x_{23}	0.436	0.288	0.382	−0.243	−0.095	−0.088
x_{24}	−0.825	−1.336	−0.824	−0.341	−0.554	0.084
x_{25}	−0.315	−0.232	−0.233	−0.093	−0.128	−0.038
x_{26}	−0.192	0.080	−0.258	0.142	0.305	0.021
x_{27}	0.443	0.564	0.427	0.290	−0.023	−0.133
x_{28}	0.277	0.534	0.358	0.174	0.343	0.101
x_{31}	−0.358	−0.230	−0.335	−0.091	0.101	−0.029
x_{32}	0.465	0.033	0.508	0.211	−0.266	0.006
x_{33}	0.449	0.396	0.261	0.208	0.040	0.210
x_{34}	−0.585	0.129	−0.490	−0.523	0.251	−0.228

Table 5 Crisp values and linguistic variables ranking

L_i	K_1	K_2	K_3	K_4	K_5	K_6
a_1	2.48	3.40	2.60	2.90	3.15	3.55
a_2	3.45	4.38	3.56	3.88	4.11	4.53
a_3	4.32	4.87	4.43	4.53	4.57	4.81
Ranking	6	1	5	4	3	2

4.2 Genetic Algorithm for Choosing Product Design

The fitness function was formulated to evaluate the product regarding three aspects: scores of Categories, Linguistic Variables Weights and Material Sustainability Criteria, as expressed in Eq. (5):

$$Max\ F_p = P_{pxj} * S_{jxk} * K_{kx1} + M_p + c \qquad (5)$$

where F = Fitness Value; P = Random Population Matrix; S = Category Scores Matrix; K = Linguistic Variables Weights Matrix; M = Materials Sustainability Value; p = Product Index; j = Total Number of Categories; k = Total Number of Linguistic Variables; $c = 20$ (Constant for non negative values).

In the genetic algorithm implemented to choose package design, each product was represented by a set of binary numbers (0 and 1), divided into two genes (Body and Lid, respectively), forming a chromosome with the size of 10 bits. In this representation, the materials are considered for each gene, with four possibilities: Body and lid made of steel, body and lid made of plastic, body made of glass and lid made of steel, and body and lid made of glass. The experiments were conducted by following the experimental design method described in [10] to analyze the influence of the variables on the results. The fitness value is considered as a response variable, and the GA parameters as input for the test (Table 6):

The crossover operator restricts the crossover point for the total number of bits on each product item. As there are two items, for the item Body, the crossover occurs between 1 and 5 bits and, for the Lid, the crossover occurs between 6 and 10. Combining all the possibilities, 16 tests were performed, each one with 20 independent rounds.

Table 6 Genetic algorithm parameters

Parameters	1	−1
Generation number	25	50
Crossover rate	0.5	0.8
Mutation rate	0.05	0.1
Elitism rate	1	1
Population size	25	50

4.3   Decode: Optimum Product Design

The optimum solutions were obtained considering three possible scenarios. The first scenario (Scenario 1) considered only costumers' preferences, while the second one (Scenario 2) considered the M values, while the third and main scenario (Scenario 3) considered costumers' preferences together with sustainability values. The representative chromosomes and their respective optimum fitness value for each scenario are shown in Table 7, and the final product designs are represented in Fig. 4.

In the Fig. 4a, for both scenarios 1 and 3 the same configuration remains; on the other hand, costumers' preferences have the highest relevance (weight) in the fitness function. Modifying the weights for costumers' preferences to 60% and the M factor to 40% in the fitness function, the choices converge to products like Scenario 2, with body and lid made of steel, but keeping the same design of Scenarios 1 and 3.

The lid of product a was drawn for this survey and apparently, it is a nonexistent item in the sweet milk packages in the Brazilian market. It shows that this technique applied for the selection of product design can be used to systematically evaluate new product designs before market launch. The most sustainable package design (Fig. 4b) is a configuration made of steel awith less material employed.

Table 7 Final product: chromosome representation

Scenario	Gene1: body	Gene2: lid	Fitness value
1	01, 010	01, 000	24.13
2	11, 110	11, 110	21.88
3	01, 010	01, 000	26.23

Fig. 4 Final package design for scenarios 1 and 3 (**a**) and scenario 2 (**b**)

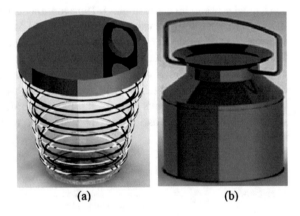

(a)   (b)

5 Conclusions

This study has shown the effective use of genetic algorithms to evaluate product design characteristics. In addition, the feasibility of incorporating sustainability criteria may improve the design process towards more sustainable products which, at the same time, satisfy costumers' preferences. The difference between virtual and real models can be softened by 3D rendering, providing the interviewee with a satisfactory evaluation of the visual aspects of product, including the type of material, and lid and body shapes. The application of the model presented, especially the idea of incorporating sustainability variables, may achieve more accurate results for different applications in different types of products, including improvements in the product development process.

References

1. Rozenfeld, Henrique, e Daniel Capaldo Amaral. Gestão de projetos em desenvolvimento de produtos. [Project Management in product development]. São Paulo: Saraiva (2006).
2. García-Arca, J., González-Portela Garrido, A. T., & Prado-Prado, J. C. "Packaging Logistics" for improving performance in supply chains: the role of meta-standards implementation. Production, 26(2), 261–272 (2016).
3. Nagamachi, M., & Lokman, A. M.. Innovations of Kansei engineering. CRC Press (2016).
4. Ushada, M., Okayama, T., Suyantohadi, A., Khuriyati, N., & Murase, H. Daily worker evaluation model for SME-scale food production system using kansei engineering and artificial neural network. Agriculture and Agricultural Science Procedia, 3, 84–88 (2015).
5. Van den Broek, P., & Noppen, J. Fuzzy weighted average: alternative approach. In Fuzzy Information Processing Society, 2006. NAFIPS 2006. Annual meeting of the North American (pp. 126–130). IEEE (2006).
6. Zhai, L. Y., Khoo, L. P., & Zhong, Z. W. A dominance-based rough set approach to Kansei Engineering in product development. Expert Systems with Applications, 36(1), 393–402 (2009).
7. Zhou, C. C., Yin, G. F., & Hu, X. B. Multi-objective optimization of material selection for sustainable products: artificial neural networks and genetic algorithm approach. Materials & Design, 30(4), 1209–1215 (2009).
8. Schütte, S. T., Eklund, J., Axelsson, J. R., & Nagamachi, M. Concepts, methods and tools in Kansei engineering. Theoretical Issues in Ergonomics Science, 5(3), 214–231 (2004).
9. Nagamachi, M. Kansei/affective engineering and history of kansei/affective engineering in the world. Kansei/Affective Engineering, 13, 1–12 (2011).
10. Montgomery, D. C., & Runger, G. C. Applied statistics and probability for engineers. John Wiley & Sons (2010).
11. Klir, G., & Yuan, B. Fuzzy sets and fuzzy logic (Vol. 4). New Jersey: Prentice hall (1995).
12. Chou, J. R. Applying fuzzy linguistic preferences to Kansei evaluation. In KEER2014. Proceedings of the 5th Kansei Engineering and Emotion Research; International Conference; Linköping; Sweden; June 11–13 (No. 100, pp. 339–349). Linköping University Electronic Press (2014).
13. Zhang, G., & Yu, E. A Novel Personalized Recommendation Method in E-business Based on Kansei Image. International Conference on Automation, Mechanical Control and Computational Engineering. Atlantis Press (2015).
14. Goldberg, D. E., & Samtani, M. P. Engineering optimization via genetic algorithm. In Electronic computation (pp. 471–482). ASCE (1986).

15. LANDIM, Ana Paula Miguel et al. Sustentabilidade quanto às embalagens de alimentos no Brasil. Polímeros, v. 26, n. SPE, p. 82–92, (2016).[in Portuguese]
16. Lin, Y. C., Lai, H. H., & Yeh, C. H. Consumer-oriented product form design based on fuzzy logic: A case study of mobile phones. International Journal of Industrial Ergonomics, 37(6), 531–543 (2007).
17. Barão, M. Z. Embalagens para produtos alimentícios [Packages for food products]. Paraná Institute of Technology-TECPAR. Curitiba, 8. (2011).
18. Bakker, C., Wang, F., Huisman, J., & den Hollander, M. Products that go round: exploring product life extension through design. Journal of Cleaner Production, 69, 10–16 (2014).
19. Jorge, N. Embalagens para alimentos [Food Packages]. São Paulo: Academic Culture: State University Paulista, Dean´s Office. (2013).
20. RECICLAGEM-CEMPRE, C. E. P. CEMPRE Review. (2013).
21. Akerman, M. Natureza, estrutura e propriedades do vidro [Nature, structure and glass properties]. Techinical Publication. Techinical Center for elaboration of Glass. Saint-Gobain, Glass-Brazil (2000).
22. Sampaio, R. Conversão da biomassa em carvão vegetal-Situação Atual com tendências 2025-Estudo prospectivo do setor siderúrgico [Conversion of biomass to charcoal-Current situation with trends 2025-Prospective study of the steel sector]. CGEE, Belo Horizonte, 13 (2008).

Entrepreneurial Ecosystems and Sustainable Entrepreneurship: Challenges Towards More Social Entrepreneurial Orientation

Mariana Pita⊙, Joana Costa⊙, and António Carrizo Moreira⊙

Abstract The purpose of this paper is to grasp the connection between entrepreneurial ecosystems and sustainable entrepreneurship and its relevance for tackling societal challenges. In particular, the work investigates if entrepreneurial ecosystems, through the lens of education and social context, simulate social entrepreneurial orientation. To accomplish the research goals, an empirical study is conducted relying on Global Entrepreneurship Monitor from 2015, where 58 countries are analyzed based on the Entrepreneurial Ecosystem Taxonomy. The results point that education and social context are supporters of social entrepreneurship, emphasizing on both institutional and social networks role. However, the findings reveal that social context tends to instigate more regular entrepreneurship when compared to social entrepreneurship, although being a driver in both cases.

Keywords Entrepreneurial ecosystems · Sustainable entrepreneurship · Social entrepreneurial orientation · Sustainability

1 Introduction

Entrepreneurship is not a new phenomenon and has gained increasing interest during the past years among academics, practitioners and policymakers due to its recognition has a relevant driver for economic growth and employment. The multidimensionality of entrepreneurship opened a large avenue for different perspectives placing entrepreneurial ecosystems and social entrepreneurship at the center of the research arena. Entrepreneurial ecosystems are composed by several interacting and interdependent elements [1, 2] aiming to nurture more favourable environments towards entrepreneurial activity. However, the significant transformations and societal challenges around the world encountered by the severe effects of economic crises are

M. Pita (✉) · J. Costa · A. C. Moreira
Department of Economics, Management, Industrial Engineering and Tourism (DEGEIT),
University of Aveiro, Campus Universitário de Santiago, 3810-129 Aveiro, Portugal
e-mail: mariana.pita@ua.pt

© The Author(s), under exclusive license to Springer Nature Switzerland AG 2021
A. M. Tavares Thomé et al. (eds.), *Industrial Engineering and Operations Management*,
Springer Proceedings in Mathematics & Statistics 367,
https://doi.org/10.1007/978-3-030-78570-3_43

pushing economies to embrace more sustainable oriented policies to tackle inequalities, raise individual's potential hoping to achieve long-term effects. Sustainable development goals, are therefore, at the core of government priorities, encompassing simultaneously social, environmental and economic concerns. Such concerns are not new, but only recently the concept of "sustainable entrepreneurship" has been discussed.

The contribution of education is generally accepted as beneficial to enhance economic development and stimulate innovative businesses, suggesting it's influence on promoting entrepreneurial mind-set. Along with education, [3] defend that the context plays a role of positive influence on developing pioneering and innovative activities, particularly with environmental responsibility [4].

Although sustainable entrepreneurship emerged as a relevant topic some years back, to the best of our knowledge, the role of education and context on social entrepreneurial orientation has not been addresses yet in prior studies.

The importance of education and the influence of social context on (social) entrepreneurial orientation constitutes a valuable research topic for three reasons: first, education is widely accepted as an undeniable condition to succeed; second, the social context as family is considered the responsible for role-modelling [5] training individuals to perform in a certain way, particularly on responsible new ventures [4].

The study uses a logistic regression and the Global Entrepreneurship Monitor—GEM—database from the year 2015, with information of 58 countries. Following the work of [6], the empirical analysis tests the effect of EE on EA and SEA, in general, and after, additional models were run, based upon sub-sampling generated by the taxonomic position of the country regarding the entrepreneurial ecosystem quality.

Several implications from both theoretical—the role of education and social context on stimulating social entrepreneurial orientation—and practical—providing insights for governmental policies adjustments—are expected with the present study.

The article is structured as follows. After the introduction, Sect. 2 incorporates theoretical contributions for study purpose, Sect. 3 exposes the methodology adopted, Sect. 4 shed the light on results and, lastly, Sect. 5 presents the main conclusions along with limitations and future avenues of research.

2   Literature Review

Recent challenges faced all over the world highlighted the emergency of new perspectives concerning social entrepreneurship. More, the acknowledgement of differences among territories regarding entrepreneurial activity and entrepreneurial ecosystems shed the light around entrepreneurial territorial patterns and its circumstances, pointing new directions towards more social entrepreneurial policies as way to create more favorable environments for social entrepreneurs. During several decades, entrepreneurship and its economic impact pushed developed economies towards new policy measures as a way to attract entrepreneurs. Recently, the recognition of social entrepreneurship as a phenomenon that combines economic utility and social efficiency, particularly among emerging economies, as showed its importance as a growth driver [7].

As indicated in the literature, the effect of social entrepreneurship can be enlarged by external factors along with institutional infrastructures [8].

According to [9] social entrepreneurship integrates a social perspective and differs from for-profit enterprises and market-oriented businesses. However, not all ecosystems are prepared to support equally social entrepreneurs as they lack appropriated mentoring processes [10] and social accelerators [11]. Entrepreneurial ecosystems studies help to describe the interconnected elements and its effect on entrepreneurial activity but commonly the focus is placed on economic oriented companies. According to [12] social entrepreneurship can be stimulated through social government policies along with social capital, placing emphasis on the plurality of elements. Additionally, as advanced by Sahasranamam and Nandakumar [13], social entrepreneurship can be enhanced through educational systems as they contribute to leverage human capital. Also, social context enables a better understanding of the social circumstances, contributing to develop social skills [14] which can stimulate social entrepreneurial orientation.

As discussed in prior studies, social entrepreneurship can be considered as hybrid organizations where business performance goals and the accomplishment of social objectives can generate a balance between economic and social impact [15]. Therefore, entrepreneurship should capture economic and social values, and following a holistic approach to the societal challenges only productive entrepreneurship sustainable development will allow to overcome inequality, poverty and unemployment gaps [9].

Entrepreneurial Ecosystems and Social Entrepreneurship literature does not fully explore its connection in terms of education and social context effects, although both dimensions are recognized as positive influencers towards entrepreneurship. Departing from this general perspective, the present work aims to provide empirical evidences around entrepreneurial ecosystems and social entrepreneurship through the lens of education and social context and deliver policy-directions to more effective social entrepreneurship strategies.

3 Methodology

The study relied on the Global Entrepreneurship Monitor (GEM) database, from 2015, and encompasses 58 countries, distributed according to the Entrepreneurial Ecosystem Taxonomy proposed by Pita and Costa [6]. The authors taxonomic perspective allows to posit countries according to their entrepreneurial ecosystem quality and entrepreneurial activity in four main groups: (1) Die-Hard (low quality of entrepreneurial ecosystem; low entrepreneurial activity); (2) Go-Getter (low quality of entrepreneurial ecosystem; high entrepreneurial activity); (3) Sugar-Coated (high quality of entrepreneurial ecosystem; low entrepreneurial activity); (4) Front-Runner (high quality of entrepreneurial ecosystem; high entrepreneurial activity).

The empirical analysis was performed relying on Binary Logistic models. Model 1 tests the effect of EE on EA and SEA, in general. Four additional models were run, based upon the sub-sampling generated by the taxonomic positioning of the country: Die-Hard (model 2), Go-Gette (model 3), Sugar-Coated (model 4), and Front-Runners (model 5)—to appraise differences of non-social businesses and social businesses. The study also includes as control variables: Education. Gender, Age, Social Context, Skill Perception, Opportunity Recognition and Fear to Fail.

4   Research Findings

The research findings for Model 1 are provided in Tables 1 and 2, respectively for Entrepreneurial Activity and Social Entrepreneurial Activity.

Results reveal that entrepreneurial ecosystems support entrepreneurship and social entrepreneurship, in all quadrants. Nevertheless, Model 1 points interesting findings concerning education and social context as an influence on starting a (social) business. Despite the recognition of education as an employment leverage, Model 1 points education as discouraging factor regarding regular entrepreneurship, while acts otherwise for social entrepreneurship. Social context appears for both types of entrepreneurship as a positive influence, despite being stronger regarding non-social entrepreneurial activity. As for other variables in use, individuals in both cases acknowledge having the necessary skills to enterprise. As for fear of failure, the

Table 1 Model 1—entrepreneurial activity

	BSTART					
	B	S.E	Wald	df	Sig	Exp(B)
Education	−0.024	0.006	18.262	1	0.000	0.976
Social context	0.626	0.016	1447.097	1	0.000	1.870
Opport. recognition	0.632	0.016	1483.547	1	0.000	1.882
Skills perception	1.353	0.016	4865.086	1	0.000	3.868
Fear of failure	−0.240	0.016	197.459	1	0.000	0.787
Gender	−0.174	0.016	118.173	1	0.000	0.840
Age	−0.108	0.006	334.277	1	0.000	0.897
Go-Getter	1.322	0.022	3654.159	1	0.000	3.749
Sugar-Coated	0.291	0.025	137.974	1	0.000	1.337
Front-Funner	1.159	0.023	2517.025	1	0.000	3.186
Constant	−3.110	0.040	6139.493	1	0.000	0.045

[a]Variable(s) entered on step 1: UNEDUC, knowent, opport, suskill, fearfail, gender, age9c, Go-Getter, Sugar-Coated, Front-Funner

Table 2 Model 1—social entrepreneurial activity

	SESTART					
	B	S.E	Wald	df	Sig	Exp(B)
Education	0.082	0.008	117.276	1	0.000	1.086
Social context	0.443	0.023	370.688	1	0.000	1.557
Oppo. recognition	0.372	0.023	266.420	1	0.000	1.450
Skills perception	0.404	0.024	273.417	1	0.000	1.498
Fear of failure	−0.124	0.023	28.899	1	0.000	0.883
Gender	−0.090	0.022	16.848	1	0.000	0.914
Age	0.058	0.008	53.419	1	0.000	1.060
Go-Getter	0.697	0.029	571.982	1	0.000	2.008
Sugar-Coated	0.133	0.032	17.745	1	0.000	1.142
Front-Funner	0.374	0.032	136.235	1	0.000	1.454
Constant	−3.912	0.053	5391.070	1	0.000	0.020

[a]Variable(s) entered on step 1: UNEDUC, knowent, opport, suskill, fearfail, gender, age9c, Go-Getter, Sugar-Coated, Front-Funner

results reinforce its negative impact on pursuing entrepreneurial endeavors in both cases. Nevertheless, is influence being minor when it comes to social entrepreneurship. The same result is verified for gender. Concerning age, while appears to be a deterring factor for regular entrepreneurship, as seen in Table 1, in the case of social entrepreneurship it seems to be the opposite.

Model 1 suggest that education has a different influence on regular entrepreneurial activity and social entrepreneurial activity, being in the first case a constraint, while for the second, an accelerator.

Model 2, 3, 4, and 5 expose the results according to the taxonomic position of countries accordingly to the Entrepreneurial Ecosystems Taxonomy [6].

Model 2 (Table 3) represents the countries positioned as Die-Hard, therefore exhibiting lower entrepreneurial activity and lower entrepreneurial ecosystem quality. For this groups, all variables hold similarly, with the exception of age. The results indicate that age is a deterring factor concerning regular entrepreneurial activity, however, for social entrepreneurship age appears as a stimulus. Regarding education, although its positive influence in both type of businesses, it's more prevalent on social oriented businesses.

Model 3 (Table 4) presents significant differences regarding the influence of education. While in prior group—Die Hard—education display a positive effect on entrepreneurial activity, for both types, in the Go-Getter group it has an opposite effect. Therefore, education for Model 3 arises as a dissuading factor in both types of ventures, with a stronger incidence on social enterprises. As for Social Context, the results are in line with Die Hard Group, showing that knowing an entrepreneur incites entrepreneurship. Lastly, age is taken as a constrain for regular enterprises, while for social enterprises possess a positive effect.

Table 3 Model 2—entrepreneurial ecosystem taxonomic position: Die-Hard

| | DIEHARD | | | | | | | |
| | BSTART | | | | SESTART | | | |
	B	S.E	Sig	Exp(B)	B	S.E	Sig	Exp(B)
Education	0.072	0.012	0.000	1.075	0.306	0.015	0.000	1.358
Social context	0.593	0.034	0.000	1.810	0.383	0.043	0.000	1.466
Opp. Recognition	0.749	0.034	0.000	2.114	0.273	0.042	0.000	1.314
Skills perception	1.538	0.043	0.000	4.654	0.380	0.044	0.000	1.462
Fear of failure	−0.361	0.036	0.000	0.697	−0.117	0.043	0.006	0.890
Gender	−0.229	0.034	0.000	0.795	−0.053	0.041	0.199	0.948
Age	−0.200	0.012	0.000	0.819	0.040	0.015	0.007	1.041
Constant	−3.177	0.077	0.000	0.042	−4.534	0.096	0.000	0.011

(p < 0.001***; p < 0.05**; p < 0.01*)

Table 4 Model 3—entrepreneurial ecosystem taxonomic position: Go-Getter

| | GO-GETTER | | | | | | | |
| | BSTART | | | | SESTART | | | |
	B	S.E	Sig	Exp(B)	B	S.E	Sig	Exp(B)
Education	−0.119	0.008	0.000	0.888	−0.170	0.012	0.000	0.843
Social context	0.592	0.028	0.000	1.807	0.547	0.042	0.000	1.727
Opp. Recognition	0.590	0.028	0.000	1.804	0.392	0.042	0.000	1.480
Skills perception	1.310	0.035	0.000	3.707	0.359	0.048	0.000	1.432
Fear of failure	−0.126	0.030	0.000	0.881	−0.188	0.044	0.000	0.829
Gender	−0.198	0.028	0.000	0.821	−0.084	0.039	0.032	0.919
Age	−0.070	0.011	0.000	0.933	0.010	0.015	0.486	1.010
Constant	−1.609	0.061	0.000	0.200	−2.392	0.086	0.000	0.091

(p < 0.001***; p < 0.05**; p < 0.01*)

In Model 4 (Table 5), both education and social context appear has positive inducers for entrepreneurs and social entrepreneurs. Nevertheless, education exhibit a higher effect among those who pursue social entrepreneurship. As for social context, the results show the opposite, revealing a lower effect when compared to regular entrepreneurship. Regarding gender, it is displayed as constrain for the two types of businesses. As for age, only appears as a deterring factor for regular entrepreneurship.

Lastly, Model 5 (Table 6) reveal that Front-Runners are positively influenced by education and social context in both social and non-social businesses. More, gender has a negative effect on pursuing entrepreneurial activity, while age only stimulus for those who are social entrepreneurial oriented.

Comparing the four taxonomic positions, the results confirm that education is a driver of regular entrepreneurship and social entrepreneurship. Education only have

Table 5 Model 4—entrepreneurial ecosystem taxonomic position: Sugar-Coated

| | SUGAR-COATED | | | | | | | |
| | BSTART | | | | SESTART | | | |
	B	S.E	Sig	Exp(B)	B	S.E	Sig	Exp(B)
Education	0.025	0.015	0.096	1.025	0.147	0.020	0.000	1.158
Social context	0.760	0.038	0.000	2.138	0.224	0.051	0.000	1.251
Opp. Recognition	0.476	0.038	0.000	1.610	0.321	0.050	0.000	1.379
Skills perception	1.458	0.044	0.000	4.297	0.419	0.051	0.000	1.520
Fear of failure	−0.205	0.038	0.000	0.814	−0.043	0.049	0.377	0.958
Gender	−0.215	0.037	0.000	0.807	−0.154	0.049	0.002	0.858
Age	−0.206	0.014	0.000	0.814	0.084	0.018	0.000	1.087
Constant	−2.664	0.091	0.000	0.070	−4.012	0.121	0.000	0.018

($p < 0.001$***; $p < 0.05$**; $p < 0.01$*)

Table 6 Model 5—entrepreneurial ecosystem taxonomic position: Front-Runner

| | FRONT-RUNNERS | | | | | | | |
| | BSTART | | | | SESTART | | | |
	B	S.E	Sig	Exp(B)	B	S.E	Sig	Exp(B)
Education	0.058	0.012	0.000	1.059	0.306	0.019	0.000	1.358
Social context	0.564	0.033	0.000	1.758	0.559	0.054	0.000	1.749
Opp.Recognition	0.671	0.033	0.000	1.957	0.427	0.053	0.000	1.532
Skills perception	1.140	0.037	0.000	3.127	0.308	0.056	0.000	1.361
Fear of failure	−0.258	0.034	0.000	0.773	−0.027	0.052	0.606	0.974
Gender	−0.069	0.032	0.028	0.933	−0.139	0.049	0.005	0.870
Age	−0.009	0.011	0.414	0.991	0.107	0.017	0.000	1.113
Constant	−2.455	0.074	0.000	0.086	−4.507	0.118	0.000	0.011

($p < 0.001$***; $p < 0.05$**; $p < 0.01$*)

a negative influence for the Go-Getter group which could be related to more fragile economies. The study points that more educated individuals tend to be more engaged in social entrepreneurial businesses suggesting that they have a stronger social orientation when compared to those who pursue regular entrepreneurship purposes. Moreover, education appears to raise alertness of individuals towards social problems, pointing education is a two-fold contributor since increase entrepreneurial activity and social entrepreneurial activity.

Regarding social context, the results reveals its positive influence towards entrepreneurship. Nevertheless, the results surprisingly point a higher effect upsetting regular entrepreneurship. This finding could suggest that economic concerns are still a priority for family and social context, despite the importance of tackling social challenges. Individuals seems to be influenced by family to pursue, in the first place, a more stable professional pathway, while education tends to turn individuals more aware of societal inequalities. This finding is corroborated by the results, as age turn individuals more prone to social enterprise. Moreover, social entrepreneurial orientation seems to be widely recognized by the society even when entrepreneurial ventures do not succeed.

5 Conclusion

The purpose of this article was to discuss the role of education and social context on entrepreneurial activity and social entrepreneurial activity among countries with different profiles. Considering the emphasis of prior studies about the importance of tackling economic, environmental and societal challenges through more social entrepreneurial orientation, the present work aims to contribute to strength this field of research and deliver empirical evidence to adjust social entrepreneurship policies. Following [13] perspective, the results reinforce the importance of education to stimulate social entrepreneurial orientation. As education helps to foster the perception of social challenges, also enables entrepreneurial activity. This finding shed the light around the role of education on developing human capital, and the need of tackling education inequalities. In the same direction, and corroborating [14] work, social context also contributes to increase social entrepreneurship, bringing a new perspective around the responsibility of individuals social and family network. Nevertheless, the results evidence that social context has a more prevalent inclination towards regular entrepreneurship when compared with social businesses, probably because is less interconnected with economic benefits.

Based on the main findings of the present study, there are three main policy recommendations. First, education policies should be expanded with entrepreneurial ecosystems policies, allowing to incorporate territorial singularities and social challenges. Second, to leverage social entrepreneurship relevance through the recognition of social success driven business models. Third, to improve support measures specifically devoted to social entrepreneurship, as a mean to capture more adequate support to social projects.

This research contributes to address entrepreneurial ecosystem literature by using an original analytic technique to evaluate the entrepreneurial ecosystem impact on social entrepreneurship. Moreover, contributes to the literature by delivering empirical evidence about the role of education and social context and its effects on entrepreneurship and social entrepreneurship, according to different four country groups. Thirdly, the results pinpoints policymakers to design effective EE policy packages to improve social entrepreneurship. Concluding, the role of EE and its effect in social entrepreneurship also needs to be rethought, opening the door for novel entrepreneurship multi-factor strategies, where academia, communities, and ecosystems work together.

Acknowledgements This work was supported by the research unit on Governance, Competitiveness and Public Policy (UIDB/04058/2020) + (UIDP/04058/2020), funded by national funds through FCT - Fundação para a Ciência e a Tecnologia.

References

1. E. Stam, 'Entrepreneurial Ecosystems and Regional Policy: A Sympathetic Critique', Eur. Plan. Stud., vol. 23, no. 9, pp. 1759–1769, 2015.
2. E. Mack and H. Mayer, 'The evolutionary dynamics of entrepreneurial ecosystems', Urban Stud., vol. 53, no. 10, pp. 2118–2133, 2016.
3. O. Toutain et al., 'Role and impact of the environment on entrepreneurial learning', Entrep. Reg. Dev., vol. 29, no. 9–10, pp. 869–888, 2017.
4. W. R. Meek, D. F. Pacheco, and J. G. York, 'The impact of social norms on entrepreneurial action: Evidence from the environmental entrepreneurship context', J. Bus. Ventur., vol. 25, no. 5, pp. 493–509, 2010.
5. L. K. Gundry and H. P. Welsch, 'The ambitious entrepreneur: ig growt strategies of women-owned enterprises', J. Bus. Ventur., vol. 9026, no. 312, pp. 453–455, 2001.
6. M. Pita and J. Costa, 'Entrepreneurial Ecosystem Quality', in 56th International Scientific Conference on Economic and Social Development, 2020, p. ahead of print.
7. M. Kabbaj, K. El Ouazzani Ech Hadi, J. Elamrani, and M. Lemtaoui, 'A study of the social entrepreneurship ecosystem: The case of Morocco', J. Dev. Entrep., vol. 21, no. 4, 2016.
8. J. Han and S. Shah, 'The Ecosystem of Scaling Social Impact: A New Theoretical Framework and Two Case Studies', J. Soc. Entrep., vol. 11, no. 2, pp. 215–239, 2019.
9. A. Singh, 'Social Entrepreneurship and Sustainable Development', Soc. Entrep. Sustain. Dev., no. March, pp. 25–40, 2020.
10. I. F. Thomaz and M. Catalão-Lopes, 'Improving the Mentoring Process for Social Entrepreneurship in Portugal: A Qualitative Study', J. Soc. Entrep., vol. 10, no. 3, pp. 367–379, 2019.
11. S. Pandey, S. Lall, S. K. Pandey, and S. Ahlawat, 'The Appeal of Social Accelerators: What do Social Entrepreneurs Value?', J. Soc. Entrep., vol. 8, no. 1, pp. 88–109, 2017.
12. I. Bozhikin, J. Macke, and L. F. da Costa, 'The role of government and key non-state actors in social entrepreneurship: A systematic literature review', J. Clean. Prod., vol. 226, pp. 730–747, 2019.
13. S. Sahasranamam and M. K. Nandakumar, 'Individual capital and social entrepreneurship: Role of formal institutions', J. Bus. Res., vol. 107, no. April 2018, pp. 104–117, 2020.
14. A. I. Journal, M. S. Nielsen, and K. Klyver, 'Meeting entrepreneurs' expectations?: the importance of social skills in strong relationships', Entrep. Reg. Dev., vol. 00, no. 00, pp. 1–20, 2020.
15. N. Franco-Leal, C. Camelo-Ordaz, J. P. Dianez-Gonzalez, and E. Sousa-Ginel, 'The role of social and institutional contexts in social innovations of spanish academic spinoffs', Sustain., vol. 12, no. 3, pp. 1–24, 2020.

TPM Adaptation as Lean Healthcare Practice to Improve the Logistics Processes of a Pharmaceutical Supply Center—PSC

L. S. de Bittencourt, A. C. Gularte, B. A. Leal, I. C. de Paula, and W. P. Bueno

Abstract Lean Healthcare (LH) is being increasingly applied in health processes, especially in higher complexity level facilities as hospitals and clinics, and less in primary care facilities. Tools as Total Productive Maintenance (TPM) can be useful to minimize process errors and bottlenecks, putting in practice a type of monitoring focused on the productivity of the organization as a whole. However, the peculiarities of health processes require adaptations in order to become efficient. This study aims at investigating the necessary adaptations of LH practices in a primary health care setting. The unit studied consists of the drug storage and distribution from a Pharmaceutical Supply Center (PSC) located in a city of the State of Rio Grande do Sul, Brazil. Health management professionals must design, manage and optimize the Pharmaceutical Assistance (PA) operations which include acquisition, storage of medicines in warehouses called PSCs. It is a single case study conducted from June 2019 to February 2020. Data collection and problem diagnosis involved qualitative interviews, chronoanalysis, process mapping and diagram of cause-effect relationships. Information collected showed eighteen problems in the PSC and losses throughout the process. One of the challenges faced throughout the paper was to adapt calculation variables used in the TPM to analyze efficiency within PSC. Such an adaptation was essential to determine the efficiency of the separation process carried out at PSC. It was not possible to complete the four steps of the Lean healthcare implementation and we suggest its completion for future works.

Keywords Total Productive Maintenance (TPM) · Lean Healthcare (LH) · Pharmaceutical Supply Center (PSC)

L. S. de Bittencourt (✉)
Federal University of Rio Grande Do Sul (UFRGS), 150 St São Luís, Porto Alegre, RS, Brazil

A. C. Gularte · B. A. Leal · I. C. de Paula · W. P. Bueno
Federal University of Rio Grande Do Sul (UFRGS), 99 Ave Osvaldo Aranha, Porto Alegre, RS, Brazil

A. M. Tavares Thomé et al. (eds.), *Industrial Engineering and Operations Management*,
Springer Proceedings in Mathematics & Statistics 367,
https://doi.org/10.1007/978-3-030-78570-3_44

569

1 Introduction

A complex municipal, state and federal management structure exists to meet the Brazilian population health needs. The Brazilian health policy called Unified Health System (UHS) defines Primary Health Care (PHC) as the first level of complexity with direct contact of the population with UHS [1]. Pharmaceutical Assistance (PA) is the term that defines the set of actions and practices to ensure the promotion, protection and recovery of individual and collective health considering medication as an essential input, aiming at its access and rational use [2]. PA comprises the Selection, Scheduling, Acquisition, Storage, Distribution and Dispensing (SPAADD) operations, known as the PA cycle, carried out under the management of the health departments in almost all of the 5000 Brazilian municipalities.

This article focuses on SPAADD cycle medication storage and distribution operations at municipal level conducted by warehouses known throughout Brazil as the 'Pharmaceutical Supply Center' (PSC) [3]. Given the importance of said operations within the SPAADD cycle, PSC must have its processes planned and performed as efficiently as possible, avoiding failures that compromise dispensing, such as medication losses and delays in the delivery to patients [4]. Furthermore, PA flows are made tangible through the product "medication" during said operations. The flows are predominantly information flows in the steps that precede storage.

Here, the processes and services optimization, typical of the knowledge treated in Production Engineering, becomes important and more specifically, the practices of Lean Healthcare (LH). Lean, also known as lean production, is a philosophy of reducing waste that does not add value to the operation of any process [5]. There has been an increase in its application in services, especially health care services, known as Lean Healthcare [6]. Pontes et al. [7] demonstrate that the LH applications in PHC processes are still reduced, as addressed in this article, with a greater predominance in hospitals and higher complexity level settings [8, 9].

The research question that arises when applying LH in PA operations and services such as PHC storage and distribution is: "what adaptations are necessary for the application of LH diagnostic tools and process improvement in this context?". In terms of reliability engineering, TPM is a subdiscipline of systems engineering that emphatically empowers the flawless use of equipment, estimating the probability and prevention of failures, so it brings security. Considering that the health environment requires maximum minimization of errors, due to the obvious impact on human life, it is believed that TPM can be an excellent tool in LH practices. Therefore, this study aims at investigating the application of TPM as a Lean Healthcare (LH) practice in a primary health care setting. The investigation unit chosen consists of the drug storage and distribution from a PSC located in a city of the State of Rio Grande do Sul, Brazil.

In this case, the theoretical contribution is to advance the adaptation and application of LH tools and practices in health service processes that are still under explored. The practical contribution is to bring better results to the PSC processes by providing tools and practices that can be replicated in similar settings.

2 Theoretical Benchmark

Organizing the PA processes, the best way possible is one of the tasks of professionals in charge of municipal and state UHS health management. This challenge involves both the use of financial resources and the continuous improvement of its processes [10]. Considering the PA cycle itself, its existence comprises the best medication management, taking into account its rational use, logistics and processes [1, 6]. The PA cycle comprises the selection, scheduling, acquisition, storage, distribution and dispensing of medicines, referred to in this text as the SPAADD cycle.

Lean means to do more with less and refers to a method that seeks to identify and eliminate waste. Waste is defined as any activity that does not add value from the customer's point of view, thus reducing the efficiency of a process and increasing its costs. There are eight possible types of waste, namely: overproduction, transportation, waiting, handling, unnecessary processing, inventory, rework or defect and talent [11].

The Lean method in health is called Lean Healthcare and consists of applying lean production tools and practices to improve the management of health systems [12]. Lean health interventions are aimed at improving the quality of health care, thus reducing waste and facilitating the flow in work processes [13–15]. The Lean Method has proven to be a method commonly used as a quality improvement approach in healthcare environments to improve provision of care [16].

The concern with reducing costs and improving processes in the health sector has motivated many studies. The study of [17] showed improved processes with a 75% performance improvement in the workspace and reduced waiting period for patient care from 2 h to 30 min. Tlapa et al. [18], shows that the focus on reducing activities without added value in LH contributes to improve the flow of patients and efficiency in hospitalization care. Gonzalez-Aleu et al. [19] analyzed the success factors in the implementation of continuous improvement projects in hospitals. LH applications found in the literature are often related to hospitals and clinics. The latter have different specifications from other health environments, such as PSCs, municipal pharmacies, basic health units (BHU) and other Primary Health Care facilities.

When assessing LH waste, analysts must use tools that include Value Stream Mapping (VSM). This is one of the most popular lean production tools used in the diagnosis of processes, which contributes to the visualization of flows and identification of losses [20]. For [21], the VSM is ideal to propose improvements from the current scenario, seeking to eliminate current waste in order to have a more efficient process.

An important advantage of VSM is to bring together, in a common language, several Lean tools that will be used together for a global benefit. For this, knowing the VSM symbology that facilitates the registration of different types of losses within the system or process under analysis is ideal [22]. What happens is that these tools had their genesis in industrial environments different from the health processes environments. Therefore, applying equations and symbols is not always obvious and they must be adjusted to fit this new context.

As in the lean manufacturing concept, total productive maintenance (TPM) appeared in the Japanese automotive industry as a way to optimize the progress of each project step. Its pillars are: (i) Autonomous maintenance: provides freedom to operators, who monitor the conditions of their own equipment and the processes around them; (ii) Collection of information for improvement: optimizes the flow of information between sectors. Thus, the manager of each area collects data that are analyzed and receive the necessary prioritization for improvements; (iii) Preventive maintenance: each sector puts into practice the necessary actions for the care with equipment, tools and machines, preventing an interruption of activities; (iv) Planned maintenance: those responsible for each process monitor the team and equipment performance accurately, preparing reports that indicate the need to implement improvements and include breaks in the schedule; (v) Maintenance of quality: the quality of industrial processes depends on a series of variables, and in this topic, the best technologies, methods and new ideas that can contribute to generate greater value in each task are considered; (vi) Capacity building and training: human capital is essential in any organization, and in order to implement process automation, it is necessary for employees to master the technologies, in addition to allowing them to act in more strategic scenarios, (vii) Occupational health and safety (OHS): in any company and, above all, in the industry, the safety of employees is paramount. To support every pillar, best practices must ensure the safety of everyone involved.

Six losses are considered in the TPM: machine breakdown, set up, short stops, speed reduction, defects and rework and performance. OEE (Overall Equipement Effectiveness) is a key performance indicator (KPI), a central component of the TPM (Total Productive Maintenance) method. It was created in 1960, by Seiichi Nakajima (TPM tenkai, JIPM Tokyo), considered as an excellence tool to monitor and improve the efficiency and effectiveness index in the manufacturing process. It is a simple and practical calculation that uses well-defined metrics and allows you to compare different machines, checking their evolution over time. The comparison between the real value and the ideal value of the machines gives rise to a percentage value, identified as OEE.

3   Method

This paper adopts the single case study as method. The unit of analysis is the PSC of a municipality in the state of Rio Grande do Sul/Brazil, with thousands of inhabitants. The PSC studied serves about 10 municipal pharmacies and 130 UBS, and supplies public and private hospitals in the municipality. Intervention occurred from June 2019 to February 2020.

Step one was to determine the perception of value by employees in the stages of the SPAADD cycle. Said definition of value and main losses perceived in their activities was carried out through qualitative interviews with the professionals responsible for the Storage and Distribution operations of the SPAADD cycle, provided for in Appendix A. The interviews were recorded and transcribed. The data collection instrument contained questions regarding: (i) the professionals' perception of value, (ii) types of Lean losses, and (iii) sub-processes and operations carried out at PSC.

The interview with the PSC professional was recorded on audio with the interviewee's consent, and the Informed Consent Form (ICF) was agreed upon and signed, and was approved by the Research Ethics Committee of the Municipal Health Secretariat with number CAEE—82,571,018.6.0000.5347, as provided for in Appendix B. Subsequently, they were transcribed and converted to a data analysis spreadsheet (Appendix C).

Then, the value stream and waste elimination were diagnosed by mapping the activities carried out during the storage, separation, checking and shipping of medicines. We used chronoanalysis to better understand and detail the logistical process, making it possible to design a flowchart of the activities, the Swim Lanes flowchart, which was prepared by the Bizagi® software—version 2019 (Appendix D).

From the information obtained in the data collection section, both in the qualitative interview and the chronoanalysis phase, we were able to identify 18 main problems that occur in the PSC processes, making it possible to design a cause-effect relationship diagram. Such a diagram allows for a broader view of the relationships between problems and their root causes, rather than analyzing the problems separately.

This survey allowed the creation of the VSM based on the results of the qualitative interview, chronoanalysis, mapping of processes and cause-effect relationships diagram. The VSM provides an analysis of the logistical processes of separation, checking and loading of medicines. The steps taken to apply the VSM were followed as described by [22].

(i)   **VSM Preparation**: familiarization with PSC processes and contextualization, based on the results obtained in the data collection, process mapping and cause-effect relationships diagram, and VSM design steps, along with the research team;

(ii) **Lead time**: taking of times and movements of each of the three processes: separation, checking and loading of medicines, through chronoanalysis data analysis, the following variables were recorded by process: Cycle time (CT), Setup time (ST), Takt-time (TKT), number of operators, number of shifts, Overall Equipment Effectiveness (OEE). Calculations as per Table 1 below were made to determine the OEE of each process.

Considering that there are peculiarities in the storage and distribution services of medicines, it was necessary to interpret the context in order to adjust the variables to it. In this study, all activities considered as losses at PSC had to be adapted to the six losses considered in the OEE: machine breakdown, setup and adjustments, speed drop, empty operation, defects and rework in the process, defects and rework in the start of production, as losses made by machines are not being treated in this study, but losses in the activities in the processes. Thus, the team involved with this project made adaptations to fill

Table 1 Formulas to calculate OEE

$$OTI = \frac{Total\ time\ available\ -time\ stop\ losses}{Total\ time\ available} \quad (1)$$

$$OPI = EWT\ x\ OSI \quad (2)$$

$$EWT = \frac{Number\ of\ processed\ products\ x\ actual\ cicle\ time}{Tempo\ total\ disponível-Tempo\ paradas\ perdas} \quad (3)$$

$$OSI = \frac{Theoretical\ cycle\ time}{Actual\ cicle\ timep} \quad (4)$$

$$PAI = \frac{Number\ of\ processed\ drugs-number\ of\ defective\ drugs\ or\ rework}{Number\ of\ processed\ drugs}$$
$$(5)$$

OTI: Operating Time Index; *OPI*: Operational Performance Index; *PAI*: Product Approval Index; *OSI*: Operational Speed Index; *EWT*: Effective Working Time

out the equations and analyzes, submitting them, to the confirmation of their validity with Lean specialists.

(iii)   **VSM Validation**: the results obtained in the VSM were validated with the PSC manager. VSM results were presented formally with suggestions for improvement in the logistical flow. VSM was designed using the *software Lucidchart®*.

4   Results

Within the conception of LH the understanding of what represents value serves as a guide for the improvement of processes. According to the Lean philosophical principles, only operations that add value can be kept. The objective described for each of the PA cycle operation, SPAADD, served as a reference for the value analysis. Said objective used as a reference was extracted from the Technical Instructions for PA Organization of the Department of Health [3]. We expect the perception of value of the professional responsible for the PSC is in line with the reference objective.

The SPAADD cycle interviewees did not fully mention the stages of the PA cycle. It is natural for the interviewees to place greater emphasis on the stage in which they are inserted, forgetting the others that make up the logistics chain. The result showed a lack of integral alignment between the interviewees' responses and the sector's objectives. The description given by the PSC interviewee, for example, was not full, as was the definition of the concept of PA. Such a misalignment can be minimized by training professionals. In particular, sector leaders need to be clear about the value proposition of the cycle stage under their coordination and in relation to the entire SPAADD system, in order to ensure: (i) delivery of results aligned with the value foreseen in their stage; (ii) disseminate the concept of value among subordinates; (iii) promote improvements in their operations towards the value desired locally and throughout the SPAADD cycle.

Regarding the Lean losses identified by the interviewee's perception, seven out of the eight types of loss (waiting time, transportation, rework, inventory, handling, defects, misuse of human resources and overproduction) were mentioned, with the exception of "loss due to overproduction", with losses due to "waiting time", "defective actions" and "loss due to inventory" predominating. Despite being just an analysis from the interviewee's perception, it served as a starting point for applying the VSM. The diagnosis made to implement the VSM provides greater precision of the real losses in the processes within the sector, bringing more depth and understanding of how the processes within the sector occur.

The purpose of the interview was, in addition to raising the perceptions of value and losses that occurred in PSC's storage and distribution operations, based on the manager's perception, how to collect information necessary to fill in the SIPOC form (Supplier, Information, Process, Output, Client). The results of this application can be accessed in Appendix C. Thus, the diagnosis proceeded through the following steps: (i) chronoanalysis; (ii) process mapping; (iii) cause-effect diagram, compiling useful information for the VSM.

Typical PSC operations include purchase, receipt, storage, separation, checking and distribution. This study will focus on separation, checking and distribution operations. The results of these applications are shown below.

(i)   **Chronoanalysis**: at PSC, the project researchers paid five on-site visits between May and July 2019 to collect the times of seven out of the twelve PSC employees responsible for the intermediate stages of the process mentioned earlier. In total, there was three hours of collection of times and movements from four employees during the separation of medicines; two hours to collect from two employees during the checking of the separate orders and two hours to collect from four employees during the loading of the trucks of the orders checked and packaged for delivery.

(ii)   **Mapping of processes**: The operation analysis phase through chronoanalysis made it possible to map all activities, identifying the operations that generate and the ones that do not generate value to the processes, carried out by the employees involved in the process of separation, checking, shipping of medicines and the administrative sector; the latter checks all orders that arrive at PSC. The flowchart of the processes main stages activities and the details of Swim Lanes general image containing the flow map are presented in Appendix D.

Medicines separation is carried out by up to 12 PSC employees since lists are divided by strategic and basic medicines with people individually responsible for each one of them. Separation is carried out with a shopping cart. These are adapted for the function as shown in Appendix E. There is an opportunity to design a "cart" that is more suitable for the function of separating or to identify another cart for separating medicines.

(iii) **Cause-effect relationship diagram**: the cause-effect relationship diagram
       made it possible to transform PSC's 18 problems into two root causes iden-
       tified as "Inadequate processes" and "Inadequate layout, infrastructure and
       furniture". The relationship between these 18 problems and their root causes
       are all connected forming a cycle that leads 'n' problems to two main root
       causes, leading to the "work overload" problem. For example, the "Insuf-
       ficient number of employees" problem with "Inadequate Process" as root
       cause is related to the "work overload" problem, which in turn is also gener-
       ated by the "Inadequate layout, lack of infrastructure, inadequate furniture"
       problem, which also creates the "Separation errors" problem, as can be seen in
       Appendix F.

4.1 Value Stream Mapping (VSM)

When diagnosing VSM, as shown in Fig. 1, it was possible to see that the flow of
information in PSC starts upon receipt of the demand for medication orders for basic
health units, district pharmacies or hospitals through PSC through its own medication
flow management system called DIS. Said demand is passed on daily to the employees
responsible for separating these drugs through lists divided between delivery points.
Meanwhile, the schedule for loading and distributing orders is forwarded weekly
following the routing of deliveries by zones and neighborhoods in the municipality.

   As for the times allocated to the separation, checking and loading operations,
it was possible to identify idleness in all of them. Since an average of 60 types of
medications are separated per person per day, which subsequently go through the
following operations, and that comparing to the cycle time, takt time and the time

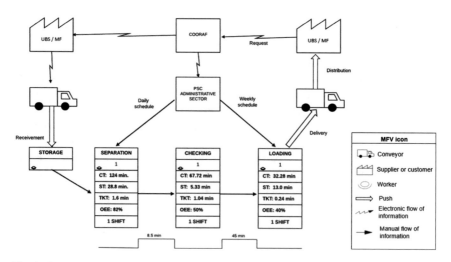

Fig. 1 Current PSC value stream map diagnosis

Table 2 Results of OEE calculation indices per PSC process

Process	Index			
	OTI (%)	OPI (%)	PAI (%)	OEE (%)
Separation	94,36	89,44	96,67	81,58
Checking	95,61	52,57	100,0[a]	50,26
Loading	98,53	40,37	100,0[a]	39,78
Limits	>90	>95	doesn't have	>85

[a]Not applicable, as there are no defective products at the conference

available to perform them, it would potentially be possible to deliver more orders. However, to recalculate the division of orders per day more effectively, it would first be necessary to eliminate or minimize the losses identified through the OEE calculation as much as possible.

According to [23], OEE values between 65 and 75% are acceptable, over 75% are accepted as very good, above 85% are rated as excellent and below 65% are considered unacceptable, and the company should implement improvement actions. Thus, for this range, the Operating Time (OTI), Operational Performance (OPI) and Product Approval (PAI) indices must be 90%, 95% and 99%, respectively.

That said, Table 2 shows that the only operation within the acceptable standards is the separation (OEE = 81.58%), with conference (50.26%) being the unacceptable process due to its low efficiency and loading (39.78%) the unacceptable and the most critical process.

In order to perform OEE calculations, the difficulties were caused by the classification of losses as required in the calculation of variables. Therefore, a table was prepared with the descriptions of the variables originally proposed for industrial processes and the team performed an analysis of these in the context of the studied context. Adaptations are described in Table 3.

In the case of loss due to "Machine breakage", for example, because it is a predominantly manual process, it was decided to consider interruption of work in this failure due to the lack or inefficiency of the instruments used for the operation. The loss due to "Setup and adjustments" was adapted considering all the adjustments made before starting the next process, such as organizing pallets and medicines. In relation to the loss "Speed drop", the employee's entire movement, such as displacement and conversations related to the activity developed, was considered a loss. The "Empty operation" is related to the interruption of the machine due to lack of raw material supply. In this context, all activities that were not part of the process, such as unnecessary stops and conversations, were considered losses. The loss due to "Defects and rework" at the beginning of production was considered an activity not relevant to the employee's function, impairing the beginning of its process.

These adaptations allowed the calculations to be performed. One expert verified the relevance of adaptations in relation to the process, but it is understood that for future applications it will be necessary to evaluate the context in more depth and

Table 3 Results of OEE loss adaptations per PSC process

OEE Losses	Adaptation	Process		
		Separation	Checking	Loading
Machine breakage	– The work is essentially manual. Few equipment and machines are used in the sector. The adaptation occurred by analyzing the instruments used in the operation – Interruption of work due to the lack of any utensil or due to the lack of capacity to perform that task		– Take pen with another colleague because it is not working	– Walk looking for information
Setup and adjustments—it is the time taken to change a running process until the next process starts	– Considering all the adjustments made before starting the next process, such as: organizing pallets, carts, workplace, organizing medicines	– Organize medicines on the cart or shelf – Move medicine box to the corner of the table – Move cart away to get around – Set up boxes on the table that has already been set up by another colleague	– Decide on which pallet to place the box – Analyze the notes on the medication list sheet, before the next conference – Prepare the boxes to place the medicines – Replace adhesive tape on devices	– Organize empty pallets

(continued)

Table 3 (continued)

OEE Losses	Adaptation	Process		
		Separation	Checking	Loading
Speed drop-Difference between the rated speed of the equipment and the actual speed	– The employee's entire movement was considered as speed drop, such as: displacement, cleaning, conversations related to the activity	– Write down inventory control sheet + calculation – Make the manual count of the medications that will be separated – Travel to the separation table – Choose and clean the table	– Observe the mural – Move pallet boxes	– Chat while taking the boxes to the truck – Chat while on the move – Conversation: exchange information with colleagues, without being the subject of the activity
Empty operation Upstream problems, parts obstruction—Machine operating empty during interruption of the feed of raw materials	– Considered activities that were not part of the process, that is, unnecessary, such as: unnecessary conversation and stops	– Unnecessary conversation (chat)	– Unnecessary conversation (chat) – Answer a question from another colleague	– Employee left his activity and did not return, without communicating the others and overloading them – Stop to sing – Wait for the truck to unload before starting to load the boxes
Defects and rework in the process	– Redo activities that were not correct	– Redo calculation; – Check basic list more than once in the same operation	– Check the medication more than once before closing the box, due to lack of attention – Separation error	– Organize the truck, close it and check if it closed – Hand paper over to another employee, as the other has run out – Carrying box, by hand, without using the cart – Pick up pallet that was halfway
Defects and rework at the start of production—from machine startup to stable production	– Considered an activity not relevant to the function of an employee, impairing the beginning of its process	– Develop an activity that is not relevant to your role, impairing the beginning of your process		

detail. Especially when considering analyzes in other PSCs in the state and in the country.

The purpose of the diagnosis is to direct the implementation of improvements. Thus, the team understood that the loading would be the first measure to optimize the PSC process, due to the low OEE value (Table 2). However, this operation is directly influenced by the amount of medications available for loading, with idleness when there are few. The group reduces its operating speed according to the time available for it. Thus, analyzing the process flow, the checking step is the one that requires adjustments, and also the one that becomes a bottleneck when restricting how many medications can be separated and checked.

When applying Lean practices in the storage area, losses were detected throughout the process. The high demand for deliveries coupled with the need to restructure work, such as the replacement of inappropriate equipment and suitable furniture for activities, contribute to increased waste. The space allocated for inventory is limited, in addition to the lack of adequate space for the checking and separation. Which makes these processes inefficient. Although it is not possible to expand or replace the physical area, given the municipality's management environment, it is understood that it is possible to make changes to the layout in order to optimize activities and reduce the burden on employees with unnecessary displacements.

The diagnosis of the VSM also made it possible to identify that an easily proposed improvement would include the redefinition of activity shifts. By reducing the time window for loading and increasing it for other operations, it will be possible to increase the size of orders served once the waiting time between the end of checking to start loading is on average 45 min, as indicated in the VSM. To increase agility in the checking and make the process more efficient, an alternative would be to check separately to avoid conversations and increase the concentration of employees.

Also, it is suggested introducing computing systems to modernize procedures and obtain more reliable data from the physical inventory to help decision making. As well as suggesting the batch number of the medication that will expire first in the packing list, thus minimizing separation errors and losses due to expiration. In addition, there is a deficiency in training. Investing in people development by providing them with information and support to become more efficient in their tasks is essential for the good performance of activities, in addition to minimizing errors and waste.

Regarding the use of Lean techniques within the health context, a tool such as OEE calculation had to be adapted because the PSC study is a service, and since the index is for machine losses. Thus, a certain difficulty was found to diagnose which process activities could be seen as OEE losses. Also, the qualitative interview transcription stage enabled a single table to be created with relevant information for diagnosing and understanding PSC processes, considering the concept of value, SIPOC, losses, problems and proposed improvements.

Whenever suggestions for improvement are proposed, there is a need to measure the gain between the "before" (as is) and the "after" (to be). A question that arises for future studies would be the use of Time-Driven Activity-Based Costing (TDABC), as a tool to identify the efficiency gains that will be obtained with Lean interventions. TDABC was created by [24] as an evolution of its predecessor, the Activity-Based

Costing (ABC) model. The authors describe this new system as "simpler, cheaper and much more powerful than the conventional ABC approach". The cost process is simplified as it no longer requires interviews and questionnaires to allocate costs to activities. According to [25], TDABC is applicable in health and can cost processes efficiently, overcoming the challenge of current cost accounting methods, thus being the solution for the health cost crisis.

5 Final Considerations

This study focused on the storage and distribution operations of a PSC. These operations are decisive for the good performance of subsequent activities in the medication chain. It was identified that the quality, efficacy and safety of the products can be affected by the lack of adequate storage conditions, either due to a lack of physical structure or non-compliance with the procedures.

The theoretical contributions of this paper include: the adaptation of OEE and TPM in the context of Lean Healthcare in a poorly explored health environment, pharmaceutical assistance, and at the primary health care level. PA is the gateway for health systems and cost reduction in this context leads to the minimization of health costs at greater complexity levels such as hospitals. As a practical contribution, it was possible to identify the different types of waste during the storage process, such as: movement, overproduction, inventory, talent, rework, transportation and waiting time, as well as helping to minimize them.

A few initiatives have been proposed to redesign the process and provide better results. Such as changing the layout of the unit, in order to have a better work environment, performance, productivity, efficiency and quality in the reception, storage, separation, checking and loading of medicines activities.

It is understood that the study carried out may be adjusted for replication purposes, with due changes based on the diagnoses themselves, in other PSCs from other municipalities, since the SPAADD cycle stages are similar in other Centers. One of the challenges faced throughout the paper was to adapt one of the tools to analyze efficiency within VSM. All activities considered losses had to be adapted to the losses considered in the OEE, as they are not machines, but losses in the processes activities. Such an adaptation was essential to determine the efficiency of the separation process carried out at PSC.

Opportunity for improvement in the current application was identified, a tool for assessing gains between the before (as is) and after (to be) scenario was proposed for improvement. The TDABC tool could bring the perspective of "costs" before and after improvements, as an indication of loss reduction. It was not possible to complete the four steps of the Lean implementation and we suggest its completion for future works. As well as evaluating the effectiveness of the actions implemented and their results and understanding the value of the improvements in the processes carried out through TDABC.

Appendix

Appendix A—Qualitative Interview Script—https://drive. google.com/file/d/1loDjbVroWm3ZEOlyFp7z9px3Eosy8ehg/ view?usp=sharing

Appendix B—Informed Consent Form (ICF)—https://drive. google.com/file/d/1AefXAfAmNa5AZqpa0V49cpSPUvcZ9 CTo/view?usp=sharing

Appendix C—PSC Diagnosis Table—https://drive.google. com/file/d/1jlpLUuv1c-Q5W3CsGIj3h_c2DxQzSF5D/view? usp=sharing

Appendix D—*Swin Lanes*—https://drive.google.com/file/d/ 16vb_SQshTH08cbh21VLVtGaFjsmTYRXi/view?usp=sha ring

Appendix E—Image of the Cart Used to Transport the Medication at PSC—https://drive.google.com/file/d/10L Tq1CB0A9-HcQkUmUTygnpSo8_FN46h/view?usp=sharing

Appendix F—PSC Problems Cause-Effect Diagram—https://drive.google.com/file/d/1LC7oeMiodHE q7NHHmAHuR9dfUNj6L_Al/view?usp=sharing

References

1. Brasil, Portaria n° 2.436, de 21 de setembro de 2017. Aprova a Política Nacional de Atenção Básica, estabelecendo a revisão de diretrizes para a organização da Atenção Básica, no âmbito do Sistema Único de Saúde. (SUS), Diário Oficial de União, Brasília, DF, n° 183, 22 de setembro de 2017 (2017).
2. Brasil, Lei n° 13.021, de 8 de agosto de 2014. Dispõe sobre o exercício e a fiscalização das atividades farmacêuticas, Diário Oficial de União, Brasília, DF, 11 de agosto de 2014 (2014).
3. Brasil, Ministério da Saúde. Secretaria de Ciência, Tecnologia e Insumos. Estratégicos, Departamento de Assistência Farmacêutica e Insumos Estratégicos, Assistência farmacêutica na atenção básica: instruções técnicas para sua organização, 2. ed, Ministério da Saúde, Brasília, DF (2006).
4. Brasil, Ministério da Saúde, Secretaria de Atenção à Saúde Departamento de Atenção Básica, Diretrizes do NASF: núcleo de apoio a saúde da família, Ministério da Saúde, Brasília, DF (2009). (Série A. Normas e Manuais Técnicos) (Cadernos de Atenção Básica, n. 27).

5. Soliman, M.S., Saurin, T.A.: Uma análise das barreiras e dificuldades de Lean Heathcare, Produção, 17(2), p. 620–640 (2017).
6. Jones, D.: Lean Enterprise Institute, Lean Institute Brasil (2015). Available at: <http://www.lean.org.br/comunidade/artigos/pdf/artigo_314.pdf >. Visited on: September, 2020.
7. Pontes, A.T., Paula, I.C., Campos, E.A.R., Lopes, E.: Análise da utilização da saúde enxuta no contexto dos serviços farmacêuticos, Sistemas & Gestão, vol.14 (2019).
8. Costa, L.B.M., Godinho Filho, M.: Lean healthcare: review, classification and analysis of literature, Production Planning & Control, 27(10), (2016).
9. D'Andreamatteo, A., Ianni, L., Lega, F., Sargiacomo, M.: Lean in healthcare: a comprehensive review, Health Policy, vol. 119, pp. 1197–1209 (2015). https://doi.org/10.1016/j.healthpol.2015.02.002
10. Brasil, Conselho Nacional de Secretários de Saúde, Assistência farmacêutica no SUS, CONASS, Brasília, DF (2007). (Coleção Progestores) (Para entender a gestão do SUS, v. 7).
11. Altman, H.: Lean: the bible: 7 Manuscripts—Lean Enterprise: Quickstart step-by-step guide to build a lean business, CreateSpace Independent Publishing Platform, Scotts Valley, Califórnia, EUA (2017).
12. Graban, M.: Lean hospitals, CRC, New York (2009).
13. Al-Hyari, K., Hammour, S.A., Zaid, M.S.A., Haffar, M.: The impact of Lean bundles on hospital performance: does size matter? International Journal of Health Care Quality Assurance, 29(8), pp. 877-894 (2016).
14. Andersen, H., Røvik, K.A., Ingebrigtsen, T.: Lean thinking in hospitals: is there a cure for the absence of evidence? A systematic review of reviews, BMJ Open, 4(1), pp. 1-8 (2014).
15. Shazali, N.A., Habidin, N.F., Ali, N., Khaidir, N.A., Jamaludin, N.H.: Lean Healthcare Practice and Healthcare Performance in Malaysian Healthcare Industry, International Journal of Scientific and Research Publications, 3(1), (2013).
16. Mazzocato, P., Stenfors-Hayes, T., Schwarz, U.T., Hasson, H., Nyström, M.E.: Kaizen practice in healthcare: a qualitative analysis of hospital employees suggestions for improvement, BMJ Open, vol. 6, e012256 (2016). doi: https://doi.org/10.1136/bmjopen-2016-012256
17. Coelho, S.M. Pinto, C.F., Calado, R.D., Marotta, E.A., Romano, E., Silva, M.B.: Lean healthcare: process improvement in a cancer utpatient chemotherapy unit, J. Innov. Heath Manag, vol. 1, pp. 1-9 (2015). doi: 10.20396 / jihm.v1i1.9305.
18. Tlapa, D. Zepeda-Lugo, C.A., Tortorella, G.L., Baez-Lopez, Y.A., Limon-Romero, J., Alvarado-Iniesta, A., Rodriguez-Borbon, M.I.: Effects of lean healthcare on patient flow: a systematic review, Value in Health, vol. 23, pp. 260–273 (2020). doi: 10.1016 / j.jval.2019.11.002.
19. Gonzalez-Aleu, F., Aken, E.M.V., Cross, J., Glover, W.J.: Continuous improvement project within Kaizen: critical success factors in hospitals, TQM J., vol. 30, pp. 335-355 (2018). doi: https://doi.org/10.1108/TQM-12-2017-0175
20. Tritos, L., Premaratn, S., Dotun, A.: Prioritizing lean supply chain management initiatives in healthcare service operations: a Fuzzy-AHP approach. In: IEEE International Conference on Industrial Engineering and Engineering Management, Cingapura (2013).
21. Maia, R.D.A., Souto, R.R., Meira, J.A., Lima, V. S.P., Oliveira, J.B.: O value stream mapping e sua relação com os princípios da abordagem enxuta: proposição de uma sistemática expandida para a gestão do lead time. In: XXX Encontro Nacional de Engenharia de Produção. Anais. ENEGEP, São Carlos (2010).
22. Rother, M., Shook, J.: Aprendendo a enxergar: mapeando o fluxo de valor para agregar valor e eliminar o desperdício, Lean Institute do Brasil, São Paulo (2003).
23. Hansen, R.C.: Eficiência global dos equipamentos: uma poderosa ferramenta de produção/ manutenção para o aumento dos lucros, Bookman, Porto Alegre (2006).
24. Kaplan, R.S., Anderson, S.R.: Time-Driven Activity-Based Costing: A Simpler and More Powerful Path to Higher Profits, Harvard Business School Press, Boston (2007).
25. Kaplan, R.S., Porter, M.E.: How to solve the cost crisis in health care, Harvard Business Review, vol. 89, pp. 46–52 (2011).

Green Innovation Ecosystems: An Exploratory Study of the Involved Actors

Arthur Marcon⬤, José Luis Duarte Ribeiro⬤, Rosa Maria Dangelico⬤, and Luca Fraccascia⬤

Abstract This article employs an innovation ecosystem approach to green innovations to understand how actors participate in the development. Thus, using a qualitative approach, we studied four cases of firms that have developed green innovations to analyze how the innovation ecosystem affects the firm. Data was collected through in-depth interviews and it was content analyzed. Our findings show that many actors can constitute innovation ecosystems. The most mentioned ones are suppliers, consultants and experts, and universities. Our findings contribute to the literature on green innovation by providing solid notions of the innovation ecosystem's impact on firms.

Keywords Green innovation · Sustainability · Innovation ecosystem

1 Introduction and Theoretical Background

Developing innovations has become a major target for industries that aim to capture returns and become technology leaders [1]. Nevertheless, innovating often requires knowledge and resources that companies do not fully comprise, thus open innovation strategies that include external actors and entities can be of great help [2]. Indeed, innovation is a systemic process and, as such, it demands that interdependencies with other players are managed in order to successfully develop new products [3]. In addition, market maintenance and success of the innovation need accompanying changes in the environment that support its existence and increase functionality and, therefore, the value delivered [1]. That is, innovations need changes that support its existence in the market, such as the development of product parts from suppliers,

A. Marcon (✉) · J. L. D. Ribeiro
Universidade Federal Do Rio Grande Do Sul (UFRGS), 90 035-190, Porto Alegre, RS, Brazil
e-mail: arthur.marcon@ufrgs.br

R. M. Dangelico · L. Fraccascia
Sapienza University of Rome, 00185 Rome, Italy

L. Fraccascia
University of Twente, 7522 NB Enschede, The Netherlands

A. M. Tavares Thomé et al. (eds.), *Industrial Engineering and Operations Management*,
Springer Proceedings in Mathematics & Statistics 367,
https://doi.org/10.1007/978-3-030-78570-3_45

partners for repair and remanufacture, and even governmental legislation changes [4]. To that end, firms have established networks of partners to enhance the breadth and depth of their inventions in innovation ecosystems [5]. Innovation ecosystems consist of a setting of multiple independent actors that jointly create value in an environment that simulates organic ecosystems where firms and other actors cooperate and compete [6, 7].

Adner and Kapoor [1] claim that the ecosystem construct helps to make interdependencies more explicit among the actors. The authors argue that the innovation ecosystem is structurally composed of the focal firm, the components of the ecosystem, the complementors (i.e., organizations that add value to the innovation of the focal firm), and the customer. Ritala and Almpanopoulou [7] and Oh et al. [8] highlight that the term "innovation ecosystem" has become popular in industry and academia. Additionally, [9] emphasizes the determining role of the institutions of innovation ecosystems. Institutions concern the routines, laws, and rules that constitute incentives and obstacles to innovation. In other words, they are the "rules of the game" [9]. Conceptually, several definitions of innovation ecosystem have been proposed (see [10]). Many theorists claim the innovation ecosystem's concept's birth belongs to [11], who proposed that a company can be viewed as a part of a business ecosystem that is not restricted to its geographical or sectoral boundaries and that is characterized by interdependence and coevolution of capabilities to increase value.

Considering the importance of innovation ecosystems in supporting and increasing the functionality of innovations, green innovation ecosystems (IEs) arise as an important topic of interest, since this green innovation demands a common joint effort from suppliers, universities, policymakers, users/customers, and complementors to succeed and make a change [12]. Green innovation entails the integration of sustainable elements into firms' new or improved products, services, processes, or practices or the development of new ones that either reduce their environmental impact of the product, neutralize it, or even cause a positive impact [13, 14].

Interest on this topic has rapidly increased and [15] report that investments in environmentally sustainable innovation are expected to reach US$10 trillion by 2020. Consequently, new markets allowing for many opportunities emerge, as well as new challenges that include changes in established markets and the need for support and participation of different actors and innovation ecosystem factors [16].

Therefore, understanding the link between the actors of this ecosystem is essential to become aware of ecosystems' boundaries and impacts. Environmentally sustainable issues are acknowledged to generally fall beyond the core activities of most firms, and even some firms that wish to develop green products do not own the necessary resources [17], which further advocates for open innovation practices. Nevertheless, how actors' dynamics occur in green IEs still remains an unanswered question in the environmentally sustainable innovation field.

Although some studies have recognized green innovation actors, they have either ignored the role of codevelopment and competition of actors within ecosystems, or they did not address the "whole picture" of the innovation ecosystem. For example, [18] only addressed how some stakeholders impact the research and development of green products and their findings only present the actors that are closest to the company, such as universities, consultancies and non-governmental organizations. Melander [14] on the other hand only addresses the integration of suppliers and customer for green product innovation which narrows the open innovation process. The focus on supplier and customer integration hinders a more profound understanding that recognizes that firms that develop green innovations are found in deeper, more complex innovation settings formed by multiple actors that are not as closely coupled with firms.

Recognizing the actors that play a role in the development of green innovation is of utmost importance as firms rely on these actors' resources and expertise on green innovation to develop green products that are more efficient from the environmental point of view and that address market needs [14]. However, being able to establish ecosystems where actors are able to coevolve and jointly deliver marketable green innovations largely depends on the effective identification of which actors integrate green IEs and which actors could help in the development of green innovations by adding resources to the ecosystem.

Thus, based on the potential contribution of the innovation ecosystem theory to the understanding of the dynamics in green innovation, this article addresses the actors of green IEs. To that end, we analyzed four different cases of firms that have developed green innovations to define actors and dynamics of green IEs. We posit that green innovations differ from regular innovations and, therefore, they create different innovation ecosystems due to the need to consider sustainability aspects during the innovation process. Responding to the call from literature, it aims to understand how green IEs are organized based on the roles played by different ecosystem actors. As such, we analyze firms that have developed green innovations using open innovation practices to understand the green IEs.

The rest of the paper is organized as follows. Section 2 concerns the methods of this research. Section 3 presents the results. The paper ends with discussion and conclusions in Sect. 4 and Sect. 5, respectively.

2 Method

Since this article aims to comprehend how green IEs are structured, we decided to use a qualitative approach to the problem, in line with the study of [19]. Thus, we conducted in-depth case studies of four companies that develop green innovations and are located in Brazil. Data were collected mainly using semi-structured interviews to gain deeper insights on the actors and dynamics of the green IEs of each company, and it was complemented using published data sources from companies to enrich the findings [20]. The questionnaire contained nine questions that addressed green innovations, ecosystem structure and governance and the impact of the ecosystem on the firm. To guarantee anonymity, firms were assigned codenames (Tractor, Heath simulation, Fashion retail, and Grain drying) and a brief description of each firm is presented next.

Table 1 Information about the cases studied

Codename	Size	Interviewee	Innovation description
Tractor	Large	Head of Product Engineering	Tractor with controlled residual emission
Health Simulation	Startup	CEO/Founder	Devices and products for health simulation such as vials, gels, and dummies
Fashion retail	Large	Sustainability Analyst	Development of a sustainable chain of fashion retail to promote green raw materials and processes
Grain drying	Startup	CEO/Founder	Development of sustainable grain drying system that eliminates current highly polluting heat sources

The company under the Tractor codename refers to one of the biggest tractor manufacturers in the world. The company has developed a low-emission tractor engine to comply with new regulations that demand the reduction of polluting gases. Health simulation is a startup that develops environmentally sustainable solutions for the simulation of health procedures for medicine and nurse students. During the development and commercialization of the products, actors had to be involved, since the company did not have all the resources and knowledge to develop and market the products. Fashion retail is the case of a major fashion Brazilian garment company. The company does not produce the clothes, rather, it just sells them, therefore, it had to engage stakeholders to create more sustainable products and greener operation. The Grain Drying case refers to a startup that has developed a greener solution to dry grains stored in silos. Table 1 summarizes the information about each case.

Interviews lasted around 75 min each and were recorded and transcribed. Transcripts were content analyzed using a coding technique to divide relevant content of transcripts into categories, as also done in the qualitative paper of [19]. After coding, we performed within cases and cross-case analyses following [21].

3 Results

A deeper understanding of the innovation ecosystem and its dynamics can only be achieved through the understanding of the actors that compose the green IEs. In this regard, our analysis shows that several are the actors play a role in green IEs. The most referenced actors are suppliers, since they directly influence the performance of the development, production, and commercialization of products. In the case of the *Tractor* firm, suppliers were included during the design of the parts of the tractor, in order to assist with the technology and the requirements necessary to allow the engine to reduce particle emission. In other cases, suppliers are responsible for the entire pool of sustainable attributes of the final product. This is the case, for example, of *Fashion Retail*, where the focal firm does not manufacture the products; nevertheless, *Fashion Retail* enforces sustainable practices for suppliers to assure that the final product is in line with the sustainable values.

> We develop suppliers to make them sustainable. We made them aware of sustainability and then we showed them how to reduce waste. (...) The first step was to promote knowledge about the importance and the impact of being sustainable and of adopting sustainable practices. We hired consultancies for the suppliers to give them knowledge about how to be sustainable and to optimize the cutting of the pieces, to diminish waste and to reduce consumption.

> (Sustainability Analyst, Fashion Retail)

Developing sustainable innovations is not a trivial task for firms, which usually lack the resources to do that. Hence, firms integrate consultancies and experts in their innovation network. Consultancies and experts span the boundary between the knowledge internally diffused and the knowledge externally available, which is not always entirely at firms' reach. Therefore, through consultancies and experts, firms can acquire knowledge related to green innovations even though the firm did not hold the entire necessary knowledge and capabilities for that. Nevertheless, it is important that firms have the necessary means to exploit the knowledge absorbed, otherwise, they may not be able to fully leverage it. The *Grain Drying* firm highlighted the lack of such mechanisms to leverage the external knowledge acquired since the startup collaborated with experts to help develop a marketable solution, but they failed to exploit it to the fullest. This is evident in their quotation:

> We were not able to do that internally. So, the practice we adopted was to have partnerships, agreements with companies that had this know-how. (...) We sought partners who gave us the technical support about an area of knowledge that we did not have. (...) So we have always looked for representative entities that had these competences, but we did not internalize the knowledge. And that was one of the pains of our project, now we see that this was a mistake...not having this intellectual capital inside the company.

> (CEO/Founder, Grain Drying Case)

The third most mentioned actor in the green IEs is the university. Universities were reported by respondents to be knowledge-rich environments where information can be achieved to improve the innovation process of firms. The *Fashion Retail* firm, for example, partnered with universities to conduct life cycle assessment of products, whereas the *Health Engineering* firm collaborated with the pharmaceutical labs of a major university to help develop its simulation products using the know-how of experts. Additionally, the *Tractor* firm required specialized laboratory tests to measure emissions and particle counting during the development of its tractor. Thus, the firm hired universities to conduct the tests and analyses in their laboratories given universities' structure and know-how of researchers.

Our findings show that Research and Technology Firms participate in the design and prototyping of products. For example, the *Tractor* firm hired technology firms to assist in the design of the new engines. According to the interviewee, partnering with technology firms is ideal for technology-related R&D issues that the focal firm is not able to handle alone, since this type of firm holds great expertise in technology development and implementation. Whereas the *Grain Drying* firm partnered with research laboratories that assisted in the testing of the solution and in measuring the results achieved during the validation of the prototypes developed. Mostly, according to the interviews, this type of actor is included in the ecosystem because of its advanced knowledge and resources to design, measure, and test solutions that will influence the innovation developed.

Complementors are the next most mentioned actors in the innovation ecosystems analyzed. Complementors, as previously defined, are actors responsible for adding value to the innovation ecosystem; however, they are not the focal firm's suppliers or direct actors. This does not mean that their role can be neglected when analyzing green IEs. The case of the *Tractor* firm is a good illustration of that. During the development of the tractor with reduced emissions, oil and gas companies were included because the new engines required fuels with specific chemical characteristics (i.e., reduced particles). Oil and gas companies had to develop new processes and systems that allowed the production of this type of fuel so that, when the tractor reached the market, the necessary asset network was ready to support its operation. The quotation presented below highlights the role of the complementor to the innovation of the *Tractor* firm:

> The oil companies had to make diesel fuels with less quantity (of particles) that forced them to have more technological refining plants, more novel processes. (...) the agricultural machinery manufacturers had to develop their machines to receive these engines. The oil companies had to make better diesel. The exhaustion systems had to be improved. The diesel filtration systems had to be improved, the systems of air filtration had to be improved (...), electronic equipment was more widely used in the machines.
>
> (Head of Product Engineering, Tractor case)

Additionally, we found that class organizations also participate in the innovation ecosystem. The *Fashion Retail* firm participates in a class organization specific for fashion firms. Through the knowledge acquired by collaborating with such an organization, the *Fashion Firm* was able to develop a sustainability standard to which its products comply and carry a label. The *Health Engineering* firm also consulted class organizations to acquire knowledge related to the regulations of the health simulation and health education sector.

Other actors in the innovation ecosystem are investors and shareholders. This was observed in the case of the *Fashion Retail*, which is a publicly-traded company that is pushed by shareholders to adopt green practices. According to the interviewee, although customers' demand for green products in Brazil is still below expected, in Europe and North America such concern is more widespread and fashion retail firms are pushed to adopt green practices and market more sustainable products. Thus, investors and shareholders press the firm to hold the same standards in Brazil.

It is important to highlight the relevant role played by regulation agencies in green IEs: they are responsible for creating policies that can enforce firms to innovate and can impact the development of the innovation. This is highlighted by the interviewee from the *Tractor* firm, which explains that the innovation developed by the firm underwent tests from the regulatory agencies to guarantee the tractor complied with the legislation on emissions.

Additionally, interviewees also reported the role played by the government in policymaking and law creation. For example, the *Health Simulation* firm developed sustainable simulation products to meet a market demand generated by a governmental law, which required health-related undergraduate courses to maintain simulation laboratories for student practice. The innovation in the *Tractor* firm was also triggered by a law that required engines to reduce the number of particles emitted in fuel burning. Nevertheless, the need for a greater incentive by the government to require environmentally sustainable practices from all the players in the market was observed. Interviewees reported that the government should push for more sustainable operations from all manufacturers, as, according to them, developing and marketing sustainable innovations can be more expensive and costly. Thus, interviewees state that as they develop green innovations, the government should require the same from competitors in order to even market competition under green conditions.

We also found that coworking spaces, innovation hubs, and incubators play a role in developing knowledge for innovation of firms. For example, the *Grain Drying* firm and the *Health Engineering* firm participated in a University's incubation program to develop marketable products and management skills. Most recently, the *Grain Drying* firm has joined a coworking space to improve its network of partners. Additionally, although such spaces and programs are more directed to startups and small firms, we found that the *Fashion Retail* firm developed a startup acceleration program to accelerate startups focused on green innovation to possibly develop new suppliers and partners.

Finally, customers play a major role in the environmentally sustainable innovation ecosystem of firms. Although the notion that the demand for sustainable products is still not as high as would be expected, firms integrate customers during the development process in order to assure that the product meets the demands and that it is aligned with the needs. This is highlighted in the following excerpt of the CEO and Founder of the *Grain Drying* firm:

> Interacting with the customer (...) this ecosystem must take you directly to the customer, to validate the solution with the customer. First, you do the back-office work, then you go back to the drawing board to develop the product.
>
> (CEO/Founder, Grain Drying Case)

Table 2 presents each actor mapped in our study, along with the number of references to the actor in the interviews and an illustrative quotation. It is important to highlight that the reference count refers to the number of different excerpts that addressed the actor and not the word count for the actor, similar to the approach used in de Medeiros [22].

Table 2 Actors, references and quotations

Actor	#	Illustrative quotation
Supplier	20	We started to promote supplier awareness about the importance and impact of being sustainable and of adopting sustainable practices. To that end, we paid consultancies to suppliers to provide information about how to be sustainable and to optimize the cutting of parts. (Fashion Retail)
Consultancies and experts	10	(…) the specialists, the experts. They were important for me to get where I needed to be. For example, in the third product, the gel, I do not understand anything about gel, so I went to a compounding pharmacy and I talked to the pharmacist. (…) I am in the prototyping phase, so I need the dealer, I need the expert, I need the supplier (Health Simulation)
Universities	9	The university is a very knowledge-fertile environment because in it I can easily find the experts that I need. Therefore, inside the university, I find the experts (…) to start a work of creation. (Health Simulation)
Research and technology firms	4	Another company is AVL. They do research in the field of engine technology, research, work and projects on this area (Tractor)
Complementors	4	The oil companies had to make diesel fuels with less quantity (of particles) that forced them to have more technological refining plants, more novel processes. (*Tractor)*
Class Organization	4	We involved class organizations formed by representatives of the textile retail sector. Through this class organization, we were even able to develop a standard (*Fashion Retail*)
Investors and shareholders	3	Sustainability within Fashion Retail comes from two main reasons: the customer asks for more and more sustainable practices and products. But a lot is due to the shareholders, mainly those from Europe and North America, and our administrative board. The shareholders of developed countries already have this concern in their countries, so they demand us to have the same concern for sustainability here. (Fashion Retail)
Regulation Agencies	2	Organizations have been created, for example, in the USA, they have the Environmental Protection Agency, there is ACABI, I think it's from California too (…) here in Brazil we have IBAMA (Tractor)
Governments	2	Health simulation is an emerging market in Brazil; it generates millions in the world. In other countries, this market (of health simulation) is very powerful, why is that? Because the student cannot touch patients before graduating. So the student goes through the whole course doing a lot of simulation. It is a legal issue. Our Brazilian curriculum still does not require this, but there is practically no health undergraduate course without simulation nowadays. (Health Engineering)

(continued)

Table 2 (continued)

Actor	#	Illustrative quotation
Coworking spaces, innovation hubs, and incubators	2	Recently I have joined one of these innovation hubs (…). It is a coworking space, but with a different proposal. It is not simply the leasing of a desk. I think that's what's making the difference. They provide you with some support. (Grain drying)
Other actors	2	We also involved startups. The idea was not to make them become our suppliers; our intention is to understand what they are doing so maybe one day they can become our suppliers in areas of support (Fashion Retail)
Customers	2	I need my customers, their opinion is very important; I am prototyping now, so I am testing the gel with customers. (Health Simulation)

4 Discussion

As discussed earlier, our study employed the innovation ecosystem theory to analyze how green innovations are developed by firms. As an open innovation approach, innovation ecosystem theory supports the notion that innovations are created from a "dialog with multiple constituents" [7]. Thus, our findings show that the alignment of values between actors is of paramount importance to any green IEs. Additionally, values must not only address the innovation itself, but also sustainability. Similar findings have been previously provided by Chen and Liu [23].

Our findings are in accordance with the work by Tsujimoto et al. [24], which is focused on discussing how actors' differing beliefs and values in the innovation ecosystem may cause unintended results at the ecosystem level. In fact, we found that firms must choose actors with similar values to be able to successfully exploit the innovation, otherwise, as the innovation process advances, such misalignment will lead to unsatisfying outcomes. To ensure the alignment with suppliers, for example, we found that firms can train them and provide consultancies. This leads to closer ties and better collaborative relationships between actors.

Additionally, using the open innovation view of green IEs, we found that some actors play a key role in adding or increasing the sustainability of the innovation. Suppliers can partly add sustainability to the final product (such as new engine parts in *Tractor*) or they can be responsible for the entire sustainability addition to the product (such as raw materials in the *Fashion Retail*). Universities can add sustainability to innovation through intellectual capital. We also found that some actors are more responsible for increasing the value of innovation, such as the complementors, but they can also be responsible for increasing sustainability. Thus, our findings are in line with results by Farla et al. [25], which highlighted the role of policymakers, customers, and other actors in the transition for sustainable societies.

5 Conclusions

This research has addressed the actors of environmentally sustainable innovation ecosystems. Our findings show that the constituents of the ecosystem play an important role in adding sustainability to the green IEs. The contribution of this article to the literature on green innovation and on innovation ecosystems theory is twofold. First, we provide a mapping of the actors of green IEs and provide a description of the role played by them. To date, although studies on green innovation have recognized the role of inbound open innovation [26], authors have not provided an in-depth description of the actors that impact the green innovation development and commercialization and, most importantly, the roles played by such actors, under the innovation ecosystem perspective.

Nevertheless, some limitations should be highlighted. First, we only address outbound-in dynamics of open innovation, and we do not study the dynamics of firms' internal knowledge that crosses boundaries to impact the ecosystem (that is, inbound-out dynamics). Thus, future research could study how firms developing and commercializing green innovations impact the other actors from the IEs through, for example, knowledge spillover that is absorbed into actors' boundaries. Moreover, future research could quantitatively study the impact of different types of actors and green IEs configurations of the outputs of innovation. More importantly, future studies can address how external resources from the ecosystem impact green IEs and how such resources are orchestrated within firms.

References

1. Adner, R., Kapoor, R., 2010. Value creation in innovation ecosystems: how the structure of technological interdependence affects firm performance in new technology generations. Strateg. Manag. J. 31, 306–333. https://doi.org/10.1002/smj.821.
2. Lee, C.-W., 2007. Strategic alliances influence on small and medium firm performance. J. Bus. Res. 60, 731–741. https://doi.org/10.1016/j.jbusres.2007.02.018.
3. Fagerberg, J., Mowery, D.C., Nelson, R.R. (Eds.), 2006. The Oxford Handbook of Innovation. Oxford University Press. https://doi.org/10.1093/oxfordhb/9780199286805.001.0001.
4. Adner, R., 2006. Match Your Innovation Strategy to Your Innovation Ecosystem. Harv. Bus. Rev.
5. Nambisan, S., Baron, R.A., 2013. Entrepreneurship in innovation ecosystems: Entrepreneurs' self-regulatory processes and their implications for new venture success. Entrep. Theory Pract. 37, 1071–1097.
6. Beltagui, Ahmad, Ainurul Rosli, and Marina Candi. 2020. "Exaptation in a Digital Innovation Ecosystem: The Disruptive Impacts of 3D Printing." Research Policy 49 (1): 103833. https://doi.org/10.1016/j.respol.2019.103833.
7. Ritala, P., Almpanopoulou, A., 2017. In defense of "eco" in innovation ecosystem. Technovation 60–61, 39–42. https://doi.org/10.1016/j.technovation.2017.01.004.
8. Oh, D.S., Phillips, F., Park, S., Lee, E., 2016. Innovation ecosystems: A critical examination. Technovation 54, 1–6. https://doi.org/10.1016/j.technovation.2016.02.004.
9. Edquist, C. 2005. "System of Innovation: Perspectives and Challenges." In The Oxford Handbook of Innovation. http://ideas.repec.org/p/cpb/discus/138.html.

10. Gomes, L.A. de V., Facin, A.L.F., Salerno, M.S., Ikenami, R.K., 2018. Unpacking the innovation ecosystem construct: Evolution, gaps and trends. Technol. Forecast. Soc. Change 136, 30–48. https://doi.org/10.1016/j.techfore.2016.11.009.

11. Moore, J.F., 1993. Predators and prey: a new ecology of competition. Harv. Bus. Rev. 71, 75–86.

12. Fagerberg, J., 2018. Mobilizing innovation for sustainability transitions: A comment on transformative innovation policy. Res. Policy 47, 1568–1576. https://doi.org/10.1016/j.respol.2018. 08.012.

13. Schiederig, Tim, Frank Tietze, and Cornelius Herstatt. 2012. "Green Innovation in Technology and Innovation Management - an Exploratory Literature Review." R&D Management 42 (2): 180–92. https://doi.org/10.1111/j.1467-9310.2011.00672.x.

14. Melander, Lisa. 2018. "Customer and Supplier Collaboration in Green Product Innovation: External and Internal Capabilities." Business Strategy and the Environment 27 (6): 677–93. https://doi.org/10.1002/bse.2024.

15. Montalvo, C., Diaz-Lopez, F., Brandes, F., 2011. Eco-innovation opportunities in nine sectors of the European Economy. Brussels.

16. Boons, F., Montalvo, C., Quist, J., Wagner, M., 2013. Sustainable innovation, business models and economic performance: an overview. J. Clean. Prod. 45, 1–8. https://doi.org/10.1016/j.jcl epro.2012.08.013.

17. Mousavi, S., Bossink, B.A.G., 2017. Firms' capabilities for sustainable innovation: The case of biofuel for aviation. J. Clean. Prod. 167, 1263–1275. https://doi.org/10.1016/j.jclepro.2017. 07.146.

18. Foster, C., Green, K., 2000. Greening the innovation process. Business Strategy and the Environment, 9, 287-303.

19. Ben Arfi, W., Hikkerova, L., Sahut, J.M., 2018. External knowledge sources, green innovation and performance. Technol. Forecast. Soc. Change 129, 210–220. https://doi.org/10.1016/j.tec hfore.2017.09.017.

20. Eisenhardt, Kathleen M. 1989. "Building Theories from Case Study Research." Academy of Management Review 14 (4): 532–50. https://doi.org/10.5465/AMR.1989.4308385.

21. Eisenhardt, K.M., Graebner, M.E., 2007. Theory Building From Cases : Opportunities and Challenges. Acad. Manag. J. 50, 25–3 https://doi.org/10.2307/20159839.

22. Medeiros, Janine Fleith de, and José Luis Duarte Ribeiro. 2017. "Environmentally Sustainable Innovation: Expected Attributes in the Purchase of Green Products." Journal of Cleaner Production 142 (January): 240–48. https://doi.org/10.1016/j.jclepro.2016.07.191.

23. Chen, Jiawen, and Linlin Liu. 2019. "Customer Participation, and Green Product Innovation in SMEs: The Mediating Role of Opportunity Recognition and Exploitation." Journal of Business Research, June. https://doi.org/10.1016/j.jbusres.2019.05.033.

24. Tsujimoto, Masaharu, Yuya Kajikawa, Junichi Tomita, and Yoichi Matsumoto. 2018. "A Review of the Ecosystem Concept — Towards Coherent Ecosystem Design." Technological Forecasting and Social Change 136 (June 2017): 49–58. https://doi.org/10.1016/j.techfore. 2017.06.032.

25. Farla, Jacco, Jochen Markard, Rob Raven, and Lars Coenen. 2012. "Sustainability Transitions in the Making: A Closer Look at Actors, Strategies and Resources." Technological Forecasting and Social Change 79 (6): 991–98. https://doi.org/10.1016/j.techfore.2012.02.001.

26. Marcon, A., de Medeiros, J.F., Ribeiro, J.L.D., 2017. Innovation and environmentally sustainable economy: Identifying the best practices developed by multinationals in Brazil. J. Clean. Prod. 160, 83–97. https://doi.org/10.1016/j.jclepro.2017.02.101.

Circular Economy and Companies: Understanding the Characteristics and the Challenge of Measurement

Marina I. Baumer-Cardoso⑩**, Lucila M. S. Campos**⑩**, and Weslynne Ashton**⑩

Abstract Companies are the main actors in the move towards a more circular and less linear economy and need to understand their role and monitor their achievements. As such, defining goals and indicators for assessing the adoption of circular economy (CE) at the company-level is essential. This paper aims to define the characteristics of the CE at the company-level and to evaluate the growing cadre of company-level CE indicators, based on these characteristics. Different from other CE micro-level reviews, this paper focuses on companies and review academic and professional practice literature. The review revealed eight current indicators developed to assess the level of CE adoption in companies, four from academia and four from practice. The indicators were evaluated according to six characteristics of CE from the literature. In general, there is inadequate coverage of holistic CE characteristics among the indicators, as well as clear differences in how academics and practitioners propose that companies approach CE. While academic literature emphasizes the need for a sustainability view for CE indicators, the practice literature is concerned to measure what represents an economic advantage for the business. As a result, none of the eight indicators addressed the CE characteristics in full, mainly due to the lack of standard approaches in water, assets management and social aspects. The circularity indicators intended for companies must be comprehensive enough to capture the characteristics of CE at the company-level while being applied to different types of industry and easy to be adopted.

Keywords Circular Economy · Circularity · Company · Business · Indicator

M. I. Baumer-Cardoso (✉) · L. M. S. Campos
Universidade Federal de Santa Catarina, Florianópolis, SC 88040-900, Brazil

W. Ashton
Illinois Institute of Technology, Chicago, IL 60616, USA

© The Author(s), under exclusive license to Springer Nature Switzerland AG 2021
A. M. Tavares Thomé et al. (eds.), *Industrial Engineering and Operations Management*,
Springer Proceedings in Mathematics & Statistics 367,
https://doi.org/10.1007/978-3-030-78570-3_46

1 Introduction

The Circular Economy (CE) appears as an alternative to satisfy the economy and the environment proposing that it is possible for nature, economy, and society to co-exist in abundance by applying the increasing knowledge of the intelligence of natural systems in designing and managing products, processes, and systems [1, 2]. In this way, the challenge of the industry is not to be less destructive, but to become a positive force. However, this does not seem to be such an easy task. Despite all the normative power of the literature and institutional pressures, companies are still strongly reluctant to fully implement CE practices [3]. The adoption of CE by companies implies the development of different strategies to improve the circularity of its production system and also cooperates with other companies in the supply chain to obtain a more effective circular pattern [2]. Companies are the lead characters in the transition of the economy, it is then fundamental to understand their role and progress [3]. A company that wants to contribute to the CE needs to understand what is really within its responsibilities and sphere of influence, to visualize its current situation, to envision what the CE means for it and also to create goals. The CE screening in companies needs to reflect all the characteristics of this new way of providing products and services, demonstrating strengths, highlighting the areas for improvement in an easy and practical way, and providing transparency to investors and customers about a company's circular economy adoption [4–7].

The CE implementation can be evaluated at the micro-level (product, company, or isolated consumer), meso level (eco-industrial parks) and macro-level (cities/regions/nations) and the success of CE policies requires efforts at these three levels [8]. Quantifying requirements of CE at all levels is critical to understanding the current situation, defining points for improvement, and supporting the transition to a CE [1, 9]. Measurement is one of the barriers pointed out in the literature on how to perform and evaluation of the transition from linear to a circular economy [1, 4, 10, 11]. The number of micro-level indicators is growing, showing there is an effort from academia and professional practice to fill this gap [9]. Several authors proposed indicators to assess the micro-level [12, 13]. As the micro-level comprises companies, products and consumers, each of these entities has its role in the transition for a new economy, which must be examined separately.

To define the requirements that involve the company-level is to translate the principles of CE to the scope of activities of the company and illustrate its responsibility in the new economy. According to the framework proposed by Elia et al. [1], the requirements to be measured embrace: (i) Reducing input and use of natural resources, (ii) reducing emission levels, (iii) reducing valuable materials losses, (iv) increasing share of renewable and recyclable resources, (v) increasing the durability of products. Although some authors identified a growing number of indicators for measure CE adoption, the development indicators applied to business are still a barrier [10, 12]. Some authors have tried to fill this gap and proposed indicators for the company-level [4–7, 14–16]. It is clear that each indicator measures different practices of CE and a consensus of what is CE at the company-level does not exist [1, 11]. Quantifying the adoption of the CE in companies is fundamental to understand the current situation, offset goals for improvement, and support the transition [1, 12, 13]. To have a robust and effective indicator system that contributes to the improvement of managerial processes of CE at the company-level, clarity on what are the CE practices for business needs to be defined first. In this sense, this research aims to define the CE characteristics for companies and to analyze whether the existing indicators are capable of assessing all the characteristics.

The paper is structure as follows. First, this section provides a short literature analysis of CE at the company-level. Section 2 details the methodology used to define the characteristics of CE at the company-level and to identify the existing indicators. The results are presented in Sect. 3, with details about the indicators founded in the literature, and with the analyses of the alignment between CE characteristics and the company-level indicators. Discussion of the results is presented in Sect. 4 followed by a conclusion in Sect. 5, which also presents directions for future research.

2 Methodological Procedures

To answer the research question mentioned above, the research method applied in this study consisted of the definition of CE characteristics at the company-level, identification of CE indicators for companies, and then analysis and discussion of the findings. The characteristics for assessing the CE level of companies were defined through an academic literature review by building on the characteristics proposed by previous CE studies. The framework proposed by Elia et al. [1] was chosen because it objectively and in a simple way encompasses various characteristics of CE at the micro-level. It has become a leading resource referenced by other authors in similar research [5, 11]. However, the literature suggests that some important characteristics of CE were not included in the mentioned framework. To fully capture the CE at the company-level, we proposed to update three of Elia's characteristics and add one more that reflects employee wellness and equity. Table 1 lists the defined six characteristics of CE at the company-level, highlighting the added parts to the framework. These defined characteristics were considered as evaluation criteria for the analysis of existing indicator systems.

The characteristics increase the share of renewable resources and fewer material/losses residuals were updated to capture water used by companies. Water is essential in many production processes, making it necessary to address this natural resource in a CE indicator for companies [15]. Companies are significant users of water locally and globally and need to be prepared for the challenges of using water in their processes. The third updated characteristic is keeping the value of products and assets. Several studies argue that companies should adopt strategies to increase the lifetime of their products [17], but barely is this said about companies adopting these strategies for their durable equipment and assets as well. It is widely understood that asset management strongly impacts finances, both in the acquisition and in its

Table 1 Characteristics of CE for company-level

Characteristic	Details
Reduced input and use of natural resources	• minimized and optimized exploitation of raw materials, while delivering more value from fewer inputs (materials, energy and water) • reduced import dependence on natural resources • efficient use of all-natural resources (material, energy, and water) • minimized overall energy and water use
Increased share of renewable and recyclable resources	• non-renewable resources replaced with renewable ones within sustainable levels of supply (material and energy) • increased share of recyclable and recycled materials • closure of material loops • sustainably sourced raw materials • increased share of reused water, with minimal or no treatment, within and outside the fence for the same or different processes • increased share of recycled resources and wastewater (treated by membrane or reverse osmosis to very high quality) within and outside the fence
Reduced emissions	• reduced emissions throughout the full material cycle through the use of less raw material and sustainable sourcing • less pollution through clean material cycles • eliminate or minimize any emission harmful to the environment, whether in the form of gases, hazardous material, toxic substances, land, and water pollution
Fewer material and *water* losses/residuals	• build-up of waste minimized • incineration and landfill limited to a minimum • dissipative losses of valuable resources minimized
Keeping the value of products, component, materials, and *assets* in the economy	• extended product lifetime keeping the value of products in use • reuse of components • value of materials preserved in the economy through high-quality recycling • extended assets lifetime through maintenance, reuse parts, and improvements • increased product-service adoption for equipment and other assets

(continued)

Table 1 (continued)

Characteristic	Details
Promote employee wellness and equity	• provide a healthy and safe work environment for the employee • provide employee satisfaction with the work environment • invest in employee training and development • committed to diversity, equity and inclusion through promoting equal opportunity-affirmative actions

maintenance. Applying the concept of keeping the value to this category also brings benefits to the company as well as to the supply chain of this asset. Finally, the last characteristic updated was reduced emissions. The purpose of the change was just to make it clearer that the emission refers to any type of toxic substance that the company may have.

Regarding the social issues, as the unit of analysis is a company, the social dimension concerns the company's employees and was developed with the regenerative concept of CE in mind. In biology, regeneration is the process of renewal, restoration, and growth that makes genomes, cells, organisms, and ecosystems resilient to natural fluctuations or events that cause disturbance or damage [18]. In this sense, regeneration suggests helping people within the company to be more resilient to changes, whether positive or negative. The people's resilience at the company-level is then achieved through promoting employee wellness and equity.

The second step comprised the identification of CE indicators for companies through a literature review. Since the CE is a concept that has received attention from academia and practice [17], in this study, the CE indicators were analyzed from both sources. The literature review was carried out following an adaptation of the methodological procedures suggested by Corona et al., Kristensen and Mosgaard [11, 17].

The academic research was investigated through two international databases Scopus and Web of Science. Only publications from peer-reviewed journals, published in English, between 2006 and May/2020 (date of this research) were considered. 2006 was chosen as the reference year because previous systematic reviews of CE indicators identified an increase in publications from 2006, with only a few studies before 2006 [17, 19]. The keywords were also based on previous publications that carried out literature reviews on the topic [9, 11, 17] and have embraced three research axes: circular economy ("Circular Economy", circularity), indicator (indicator*, indices, index, measur*, assess*, tool, indicator) and company (company, companies, industry, industries, business, businesses), which were searched in titles and abstracts for the purpose of selecting qualified literature.

Since the professional practice publications are not published in the same databases of academic works, these publications were investigated by exploring the websites of organizations engaged in the CE and by searching for studies, reports, tools using Google as a search engine. The keywords for the search involved the words "circular economy" and "indicator". Others literature reviews papers also were used as a reference to find grey literature indicators. All the selected publications were analyzed considering if they (i) propose a CE measurement, (ii) focus on the company and (iii) were not specific for a sector or industry. According to this and the searching criteria described before, 8 indicators were identified, 4 from academia literature and 4 from practical literature. Section 4 details the indicators founded.

3   Company-Level Indicators

The review found eight indicator systems that are focused on CE at the company-level, as presented in Table 2. The indicators are Sustainable Circular Index (SCI) [6], Expanded Zero Waste (EZW) [16], Circularity Measurement Toolkit (CMT) [5], Indicators for Organizations considering Sustainability and Business Models (IOSBM) [4], Material Circularity Indicator (MCI) [20], Circularity Facts (CF) [14], Circulytics (CYTICS) [7], and Circular Transition Indicators v1.0 (CTI) [21]. The indicators were categorized based on the number of indicators considered, the type of indicators used (quantitative and/or qualitative), and the type of the final representation of the indicator used: (1) single final number or another grade that represents CE, or (2) a set of indicators. Information about the final representation of the indicator was included for the type 1 indicator.

The relatively low number of indicators found (8) indicates that research on CE indicators at the company-level is currently underdeveloped. Other literature reviews at the micro-level support the same result [9, 11, 17]. Professional practical contributions represent half of the total of indicators founded. The shift of CE moving from a strategic to a practical level may be the reason for the increased focus on micro-level indicators [9, 17].

Table 2   Summary of the indicators founded that focus on CE at the company-level.

Details	SCI	EZW	CMT	IOSBM	MCI	CF	CYTICS	CTI
A/P	A	A	A	A	P	P	P	P
Author	[6]	[16]	[5]	[4]	[20]	[14]	[7]	[21]
Type	1 quant	2 quant, qual	1 qual	2 quant, qual	1 quant	1 quant, qual	1 quant	2 quant, qual
Number of indicators	17	31	36	26	2	9	30	6
Final representation	0–1	–	0–8	–	0–1	0–100%	A + to E	–

Academic (A) and practice (P). Type: (1) single indicator/index; (2) indicators set; Quant = quantitative indicators, Qual = qualitative indicators

All eight indicators include quantitative indicators, and five also consider qualitative indicators. The majority of the indicators (five) proposes to assess CE at the company-level as a single number or grade (that is, an index). The orientation towards an index value can be understood as a tendency to want an indicator that can enable benchmarking within companies and rankings across companies.

Although all the indicators state that are they developed for any type of company, most were concentrate on manufacturing firms. Only the Sustainable Circular Index (SCI) indicator declares that is applied only for manufacturing industries and suggests each company uses the Delphi method to define the optimal weight for each indicator. Expanded Zero Waste (EZW) indicator proposes several indicators but leaves it up to the company to select what makes sense to their business. Circularity Measurement Toolkit (CMT) indicator also is flexible as a company is given the option to exclude situations that do not apply to it. Circulytics (CYTICS) defines the indicators considering the industry (manufacturing, service, waste and wastewater treatment, energy provider, financial institution). For the Circular Transition Indicators (CTI) indicator, of the six defined indicators, three are optional. The flexibility in responding to the indicators shows the concern for developing indicators that can be applied across all types of companies.

3.1 Alignment Between CE Characteristics and the Company-Level Indicators

Table 3 presents an overview of the characteristic's coverage by the indicators. Each characteristic is analyzed separately below.

Reduced input and use of natural resources: It should be noted that a CE not only closes the cycle but is also concerned with 'decreasing the flow' of resources

Table 3 Characteristic's coverage by the indicators analyzed. X = fully attends the characteristic, and (X) = partially attends the characteristic

Characteristics	SCI	EZW	CMT	IOSBM	MCI	CF	CYTICS	CTI
Reduced input and use of natural resources	(x)	x	x	x				
Increased share of renewable resources	(x)	(x)	x	x	(x)		(x)	(x)
Reduced emissions	x	(x)	x	x				
Fewer material and water losses/residuals	(x)	(x)	(x)		(x)	(x)	(x)	(x)
Keeping the value of products/assets	(x)	x	(x)	(x)	(x)	(x)	x	(x)
Promote employee wellness and equity	x	(x)	(x)	(x)			(x)	

and 'restricting the flow of resources' [22]. Although resource-efficiency is a well-known strategy to use fewer inputs in products/processes, it is not prioritized in CE indicators [17]. Only EZW, CMT, and Indicators for Organizations considering Sustainability and Business Models (IOSBM) involve indicators about reducing the amounts of material, energy, and water consumed. SCI measures the amount used for energy and water and none of the professional practice indicators include reduction of inputs.

Increased share of renewable and recyclable resources: The source type is a common measurement in almost the indicators founded, with the percentage of renewable sources used in relation to the total amount used for input (energy, material, or water) as the prevalent way to measure. Energy and material renewable resources are measured by all indicators. However, the indicators fail when analyzing the source for water. SCI and MCI assess the resource only for the material. EZW, CYTICS and CTI cover material and energy resources. CMT covers resources for all inputs, including water assessed as reused from another company. For IOSBM, the water assessment is not so explicit, but it is understood that it includes the reuse of water by other companies.

Reduced emissions: Although GHG emissions reduction may be a consequence of adopting renewable sources [22], it is not the only type of emission harmful to the environment. None of the professional practice indicators embrace the topic of reduced emissions in their indicators and it can only be found in the academic indicators. EZW proposes to avoid GHG emissions alone and it does not mention any other toxic substance. CMT goes deeper and also considers the reduction of pollutants as fertilizers, pesticides, petrol, diesel, and natural gas. IOSBM measures the reduction of toxic substances and SCI in the reduction of hazardous waste, which both can be understood as pollutants in general.

Fewer material and water losses/residuals: This characteristic aims to cover the quantity reduction (less amount of waste in general) and also the type reduction (less landfill or incineration), following the same premise for input. Most of the indicators (MCI, SCI, CYTICS, CTI) measure only the type reduction for materials, not water. EZW and Circularity Facts (CF) cover the amount and the type reduction but only for materials. Only CMT measures the type reduction for materials and water and examines if the water is treated internally or sent to another company. Although the characteristic of waste is strongly addressed by the indicators, in general, they fail in measuring the amount reduction of waste and also wastewater.

Keeping the value of products, components, materials, and assets in the economy: Keeping the value of products and equipment follows the same situation as the previous characteristics. Most of the indicators cover strategies to prolong the product life cycle, such as recycling, biochemical feedstock use, composting, anaerobic digestion, biofuel, byproduct use/sale, and reuse. However, indicators often focus only on the technological cycle. CF and CYTICS are the only indicators that explicitly prepare their indicators to include biological cycles. Concerning equipment, the indicators neglect to analyze how companies can also prolong the lifetime of their equipment and other durable assets. Only EZW and CYTICS regard this issue. CYTICS questions if the company has a plan in place for the end of life of

PPE assets (property, plant, and equipment: physical assets with a use period of one year or more) that adheres to circular economy principles. EZW measures the percentage of products or equipment reused, as well as economic indicators, such as the Fair market value (FMV) of reused equipment, FMV of repurposed equipment and long term ROI of upgraded equipment/systems.

Promote employee wellness and equity: The promotion of employee wellness and equity is well covered by SCI, precisely because it is an indicator concerned with the vision of sustainability. The other indicators cover only some characteristics in this area, such as training (SCI, EZW, CMT, CYTICS, IOSBM), communication (EZW, CYTICS), and job creation through CE practices (EZW, IOSBM). EZW reinforces the importance to have employee engagement indicators based on principles of "inform, educate, empower, and recognize" [16]. The indicator also suggests that the company measure other social impacts like scientists supported, students benefitted, and NGOs served. Training is a well-known indicator in the social dimension, especially in sustainability works.

4   Discussion

Indicators are essential to document, measure progress and set goals towards CE. Currently, indicators for CE at the micro-level are less developed [9, 11, 17], and even less attention is paid to the company-level. This review defined the CE characteristics at the company-level and presented an overview and categorization of eight company-level indicators. Although the majority of micro-level indicators originate from academic publications [9, 11, 17], in this review the number of publications demonstrates interest from researchers and practice. And both academy and practice similarly cover aspects of the increased share of renewable resources, the fewer material, and water losses/residuals, and keeping the value of products and equipment.

The reduction of inputs is one of the least addressed CE characteristics. While the main focus of measurement is on the type of source and waste generated, monitoring the reduction in the amount of input in a company is something still underdeveloped. CE is not only about choosing the best sources of input but also about using them responsibly and consciously [2]. Therefore, companies need to be more aware of the quantity needed for their production.

The characteristics of increasing the share of renewable sources and having fewer material losses are often addressed within the indicators. Material and energy are well covered and have a common measurement indicator, as a percentage or ratio of the renewable source/waste to the total amount. However, the indicators fall short when analyzing both source and disposition of water. The circularity of water use is not commonly assessed in the indicators analyzed. Even though water is a common input for all types of companies and there are significant regional limitations on freshwater availability, this concern has failed to materialize in indicators [9, 11, 17]. WBCSD [21] already recognizes the water importance and declared that they intend to include a water indicator in an updated version of their CTI indicators in the year 2021.

Emissions reduction is widely addressed in the CE studies. However, it is noticeably absent from indicators, even though there are common methods of measurement. One reason may be that emission reduction is the consequence of utilizing a renewable source of energy/material [22]. However, the company may have other forms of emission, and it is understood that the emission needs to be monitored closely and individually.

The strategies of keeping the value of products are widely addressed by the indicators, even though all of them do not address both the technical and biological cycles. As most indicators are based on manufacturing industries, industries with materials from the biological cycle may not be able to use the indicators. Manufacturing companies consume more non-renewable materials and receive more attention in current indicators. But companies with biological materials also deserve attention when developing an indicator for CE because, despite having the advantage of already using a renewable source of material, these same companies may not have practices related to the amount consumed, use of energy and water, or even the best waste disposal practices.

Regarding the indicators for employee wellness and equity, training is the most common indicator. Training is key to engage employees and make the CE actions happen [16]. Even though new jobs may emerge within a company due to the adoption of CE practices, jobs can also occur due to the creation of new companies in the supply chain when open loops emerge. Although job creation can be expected to be local at the micro-level [17], it is not clear yet if this situation would happen inside a company or because the supply chain would demand new business models, which may justify measuring job creation at a meso or macro level.

Concerning the total number of indicators, it is possible to identify that indicators from the academy (SCI, EZW, CMT, IOSBM) have a greater number of indicators in relation to indicators by professional practice literature (MCI, CF, CYTICS, CTI). This is because academic indicators cover more characteristics of CE than the professional practice literature and have shown concern with going beyond the CE and have a more strategic vision thinking about the ultimate goal of sustainability. Only the indicators proposed by the academy reinforce the three pillars of sustainability when approaching CE at the company-level [4–6, 16]. The limited inclusion of the three dimensions of sustainability in indicators at the company-level was also reported in the work of [17]. Even though the indicators derived from professional practice partially address some economic and social characteristics, it is not the purpose of these works [7, 14, 20, 21]. It appears that these business-led indicators are more focused on measuring what can be translated in monetary values and represents an economic advantage for the business, like reduce inputs of material and energy, and less waste. Kristensen and Mosgaard and Geissdoerfer et al. [17] and [23] also highlight in their works that the CE aims to benefit economic actors above to equally benefit also the environment and society, which is the sustainability proposal.

Even though not reaching 100% coverage in all characteristics, the analysis shows that the authors are concerned with developing indicators that seek the greatest possible coverage when it comes to measuring the degree the CE in companies. However, the small number of indicators found shows the low maturity of CE indicators for companies. If CE is not measured in companies, it is not manageable and makes it difficult to adopt. In this sense, indicators aimed at measuring the level of adoption of CE in companies are essential to support the transition to CE.

5 Conclusions

This research aims to define the CE characteristics for companies and to analyze whether existing indicators are adequately covering these characteristics. This paper contributes to the definition of the CE characteristics at the company-level and then the analysis of the existing indicators. The results demonstrated that the existing indicators fulfill the majority of the requirements but fail in indicators aimed at reducing input resources in terms of the amount used, the consumption and disposal of water, the control of greenhouse gas emissions, the management of the company's assets company focused on increasing lifetime, and indicators focused on employee wellness and social equity. As a result, none of the indicators met all the characteristics of the CE defined for the company-level, which suggests there is a significant gap for guiding companies on how to adopt the CE and measure its performance. Also, it is clear that academy and professional practice follow different paths to measure circularity in companies. Companies are the main actors for the transition to CE and it is essential to develop simple but robust indicators, easy to apply, and applicable across any industry segment.

This research has limitations related to the use of search engines and the methodological choices concerning the selected search strings, filters, and databases. Therefore, an agenda for future work is defined here: (i) to investigate and define the best indicators for those CE characteristics not well covered by current indicators; (ii) to develop indicators capable of broadly capturing the CE key characteristics defined in this work, which would be easy to apply and flexible for the different types of business, including business with technological and biological materials; (iii) to explore the boundaries of job creation at the company, supply chain and region-levels of a CE perspective and how to measure it; as well as investigate deeper how the economic pillar of sustainability transcribe at the company-level, could also be suggestions for future studies.

Acknowledgements This study was financed in part by the Conselho Nacional de Desenvolvimento Científico e Tecnológico of Brazil (CNPq) and the Coordenação de Aperfeiçoamento de Pessoal de Nível Superior Brasil (CAPES) – finance code 001.

References

1. Elia, V., Gnoni, M.G., Tornese, F.: Measuring circular economy strategies through index methods: A critical analysis. J Clean Prod, 142, 2741–2751 (2017).
2. EMF: Towards a Circular Economy—Economic and Business Rationale for an Accelerated Transition. (2015).
3. Linder, M., Williander, M.: Circular Business Model Innovation: Inherent Uncertainties. Bus. Strateg. Environ. 26, 182–196 (2017).
4. Rossi, E., Bertassini, A.C., Ferreira, C.S., do Amaral, W.A.N., Ometto, A.R.: Circular economy indicators for organizations considering sustainability and business models: Plastic, textile and electro-electronic cases. J Clean Prod 247 (2020).
5. Garza-Reyes, J.A., Valls, A.S., Nadeem, S.P., Anosike, A., Kumar, V.: A circularity measurement toolkit for manufacturing SMEs. Int J Prod Res, 57, 7319–7343 (2019).
6. Azevedo, S., Godina, R., Matias, J.: Proposal of a Sustainable Circular Index for Manufacturing Companies. Resources, 6 (2017).
7. EMF: Circulytics - Method Introduction (2020).
8. Kirchherr, J., Reike, D., Hekkert, M.: Conceptualizing the circular economy: An analysis of 114 definitions. Resour Conserv Recycl, 127, 221–232 (2017).
9. Saidani, M., Yannou, B., Leroy, Y., Cluzel, F., Kendall, A.: A taxonomy of circular economy indicators. J Clean Prod, 207, 542–559 (2019).
10. Galvão, G.D.A., Nadae, J., Clemente, H.D., Chinen, G., Carvalho, M.M.: Circular Economy: Overview of Barriers. Procedia CIRP, 73, 79-85 (2018).
11. Corona, B., Shen, L., Reike, D., Carreón, J.R., Worrell, E.: Towards sustainable development through the circular economy—A review and critical assessment on current circularity metrics. Resour Conserv Recycl, 151 (2019).
12. Cayzer, S., Griffiths, P., Beghetto, V. Design of indicators for measuring product performance in the circular economy. Int J Sustain Eng, 10, 289–298 (2017).
13. Linder, M., Sarasini, S., van Loon, P.: A Metric for Quantifying Product-Level Circularity. J Ind Ecol, 21, 545–558 (2017).
14. UL: Standard UL3600. https://industries.ul.com/environment/certificationvalidation-marks/circularity-facts-program. Accessed 6 Jan 2019 (2019).
15. WBCSD: Business guide to circular water management: spotlight on reduce, reus and recycle. (2017).
16. Veleva, V., Bodkin, G., Todorova, S.: The need for better measurement and employee engagement to advance a circular economy: Lessons from Biogen's "zero waste" journey. J Clean Prod, 154, 517–529 (2017).
17. Kristensen, H.S., Mosgaard, M.A.: A review of micro level indicators for a circular economy e moving away from the three dimensions of sustainability? J Clean Prod, 243, (2020).
18. Ichihashi, Y., Hakoyama, T., Iwase, A., Shirasu, K., Sugimoto, K., Hayashi, M.: Common Mechanisms of Developmental Reprogramming in Plants—Lessons From Regeneration, Symbiosis, and Parasitism. Front Plant Sci, 11, 1–10 (2020).
19. Merli, R., Preziosi, M., Acampora, A.: How do scholars approach the circular economy? A systematic literature review. J Clean Prod, 178, 703–722 (2018).
20. EMF: Circular Indicators: An approach to measuring circularity. Methodology. (2015).
21. WBCSD: Circular Transition Indicators. (2020).
22. EMF: Renewable Materials for a Low—Carbon. (2018).
23. 23. Geissdoerfer, M., Savaget, P., Bocken, N.M.P., Hultink, E.J.: The Circular Economy – A new sustainability paradigm? J Clean Prod, 143, 757–768 (2017).

Reviewing Photovoltaics and Electric Vehicles: Synergy and End of Life Management

Rafael Marcuzzo⬤, **Lucila M. S. Campos**⬤,
Mauricio Uriona Maldonado⬤, **and Caroline Rodrigues Vaz**⬤

Abstract Energy supply and mobility are necessary functions of economic development and have become global latent themes concerning sustainability. In the electricity sector renewable energy technologies, such as solar photovoltaic (PV), have grown significantly over its competitors in terms of investment in the past 20 years. Still, adoption of the technology has not grown quickly enough to decarbonize the sector. In the mobility sector, however, combustion engines have been replaced with electric vehicles (EV), illustrating a greater potential to mitigate emission of polluting gases. Still, the production of EV technology can yield high emissions. Instead of either technology working separately, both PV and EV may be able to cooperate in a manner where PV systems power clean electricity to produce and charge EV batteries until they return to a stationary storing system (i.e., second life) for PV systems. The following study investigates the use of EV batteries for PV systems, assessing the possible toxic effects on the environment from manufacturing, using, and disposing the technology. A review of peer-reviewed articles shed light on possible hazards of clean technologies in the energy and mobility sectors. Findings suggest the diffusion of clean technologies must be accompanied by policies that guarantee responsible disposal of them. The study concludes that advanced technological mechanisms for renewable energy may have significantly negative and sometimes obscured environmental impacts, which must be considered in future policies advocating for those technologies.

R. Marcuzzo
Graduate Program in Production Engineering, Universidade Federal de Santa Catarina, Florianópolis, Brazil

L. M. S. Campos (✉) · M. U. Maldonado
Department of Production Engineering and Systems, Universidade Federal de Santa Catarina, Florianópolis, Brazil
e-mail: lucila.campos@ufsc.br

M. U. Maldonado
e-mail: m.uriona@ufsc.br

C. R. Vaz
Department of Textile Engineering, Universidade Federal de Santa Catarina, Blumenau, Brazil
e-mail: caroline.vaz@ufsc.br

Keywords Photovoltaics · Electric Vehicles · End of life battery management

1 Introduction

Growing energy needs and depletion of fossil fuel resources require more searching for sustainable alternatives, including renewable energy sources and storage technologies. It worries that significant decreases in energy use, as for instance electricity, in a short period of time would be difficult to achieve in modern society [1, 2].

From the mobility point of view, the traditional vehicles powered by internal combustion have also received more attention from an environmental perspective in the last years. Regarding only transport and electricity generation together they are responsible for more than 50% of the carbon emissions associated with combustion, but this percentage can decrease rapidly with the generation of solar photovoltaic (PV) and the use of electric vehicles (EV) [3].

Regarding the reduction of greenhouse gas emissions, such as CO_2, if the country has more renewable energy sources in the matrix the more environmental advantages are obtained through replacing vehicles powered by internal combustion engines with EVs. By the way, if a country has a thermal power as a base source in its electricity matrix, for example, the emission reduction resulting from the replacement of vehicles powered by an internal combustion engine with EV can be canceled by the plants, because more electricity will need to be generated to meet EV energy demand [4, 5].

One of the Achilles' heels in cleaner technologies and energy industries, which try to achieve stability in socio-technical regimes, is found in storage systems, as for instance electrochemical batteries [6]. Batteries are important elements for transitions in these great social functions, energy and mobility, however, technological incipience and chemical characteristics make batteries both a savior and a villain at the same time.

In this manuscript it is discussed the PV and EV technologies, both related to end of life management. A literature review in the main scientific databases provided the structuring of the knowledge update according to the specificities of the two industries (PV and EV) and also in their integration (PV storage systems with EV's end of life batteries). This article is organized in this introduction which is followed by section two on methodological procedures, and sections three and four present the results and discussions, and final considerations, respectively.

2 Methodology

To start the literature review, first there was the creation of a database of scientific documents on PV, EV and end of life, preliminary random searches were carried out in order to identify some more relevant works [1, 7–11] and visualize the research field from an already formed perspective in order to add something to the existing knowledge. This initial process was also crucial in dimensioning search strings, which, after improvements and tests and using Boolean operators, was defined as: "end of life" AND "solar PV" OR "solar cell" OR "solar panel*" AND "electric vehicle*", in the spectrum of titles, abstracts and keywords. Temporality considered the entire available period of the databases, with the selected results ranging from 2000 to 2019.

First selection resulted in 55 complete articles found—Web of Science (14), Scopus (26) and Google Scholar (15) databases—and organized in the reference manager software EndNote®. After reading the abstracts, there were still documents unrelated or not being complete article type or not in English. At the end of these filters, 27 articles remained.

The 27 selected articles were read completely for collecting useful information. Still, this complete reading process demonstrated that some materials were not adhering to the theme (very technical or theoretical), resulting in a database of 17 articles aligned and available for the final analysis. Microsoft Excel® software was used for collecting and organizing information about objectives, methods, results and contributions and limitations of the articles. Much information was analyzed and discussed in an organized way to offer an understanding of the theme, considering the systematization used.

3 Results and Discussions

This section presents the findings of the content review of the portfolio of selected articles on PV, batteries and EV from the perspective of the environmental impact of manufacturing, using and end of life of the equipment. It allowed the identification of focus patterns of the analyzed studies, organized in four sections: (i) EV's batteries, (ii) PV, (iii) EV's batteries as PV's storage systems, and (iv) sustainability energy transitions.

3.1 Electric Vehicles and Batteries: Heroes or Villains?

The technological revolution of the last centuries was mainly fueled by variations in the combustion reaction, the fire that marked the dawn of humanity. But it comes at a price: burning brings emissions of carbon dioxide, contributing to global climate changing. For the sake of future generations, there is an urgent need to reconsider the use of energy in everything from barbecues to jet planes and power plants [10]. But how can scientists achieve the performance required by each application? With many other portable devices being developed, how batteries will power them within more sustainable perspectives?

Questions about the sustainability of energy and mobility abound, but the use of EVs is an alternative that is evaluated as positive. At least, it is beneficial compared to the use of the fossil fuel mechanism used on a large scale in our society in transportation. Is this candidate for mitigating the accelerating effects of global warming so good? Well, the fact is that EVs are hostage to the use of energy storage systems, which deal with many toxic substances. Besides that, evaluation of expected impacts of the introduction of EVs in the electrical system also found that the disadvantages lie in the high cost of the batteries, in the relatively long charging time and in the accuracy of the useful life of the latter [4].

Being able to predict the lifespan of batteries is of paramount technical and commercial importance for planning systems, selecting the most suitable battery, determining operating conditions and battery replacement intervals [9]. They also affirm that a successful forecast requires knowledge of the aging processes within a battery that lead to a loss of performance and the stress factors that induce aging and influence the rate of aging.

Electrochemical cell manufacturing is responsible for the main energy and environmental impact of battery manufacturing. The analysis of this step, which represents a key part of the battery's life cycle, demonstrates that the total impact of the battery can be reduced by increasing efficiency and decreasing the carbon emissions of the manufacturing process. Measurements of energy and environmental impacts of sodium and nickel chloride batteries resulted in the operating stage having the greatest energy impact (55% and 70% of the total) and the manufacturing stage, particularly of electrochemical cells, contributing with the greatest environmental impact (more than 60% of the total) [12]. These numbers can be optimized by reducing direct energy consumption or increasing energy from renewable sources.

The use of lithium is the majority as a chemical component of batteries and the great demand for this element could be met by recycling, which has already proved its value with lead-acid batteries. However, it is not certain whether the next generation of batteries can be successfully integrated into an energy market that is currently linked to global warming, as its construction process is dependent on electrical matrixes that are sometimes non-renewable, in addition to the toxicity of the improper disposal of battery and waste from battery production processes [10].

More generally, energy and environmental sustainability of lithium-ion batteries regarding elemental abundance, toxicity, synthetic methods and scalability, highlight the importance of battery recycling, especially because society is more aware than ever about sustainability issues and less willing to compromise future technologies. Scientists even share a certain alignment and commitment to evolve research with the appeal of sustainability, but the discrepancies are salient. However, in fact sustainability is being considered as an extra dimension, in addition to the structure, composition and morphology in the design of new materials [1].

More precisely, it takes more than 400 kWh to make a 1 kWh Li-ion battery, resulting in the emission of around 75 kg of CO_2—as much as burning 35 L of gasoline. In comparison, the production of 1 kWh of electricity from coal produces about 1 kg of CO_2. Although the CO_2 generated by the manufacturer of batteries is undesirable, the energy cost associated with use is still a major concern [1].

Concerns about the increasing usefulness of batteries have led to the use of partially degraded lithium batteries used first in automotive applications (also known as second life batteries), which are becoming more available for secondary applications as the market share of plug and hybrid electric vehicles grows [13]. The EVs' recovered batteries have technological, economic and environmental opportunities to improve energy efficiency of systems within supply and storing. Batteries can be reused in stationary applications (as will be discussed in Sect. 3.3) as part of a smart grid, for example, to provide energy storage systems for charge leveling, and increasing life of residential or commercial energy systems [14].

Increasing batteries' lifetime is the key issue to extend its use. With the intense process of charging and discharging, batteries reduce its life, which is a major problem in terms of environmental degradation and price. Other concerns are also considered in research on batteries such as cost, size (weight), energy density and recharging capacity [4].

Despite the accurate forecasting of a battery's life requiring knowledge of aging processes and the availability of battery models, user requirements, operating regimes and operating conditions of batteries should be linked to aging processes and loss of performance. Examples of criteria linked to these requirements are the user's load profile, energy availability over time, ambient temperature and installation conditions, design and battery size [9].

Undoubtedly batteries are an essential part of energy transition and can bring advances in protecting the environment through increasing electric vehicles on road transport. However, there is need for precaution because increasing electric vehicles may increase the need for generation coming from fossil fuels, in the sense of charging the batteries will require more electricity from the grid (it could be from renewables only, but most electrical markets and systems are still having initial contact with this option). The good notice is that there is already knowledge on how to move forward this conflict: first is that CO_2-free electricity has to be used to manufacture batteries, second is that batteries' energy efficiency of has to reach levels of commercial viability, and third is that recycling systems have to be more intense in helping to decrease costs of raw material in manufacturing chains [1].

Performing tasks to align energy systems to societies and technologies is not easy, but supporting EV diffusion can provide a path to cleaner electricity by leveling the energy demand cycle and lessening the need to add generation to the grid (as opposed to the previous paragraph), but it depends on a costlier process. EVs can be recharged during the night, outside peak hours, offering less risk of an increase in demand and delivering power to the grid during times of greater demand thanks to batteries, but for that investments in smart networks and systems are required, changing habits and creation of strong policies [4].

In a nutshell, detailed models that describe how to provide alignment with sustainable practices regarding constructive and operational technical and economical issues are needed to expand the discussion on how to address the still high cost of batteries. Nevertheless, fame and fortune await anyone who proposes a viable alternative to fossil fuels [9, 10].

3.2 Sustainable and Secure Photovoltaics

Photovoltaic solar energy appears to contribute to reducing greenhouse gas emissions and sustainable development. However, considering the prediction of its rapid growth, some situations shall jeopardize its benefits, as for instance the use of potentially toxic substances and manufacturing processes that present health and safety problems. The data gaps in waste management of photovoltaic panels concern not only the end of life phase, but also the manufacture of defects and the generation of waste in the production process [15].

The disposal of photovoltaic modules (which useful life is around 30 years) must be done in the most environmentally friendly way possible and for that there must be specific and rigorous legislation and inspection. Environmental regulations can determine the cost and complexity of dealing with end of life photovoltaic modules, as long as they are characterized as hazardous, which would generate special requirements for material handling, disposal and maintenance of records and reports. This would increase the cost of disposal, however the improper disposal in municipal landfills is much worse, due to the existence of quantities of regulated materials in the modules, such as Cadmium, Lead and Selenium [7].

There are also information deficits about some sensitive life cycle indicators and environmental impacts of photovoltaic modules, along with incomplete information on toxicological data and studies of exposure of workers to different chemical and physical risks. Due to technological advances in the photovoltaic industry it is possible to reduce greenhouse gas emissions and comply with current principles of sustainable development. However, in order to achieve these global sustainability ambitions—and create and implement instruments for this (to measure and to analyze energy systems and its real impacts)—joint efforts by manufacturers, workers, scientists and government officials are essential [15].

A study to test the impacts of photovoltaic energy compared the technologies of traditional public lighting connected to the central electrical grid with an autonomous photovoltaic system (highlighting the expensive process of expanding the grid to reach rural areas and highlighting the priority of analyzing the location and climatic conditions, since renewable energy generation systems are sensitive to the environment) in Lebanon. The traditional system has less environmental impact than the autonomous solar system considering raw material extraction and production phase, mainly due to the presence of lead and other electronics used in the solar system. However, that difference is compensated because the traditional system consumes a significant amount of energy from the power grid during the use phase [16].

Within solar cell manufacturing process the impactful categories most affected (more than 90%) are mineral extraction (due to the copper used in mining and refining operations), non-renewable energy (due to the use of natural gas, coal and crude oil), respiratory inorganics (due to particulates produced by the combustion of oil and diesel for electricity production), and carcinogens (due to atmospheric emissions, for example, of hydrocarbons, caused by natural gas and oil production) [16].

In addition, the installation, maintenance and dismantling of photovoltaic systems pose risks to the safety of workers (for example, falls from height, ergonomic risks and injuries related to the handling of solar panels and risks of electric shock). The electrical risks are the most complex, involving both the direct current circuit, associated with the photovoltaic panels and their wiring complex, and the alternating current circuit, associated with the inverters and their cabling to the public electrical network or batteries and generators [15]. Regarding any aspect of these analyzed photovoltaic issues, awareness of the best organizational and environmental practices is necessary [7, 17].

3.3 Second Life Batteries

Batteries are one of the most expensive components of electrical systems and can make projects unfeasible. Nevertheless, many types of autonomous photovoltaic systems utilize batteries to mitigate irregularities in solar irradiation (intermittency and seasonality) [8]. However, the storing possibility is not too affordable to the most prosumers and a rapid increase in the uptake of domestic photovoltaic solar energy without storing system raises concerns about intermittent exports of electricity to the grid and related balancing problems [5]. A micro generation system that combines, for example, photovoltaic solar energy, heat and power plants and battery storage could potentially alleviate these concerns while improving energy self-sufficiency in homes. In doing so, the micro generation system would provide significant improvements in all environmental impacts compared to conventional energy supply, ranging from 35% for fossil fuel depletion to 100% for terrestrial ecotoxicity, regarding operational phase [18].

From the point of view of performance and the energy return from the manufacturing process, different types of batteries (lithium, sodium, nickel and lead-acid) were analyzed in terms of charging and discharging times, lifetime, gravimetric density and requirements for their manufacture and operation. The photovoltaic system analyzed comprised batteries, solar panels, modules, inverters and supports. Results showed that lead-acid and nickel–cadmium batteries are the most used in photovoltaic systems and the production and transport of batteries contribute between 24–70% of the requirements of photovoltaic systems with storage [8].

From another perspective, analyzing batteries, with a capacity three times greater than the consumption capacity, the energy return factor for the photovoltaic system varies from 2, 2 to 10 (energy units), which is a very large variation and can imply wasteful. However, for any analysis of this sphere it is important to characterize the context of operational conditions, which resources need to be minimized, which alternative technologies and which general parameters of the energy system, as renewable and non-renewable can be combined as long as conflicting interests linked to supply and demand profiles are mitigated [19].

Consumer behavior profiles play an essential role in the design of photovoltaic solar energy storage systems, as they are reflected in the battery operation profile.

Account must be taken regarding technical and economic parameters of the photo-voltaic and battery system, as for instance the consumer's time load and the photovoltaic production profiles, as well as existing regulatory framework [20].

The highest impact of batteries on its prices results from the duration of charging and discharging, which increases with battery size (on the other hand, smaller batteries have longer operating times at full power). In addition, most part of the time (approximately 80–95%) batteries are on standby (fully charged or discharged) for a system less than or equal to 10 kWp. The amount of time when that battery is operated in charge or discharge mode is limited to the remaining period (5–20%). On the other hand, battery components, such as electrodes, substrates and the manage-ment system, contribute to several potential environmental impacts. However, it is the stages of battery use and reuse that dominate energy demand and atmospheric emissions [20].

Using second life batteries for solar energy storing can make photovoltaics more attractive to many regions, as much as reducing its costs. The cost of such a system using a second life battery is estimated to be 50% less than a system with the same performance that uses new lithium batteries [13]. Besides, second life batteries built from partially degraded battery cells have greater internal resistance compared to a new battery. Also, the second use of batteries after its initial capacity is exhausted (around 80–90% of the initial capacity) can be achieved by remanufacturing and recycling the old ones [4].

Identifying individual cells capacity and applying battery management tech-niques, a second life battery pack can be assembled to perform as well as new batteries (with the same affordable capacity) but cheaper than them. It has already proved the application of a second life battery in an off-grid energy storage system, such as a solar panel charging station, is feasible and economical compared to systems using new lithium-ion batteries [13].

A study also evaluated the life cycle of a lithium-ion battery used in an electric vehicle and then reused in a stationary application. Results indicate the manufac-turing phase of the Li-ion battery dominates environmental impacts, but the cascading system appears to be significantly beneficial in consuming clean energy sources for both use and reuse, causing reductions in global and local environmental stress. Lithium-ion batteries present opportunities to power mobility and stationary appli-cations in the necessary transition to cleaner energy as long as the health of the battery is adequate [14].

In Charles et al. [21] there has been an investigation into which battery technology would be most suitable for sustainable solar energy storage in small-scale domestic use in rural Africa. Its questions permeated whether lead acid would be the best choice within a circular economy, as there are currently no African facilities for recycling Li-ion, with little opportunity to retain material value from these batteries in the region.

In fact, both hazardous materials Li-ion and Pb-acid pose risks at the end of a battery's life and high recycling costs increase the likelihood of inadequate end of life management, with resulting impacts on populations and the environment. Despite lower efficiency and shorter service life, Pb-acid batteries (more readily available and less expensive), are the most sustainable current choice for small-scale home photovoltaic systems (<10kWp). Despite the relatively low costs, the proposed system is still not profitable over its 20 years of life; however, the adoption of circular economy practices has great potential to improve this [21].

It is noteworthy that in the interaction between solar photovoltaic and energy storage systems, the longevity of the latter is a key aspect to be improved (with less environmental impact). Business models for maximizing the return on batteries at the end of life, rental and return schemes and deposit schemes are also very important. Nevertheless, there is a call for an additional basic education and training package on the benefits of solar energy, risks associated with technologies and operation, proper maintenance and replacement of components, complete monitoring and analysis of system performance for forecasting and detecting problems and failures [4, 13].

3.4 Sustainability Energy Transitions

In view of the characteristics previously mentioned about individual technologies and their integration, the ways to enable them in a sustainable manner in the energy and transport sectors are arduous, due to the toxic properties of the materials. Thus, these technologies still depend on synergy with each other and on the right decisions to be inserted in a sustainable manner on a large scale [1]. Although the use of electric vehicle batteries in solar energy storage stations is a more sustainable activity, there are still a number of improvements and different developments (social, economic, political) to help transition to greater sustainability in these sectors.

The increase in the number of batteries is linked to the increase in electric vehicles, which also depend on improvements in storage technologies, among other factors such as geographic extension. Countries with a less extensive geographical area have advantages in the transition to the use of EV, as they do not have transit capacity for long periods. Countries where smart grids already exist are even more conducive to success in the diffusion of EVs, but it still depends on government and technological incentives, in addition to greater consumer awareness [4].

Another relevant point to understand the sustainability of the transition of these technologies is to keep in mind that, on the one hand, it is possible to reduce the use of fossil fuels in the transport sector (in addition to energy), but on the other hand, it may require more energy from network to supply batteries for these vehicles. Solutions to mitigate this effect are to supply the central power grid with excess energy from electric vehicle batteries (vehicle-to-grid) which could add capacity to the power grid during peak hours, which would reduce the need to build new power plants. Another point is that the energy demand for charging the batteries can come from the grid when the demand is low (charging the batteries during the night). In short, there are advantages that go beyond the criticisms associated with the construction and operation of electrochemical and electronic elements needed in these technologies [11, 22].

Although the potential benefits of an energy transition remain significant, they may not accumulate without social conflict. Moving car pollution to distant power plants, which produce energy for EVs, can ensure that the negative externalities of energy production do not affect city dwellers. But such a movement could polarize relations between rural and urban communities or different economic classes of people [4, 11]. Besides, many of the most significant players in the existing energy and transport infrastructure have a vested interest in maintaining the status quo. Fortunately, the opposition does not always prevent technologies from appearing on the market and reconfiguring current social and technological regimes.

4 Final Considerations

This study carried out a literature review in the main scientific databases and provided a portfolio of knowledge structured according to the specificities of the PV and EV industries and also in their integration with the use of batteries for a more sustainable transition for both.

PV storage by reusing EV second life batteries modifies the impact that EV and batteries can have on the environment and society. EV can demonstrate more benefits and consequently diffusion capacity due to the reuse of their batteries in PV systems, also helping with the problem of intermittency of PV.

Important barriers to the spread of EV are not only technical, but political and socio-cultural as well. The effects of manufacturing processes, use and disposal of clean technologies influence the economic configuration of its products and the decision-making process for adopting them. The misunderstanding of facts and data of toxicity in the entire productive chain of technologies should be better discussed and treated, in order to expand the knowledge available to consumers.

Advanced technological mechanisms for renewable energy may have significantly negative and sometimes obscured environmental impacts, which must be considered in future policies advocating for those technologies. As support for the continuation of discussions on this topic, some research gaps are presented as necessary for further investigation. First, the political influence of major players in the stakeholder industries in the energy and transportation sectors; second, the dissemination of specific knowledge on all technical and regulatory processes linked to the cleaner technology market; third, clean technologies have to be accompanied by policies that guarantee future treatment of them.

The limitations of this research are the choice of the string and the databases. An expansion in strings and databases could demonstrate different and equally important studies. Search terminology and configuration of filters are also pointed out as potential limiters, being recommended to be modified and tested for comparison of results.

Acknowledgements This study was financed in part by the Conselho Nacional de Desenvolvimento Científico e Tecnológico of Brazil (CNPq) and the Coordenação de Aperfeiçoamento de Pessoal de Nível Superior Brasil (CAPES)—finance code 001.

References

1. Larcher, D., & Tarascon, J.-M. (2015). Towards greener and more sustainable batteries for electrical energy storage. Nature chemistry, 7(1), 19.
2. Coram A., Katzner, D. W. Reducing fossil-fuel emissions: dynamic paths for alternative energy-producing technologies. Energy Econ 2018; 70:179–89. https://doi.org/10.1016/j.eneco.2017.12.028.
3. Erickson, L. E., & Jennings, M. (2017). Energy, transportation, air quality, climate change, health nexus: Sustainable energy is good for our health. AIMS public health, 4(1), 47.
4. Teixeira, A. C. R., da Silva, D. L., Neto, L. D. V. B. M., Diniz, A. S. A. C., & Sodré, J. R. (2015). A review on electric vehicles and their interaction with smart grids: the case of Brazil. Clean Technologies and Environmental Policy, 17(4), 841–857.
5. Kourkoumpas, D. S., Benekos, G., Nikolopoulos, N., Karellas, S., Grammelis, P., & Kakaras, E. (2018). A review of key environmental and energy performance indicators for the case of renewable energy systems when integrated with storage solutions. Applied Energy, 231, 380–398. https://doi.org/10.1016/j.apenergy.2018.09.043.
6. Van der Kam, M. J., Meelen, A. A. H., van Sark, W. G. J. H. M., & Alkemade, F. (2018). Diffusion of solar photovoltaic systems and electric vehicles among Dutch consumers: Implications for the energy transition. Energy research & social science, 46, 68–85.
7. Fthenakis, V. M. (2000). End of life management and recycling of PV modules. Energy Policy, 28(14), 1051–1058.
8. Rydh, C. J., & Sandén, B. A. (2005). Energy analysis of batteries in photovoltaic systems. Part I: Performance and energy requirements. Energy Conversion and Management, 46(11–12), 1957–1979.
9. Wenzl, H., Baring-Gould, I., Kaiser, R., Liaw, B. Y., Lundsager, P., Manwell, J., Svoboda, V. (2005). Life prediction of batteries for selecting the technically most suitable and cost effective battery. Journal of Power Sources, 144(2), 373–384.
10. Armand, M., & Tarascon, J. M. (2008). Building better batteries. Nature, 451(7179), 652.
11. Sovacool, B. K., & Hirsh, R. F. (2009). Beyond batteries: An examination of the benefits and barriers to plug-in hybrid electric vehicles (PHEVs) and a vehicle-to-grid (V2G) transition. Energy Policy, 37(3), 1095–1103.
12. Longo, S., Antonucci, V., Cellura, M., & Ferraro, M. (2014). Life cycle assessment of storage systems: The case study of a sodium/nickel chloride battery. Journal of Cleaner Production, 85, 337–346. https://doi.org/10.1016/j.jclepro.2013.10.004.
13. Tong, S. J., Same, A., Kootstra, M. A., & Park, J. W. (2013). Off-grid photovoltaic vehicle charge using second life lithium batteries: An experimental and numerical investigation. Applied Energy, 104, 740–750.
14. Ahmadi, L., Young, S. B., Fowler, M., Fraser, R. A., & Achachlouei, M. A. (2017). A cascaded life cycle: reuse of electric vehicle lithium-ion battery packs in energy storage systems. International Journal of Life Cycle Assessment, 22(1), 111–124. https://doi.org/10.1007/s11367-015-0959-7.
15. Bakhiyi, B., Labreche, F., & Zayed, J. (2014). The photovoltaic industry on the path to a sustainable future - Environmental and occupational health issues. Environment International, 73, 224–234. https://doi.org/10.1016/j.envint.2014.07.023.
16. Tannous, S., Manneh, R., Harajli, H., & El Zakhem, H. (2018). Comparative cradle-to-grave life cycle assessment of traditional grid-connected and solar stand-alone street light systems:

A case study for rural areas in Lebanon. Journal of Cleaner Production, 186, 963–977. https://doi.org/10.1016/j.jclepro.2018.03.155.

17. D'Adamo, I., Miliacca, M., & Rosa, P. (2017). Economic feasibility for recycling of waste crystalline silicon photovoltaic modules. International Journal of Photoenergy, 2017. https://doi.org/10.1155/2017/4184676.

18. Balcombe, P., Rigby, D., & Azapagic, A. (2015). Environmental impacts of microgeneration: Integrating solar PV, Stirling engine CHP and battery storage. Applied Energy, 139, 245–259.

19. Rydh, C. J., & Sandén, B. A. (2005a). Energy analysis of batteries in photovoltaic systems. Part II: Energy return factors and overall battery efficiencies. Energy Conversion and Management, 46(11–12), 1980–2000.

20. Linssen, J., Stenzel, P., & Fleer, J. (2017). Techno-economic analysis of photovoltaic battery systems and the influence of different consumer load profiles. Applied Energy, 185, 2019–2025.

21. Charles, R. G., Davies, M. L., Douglas, P., Hallin, I. L., & Mabbett, I. (2019). Sustainable energy storage for solar home systems in rural Sub-Saharan Africa—A comparative examination of lifecycle aspects of battery technologies for circular economy, with emphasis on the South African context. Energy, 1207–1215. https://doi.org/10.1016/j.energy.2018.10.053.

22. Telaretti, E., Graditi, G., Ippolito, M. G., & Zizzo, G. (2016). Economic feasibility of stationary electrochemical storages for electric bill management applications: The Italian scenario. Energy Policy, 94, 126–137. https://doi.org/10.1016/j.enpol.2016.04.002.

Entrepreneurial Resilience and Gender: Are They Connected? Contributions Toward Entrepreneurship Policy-Package

Mariana Pita🆔 and **Joana Costa**🆔

Abstract Gender issues are increasingly studied in entrepreneurship along with the debate about the centrality of resilience in both individuals and organizations. However, the analysis of entrepreneurial initiative based on genderized resilience is a novel insight. Prior studies point out that, against the odds, in more vulnerable environments, individuals tend to be more resilient and exhibit higher levels of entrepreneurial initiative. Considering previous studies and its remarks about the importance of self-efficacy, self-determination, self-regulation, and social environment to determine a resilient profile, this work bridges the existing theoretical contributions with empirical findings, using resilience and entrepreneurship as the ground research field. Empirical results evidence the role of resilience and opens the discussion around its importance through a genderized perspective along with its effect on the entrepreneurial initiative. The existence of a gendered connection between individual resilience and entrepreneurial initiative emerges as a relevant insight. These findings provide valuable information to policymakers and practitioners understanding and designing policy packages which better suit gender singularities.

Keywords Gender · Resilience · Entrepreneurship · Environment · GEM

1 Introduction

The uncertainty lived around the world and emergence of complex challenges, namely, economic, environmental and societal, contributed to increasing the interest of promoting more capable and equal societies. In the same line, the European Commission [1] endorses the necessity to develop business skills and entrepreneurial attitudes as creativity, resilience, responsibility, risk tolerance and team engagement,

M. Pita (✉) · J. Costa
Department of Economics, Management, Industrial Engineering and Tourism (DEGEIT),
University of Aveiro, Campus Universitário de Santiago, 3810-129 Aveiro, Portugal
e-mail: mariana.pita@ua.pt

© The Author(s), under exclusive license to Springer Nature Switzerland AG 2021
A. M. Tavares Thomé et al. (eds.), *Industrial Engineering and Operations Management*,
Springer Proceedings in Mathematics & Statistics 367,
https://doi.org/10.1007/978-3-030-78570-3_48

to facilitate the transformation of ideas into action and foster employment opportunities. In this setting, women are taken as the large pool of entrepreneurial potential in Europe, therefore, paths to entrepreneurship should be sensitive to diversity. Entrepreneurship should not be stigmatized, encouraging individuals to be resilient and strive even in less favourable environments.

Recent studies point more developed economies as those with better organizational conditions, and consequently, more open and flexible toward gender equality. Nevertheless, business environments are a men field [2] which is corroborated by several studies pointing women as a minor workforce. Also, other findings point that females are still socially pressured to answer multiple roles, in a different extant compared to men, jeopardizing their initiative to enterprise.

Gender disparities are internationally recognized as an important challenge and is no exception to the entrepreneurship field. As a consequence, most of the countries are committed to attain gender equality through the reinforcement of entrepreneurial spirit in general, raising balanced entrepreneurial opportunities, competitiveness, employment and innovation [3, 4]. Despite the efforts to plain gender issues, entrepreneurial initiative is not equitable, evidencing a gendered phenomenon concerning labour market [5] and self-employment. Particularly in entrepreneurship studies, it is common to see gender as a supplementary research perspective neglecting its importance [6].

According to [7] gender, and entrepreneurship should be decomposed. Previous research focuses on gender as a variable to explain the differences around entrepreneurial initiative, focusing on individual characteristics. This perspective offers a limited comprehension about the determinants of entrepreneurship since lacks an exogenous analysis on the context where individuals are rooted.

The question posed by Marlow [8] opens the floor for an ultimate debate on how entrepreneurship fit women and not how to fit women into entrepreneurship. Currently, entrepreneurship, still suffers for a masculinized perspective entailing a stereotyped vision of the entrepreneur and new venture leadership, associating male figure to a more resilient profile and roughness, assuming a heroic expression [9].

The entrepreneur is considered someone resilient and skilled to overcome wicked challenges. There is a gender division in this matter and women, nevertheless, are pointed as more resilient and exhibiting higher levels of entrepreneurial initiative, depending on the context and country development stage [10].

The role of resilience and its connection to gender and entrepreneurship are, therefore, a topic of paramount relevance considering three motives (i) resilience is an emergent field of research within entrepreneurship (ii) the influence of resilience on entrepreneurial initiative among women needs to be fully comprehended, (iii) the design of effective entrepreneurship policies depends on the clarification of factors that ignite women entrepreneurship.

This work aims to identify the factors that enhance or dissuade women to act entrepreneurially and capture empirical evidence on the role of resilience in entrepreneurial initiative through a comparative perspective between women and men. Considering the research purposes, several proxies were explored, such as self-efficacy, self-determination, self-esteem, and social environment to explain the entrepreneurial initiative phenomenon. Additionally, several control variables were included as age, family environment, self-regulation, and quality of life.

The contributions obtained are relevant to academics, practitioners and policy-makers since denotes the importance of entrepreneurial exposure and role-modelling as enhancers for women to choose an entrepreneurial career, influencing the design of programs and policy packages devoted to stimulating entrepreneurship.

This paper is structured in five sections. After the introduction of the topic, the second part presents the theoretical background. Section three discloses the methodology followed by the results and discussion. The last section offers an overall perspective of the study along with limitations and future lines of research.

2 Entrepreneurship and Resilience: Is There a Genderization?

Entrepreneurship research has been growing in the past decades, which has stimulated different research questions and perspectives around the phenomenon. The gender issues have become a prominent concern globally pushing United Nation's Sustainable Development Goals (SDGs) to recognize the empowerment of women has a priority in all dimensions. The contribution of women for economic growth and innovation [11] is undeniable but the strong normative imposed by cultural patterns has prejudice the women career choices, affecting their willingness to enterprise.

A recent report devoted to analysing the entrepreneurial initiative among several countries indicated that women are more entrepreneurial when comparing previous years [12]. However, the data confirms that women are still behind men concerning entrepreneurship, with only 7 female entrepreneurs for every 10 male entrepreneurs.

The Global Entrepreneurship Monitor (GEM), acknowledges only 6 countries where entrepreneurial initiative is balanced in a gender perspective, namely, Indonesia, Thailand, Panama, Qatar, Madagascar, and Angola [13]. Considering the importance of gender issues, GEM assessed women entrepreneurial initiative using different variables achieving surprising results. Considering the timeframe of 2017/2017, GEM reported that female entrepreneurial initiative increased among 63 countries, reducing the gender gap by 5% [11]. These results suggest a positive evolution on female entrepreneurship with more women choosing an independent career. According to [9], the job market is informally ordered, with activity division oriented by gender. The discourses prescribe related roles marking entrepreneurship as a less desirable career choice for women. Certainly, support policy measures toward female entrepreneurship [14] could influence the partition of the phenomenon.

The literature reveals that women tend to have limitations in pursuing job market opportunities compared to men, and a similar situation occurs when women attempt to develop an entrepreneurial career [15]. Although higher levels of education are relevant for women to achieve success in their entrepreneurial ventures, when it comes to other jobs, the stereotype and gender discrimination reduce women opportunities.

The gender gap has a profound impact on female entrepreneurship [16], but other factors contribute to influencing negatively their initiative of creating new businesses. Previous studies consider men more entrepreneurial [17] but women also exhibit determination to enterprise. Due to several limitations related to age, education, culture or social condition, women are frequently disconnected to the business context [18]. This scenario is more evident in less developed countries, where the occurrence of entrepreneurship is mostly induced by necessity reasons with women being excluded from men jobs. Women recognize the lack of education, the limited business experience and the scarcity of training opportunities as barriers to success as entrepreneurs [19], particularly compared to males. Despite women and men exhibit similar levels of formal education, there is a significant difference concerning self-professional experience leaving women behind in this matter [20]. To overcome the lack of entrepreneurial competences or experience, the role of the family context is crucial since can become the needed support for women to become more prepared to act in the business setting [21].

Other barriers are pointed out as inhibitors for women to enterprise, namely, the family obligations as the care for children's and household [22]. Women are demanded to fulfil different roles and social expectations, turning challenging confrontation with leading masculine contexts [23].

Entrepreneurs "*perceive new business opportunities, organize businesses where none existed before, direct these businesses by using their own and borrowed fund, take the associated risks, and enjoy profit as rewards for their efforts*" [15]. Since entrepreneurship involves risk, those who fear failing tend to avoid self-employment experiences. In terms of gender, women are more reluctant to risk compared to men, struggle to recognize opportunities [24] and fail in persisting a business [18].

The entrepreneurial journey is commonly difficult, pushing entrepreneurs to constantly adapt and manage their resources as a response to the changing setting. The literature confirmed that women face higher barriers to establish a new venture compared to men, meaning that women are resilient as they need to overcome particularly demanding circumstances to enterprise [25]. Therefore, resilience is a relevant skill for entrepreneurs to be successful, along with other competences related to leadership, communication, innovation, or networking [26].

The studies devoted to exploring the relationship between entrepreneurship and resilience, exposed contradictory results when comparing women and men, evidence an existent gender gap [25]. To the best of our knowledge, research on entrepreneurship and resilience lacks empirical findings. Departing from the latest contributions and considering that resilience research is very fragmented and offers a detached set of clusters focusing resilience at different levels [2], the present work is devoted to grasp the relation between resilience, gender, and entrepreneurship through the examination of individual performance and social and environmental axes. Resilience is a relevant entrepreneurship research topic as is repeatedly noticed as something positive or desirable, since separate 'winners' and 'losers', unequally distributed but rather largely determined by different factors [27].

3 Methodology

3.1 Methodological Options

To address the study purposes, it was used the GEM dataset from 2016, at the individual level, comprising a sample of 193.766 individuals. The sample used corresponds to the most up-to-date database comprising the Adults Population Survey (APS) from 2016. According to [13], GEM objective is to track the characteristics and motivations of individuals in pursuing entrepreneurial initiatives as well to assess its determinants. GEM offers a solid framework and corresponding data to explore the empirical application of theoretical contributions for gender and resilience in an organizational environment as proved by several studies. Therefore, the entrepreneurial initiative will be considered as the dependent variable (corresponding to those who intend or are involved in starting a business).

Since entrepreneurship is seen as a 'man world' and given the importance of understanding what influences women to pursue such challenging pathways, particularly when facing more adverse environments (here, as countries with lower levels of quality of life), research-grounded on GEM could contribute to overcome the lack of empirical findings and point possible alternatives to measure the relation between resilience and gender.

The empirical analysis is conducted within two steps: the first presents the descriptive results; on the second, different models are tested considering a multiplicity of sub-samples (males, females, all sample, all sample and gender effect, all sample and life motives). The estimations rely on regression procedures to calculate the resilience of men and women concerning entrepreneurial initiative and explore the possible genderization of the phenomenon.

3.2 Variables

The current study explores the relationship between gender and resilience since resilience is foremost of the times considered as a masculine attribute, and frequently related to the ability to overcome obstacles. For entrepreneurship research, resilience is understood as the ability to pursue entrepreneurial experiences, therefore, measured by the entrepreneurial initiative. The complete sample was used and the dimensions of gender, social environment, self-efficacy, self-esteem, self-determination, age, family environment, and quality of life were pondered as predictors of entrepreneurial initiative.

4   Results and Discussion

4.1   Descriptive Analysis

The complete sample is composed of a total of 193.766 individuals, with a similar distribution between women (50%) and men (50%). Considering age, the individuals between 25 and 34 years old are the most represented (with 22.65%), followed by those with 35–44 (21.29%). Younger individuals characterize 15.27% of the sample, and more mature individuals, between 45 and more than 65 years old, represent near 40%.

Concerning quality of life (proxied by country stage of development), the sample reports that only 8.06% of individuals are from less developed countries, which means countries with poor living conditions, and in the opposite, 48.61% of the sample come from more developed environments (innovation-driven).

Regarding resilience (entrepreneurial initiative), only 14.14% of respondents are involved in starting a business, which shows a moderate activity when it comes to starting a business. From those who decide to enterprise, men are more active compared to women with 16.37%. Interestingly, although the difference of entrepreneurial activity is testified as minor, when it comes to the other variables, such as social influence, self-determination, self-esteem, or self-efficiency, there is a stronger variation between gender.

In terms of the social environment (proxied by social context), only 38.39% report knowing someone who already enterprise. From those who contact directly with entrepreneurs, the man reveals a higher exposure with 41,81%. When it comes to self-efficacy (proxied by Skills Perception), surprisingly, half of the sample reveals having the skills to enterprise (50.77%). Regarding self-esteem (proxied by fear of failure), almost 60% declared that fear of failure would not prevent them from pursuing an entrepreneurial journey (58.47%), against 40% of individuals with lower self-esteem (41.53%). Only 26.63% of the sample exhibits graduate education, but surprisingly the differences between gender among more educated people are not significant, with man revealing 27.03% and women 26.23%. Self-determination (proxied by education) evidences the need for a proper educational background in a similar proportion between men and women.

4.2   Empirical Results

Tables 1 and 2 present the results of regressions and multiple models tested. Model 1 presents the estimations (regression analysis) for determinants of entrepreneurial initiative among a sub-sample, considering only males. Model 2 follows the same procedure implemented in Model 1, considering the sub-sample females. To establish a comparison between groups and understand the magnitude of effects, Model 3 is conducted with the complete sample. To capture the effect of self-esteem on women,

Table 1 Logistic regression–model 1, model 2, model 3

	Model 1 (Males)		Model 2 (Females)		Model 3 (Males & Females)	
	B	Exp(B)	B	Exp(B)	B	Exp(B)
Social environment	0.683***	1.979	0.707***	2.028	0.694***	2.001
Self-efficacy	1.363***	3.906	1.490***	4.439	1.425***	4.156
Self-esteem	−0.249***	0.779	−0.278***	0.757	−0.263***	0.769
Self-determination	0.043***	1.044	0.017**	1.017	0.031***	1.031
Age	−0.135***	0.874	−0.113***	0.893	−0.125***	0.882
Family environment	0.039***	1.039	0.033***	1.033	0.036***	1.037
Gender	–		–	–	−0.192***	0.825
Gender & Self-esteem	–		–	–	–	–
Self-regulation	–		–	–	–	–
Life quality	−0.582***	0.559	−0.674***	0.509	−0.622***	0.537
	–		–		–	
Constant	1.208	0.299	1.274	0.280	1.156	0.315

(p < 0.001***; p < 0.05**; p < 0.01*)

Table 2 Logistic regression–model 4, model 5

	Model 4(Males & Females)		Model 5 (Males &Females)	
	B	Exp(B)	B	Exp(B)
Social environment	0.694***	2.001	0.278***	1.320
Self-efficacy	1.424***	4.155	0.370***	1.447
Self-esteem	−0.242***	0.785	−0.212**	0.809
Self-determination	0.031***	1.031	−0.065**	0.937
Age	−0.125***	0.882	0.015	1.015
Family environment	0.036***	1.037	0.004	1.004
Gender	−0.176***	0.838	0.210**	1.233
Gender & Self-esteem	−0.046	0.955	0.367**	1.444
Self-regulation	–	–	−0.30	0.970
Life quality	−0.622v	0.537	−0.273***	0.761
	–		–	
Constant	1.163	0.313	3.008***	20.251

(p < 0.001***; p < 0.05**; p < 0.01*)

Model 4 presents a new variable of interaction. Lastly, model 5 assess the significance of life motives toward entrepreneurship, considering men and women.

Model 1 reports all variables in use as significant for entrepreneurial activity. However, self-esteem presents a negative effect, along with age and quality of life.

The fear of failure deters men to pursue entrepreneurial endeavors, but in contrast, education has a positive effect on the entrepreneurial initiative, although with a minor effect compared to self-efficacy. Such results show that for men, even lacking educational background, being confident in their knowledge and abilities is more relevant to enterprise. The existence of a social environment where the role model occurs is also influenced man's desire for starting a business. With a moderate effect, the household also has a positive significance in entrepreneurial activity. Such finding could be explained by the need to enterprise, grounded on survival motives. Regarding the quality of life, proxied by the country development stage variable, the results point a negative effect on the entrepreneurial initiative. Although such countries offer a more prosperous and stable environment, due to secure economic and employment levels, individuals are less resilient because they tend to riskless. Therefore, more resilient individuals exhibit higher levels of self-efficacy as they are more exposed to entrepreneurial role modelling through the social environment.

Model 2, report differences compared to Model 1, but only in the magnitude of the effects. For instance, the social context is evidenced as more significant for women, emphasizing its importance to choose an entrepreneurial career. Following previous studies, women who enterprises exhibit stronger self-efficacy. Such findings suggest that women need to be more confident to enterprise compared to men since they are more conservative in professional pathways.

For instance, when comparing the family environment, it's possible to recognize a major influence among man. This fact could be related to the traditional role of man as the family supporter. The result could also be analysed simultaneously with quality of life among women, which suggests that women in more developed countries enterprise less than men.

This goes in line with literature that emphasizes the social role of women and their need to grasp for stable positions and protect the family. However, on the contrary, women in adverse environments tend to enterprise more, shortenning the gender differences. Model 3 corroborates the prior models and reveals gender as a deterring factor.

Model 4 acknowledges the relevance of the interaction variable, evidencing that gender and self-esteem have a negative effect on resilience, placing women as less entrepreneurial. Lastly, Model 5 discloses a significant difference once compared to other analyses, namely, the significance of self-determination, suggesting that having higher competences discourages women from the enterprise. Concerning the other variables, the results are aligned with the main findings.

Generally, the empirical results reveal a degenderized phenomenon and posit resilience as a similar behaviour between men and women. However, despite the similarities concerning the influence of factors, the effect is more significant for women, except family environment and self-esteem. It should be also noted that self-efficacy is the most relevant ignition factor toward entrepreneurship, followed by the social environment. Also, self-determination (proxied by education) evidences a short impact on the entrepreneurial initiative (resilience) emphasizing the relevance of other variables. The differences reported, in terms of magnitude, answers the initial question and allow to support that entrepreneurial resilience is not a genderized phenomenon, since man and women in analogous conditions tend to act similarly.

5 Conclusion

This paper discusses the relationship between resilience, gender, and entrepreneurship, offering new insights to theoretical contributions grounded on empirical findings. Since the participation of women in the job market as entrepreneurs is lagging in many countries, more contributions are needed to appraise how gender influences the creation of new ventures. The initiative of women toward entrepreneurship is considered resilience, since they act under less favourable circumstances. Therefore, women that choose to enterprise are more resilient when compared to those who do not pursue an entrepreneurial career.

Previous studies show that the entrepreneurial initiative among women is lower if compared to men [18]. Although the results of this study point in the same direction, additional factors emerge as enhancers of entrepreneurial initiative among women, such as self-efficacy and social environment.

The empirical findings prove that entrepreneurial initiative is stimulated when individuals possess knowledge, skills, and are exposed to role models. For women, the contact with entrepreneurs increases the desire to purse an entrepreneurial career and raise their confidence.

Concerning education, the results are sensitive following the perspective of [19], demonstrating the importance of education to intensify business skills and initiative [28]. Observing the results and understanding the importance of self-efficacy, more business programs should be offered to women. The importance of social context is also denoted as a relevant driver of entrepreneurship, with similar impact in both genders [29].

In sum, the results of this paper challenges gender inequalities concerning entrepreneurship, revealing the need to develop an appropriate policy package to support entrepreneurship, driven by gender specificity. Consequently, policy measures should recognize successful entrepreneurs, namely, women, to enhance the role model effect. Simultaneously, the role of women in society should be deconstructed and alleviated to lessening the social pressure faced when it comes to professional diversity pathways. Moreover, a combination of sustainable educational policies oriented to business, along with the raise of confidence through literacy and role modelling, will allow to increase significantly the entrepreneurial initiative, particularly in more developed countries.

Finally, the findings point new avenues of entrepreneurship research by focusing on resilience and gender, particularly using endogenous and exogenous perspectives.

Acknowledgements This work was financially supported by the Project BOT-Learning as a modern teaching method of GEN Z (2020-1-PL01-KA203-081777), funded by European Program ERASMUS+.

References

1. European Commission, 'ENTREPRENEURSHIP 2020 ACTION PLAN Reigniting the entrepreneurial spirit in Europe', *COM 795 Final*, no. Brussels, 9 January 2013, pp. 1–33, 2013 [Online]. Available: http://eur-lex.europa.eu/legal-content/EN/TXT/PDF/?uri=CELEX: 52012DC0795&from=EN.
2. L. Branicki, V. Steyer, and B. Sullivan-Taylor, 'Why resilience managers aren't resilient, and what human resource management can do about it', *Int. J. Hum. Resour. Manag.*, vol. 30, no. 8, pp. 1261–1286, 2019. https://doi.org/10.1080/09585192.2016.1244104.
3. W. Adema et al., 'Enhancing Women's Economic Empowerment through Entrepreneurship and Business Leadership in OECD Countries', 2014.
4. C. A. Ennis, 'The Gendered Complexities of Promoting Female Entrepreneurship in the Gulf', *New Polit. Econ.*, pp. 1–20, 2018. https://doi.org/10.1080/13563467.2018.1457019.
5. D. Kelley et al., 'Global Entrepreneurship Monitor Women' s Entrepreneurship 2016 / 2017', pp. 1–93, 2017.
6. C. Holmquist and E. Sundin, 'Is there a place for gender questions in studies on entrepreneurship, or for entrepreneurship questions in gender studies?', *Int. J. Gend. Entrep.*, vol. 12, no. 1, pp. 89–101, 2020. https://doi.org/10.1108/IJGE-05-2019-0091.
7. C. G. Brush, P. G. Greene, and F. Welter, 'The Diana project: a legacy for research on gender in entrepreneurship', *Int. J. Gend. Entrep.*, vol. 12, no. 1, pp. 7–25, 2020. https://doi.org/10. 1108/IJGE-04-2019-0083.
8. S. Marlow, 'Gender and entrepreneurship: past achievements and future possibilities', *Int. J. Gend. Entrep.*, vol. 12, no. 1, pp. 39–52, 2020. https://doi.org/10.1108/IJGE-05-2019-0090.
9. H. Witmer, 'Degendering organizational resilience—the Oak and Willow against the wind', *Gend. Manag.*, vol. 34, no. 6, pp. 510–528, 2019. https://doi.org/10.1108/GM-10-2018-0127.
10. J. Costa and M. Pita, 'Appraising entrepreneurship in Qatar under a gender perspective', *Int. J. Gend. Entrep.*, vol. ahead-of-p, no. ahead-of-print, pp. 1–25, 2020. https://doi.org/10.1108/ IJGE-10-2019-0146.
11. D. J. Kelley et al., 'Global Entrepreneurship Monitor: Women ' s Entrepreneurship 2016 / 2017 Report', 2017.
12. V. K. Gupta and N. M. Bhawe, 'The Influence of Proactive Personality and Stereotype Threat on Women's Entrepreneurial Intentions', *J. Leadersiph Organ. Stud.*, vol. 13, no. 4, pp. 73–85, 2007.
13. N. Bosma and D. Kelley, 'Global Entrepreneurship Monitor: 2018/2019 Global Report', 2019.
14. F. Welter, 'Contexts and gender – looking back and thinking forward', *Int. J. Gend. Entrep.*, vol. 12, no. 1, pp. 27–38, 2020. https://doi.org/10.1108/IJGE-04-2019-0082.
15. Hossain, K. N. and A. Zaman, and R. Nuseibeh, 'Factors influencing women business development in the developing countries', *Int. J. Organ. Anal.*, vol. 17, no. 3, pp. 202–224, 2009.
16. M. E. Heilman, 'Description and Prescription : How Gender Stereotypes Prevent Women ' s Ascent Up the Organizational Ladder', vol. 57, no. 4, pp. 657–674, 2001.
17. K. Fellnhofer and K. Puumalainen, 'Can role models boost entrepreneurial attitudes?', *Int. J. Entrep. Innov. Manag.*, vol. 21, no. 3, pp. 274–290, 2017. https://doi.org/10.1504/IJEIM.2017. 083476.
18. M. Pines, M. Lerner, and D. Schwartz, 'Gender differences in entrepreneursip: Equality, diversity and inclusion in times of global crises', *Equal. , Divers. Incl. An Int. J.*, vol. 29, no. 2, pp. 2186–198, 2010.
19. J. Raghuvanshi, R. Agrawal, and P. K. Ghosh, 'Analysis of Barriers to Women Entrepreneurship: The DEMATEL Approach', *J. Entrep.*, vol. 26, no. 2, pp. 220–238, 2017. https://doi.org/10. 1177/0971355717708848.
20. S. Marlow, C. Henry, and S. Carter, 'Exploring the impact of gender upon women's business ownership', *Int. Small Bus. J.*, vol. 27, no. 2, pp. 139–148, 2009.
21. L. K. Gundry and H. P. Welsch, 'The ambitious entrepreneur: ig growt strategies of women-owned enterprises', *J. Bus. Ventur.*, vol. 9026, no. 312, pp. 453–455, 2001.

22. T. Mazzarol, T. Volery, N. Doss, and V. Thein, 'Factors influencing small business start-ups: A comparison with previous research', *Int. J. Entrep. Behav. Res.*, vol. 5, no. 2, pp. 48–63, 1999. https://doi.org/10.1108/13552559910274499.

23. N. Patterson, S. Mavin, and J. Turner, 'Envisioning female entrepreneur: Leaders anew from a gender perspective', *Gend. Manag.*, vol. 27, no. 6, pp. 395–416, 2012. https://doi.org/10.1108/17542411211269338.

24. M. Minniti and C. Nardone, 'Being in Someone Else's Shoes: the Role of Gender in Nascent Entrepreneurship', *Small Bus. Econ.*, vol. 28, pp. 223–238, 2007. https://doi.org/10.1007/s11187-006-9017-y.

25. J. C. Ayala and G. Manzano, 'The resilience of the entrepreneur. Influence on the success of the business. A longitudinal analysis', *J. Econ. Psychol.*, vol. 42, no. November 2011, pp. 126–135, 2014. https://doi.org/10.1016/j.joep.2014.02.004.

26. J. C. Jordan, 'Deconstructing resilience: why gender and power matter in responding to climate stress in Bangladesh', *Clim. Dev.*, vol. 11, no. 2, pp. 167–179, 2019. https://doi.org/10.1080/17565529.2018.1442790.

27. F. Liñán, 'Skill and value perceptions: how do they affect entrepreneurial intentions?', *Int. Entrep. Manag. J.*, vol. 4, pp. 257–272, 2008. https://doi.org/10.1007/s11365-008-0093-0.

28. Fayolle and B. Gailly, 'The Impact of Entrepreneurship Education on Entrepreneurial Attitudes and Intention: Hysteresis and Persistence', *J. Small Bus. Manag.*, vol. 53, no. 1, pp. 75–93, 2015. https://doi.org/10.1111/jsbm.12065.

29. O. Toutain *et al.*, 'Role and impact of the environment on entrepreneurial learning', *Entrep. Reg. Dev.*, vol. 29, no. 9–10, pp. 869–888, 2017. https://doi.org/10.1080/08985626.2017.1376517.

Printed in the United States
by Baker & Taylor Publisher Services